D0764068

Venomous Animals and Their Venoms

VOLUME III

Venomous Invertebrates

Contributors to This Volume

Lucien Balozet
Wolfgang Bücherl
D. de Klobusitzky
Alvaro Delgado
Carlos R. Diniz
J. R. do Valle
E. Habermann
Bruce W. Halstead
Werner Kloft
Donald F. McMichael
Zvonimir Maretić
Ulrich W. J. Maschwitz
F. A. Pereira Lima
Hugo Pesce
Zuleika P. Picarelli
A. Rotberg
Pablo Rubens San Martin
S. Schenberg

VENOMOUS ANIMALS AND THEIR VENOMS

Edited by

WOLFGANG BÜCHERL
INSTITUTO BUTANTAN
SÃO PAULO, BRAZIL

ELEANOR E. BUCKLEY
WYETH LABORATORIES
PHILADELPHIA, PENNSYLVANIA

VOLUME III *Venomous Invertebrates*

ACADEMIC PRESS New York · London

1971

ACADEMIC PRESS, INC.
111 Fifth Avenue, New York, New York 10003

United Kingdom Edition published by
ACADEMIC PRESS, INC. (LONDON) LTD.
24/28 Oval Road, London NW1 7DD

LIBRARY OF CONGRESS CATALOG CARD NUMBER: 66–14892

PRINTED IN THE UNITED STATES OF AMERICA

Contents

List of Contributors ix

Preface xi

Contents of Other Volumes xiii

Introduction xix

VENOMOUS INSECTS

Chapter 44. Morphology and Function of the Venom Apparatus of Insects—Bees, Wasps, Ants, and Caterpillars

ULRICH W. J. MASCHWITZ AND WERNER KLOFT

I. Introduction 1
II. The Sting Apparatus of Aculeate Hymenoptera 2
III. The Urticating Hairs of Lepidopterous Larvae 38
References 56

Chapter 45. Chemistry, Pharmacology, and Toxicology of Bee, Wasp, and Hornet Venoms

E. HABERMANN

I. Introductory Remarks 61
II. Biochemistry and Pharmacology of Single Venom Constituents 63
III. Possible Hazards for Man and Their Prevention 85
References 89

Chapter 46. The Venomous Ants of the Genus Solenopsis

PABLO RUBENS SAN MARTIN

I. Introduction 95
II. Systematics 98
III. Biological Cycle of *Solenopsis saevissima richteri* 100
References 101

Chapter 47. Pharmacological Studies on Caterpillar Venoms

ZULEIKA P. PICARELLI AND J. R. DO VALLE

I. Historical Data on Lepidopterism 103
II. Urticating Caterpillars 104
III. Venom Apparatus 105
IV. Pharmacology of Extracts from Caterpillar Setae 108
References 117

Chapter 48. Poisoning from Adult Moths and Caterpillars

HUGO PESCE AND ALVARO DELGADO

I. Introduction 120
II. Lepidopterism 120
III. The Poisonous Lepidoptera 120
IV. Ecology 125
V. Symptomatology of Lepidopterism 129
VI. Prevention and Treatment of Lepidopterism 131
VII. Erucism 133
VIII. Poisonous Caterpillars 133
IX. Ecology 145
X. Symptomatology of Erucism 149
XI. Prevention and Treatment of Erucism 153
References 155

Chapter 49. Lepidopterism in Brazil

A. ROTBERG

I. History 157
II. Species 158
III. Clinical Symptoms 159
IV. Pathology 167
V. Treatment 167
References 167

VENOMOUS CENTIPEDES, SPIDERS, AND SCORPIONS

Chapter 50. Venomous Chilopods or Centipedes

WOLFGANG BÜCHERL

I. Introduction 169
II. Description, Classification, Distribution, and Biology of Scolopendromorpha 172
III. Venom Apparatus and Toxicity of Scolopendromorph Venoms 190
References 195

Chapter 51. Spiders

WOLFGANG BÜCHERL

I. Introduction 197
II. The Morphology of Spiders 199
III. The Classification of Venomous Spiders 209
IV. Description, Distribution, and Biology of Dangerous Species 223
V. Defense against Dangerous Spiders 246
VI. The Venom Apparatus of Spiders 247
VII. Extraction and Toxicity of Spider Venoms 262
References 273

Chapter 52. Phoneutria nigriventer Venom—Pharmacology and Biochemistry of Its Components

S. SCHENBERG AND F. A. PEREIRA LIMA

I. Crude Venom Pharmacology 280
II. Biochemistry of the *Phoneutria nigriventer* Venom 287
References 297

Chapter 53. Latrodectism in Mediterranean Countries, Including South Russia, Israel, and North Africa

ZVONIMIR MARETIĆ

I. Introduction and Symtomatology 299
II. Kinds of *Latrodectus* in Mediterranean Areas 301
III. History 302
IV. Epidemiology of Latrodectism 304
V. Spread of Latrodectism in Single Mediterranean Countries 306
VI. Folklore Treatment of Latrodectism in Mediterranean Areas 308
References 309

Chapter 54. Chemical and Pharmacological Properties of Tityus Venoms

CARLOS R. DINIZ

I. Introduction 311
II. Toxicity and Symptoms of Intoxication 312
III. Chemistry of the Venoms 312
IV. Pharmacological Properties 313
V. Conclusion 314
References 314

Chapter 55. Classification, Biology, and Venom Extraction of Scorpions

WOLFGANG BÜCHERL

I. Introduction 317
II. Morphology, Classification, and Distribution 318
III. Biology of Scorpions 333
IV. Extraction and Toxicity of Scorpion Venoms 341
References 346

Chapter 56. Scorpionism in the Old World

LUCIEN BALOZET

I. History 349
II. Geographic Distribution of Dangerous Scorpions 351
III. The Venom 352
IV. Experimental Study of Toxicity 359
V. Envenomation in Man 360
VI. Pathological Physiology 362
VII. Treatment 365
References 368

VENOMOUS MOLLUSKS

Chapter 57. Mollusks—Classification, Distribution, Venom Apparatus and Venoms, Symptomatology of Stings

DONALD F. MCMICHAEL

I. Introduction 373
II. Poison Cone Shells 374
III. Octopuses 384
References 391

VENOMOUS COELENTERATES, ECHINODERMS, AND ANNELIDS

Chapter 58. Venomous Coelenterates: Hydroids, Jellyfishes, Corals, and Sea Anemones

BRUCE W. HALSTEAD

I. Introduction 395
II. List of Representative Venomous Coelenterates 398
III. Nature of Coelenterate Venoms, Medical Aspects 407
References 416

Chapter 59. Venomous Echinoderms and Annelids: Starfishes, Sea Urchins, Sea Cucumbers, and Segmented Worms

BRUCE W. HALSTEAD

I. Echinoderms 419
II. Venomous Annelids 433
References 440

Chapter 60. Animal Venoms in Therapy

D. DE KLOBUSITZKY

I. Introduction 443
II. Snake Venoms 444
III. Toad Venoms 458
IV. Spider Venoms 461
V. Bee Venom 461
VI. Venom Combinations 466
VII. Appendix: Snake Venoms as Anticoagulants 467
References 470

Author Index 479

Subject Index 497

List of Contributors

Numbers in parentheses indicate the pages on which the authors' contributions begin.

Lucien Balozet,[1] Institut Pasteur, Algiers, Algeria (349)

Wolfgang Bücherl, Instituto Butantan, São Paulo and Conselho Nacional de Pesquisas, Rio de Janeiro, Brazil (169, 197, 317)

D. de Klobusitzky, Cx. p. 1036, São Paulo, Brazil (443)

Alvaro Delgado,[2] Facultad de Medicina, Universidad Nacional Mayor de San Marcos, Lima, Peru (119)

Carlos R. Diniz, Departamento de Bioquimica, Faculdade de Medicina, Universidade Federal de Minas Gerais, Belo Horizonte, Brazil (311)

J. R. do Valle, Departamento de Bioquimica e Farmacologia, Escola Paulista de Medicina, São Paulo, Brazil (103)

E. Habermann,[3] Institut für Pharmakologie und Toxikologie der Universität Würzburg, Würzburg, Germany (61)

Bruce W. Halstead, World Life Research Institute, Colton, California (395, 419)

Werner Kloft, Institut für Angewandte Zoologie der Universität Bonn, Federal Republic of Germany (1)

Donald F. McMichael,[4] Curator of Mollusks, The Australian Museum, Sydney, Australia (373)

Zvonimir Maretić, Medicinski Centar, Pula, Croatia, Yugoslavia (299)

Ulrich W. J. Maschwitz, Zoologisches Institut der Universität, Frankfurt am Main, Federal Republic of Germany (1)

Francisca Augusta Pereira Lima, Serviço de Fisiologia do Instituto Butantan, São Paulo, Brazil (279)

Hugo Pesce, Facultad de Medicina, Universidad Nacional Mayor de San Marcos, Lima, Peru (119)

Zuleika P. Picarelli, Departamento de Bioquimica e Farmacologia, Escola Paulista de Medicina, São Paulo, Brazil (103)

A. Rotberg, Escola Paulista de Medicina, Sâo Paulo, Brazil (157)

Pablo Rubens San Martin, Museo Nacional de Historia Natural, Montevideo, Uruguay (95)

Saul Schenberg, Serviço de Fisiologia do Instituto Butantan, Sâo Paulo, Brazil (279)

[1] Present address: Le Rouvre, 43, Traverse du Commandeur, Marseille, France
[2] Present address: Instituto de Medicina Tropical "Daniel Alcides Carrion," Lima, Peru.
[3] Present address: Pharmakologisches Institut der Justus Liebig-Universität Giessen, Federal Republic of Germany.
[4] Present address: National Parks and Wildlife Service, Sydney, Australia

Preface

The modern trend in the study of the wide field of venomous animals and their venoms is directed toward basic research that emphasizes zoological ecology, biochemistry, pharmacology, and immunobiology. The increasing importance of this development, stimulated also by the political and industrial expansion into the undeveloped areas of the tropics, is reflected by the great number of publications on venoms of animal origin. Every year about 10,000 papers are published on this subject, scattered in hundreds of journals in many languages, thus making it impossible for the individual scientist to keep abreast of new developments.

The present treatise is an attempt to offer, for the first time, a comprehensive presentation of the entire field of the venomous members of the animal kingdom, of the chemistry and biochemistry of the venoms, of their pharmacological actions and their antigenic properties. The medical aspects, both symptomatology and therapy, are included. The work is the result of close cooperation of seventy-two scientists from thirty-two countries on all continents. The authors are highly qualified specialists in their specific areas of research; their concerted efforts make this work one of unusual scope and depth.

Volume I of this three-volume work is devoted to venomous mammals and begins the extensive section on snakes. Volume II continues the discussion on snakes and includes the saurians, batrachians, and fishes. The venomous invertebrates, such as insects, centipedes, spiders, and scorpions, venomous mollusks, and marine animals, are presented in Volume III.

The interdisciplinary aspects of the subject necessitated assigning several chapters to a single group of animals and offering separate sections covering the zoological, chemical, and biomedical points of view.

It is hoped that these volumes will be valuable reference works and stimulating guides for future research to all investigators in the field; they will also serve the needs of physicians and veterinarians seeking information on the injuries caused by venomous animals. The volumes should also facilitate the teaching of this important topic and should prove a welcome source of instruction to students and to the large group of laymen interested in this fascinating field of natural science.

The editors wish to thank the authors for their cooperation and for generously contributing the results of their work and experience.

Our thanks are also due to the staff of Academic Press for helpful advice, patience, and understanding.

We cannot conclude this preface without expressing our gratitude to Professor Dionysio de Klobusitzky who conceived the idea of this treatise and outlined its initial organization.

WOLFGANG BÜCHERL
ELEANOR E. BUCKLEY

Contents of Other Volumes

Volume I
VENOMOUS VERTEBRATES

Development of Knowledge about Venoms
CHAUNCEY D. LEAKE

VENOMOUS MAMMALS

Chapter 1. **The Platypus (*Ornithorhynchus anatinus*) and Its Venomous Characteristics**
J. H. CALABY

Chapter 2. **Classification, Biology, and Description of the Venom Apparatus of Insectivores of the Genera *Solenodon, Neomys,* and *Blarina***
GEORGE H. POURNELLE

Chapter 3. **Chemistry and Pharmacology of Insectivore Venoms**
MICHALINA PUCEK

VENOMOUS SNAKES

Chapter 4. **Karyotypes, Sex Chromosomes, and Chromosomal Evolution in Snakes**
WILLY BEÇAK

Chapter 5. **Extraction and Quantities of Venom Obtained from Some Brazilian Snakes**
HELIO EMERSON BELLUOMINI

Chapter 6. **The Protein and Nonprotein Constituents of Snake Venoms**
ANIMA DEVI

Chapter 7. **Enzymes in Snake Venoms**
N. K. SARKAR AND ANIMA DEVI

Chapter 8. **Bradykinin Formation by Snake Venoms**
CARLOS R. DINIZ

Chapter 9. **Coagulant, Proteolytic, and Hemolytic Properties of Some Snake Venoms**
G. ROSENFELD, L. NAHAS, AND E. M. A. KELEN

VENOMOUS SNAKES OF THE WORLD

Chapter 10. **Methods of Classification of Venomous Snakes**
KONRAD KLEMMER

Chapter 11. **Venomous Sea Snakes (Hydrophiidae)**
MICHEL BARME

VENOMOUS SNAKES OF CENTRAL AND SOUTH AFRICA

Chapter 12. **Classification and Distribution of European, North African, and North and West Asiatic Venomous Snakes**
KONRAD KLEMMER

Chapter 13A. **Chemistry and Biochemistry of the Snake Venoms of Europe and the Mediterranean Regions**
P. BOQUET

Chapter 13B. **Pharmacology and Toxicology of Snake Venoms of Europe and the Mediterranean Regions**
P. BOQUET

Chapter 14. **Symptomatology, Pathology, and Treatment of Bites by Near Eastern, European, and North African Snakes**
S. GITTER AND A. DE VRIES

Chapter 15. **The Venomous Snakes of Central and South Africa**
DONALD G. BROADLEY

Chapter 16. **The Venoms of Central and South African Snakes**
POUL AGERHOLM CHRISTENSEN

Chapter 17. **The Symptomatology, Pathology, and Treatment of the Bites of Venomous Snakes of Central and Southern Africa**
DAVID S. CHAPMAN

VENOMOUS SNAKES OF EAST ASIA, INDIA, MALAYA, AND INDONESIA

Chapter 18. **The Venomous Terrestrial Snakes of East Asia, India, Malaya, and Indonesia**
ALAN E. LEVITON

Chapter 19. **Chemistry and Biochemistry of the Venoms of Asiatic Snakes**
B. N. GHOSH AND D. K. CHAUDHURI

Chapter 20. **Symptomatology, Pathology, and Treatment of Land Snake Bite in India and Southeast Asia**
H. ALISTAIR REID

Volume II

VENOMOUS VERTEBRATES

VENOMOUS SNAKES

Chapter 21. **Pharmacology and Toxicology of the Venoms of Asiatic Snakes**
D. K. CHAUDHURI, S. R. MAITRA, AND B. N. GHOSH

Chapter 22. **The Story of Some Indian Poisonous Snakes**
P. J. DEORAS

Chapter 23. **The Venomous Snakes of Australia and Melanesia**
HAROLD G. COGGER

Chapter 24. **The Pharmacology and Toxicology of the Venoms of the Snakes of Australia and Oceania**
E. R. TRETHEWIE

Chapter 25. **The Pathology, Symptomatology, and Treatment of Snake Bite in Australia**

E. R. TRETHEWIE

Chapter 26. **Classification, Distribution, and Biology of the Venomous Snakes of Northern Mexico, the United States, and Canada: Crotalus and Sistrurus**

LAURENCE M. KLAUBER

Chapter 27. **The Coral Snakes, Genera Micrurus and Micruroides, of the United States and Northern Mexico**

CHARLES E. SHAW

THE CHEMISTRY, TOXICITY, BIOCHEMISTRY, AND PHARMACOLOGY OF NORTH AMERICAN SNAKE VENOMS

Chapter 28. **The Chemistry, Toxicity, Biochemistry, and Pharmacology of North American Snake Venoms**

ANIMA DEVI

Chapter 29. **Comparative Biochemistry of Sistrurus Miliarius Barbouri and Sistrurus Catenatus Tergeminus Venoms**

CARLOS A. BONILLA, WAYNE SEIFERT, AND NORMAN HORNER

Chapter 30. **Neotropical Pit Vipers, Sea Snakes, and Coral Snakes**

ALPHONSE RICHARD HOGE AND SYLVIA ALMA R. W. D. L. ROMANO

Chapter 31. **Lethal Doses of Some Snake Venoms**

D. DE KLOBUSITZKY

CHEMISTRY AND PHARMACOLOGY OF THE VENOMS OF BOTHROPS AND LACHESIS

Chapter 32. **Chemistry and Pharmacology of the Venoms of Bothrops and Lachesis**

E. KAISER AND H. MICHL

Chapter 33. **Intermediate Nephron Nephrosis in Human and Experimental Crotalic Poisoning**

MOACYR DE F. AMORIM

Chapter 34. **Symptomatology, Pathology, and Treatment of Snake Bites in South America**

G. ROSENFELD

VENOMOUS SAURIANS AND BATRACHIANS

Chapter 35. **The Biology of the Gila Monster**

ERNEST R. TINKHAM

Chapter 36. **The Venom of the Gila Monster**

ERNEST R. TINKHAM

Chapter 37. **Venomous Toads and Frogs**

BERTHA LUTZ

Chapter 38. **The Basic Constituents of Toad Venoms**

VENANCIO DEULOFEU AND EDMUNDO A. RÚVEDA

Chapter 39. **Chemistry and Pharmacology of Frog Venoms**

JOHN W. DALY AND BERNHARD WITKOP

Chapter 40. **Collection of Toad Venoms and Chemistry of the Toad Venom Steroids**

KUNO MEYER AND HORST LINDE

Chapter 41. **Distribution, Biology, and Classification of Salamanders**

WOLFGANG LUTHER

Chapter 42. **Toxicology, Pharmacology, Chemistry, and Biochemistry of Salamander Venom**

GERHARD HABERMEHL

VENOMOUS FISHES

Chapter 43. **Venomous Fishes**
BRUCE W. HALSTEAD

Introduction

The so-called venomous animals described in these volumes possess at least one or more venom glands and mechanisms for excretion or extrusion of the venom, as well as apparatus with which to inflict wounds or inject the venomous substances. The venom often may be injected at will. These animals have been characterized by several authors as being "actively venomous." The "passively venomous" species have venom glands and venom-excreting ducts, but lack adequate apparatus for inflicting wounds or injecting venom (toads, frogs, and salamanders).

In their struggle for life, all venomous animals seem to be rigorously extroverted. Their energies are directed against the other animal and vegetable organisms in their environment. All the venom glands of these animals are of the exocrine type. Their venoms produced by special epithelial cells and stored in the lumina of glands are always extruded to the outer world, generally by biting or stinging such as is the case with shrews, serpents, saurians, some fishes, stinging social insects, scolopendrids, spiders, scorpions, molluscs, some echinoderms, and worms. Other animals envenomate their victims by direct bodily contact such as is true with caterpillars, a few moths, with several representatives of Coleoptera (*Paederus*, etc.), certain hydroids, jellyfishes, sea anemones, urchins, cucumbers, starfishes, and a few marine worms. The venomous compounds of toads and salamanders act generally in direct contact with mucous membranes (eyes, lips, or throats). All venomous animals possess characteristics which distinguish them from other members of the animal kingdom. Often venomous animals are hunters, predators, solitaries, and also enemies of other members of the animal kingdom. There are exceptions, of course, such as the social Hymenoptera.

The wounding apparatus is located on the head, on the hind portion of the body, or over the entire exposed surface of the animal. In shrews, serpents, Gila monsters, and some molluscs the venom apparatus is inside the mouth. The venom glands are in fact salivary glands; the bite is inflicted by modified teeth equipped with venom canals. In scolopendrids and spiders, the venom system is situated outside the mouth, but is in close proximity to it, and is designed for protection and acquisition of food. A strange situation is present in the scorpion: the venomous mechanism is found in the last segment of the body, "in cauda venenum." In fact, the scorpion sting must be considered "peribuccal." The scorpion is able to move its tail sufficiently far in front of its head to kill prey before eating it.

In the venomous Hymenoptera, such as ants, bees, wasps, and hornets, the wounding apparatus and the venom glands are also situated in the last segments of the abdomen, far from the mouth. The stinging mechanism often may function primarily as an ovipositor, having no connection with the mouth, and its venom-injecting function may only be secondary (honeybee queen).

In some venomous fishes and bristleworms, the venomous organs may be distributed over certain exposed portions of the body or may cover more or less the entire body surface, as in caterpillars, some echinoderms, and coelenterates, with no relation to the mouth.

The location of the venom system and the transformation of certain organs into venom-conducting channels may lead us to theorize as to the significance of venom in the animal kingdom. Why do venomous animals exist? What is the primary function of venom? Are venoms present principally for digestion of food, and is the wounding apparatus intended for self-defense and even attack in the never ending struggle for survival? Is the stinging designed mainly for oviposition, or for defense and attack, and is it combined with the mechanism for obtaining food and provision for offspring, as is true for all the solitary wasps? Thus, the role of venoms immediately appears very complex.

Shrews, serpents, scolopendrids, spiders, scorpions, solitary wasps, and some coelenterates are exclusively carnivorous, but they never feed on an animal that is already dead. They are predators and active hunters, and they capture and kill their prey. The social wasps, bees, hornets, and caterpillars are exclusively herbivorous; other venomous animals may be omnivorous, i.e., they will feed on creatures that have died of other causes.

The venom and wounding apparatus must also be considered in relation to sex, particularly in venomous adult insects such as bees, wasps, and hornets. Only the adult female Hymenoptera are poisonous, not the adult males. In all other venomous animals both sexes may be equally poisonous, or the males, which often are much smaller, may do less serious harm, as is true of most spider species.

Consideration of the localization of the venom apparatus, the mode with which these animals take their prey or their food, and the fact that often only one sex bears a venom-conducting apparatus may guide us to another very important question: For what purpose is the venom used?

Toads, venomous frogs, salamanders, and other "passively" venomous animals certainly may use their toxic products for self-defense. Often these animals may not rely entirely on their venomous power, but may prefer to use other protective methods such as mimicry, flight, and concealment. Caterpillars and other Lepidoptera larvae are also in this category. The latter procure food from plants, and desire peace from other animals.

One habit of several solitary wasps is rather curious: They use their stinging apparatus to paralyze spiders and other insects. Then they bring the prey to the nest, deposit an egg over the body of it, and close the orifice of the nest. The wasp larva, hatched a few weeks later, thus is provided with fresh food. These wasps possess a nerve- or muscle-paralyzing venom with long-lasting effect and they may attack in order to protect their offspring. The social Hymenoptera, such as the bee, wasp, and hornet, may use the venom apparatus primarily for defense against enemies, even against other groups of the same family. Also, they may attack and kill, e.g., the females of bees kill the male after fertilization of the new queen. A newly hatched queen bee kills all the other queens present in the hive. Thus the stinging apparatus and venom have both defensive and offensive functions. The venomous fishes, coelenterates, and echinoderms, as well as the bristleworms, use the wounding mechanism for self-defense.

It is curious that in all these animals—toads, salamanders, bees, wasps, hornets, caterpillars, some fishes, molluscs, sea cucumbers, urchins, starfishes, sea anemones, and jellyfishes—the venom and the biting or stinging system have nothing to do with the acquisition of food. Consequently, the venom apparatus will have nothing in common with the digestive or salivary organs.

In scolopendrids, spiders, scorpions, venomous snakes, Gila monsters, and venomous shrews, the venom apparatus and the wounding system are designed primarily for food acquisition, and not so much for the predigestion of food. This is especially true of scorpions. Their venoms are paralyzing, not digestive agents. They use the sting only when the prey is large and vigorous in defending itself, as spiders. Small animals are captured directly with the pedipalps, and immediately killed and eaten; the sting is not needed. Scolopendrids and spiders use the wounding apparatus in two ways: to hold the prey and introduce it into the mouth, or, when resistance is offered, to inject and kill the prey with the venom. The salivary function of venom in scolopendrids, spiders, and especially in scorpions may be questionable.

The situation appears to differ with snakes, venomous saurians, and Insectivora. Since the venom glands and the venom-injecting apparatus are found in the mouth, with phylogenetic transformation of a few teeth, and the glands may be true salivary glands, with or without digestive ferments and enzymes, one might think that the main purposes of the venom mechanism are the capture of prey and the partial breakdown of body tissues. On the other hand, it is also true that venomous snakes may be force-fed with rats, birds, and other small animals, which they do not envenomate but which they digest very well. Without the venom apparatus it may be very difficult or even impossible for them to obtain their food. Venom may also activate the digestive processes in some manner, but probably it is not necessary for this purpose. Scolopendrids, scorpions, spiders, snakes, venomous saurians, and shrews may be considered primarily of the offensive type, their venom apparatus being used for the capturing of food; secondarily, of course, they use such apparatus for self-defense.

Exact knowledge of the biological habits of venomous animals would provide more accurate answers as to the real purpose of venoms. Too little is known about this broad subject.

Another very important issue to be clarified concerns the intensity of action of the venoms of all species. For example, a venom of one species of snake may be several times more active in rats, mice, and birds than in other animals. Human beings are extremely sensitive to certain animal venoms. One-tenth of one milligram of Loxosceles venom may seriously endanger human life. It is conservatively estimated that 40,000 to 50,000 people throughout the world may be killed every year by accidental contact with venomous animals. Every scientific effort must be directed toward the prevention of this tragedy.

WOLFGANG BÜCHERL

VENOMOUS INSECTS

Chapter 44

Morphology and Function of the Venom Apparatus of Insects—Bees, Wasps, Ants, and Caterpillars*

ULRICH W. J. MASCHWITZ AND WERNER KLOFT

ZOOLOGISCHES INSTITUT DER UNIVERSITÄT FRANKFURT AM MAIN AND INSTITUT FÜR ANGEWANDTE ZOOLOGIE DER UNIVERSITÄT BONN, GERMANY

I. Introduction 1
II. The Sting Apparatus of Aculeate Hymenoptera 2
A. Introduction 2
B. Morphology 3
C. Function 25
D. Evolutionary Peculiarities 35
III. The Urticating Hairs of Lepidopterous Larvae 38
A. Scope of the Problem 38
B. Types of Urticating Hairs and Spines 42
C. Ecological Importance of Urticating Hairs and Spines 55
References 56

I. INTRODUCTION

The evolutionary origin of animal venoms might be directed by two principal needs of the organism: nutrition and defense. The urticating hairs of the slow-moving phytophagous caterpillars seem to have only protective function. On the other hand, the venom apparatus of the quick-moving aculeate hymenopterans has its origin in the needs of nutrition. The adults of this insect group, with the exception of many ants, feed mainly on carbohydrates of plant origin. However, most supply their brood with protein from animals captured with the aid of a sting apparatus, which in higher Hymenoptera, also serves for defense. The sting was modified in many ways during

* We would like to thank Mr. John Kefuss, University of Frankfurt, for the critical reading of the manuscript.

the course of evolution and became one of the most important body structures of this insect order.

II. THE STING APPARATUS OF ACULEATE HYMENOPTERA

A. Introduction

The ability of many hymenopterans to sting their victims painfully led many early scientists to investigate the sting apparatus. In the mid-eighteenth century, Reaumur (1740) and Swammerdam (1752) described the very complicated venom apparatus of these insects. However, not all groups of the order Hymenoptera are able to sting. Gerstaecker (1867) subdivided the hymenopterans into two groups, the Symphyta and Apocrita. The more primary phytophagous symphytes are no monophyletic unit (Ross, 1937; Oeser, 1961). The apocrites can be split into two suborders, the Terebrantia (ichneumons and gall wasps) and the Aculeata (sting wasps). The aculeates are of monophyletic origin. The origin of the terebrantes is not clear.

The symphytes do not have the ability to sting prey or enemies but the female possesses a stinglike ovipositor at the end of the abdomen. With the help of this apparatus some groups can bore holes into plant substrates in order to deposit eggs. We know from the comparative morphological investigations of Lacaze-Duthiers (1849) that this organ is homologous to the ovipositor of the terebrantes and the venomous sting apparatus of the aculeates. This explains why only female hymenopterans are able to sting. This chapter will be mainly concerned with the latter group, chiefly with the *Aculeata sensu stricto*, because some Bethyloidea also use their sting for oviposition. The following superfamilies (Imms, 1957) are treated: Formicoidea, Pompiloidea, Vespoidea, Scolicoidea, Sphecoidea, and Apoidea. Most of these groups use their sting for predation as well as for defense. The modified ovipositor is no longer capable of laying eggs (exception, Sapygidae). These are issued at the base of the sting (Bischoff, 1927). Some groups have a reduced sting apparatus, such as different ants (Forel, 1878; Foerster, 1912) and the meliponines (von Ihering, 1886), thus they are again unable to sting.

Investigations of the details of the venomous apparatus have been made by numerous authors. Mostly studied is the honeybee sting, whereas other groups such as Mutillidae and Pompilidae have been more or less neglected. Of those sting apparatus that have been studied, the chitinous parts are best known. Glands, muscles, and innervations have not been sufficiently investigated. Therefore a complete description of the total apparatus cannot as yet be given.

B. Morphology

I. The Nonglandular Parts of the Sting Apparatus (especially according to Oeser, 1961)

a. Position. In the more primary Hymenoptera with short ovipositor, the sting apparatus hangs completely free at the posterior end (Fig. 1). This

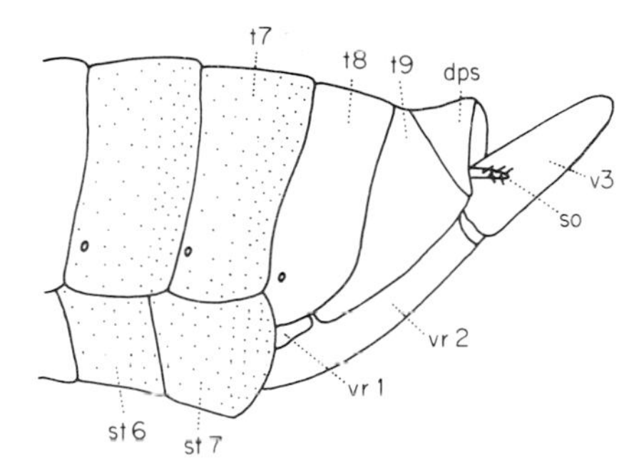

FIG. 1. Lateral view of the free ovipositor of a symphyte (*Cephus pygmaeus*) (according to Oeser, 1961). See Fig. 3 for abbreviations.

free position is found in the symphytes and in most of the terebrantes, except the Proctotrupoidea (Microphanurus); in this group parts of the ovipositor are withdrawn into the abdomen. In the *Aculeata s. str.*, the whole sting apparatus lies in the interior of the abdomen (Figs. 2 and 3). It is

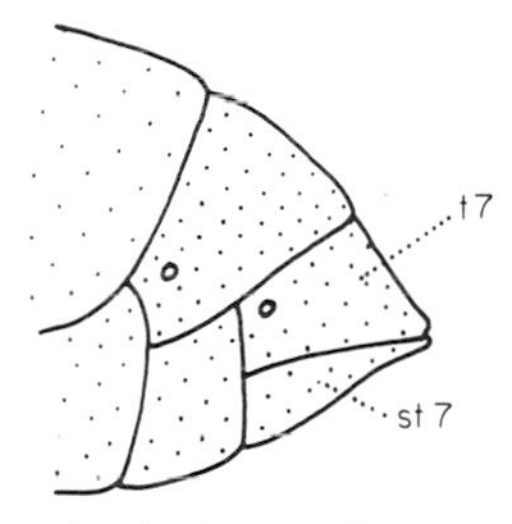

FIG. 2. Lateral view of the abdominal end of an aculeate (*Apis mellifica*). See Fig. 3 for abbreviations.

enclosed in a chamber at the posterior end of the abdomen which is formed by the seventh abdominal tergite and sternite. The seventh abdominal segment represents consequently the end of the body.* In two families of the bethylids (Chrysididae and Cleptidae) additional segments are drawn

* Abdominal segment No. 1 (epinotal segment) is found added to the thorax in the Apocrita.

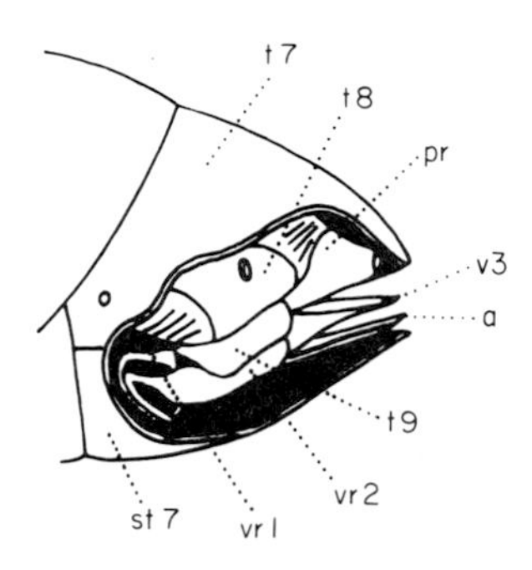

FIG. 3. Position of the sting apparatus in the abdomen of an aculeate (*Apis mellifica*) a, Aculeus; dps, dorsal proctiger sclerite; pr, proctiger; so, socius; st 6 and 7, sternum Nos. 6 and 7; t 7, 8, and 9, tergum Nos. 7, 8, and 9; v3, valvula No. 3; vr1, and 2, valvifer Nos. 1 and 2 (after Oeser, 1961).

into the interior of the abdomen. Most of the chrysidids form their sting chamber with the sclerites of segment 4. In the bethylid *Cephalonomia* however a small part of the eighth tergite is visible. This aculeate genus possesses a sting chamber form that is transitional between the terebrantes and aculeates *s. str.* The process of invagination of an organ into the body's interior is called "internalization" and may be found in plants and animals, indicating a more developed stage of evolution (Remane, 1952). Concomitant with this sting internalization are morphological alternations of the indrawn chitinized parts. In the aculeates the seventh abdominal segment, which terminates the abdomen in a more or less conical form, is basically unaltered; the eighth segment forms parts of the sting. In the sting chamber there is found, besides the sting apparatus, a dorsal lobe, the proctiger on which the anus ends. The vagina opens beneath the sting so that the sting chamber is also a cloacal chamber.

The sting of the aculeates can be divided into two functionally different parts: one of them is the glandular part, in which, among other substances, the venom is produced. The other has very complicated chitinous and muscular structures which serve in the ejection of venom and the protrusion and injection of the sting.

b. The Skeleton. The sting apparatus of the Hymenoptera is derived from a type of egg-laying apparatus which is found in most of the pterygote insects, the so-called orthopteroid ovipositor (Fig. 4). Segment 8 forms above its scalelike sternum (subgenital plate) a pair of triangular plates, called the first valvifers, which bear the first valvulae. At segment 9 the posterior valvifers with the second valves are situated. These are movable, interlocked with the first ones, and are enclosed by the third valvulae, which arise from

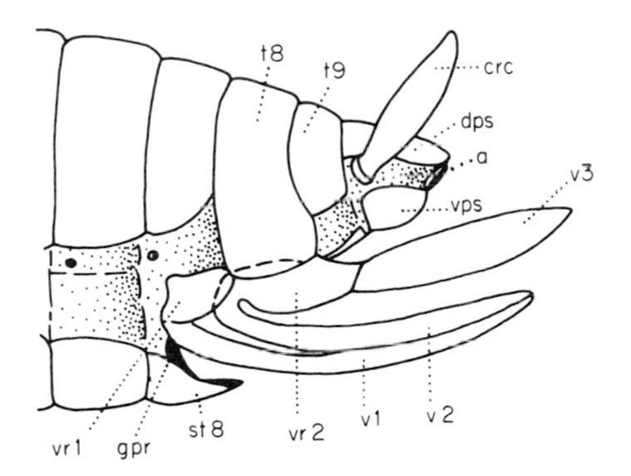

FIG. 4. Lateral view of the orthopteroid ovipositor. a, Anus; crc, cercus; dps, dorsal proctiger sclerite; gpr. gonopore; st 8, sternum No. 8 (subgenital plate); v1, 2, 3, valvula Nos. 1, 2, 3; vps, ventral proctiger sclerite; vr1, and 2, valvifers Nos. 1, and 2 (after Weber, 1954).

segment 9. Consequently the hymenopteran sting originates in the tergites and sternites of the eighth, ninth, and, following more recent studies (Trojan 1936; Flemming 1957), tenth abdominal segments and their appendices. This is substantiated by a great number of comparative morphological and ontogenetic studies of the chitinous parts, their muscles, and innervation. Ontogenetic studies were performed by Ouljanin (1872), Kraepelin (1873), Dewitz (1875, 1877), Beyer (1891), Kahlenberg (1895), and Zander (1899, 1922); comparative morphological studies of the chitinous parts, muscles, and innervation have been made by Lacaze-Duthiers (1849, 1850), Kraepelin (1873), Zander (1899), Betts (1922), Morison (1928), Snodgrass (1931, 1933, 1935), Trojan (1936), Rietschel (1938), Flemming (1957), and Oeser (1961).

The question of whether or not the appendices of the genital segments correspond to transformed extremities or are new formations is not yet clear and will not be discussed here. [For reviews see Zander (1899) and Oeser (1961).]

The correct nomenclature of the single sting parts is a difficult question. In the literature there are many different terms, partly derived from position, form, or function, partly based on supposed homologous interpretations (see Oeser, 1961). At the present stage of study, the terminology should follow, according to Oeser, the function or position of the sclerites (e.g., sting sheath, outer plate). A characterization of the parts in relation to their form seems to be unsuitable because of their great variation. If we employ, nevertheless, the most usual trivial names, we do so because they are well known and widely used. To discard them completely would cause further confusion.

Two functionally different sets of skeletal parts can be distinguished in the sting (Snodgrass, 1956). The first has a basal position and consists of a system of muscle-bearing plates, which serve for protrusion and movement of the sting and fix it in the sting chamber. They originate from tergite 9 and the two valvifers. The second set of chitinous parts, the long and pointed aculeus, is the real boring and stinging instrument. It is protruded when stinging and serves to inject venom. It consists of two pairs of valves. Both chitinous sets are connected in an unarticulated way by two pairs of curved processes of the second valves, the rami. If the proctiger which derives, as Snodgrass (1933) indicates, from segment 10, or according to others (Weber, 1954) from segment 10 plus 11, is fused with the ninth tergite, it is considered as part of the sting also. Because of the lateral symmetry, it is sufficient to describe only one of the paired skeletal elements.

The first proximal valvifer, which is a part of the moving apparatus, is called the triangular plate [fulcral plate, *pièce triangulaire*, Winkel; Fig. 5 (b), vr1]. It is derived from sternite 8, and articulates dorsoapically with

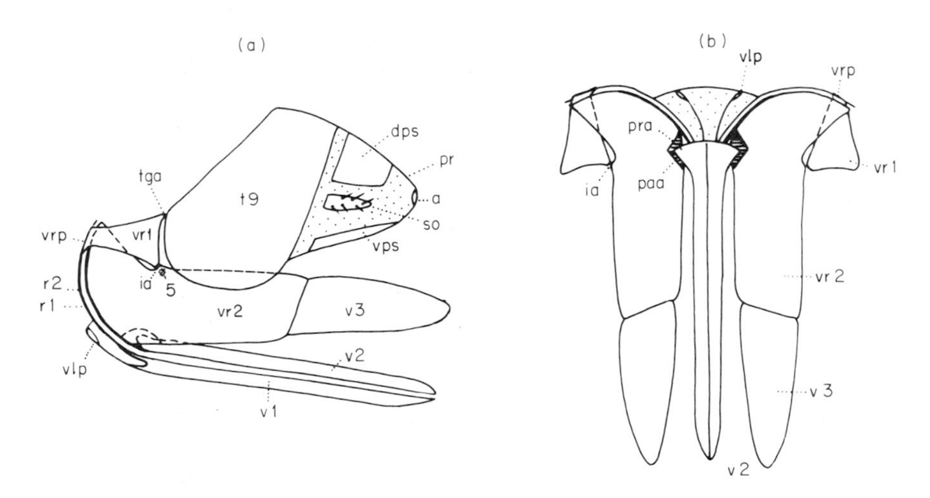

FIG. 5. General plan of a hymenopteran sting apparatus. (a) Lateral view; (b) dorsal view. a, Anus; dps, dorsal proctiger sclerite; ia, intervalvifer articulation; paa, pars articularis; pr, proctiger; pra, processus articularis; r1, r2, ramus of the first and second valvula; s, sensory bristles; so, socius; t9, tergum No. 9; tga, tergovalvifer articulation; v1, 2, 3, valvula Nos. 1, 2, 3; vlp, ventral processus; vps, ventral proctiger sclerite; vr1, 2, valvifer Nos. 1 and 2; vrp, valvifer processus (after Oeser, 1961).

tergite 9, the quadrate plate [tergovalvifer articulation, Fig. 5 (a), tga]. Furthermore, the anterior valvifer is connected with the second valvifer, the oblong plate, by a ventroapical articulation point [intervalvifer articulation, Fig. 5 (b), ia]. The ramus of the first valve [Fig. 5 (a), r1] is inserted at the rostral edge of the first valvifer. Contrary to the description of some authors (Lacaze-Duthiers, 1849; Sollman, 1863; Leuenberger, 1954), the connection of the two parts is rigid.

The second distal valvifer [Fig. 5 (b), vr2], also a plate of the moving apparatus, is called the oblong plate because of its shape, or because of its position, the inner plate (*écaille latralé,* lamina interna). It is strengthened by ridges. The ramus of the second valve is connected rigidly with the rostral edge of the second valvifer. Furthermore, there is a double articulation between the base of the second valvifers and the base of the second valves [Fig. 5 (b), paa and pra].

The pars articularis consists of a strongly sclerotized region at the inner side of the oblong plates. Therein rest the two processi articulares of the second valvulae which are hinged to it by a membranous cuticle. Because of this basal articulation, the sting can be moved up and down in the median line by antagonistic muscles. In the aculeate family of the Pompilidae this articulation is strongly reduced; in the Cleptidae it can no longer be found. A characteristic of this area in the aculeate's oblong plates is an indentation of the basal part containing the pars articularis. Oeser named it the incisura postarticularis (see Fig. 10, ipo); its function is unknown. In this way a sclerotized tongue is formed ventral to the incisura. It is a part of the oblong plate, as Oeser demonstrated, and not a newly formed lamina between valvifers and valves or even a part of the second valve, as several authors have supposed (Kraepelin, 1873; Carlet, 1890; Snodgrass, 1933; and Trojan, 1936).

Furthermore, the oblong plate can have one more indentation in front of the pars articularis, the incisura praearticularis (Oeser, 1961). In the aculeates it is rather unobtrusive in comparison to the incisura postarticularis. It facilitates the flexation of the rami, coming from the valvifers, along the processus articularis.

Above the second valvifer, connected with it by an intersegmental membrane, a further sclerotized part of the moving apparatus can be found, the strongly transformed ninth tergite [Fig. 5 (a), t9]. Its hemitergites are called the quadrate plates (outer plate, *écaille anale*, epipygium), though such a form seldom can be found. The dorsal connection of the two plates is reduced to a thin presegmental bridge or even is simply membranous. At its dorsal side the quadrate plate carries a rigid triangular apodeme, on which the muscles are inserted [see Fig. 10 (a), ap] (Rietschel, 1938). The proctiger is often connected with the ninth tergite carrying divided or undivided sclerites [Fig. 5 (a), vps]. The bristled appendices, called socii which can be found in the symphytes (Fig 1, so) and terebrantes, are lacking in the Aculeata.

Parts of the actual piercing instrument, the first valvulae [Figs. 5 (a) and 6, v1], are appendices of the first valvifers. The root of the first valve is

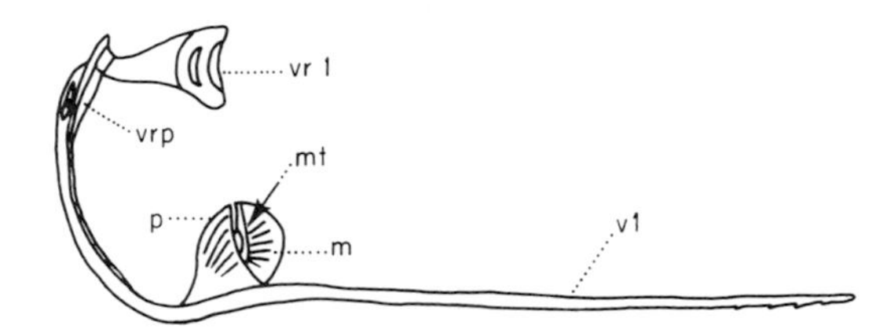

FIG. 6. Lancet with valve (*Apis mellifica*). m, Membranous appendix; mt, membrane support; p, chitinous pouch; v1, lancet (first valvula); vrp, valvifer processus; vr1, triangular plate (first valvifer) (after Oeser, 1961).

formed by a rostrally directed ventral process, the ramus. At the base of the first valves in many hymenopterans a ventral processus is found. This primitive structure, which is bifurcated, is mostly reduced in the aculeates. The ramus is curved and, as mentioned above, connected rigidly with the corresponding valvifer processus (Fig. 6, vrp). The first valves are long chitinous parts which are not attached to each other and often bear teeth and barbs at the distal end. Because of their function they are called lancets (stylets, anterior gonapophyses, *Stechborsten*). Their interior is hollow, but does not serve for venom transport. The lancets of most of the aculeates (with the exception of the Sapygidae, Pompilidae, and Vespidae) bear valvular lobes (Fig. 6). These consist of a chitinous processus (mt) on which are found two membranous appendices (m). A lateral chitinous pouch (p) (Apis) may be found also. These valves serve for transport of venom (or eggs). Some terebrantes have more simply constructed valves.

The lancets are held close against the undersurface of the second valvulae (Fig. 5, v2) by grooves that fit over tracklike ridges of the latter (Fig. 7). The second valvulae are fused into a single unit (stylet, lance, *gorgeret*,

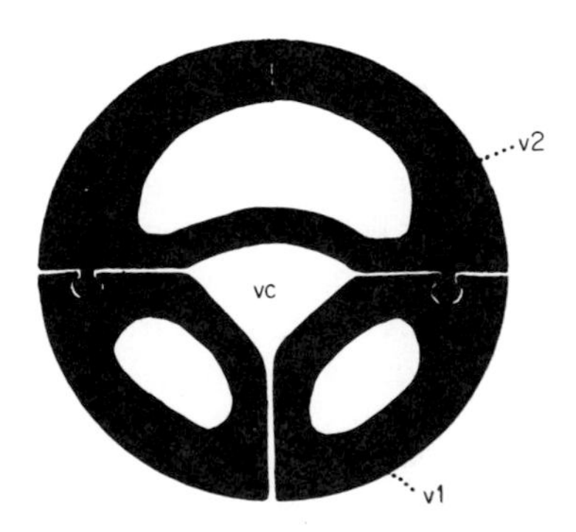

FIG. 7. Cross section of the aculeus. v1, Lancet (first valvula); v2, stylet (second valvula); vc, venom duct (after Oeser, 1961).

gonapophyses, *Stachelrinne*). In the aculeates the fusion of both second valves is total while in the symphytes and partly in the terebrantes the base is more or less disconnected. In many aculeates the basal part of the stylet is expanded in a bulblike enlargement [see Fig. 9 and 10 (a)]. (For inner structures of this bulbus, see Section II,B,2 and Fig. 9.) On the second valvulae, distal teeth and barbs can be found. The base of the stylet bears the processus articularis (Fig. 10, pra). The stylet is ventrally arched in between two ridges forming a groove. Between the stylet groove and the lancets a tube is formed, in which the venom (or the egg) is transported outward. The lancets can move independently of each other. The interlocking device is continued to the basal rami of the valves. Similar to the lancets, the stylet is hollow and provided with tracheas and a nerve (Janet, 1898; Trojan, 1922; Flemming, 1957).

The furcula (*Gabelbein*) also belongs to the movable part of the sting apparatus and is located on the bulbus (see Fig. 10, f). It possesses a more or less bifurcated anterior and an unpaired posterior end. Some important sting muscles end here which originate at the oblong plates and play a part in the up- and down-movements of the sting. According to Trojan (1936) and Flemming (1957), the furcula is an apodemal derivation of sternite 10. This is contrary to Zander (1899), who derived the furcula from sternite 9.

This chitinous structure together with the incisura postarticularis, the lack of socii, and of apophyses of tergite 9 form a combination of characteristics that is found only in the Aculeata (Bethyloidea and *Aculeata s. str.*) and mark these as a monophyletic group. The formation of a natural system based on comparative morphological studies of the sting apparatus therefore is possible, though several authors, according to Oeser (Lacaze-Duthiers, 1850 and Berland and Bernard, 1951) have thought that the sting would be phylogenetically too plastic for this purpose.

The third valvulae (gonostyli, posterior lateral gonapophyses) belong neither to the moving apparatus nor to the piercing instrument. They originate at the apical end of the second valvifers (Fig. 5, v3) and consist of one segment (e.g., Apidae, Vespidae), or two segments (e.g., Pompilidae, Mutillidae, Tiphiidae). They wrap up the resting aculeus with their concave inner side and thus are called sting sheaths.

The eighth tergite is not part of the aculeate sting apparatus. It is inside the sting chamber, and represents the point of muscle and intersegmental membrane attachment on which the sting is fixed. It bears the last spiracles and is called therefore the spiracular plate. In more primitive aculeates it consists of a single part (*Ammophila, Polistes*). In higher aculeate forms, a sclerotized premarginal ledge connects the two hemitergites (*Vespa*) or the latter are even separated into two independent parts (Apidae; Zander, 1899; Rietschel, 1938).

c. Muscles and Innervation. i. General view. In order to understand the function of the chitinous sting parts in the stinging act, it is necessary to know the musculature. Since the parts of the sting apparatus are derived from segments and their appendices (Lacaze-Duthiers, 1849; Ouljanin, 1872; Kraepelin, 1873), muscles and innervation furthermore are important for homologization. Therefore they have been studied by several authors (muscles; Betts, 1922; Morison, 1928; Snodgrass, 1933, 1935, 1942; Trojan, 1936; Rietschel, 1938; muscles and innervation: Rehm, 1940; Flemming, 1957). Almost all investigators restricted themselves to the honeybee. Other bees and some wasps were only treated for comparison purposes (Rietschel, 1938; Flemming, 1957). Of other families only the Formicidae were studied (Foerster, 1912; Callahan *et al.*, 1959; Hermann and Blum, 1966). Their muscles correspond largely to those of the honeybee.

ii. Muscles. Description and classification of the musculature follow essentially the results of Snodgrass (1933), Rietschel (1938), and Flemming (1957). The body muscles can be divided into four groups: (a) dorsal intersegmental muscles; (b) ventral intersegmental muscles; (c) dorsoventral or lateral muscles (intrasegmental muscles); (d) spiracular muscles. A survey of all these muscles is given in Table I (Snodgrass, 1933; Rietschel, 1938).

iii. Innervation. The central nervous system of the aculeates from which the innervation originates possesses only a few fused ganglia. The honeybee worker has five abdominal ganglia. In the fourth ganglion are fused segmental ganglia 6 and 7; in the fifth are fused the remaining ganglia, (8–11). In spite of fusion, the segmental nerves (Flemming, 1957; Rehm, 1940) have maintained their original segmental connections. With the exception of No. 11, all other nerves which emerge from the fused terminal ganglion complex are paired (Fig. 8). The segmental nerve of each side splits into two secondary branches, which innervate the corresponding intra- and intersegmental muscles of their segment.

The honeybee queen has only four abdominal ganglia. Their sting innervation has been studied intensively by Ruttner (1961). In *Vespa* workers and queens five ganglia can be found as in the worker bee; in this genus however the segmental paired nerves 7–11 originate in the terminal ganglion complex.

d. Sensory Organs of the Sting Apparatus. Sensory bristles can be found in the chitinous parts of the moving apparatus. The oblong plates of all hymenopterans bear bristles at the articulation point near the triangular plate [Figs. 5 (a) and 10 (a), s]. In the nonaculeates the bristles are dispersed (Soliman, 1941; Cohic, 1948; Oeser, 1961). In the aculeates, however, they form a distinct bristle field, a general characteristic of this group, as Oeser demonstrated. He found further sensory bristles in varying number and distribution at the rami of the stylet, where these are fused with the oblong

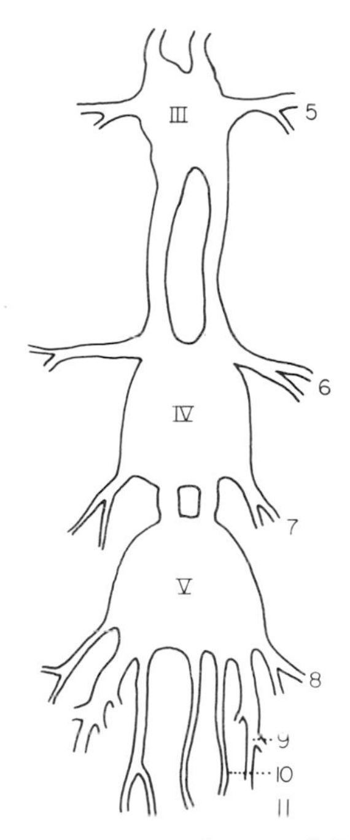

FIG. 8. Abdominal central nervous system of *Apis mellifica* (worker). III, IV, V, Ganglion Nos. 3, 4, and 5. 5–11 Segmental nerves Nos. 5–11 (after Flemming, 1957).

plates. Similar bristles were found in ants (Janet, 1898; Callahan *et al.*, 1959). At the basal articulation, bristles can be found at the base of the stylet and on the oblong plates (Oeser, 1961; Callahan *et al.*, 1959). As a rule, sensory bristles can be found on the sting sheaths. In *Philanthus triangulum* (Sphecidae) a distinct organ exists which is provided with a large nerve fiber (Rathmayer, 1962).

Sensilla campaniformia have been described on the stylet and the lancets of the honeybee by Trojan (1922), in *Philanthus triangulum* by Rathmayer (1962), and in *Myrmica* by Janet (1898).

The function of all these sense organs has not yet been clearly demonstrated. It must, however, be supposed that they provide proprioreceptive information on the position of the single parts to each other respective, of the degree of their deflection. The sheath lobes of the sting in terebrantes and aculeates were long supposed to function as sensory organs for localization of the oviposition or stinging spot (Imms, 1919; James, 1926; Hanna, 1934; Soliman, 1941; Bucher, 1948; Zander, 1922; Weinert, 1920; Weber, 1933). Recently Rathmayer (1962) has brought forth strong evidence for such touch sensory function in *Philanthus triangulum*. He observed that this sphecid

TABLE I

MUSCLES OF THE STING APPARATUS OF THE HONEYBEE[a]

Muscles	Muscle No.	Origin of the muscle	Insertion of the muscle
Intersegmental muscles of abdominal segments VII/VIII			
A. Dorsal muscles			
musc. adductor dorsalis internus	1	Behind apodeme of tergum VII	At apodeme of spiracle plate
musc. adductor dorsalis externus	2	Close at origin of m.1	At premarginal ridge of spiracle plate
m. abductor dorsalis	3	Caudal margin of tergum VII	Apodeme of spiracle plate
B. Ventral muscles			
m. adductor ventralis externus	8	Apodeme of sternum VII	Anterior end of side lobe of spiracle plate
Muscles of segment VIII			
A. Dorsoventral muscles			
m. adductor lateralis medius	14	Side margin of spiracle plate	Apodeme of triangular plate
B. Spiracle muscles			
m. occlusor spiraculi	12	One end of closing valve of spiracle	Other end of spiracle closing valve
m. dilatator spiraculi	13	Lower end of spiracle valve	Anterior part of side lobe of spiracle plate
Intersegmental muscles of abdominal segments VIII/IX			
A. Dorsal muscles			
m. adductor dorsalis internus	11	Apodeme of spiracle plate	Upper margin of apodeme of quadrate plate
m. adductor dorsalis externus	10	Dorsal at apodeme of spiracle plate	Middle of quadrate plate's apodeme base

B. Ventral muscles reduced			
Muscles of segment IX			
A. Dorsoventral muscles			
m. abductor lateralis	17a and 17b	Posterior margin of quadrate plate's apodeme	End of stylet ramus and anterior end of oblong plate
m. adductor lateralis posterior	18	Inner face of quadrate plate	Posterior end of oblong plate
Intersegmental muscles of abdominal segments IX/X			
B. Ventral muscles			
m. adductor ventralis externus	19a	Inner face of oblong plate	Furcula
m. abductor ventralis	19b	Middle of oblong plate at articulation process with triangular plate	Furcula
m. of uncertain origin	20	Upper part of stylet ramus	Processus articularis of stylet
Muscles of segment X			
Rectal muscles			
m. dilatator primus recti	21	Posterior end of apodeme of quadrate plate	Lateral at proctiger
m. dilatator secundus recti	22	Close at No. 21	Middle of proctiger

[a]After Snodgrass (1935), Rietschel (1938), and Flemming (1957). The muscles of the closing mechanism in the stylet (see Section II, B, 2) are not yet homologized.

always stings its victim in certain nonsclerotized cuticular spots which are likely discovered by the sensory organs of the sheaths.

2. Special Peculiarities of Some Aculeate Stings

The best-known stinging aculeates are representatives of the genus *Apis,* the honeybees, and the genus *Vespa*, the wasps and hornets (Apidae, Vespidae). Their sting apparatus corresponds to the normal type. A series of remarkable peculiarities, however, can be found. The genus *Formica* of the Formicidae, to which belong the well-known mound-building ants of the Palearctic region, possess a vestigial sting which is no longer suitable for stinging. In the more primitive wasp sting the spiracular plates are rigidly connected. Because of this, in combination with strong muscles, the wasp sting is very movable. The barbed sting of *Vespa* possesses sharp edges at the end and at the margins of the stylet which are capable of severing fibers. When the sting is retracted from the wound of the enemy these sharp edges prevent the sting from being caught by the skin of the opponent (Rietschel, 1938).

In *Vespa* the two lancets differ in their structure (Schlusche, 1936). The left one is thickened in the last third, but not the right one. Both have two thin lamina located caudally, which are directed against the stylet. They are larger on the left lancet and overlap each other. In this way a tube is formed which conducts the venom to the end of the sting. At the tip, the lamina narrow again, so that a cleft is formed from which the venom issues.

Special structures are found in the bulb of vespids. The lack of the lancet valves has been mentioned. The venom channel enters the interior of the bulb and goes through it as a distinct tube till it ends free between stylet and lancets. A locking apparatus of the venom duct could not be found (Fig. 9; Schlusche, 1936).

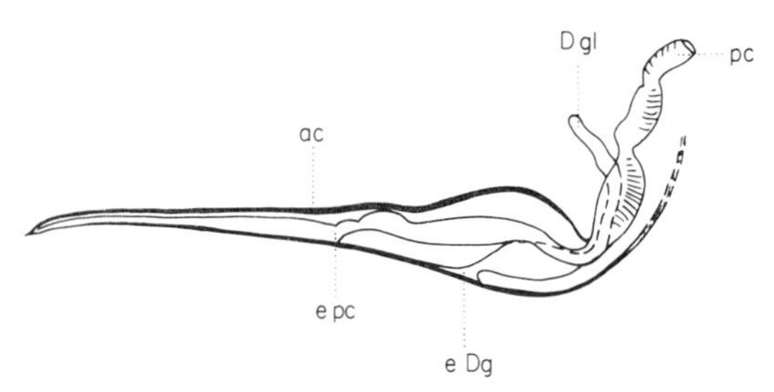

FIG. 9. Longitudinal section of the aculeus of *Vespa*. ac, Aculeus; D gl, Dufour gland; e Dg, end of the Dufour gland; e pc, end of the venom duct; pc, venom duct (after Schlusche, 1936).

In the Apidae the spiracular plates are much more separated than in other families and the separation is greatest in the worker honeybee. Here the hemitergites are completely separated, partly reduced (Zander, 1899) and the quadrate plates are divided. These separations and simultaneous reductions of muscles are the reason why a honeybee cannot move its sting other than forward and backward and why it easily loses its sting during the stinging act (see Section II,D,1). Furthermore, the Apidae have a membrane stretched over the bulb which connects the margins of the oblong plates with the furcula (Kraepelin, 1873). It is well developed and strongly bristled in the honeybee worker (*Stachelrinnenwulst* of the German authors). This development is probably correlated with its storing function for the alarm substance (see Section II,C,3). In the honeybee queen these extreme reductions and special developments are lacking. Furthermore, still other differences between the unmodified queen bee sting and the derived worker bee sting can be found, so that a characteristic and obvious caste dimorphism of the honeybee sting apparatus exists. This is demonstrated in Table II. The strong specialization of the worker bee sting presumably may be derived from its special function as defense organ against vertebrate enemies (see Section II,D).

TABLE II

CASTE DIMORPHISM OF THE HONEYBEE STING[a]

Sting part	Structure in the queen bee	Structure in the worker bee
Whole sting	Big, strong	Smaller, delicate
Stylet	Saberlike, curved down	Straight
Lancet	Three barbs, otherwise like worker bee	Ten strong barbs
Spiracular plate	Strong, fixed fast with the other plates	Weak, loosely fixed at the sting (preformed breaking point)
Quadrate plate	Normal as in other Apidae	A little reduced
Furcula	Especially unpaired stem, big and strong	Weakly developed
Membrane stretched over bulb	Membranous as in other Apidae	Very strongly developed with many bristles
Sting sheath	Strong, curved, strongly sclerotized as in other Apidae, strongly bristled, especially at the top; thick nerve	Flabby, straight, only at a narrow strip some sclerotization, bristles almost fully reduced; nerve reduced
Muscles	Normal as in other Apidae	Spiracle plate muscles especially reduced (Nos. 10 and 11)
Venom gland	Larger as in worker bee, short unpaired, long paired tube	Short; long unpaired, very short paired tube
Dufour gland	Larger as in worker bee	Relatively small
Sting sheath gland	Narrow palisade epithelium under strong sclerotized chitinous strip	Almost the whole circuit with palisade epithelium

[a] After Zander (1922) and Maschwitz, unpublished.

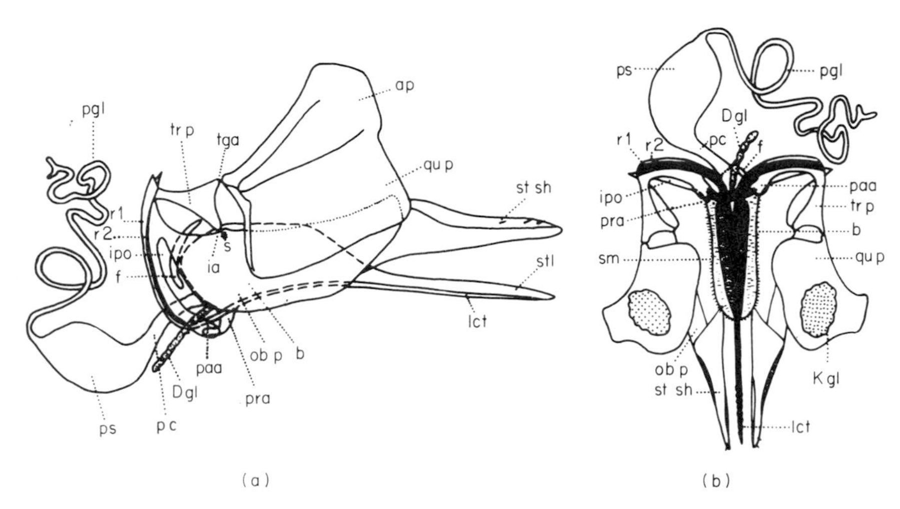

FIG. 10. Sting apparatus of the honeybee worker. (a) Lateral view (modified from Oeser, 1961). (b) Ventral view (combined after Snodgrass, 1935; Oeser, 1961; and our own preparation). ap, Apodeme; b, bulb; D gl, Dufour gland; f, furcula; ia, intervalvifer articulation; ipo, incisura postarticularis; K gl, Koshewnikow gland; lct, lancet; ob p, oblong plate; paa, pars articularis; pc, venom duct; pgl, venom gland; pra, processus articularis; ps, venom reservoir; qu p, quadrate plate; r1, ramus of the lancet; r2, ramus of the stylet; s, sensory bristles; sm, setose membrane; st sh, sting sheath; stl, stylet; tga, tergovalvifer articulation; tr p, triangular plate.

The worker bee sting is shown in Fig. 10. The interior of the sting bulb of the bee is rather complicated. In it the venom duct can be closed and opened by means of a locking apparatus (Trojan, 1922). The venom channel into the bulb narrows to a thin horizontal cleft which is normally almost shut. Two closing muscles originating at its dorsal side and inserted downward in the bulb are able to pull down the roof of the cleft and in that way shut it completely. A thick muscle inserted at the ventral side of the cleft opens, when contracted, by pulling down its underside (Trojan, 1922) so that the venom can be pumped through.

Similar locking structures are known in Myrmicinae ant stings (Janet, 1898; Callahan *et al.*, 1959). Ants show a great variety of sting modification. Their morphology has been studied by a series of authors (Meinert, 1860; Dewitz, 1877; Forel, 1878; Janet, 1898; Zander, 1899; Pavan and Ronchetti, 1955; Callahan *et al.*, 1959; Blum and Callahan, 1963). A detailed comparative morphology was worked out by Foerster (1912). According to the latter and to Forel (1878), the Ponerinae possess a strong, well-developed sting. The Pseudomyrmicinae (Blum and Callahan, 1963) and partly the

Myrmicinae (Foerster, 1912)* also have well-developed stings. In the latter subfamily, however, many genera with sting apparatus reductions are found. Their sting parts are relatively smaller and partly reduced and these genera can no longer sting. The Dorylinae possess more or less atrophied stings (Foerster, 1912; Stumper, 1960). They are still more reduced in the Dolichoderinae. In this subfamily tiny sting organs exist. Here, as in some Myrmicinae genera, the oblong and quadrate plates are reduced with exception of their thickened margins. The tip of the aculeus also is atrophied. The most stunted stings can be found in Camponotinae. Their chitinous relicts serve only as support for the venom channel.

The modified and reduced sting muscles close the venom duct. Dewitz (1877) identified the rudiments of the sting of *Formica* as homologous with the sting parts of *Myrmica* and *Apis*. He supposed, however, that it was an organ which stopped developing at a primitive stage. This error was corrected in the comparative morphological study of Forel (1878) and the ontogenetic researches of Beyer (1891). Foerster investigated a series of genera and species. He found that *Lasius* and *Plagiolepis* have the most rudimentary stings; less reduced are the sting apparatus of *Camponotus* and *Polyergus* and least reduced is that of *Formica*. The sting rudiment of *Formica* is shown in Fig. 11.

The chitinous rudiment of the stylet (Fig. 11, stl) is found in the middle. It is connected with the rudiments of the oblong plates; the connection with the lancets by means of a fold is apparently absent. The lancets' valves are lacking; the lancet tips are bulblike and thickened; furcula and quadrate plates are rudimentary.

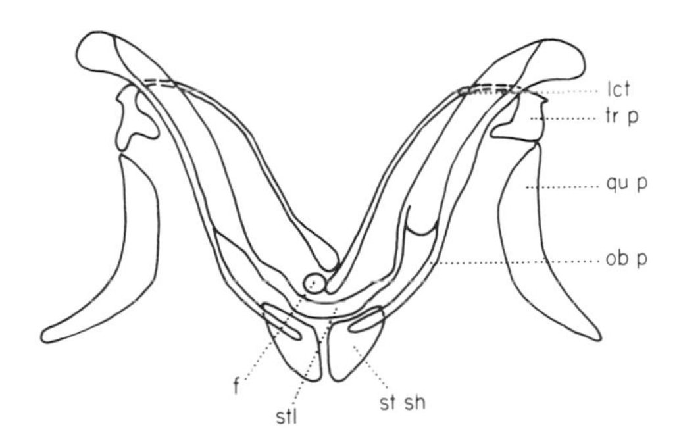

FIG. 11. Reduced sting apparatus of *Formica pratensis*. f, Furcula; lct, lancet; ob p, oblong plate; qu p, quadrate plate; st sh, sting sheath; stl, stylet; tr p, triangular plate (after Foerster, 1912).

* For modern subfamily classification of ants, see Brown (1954).

3. Venom Glands

a. General View. The glands of the aculeate sting apparatus may be derived phylogenetically from ectodermal glands of the ovipositor. In symphytes, as far as is known, they produce no venom; in aculeates and in many terebrantes they have changed their function and became accessory glands of the venom apparatus, with corresponding functions.

Up to now there have been very few comparative morphological studies on sting glands (Bordas 1895, 1897; Robertson, 1968). The primitive groups especially have been neglected. Even in the best known aculeates, only some of the families and their genera have been studied. Furthermore, the known glands are formed very differently in various groups. Therefore a general morphological review is not yet feasible. The functions of the sting glands, with the exception of the venom gland, are mostly unknown.

Several glands can be distinguished in the sting apparatus. Two are usually present, the venom gland *s. str.* and one accessory gland which is called the Dufour gland. In addition to these glands, others can be found: The Koshewnikow gland, the Bordas gland, and the sting sheath gland. In the phylogenetically advanced ant subfamily of the Dolichoderinae and partly in the Meliponinae (Apidae) these glands are more or less reduced. Instead, other glands have taken over their combat function; in the Dolichoderinae it is the newly formed anal gland (Forel, 1878) or even the mandibular gland in some Meliponinae (*Oxytrigona,* Maidl, 1934; Lindauer, 1957) (for ant mandibular glands with combat function see Pavan, 1956; Ghent 1961; Blum *et al.*, 1968; and Blum *et al.*, 1969).

b. The Venom Gland. The largest sting gland is the venom gland, which was discovered by Swammerdam (1752). It is situated between the rectum and the vagina, and ends in the aculeus. Three different regions may be distinguished. The secreting part of the gland consists of tubes which produce the venom to be conducted into a reservoir. From the latter a canal carries it into the aculeus. The gland cells of the secreting tubes are situated cylindrically in one or more layers under a basal membrane. From these begin small chitinous canaliculi which penetrate the cuticular intima and end in the lumen of the tube (Leydig, 1859; Bordas 1894; Forel, 1878; Heselhaus, 1922; Pawlowsky, 1927). The beginning of this organelle is widened to an ampulliform or dendritically branched vesicle in the gland cell (Pawlowsky, 1914; Autrum and Kneitz, 1959). Investigations on the fine structure of the aculeate venom gland cells are lacking. They are probably not very different from the venom gland cells of the terebrantes which have been described by Ratcliffe and King (1967) in *Nasonia vitripennis.* The structure of the latter is typical for protein-producing cells with a densely packed endoplasmatic reticulum and numerous ribosomes (Fig. 12.) The venom

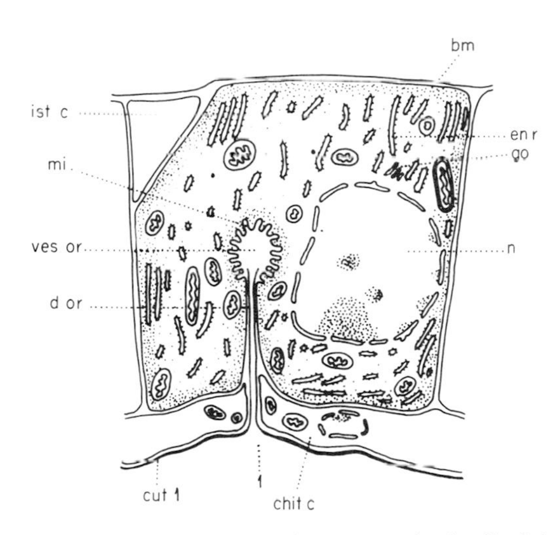

FIG. 12. Diagram of the ultrastructure of a venom gland cell of the chalcid *Nasonia vitripennis* (after Ratcliffe and King, 1967). bm, Basal membrane; chit c, chitinogenous cell; cut l, cuticular lining; d or, duct of organelle; en r, endoplasmatic reticulum; go, Golgi vesicles; ist c, interstitial cells; l, lumen of the secretion tubus; mi, microvilli; n, nucleus; ves or, vesicular organelle.

gland cells may belong to the Leydig gland type, one active gland cell originating from four cells (Leydig, 1859; Lukoschus, 1962b). Morphology of the venom gland varies widely, not only between the families, but even between genera and species. Some forms are shown schematically in Fig. 13. Pawlowsky (1927) gave a tentative description of three gland types according to the number of gland tubes, location of orifices, and structure of the reservoirs: (1). Braconid type: Numerous gland tubes end basically in the reservoir which has muscles but no glandular elements. (2). Vespid type (Vespidae, Pompilidae): Two venom gland tubes end in a distinct spherically formed reservoir whose wall has a strong muscle layer but no glandular elements. (3). Apid type (Apidae, Sphecidae, Mutillidae, Scoliidae): An unpaired venom gland is widened into a saclike reservoir which contains glandular elements but no muscles.

Without going into details, one can say that already in the braconids this gland type may strongly diverge (Robertson, 1968). The glands of the other terebrantes may also be differently structured. At best one can speak of a vespid type. All previously investigated Vespidae and Pompilidae correspond to this type. The reservoir is characteristically covered with strong transversal muscles which are spread out between four chitinous ridges of the wall. The gland tubes end divided in the reservoir apex. Here they fuse into a single glandular tube which is invaginated into the cavity of the reservoir (Robertson

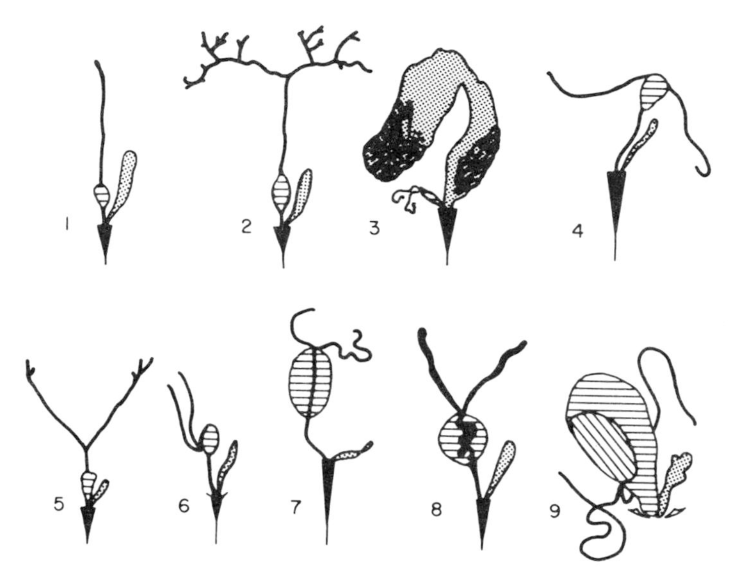

FIG. 13. Venom and Dufour glands of the sting apparatus of some aculeate families. Left, venom glands; right, Dufour glands. Apidae: 1, *Osmia spinulosa* (Pawlowsky, 1927); 2, *Bombus distinguendus* (Pawlowsky, 1927); 3, *Anthophora acervorum* (Heselhaus, 1922); Sphecidae: 4, *Tachysphex nitidus* (Rathmayer, 1962); Mutillidae: 5, *Mutilla maura* (Pawlowsky, 1927); Scoliidae: 6, *Scolia villosa* (Pawlowsky, 1927); Vespidae: 7, *Vespa germanica* (Maschwitz, 1964); Formicidae: 8, *Myrmica laevinodis* (Forel, 1878); 9, *Formica polyctena* (Maschwitz, 1964).

1968). In the third group, all the characteristics given by Pawlowsky vary so strongly that it is not possible to speak of a type. In most cases the glandular tubes are originally paired before joining (Sphecidae, Scoliidae). In the Thynnidae they end evenly divided from each other in the venom reservoir. Only in the Apidae the tubes are more or less distally fused and in extreme cases (*Osmia, Megachile*) unpaired. The single tubes may be multiple branched. The glandular tubes may not necessarily end apically in the reservoir, for instance in *Campsomeris* (Scoliidae, see also Formicidae). That the gland tube widens gradually into a reservoir which in its beginning part contains a glandular epithelium, may only hold true for the Apidae. The reservoir gland cells may have a different structure as the tube gland cells (Heselhaus, 1922; Kerr and De Lello, 1962). Heselhaus states that they have no canaliculi. In *Apis*, according to Autrum and Kneitz (1959), these cells are lacking. In *Scolia* the gland tube is invaginated into the reservoir. The reservoirs are saclike or spherically shaped. As a rule they possess a muscle layer which sometimes is strongly developed (*Diamma, Campsomeris*, Scolicoidea; Robertson 1968). It is lacking only in the Apidae. The reservoir ends in a duct leading to the aculeus. Its length differs widely in different species. The duct walls have no muscles and are stiffened by irregular chitinous folds. In the Apidae (and Myrmicinae) a locking mechanism could be found which is lacking in wasps (see Section II, B, 2). As a common characteristic of type III it can be said that the glands of these groups are emptied as a rule (exception Meliponinae) by a pumping mechanism. Emptying by a pump mechanism is also true for most of the ants. Because of their specialities the venom glands of this family, and also of the bee subfamily Meliponinae, will be considered separately.

i. Meliponinae venom glands (Kerr and De Lello, 1962). Although the Meliponinae no longer possess a functional sting, in some of the primitive species (*Meliponula bocandei, Trigona freiremaiai*) the reservoir of the venom gland is strongly enlarged and occupies the last third of the abdomen. The tube of the original venom gland has disappeared or exists only as a rudiment (*Melipona quadrifasciata*). The sac surface encloses a folded gland epithelium which may be homologous to the reservoir gland cells of other Apidae. In higher Meliponinae the sac is smaller (*Trigona droryana*), rudimentary (*Trigona tataira, T. spinipes, Melipona marginata, M. quadrifasciata*), or lacking in the worker bees (*Trigona postica, T. xanthotricha, Lestrimelitta limao*). In the species where the worker glands are reduced or completely lacking, the females have less reduced glands. Secretions of the sac glands are lipophil and nonmiscible with water.

ii. Ant venom glands. The venom glands are well developed in the ant subfamilies Myrmeciinae, (Cavill *et al.*, 1964), Ponerinae (Forel, 1878; Whelden, 1960; Hermann and Blum, 1966; Robertson, 1968), Pseudomyrmicinae (Blum and Callahan, 1963), and partly in the Myrmicinae (Forel, 1878; Janet, 1898; Callahan *et al.*, 1959). The reservoir is mostly oval in form and with exception of some Myrmicinae, it is surrounded by a muscle layer. In primitive genera (*Myrmecia, Bothroponera*) the proximal part of the gland tube is situated between these muscles and the wall of the reservoir (similar as in, e.g., *Scolia*) before it ends in the lumen. The two usually unbranched gland tubes unite often before ending apically or laterally in the reservoir. It is typical that the gland tube does not end directly in the reservoir. It is more or less invaginated and forms in the reservoir lumen a double-walled duct which may be wrinkled. Before its orifice the duct may be thickened by a great number of gland cells (venom gland with button, Forel 1878. Fig. 13, No. 8). While in *Myrmica* all gland cells are formed in the same way (Forel, 1878), in *Solenopsis saevissima* the gland cells of these two regions differ histologically (Callahan *et al.*, 1959). At the gland tube end in *Solenopsis* a filter-like structure was described. In the ant subfamilies of Dorylinae and Dolichoderinae the same gland type is found (Forel, 1878). Its size however is greatly reduced.

Strongly divergent are the formic acid-producing venom glands of the ant subfamily Camponotinae (Forel, 1878; Fröhlich and Kürschner, 1964, Fig. 13, No. 9). The secreting gland tubes lie on a oval reservoir which ends in a short and broad emptying duct. The chitinous remainders of the sting strengthen this orifice. Around a tunica propria thin muscles are situated which surround the gland tubes and the venom sac. The secreting elements of the gland consist of two parts. They begin with two free gland tubes which penetrate the tunica, fuse, and end in an oval white cushion of clustered tubes laying loose on the reservoir wall. The cushion consists of a long, densely clustered tube which ends in the middle of the reservoir's lumen. The convoluted tube which is well supplied with trachea may be unramified (*Camponotus, Lasius*) or may bear blind-ending branches (*Formica, Polyergus, Cataglyphis*). Cushion tissue and free gland tissue differ histologically. The function of the single gland parts is still unknown. The gland system may be very large. In *Formica rufa* its length is two thirds that of the gaster length. Queen glands are smaller than those of the workers.

4. The Dufour Gland

Dufour (1841) described another gland at the sting apparatus, ventral to the venom gland. It can be found generally in the Apocrita (Forel, 1878; Bordas, 1895; Pawlowsky, 1927; Heselhaus, 1922). In the Symphyta a homologous gland seems to exist (Robertson, 1968). The Dufour gland is of a more simple structure than the venom gland and consists of a simple tube which secretes and stores its products at the same time. The gland cells are located mostly in one layer on the basal membrane. They secrete directly into the lumen, which is lined with a cuticular intima. Generally its folded surface is covered by a thin muscle layer. The gland seems to possess, with exception, of the Vespidae (Schlusche, 1936), no specific muscular closing mechanism. Normally the gland is considerably smaller than the venom gland (Fig. 13). In the Apidae its size varies greatly. While it is lacking in some Meliponinae (Kerr and De Lello, 1962), in some solitary bees it may be many times larger than the venom gland and may contain numerous saclike pouches [Fig. 13, No. 3 (*Anthophora*); Heselhaus, 1922]. Also in the ant *Cremastogaster scutellaris* the Dufour gland is many times larger than the venom gland (Maschwitz, unpublished). With the exception of the Vespidae the Dufour gland ends inside of the aculeus. In *Vespa* the gland tube penetrates into the bulb, perforates its underside, and freely ends in the sting chamber (Fig. 9). According to Trojan (1930), that is true for *Apis* also. This could not be confirmed by Callahan *et al.*, (1959) and Robertson (1968). In the honeybee the Dufour gland approaches the bulb, takes its course along its underside between the dorsal venom channel and a ventral membrane which is spread out between the beginning of the second rami [Fig. 10 (b)]. Beyond the venom duct orifice it ends in the aculeus. In myrmicine ants (*Myrmica* and *Solenopsis*) the Dufour gland ends within the aculeus also (Janet, 1898, and Callahan *et al.*, 1959), while in Camponotinae, owing to the strong reduction of the sting, it freely ends under the venom duct.

There are many speculations over the general function of this gland. Dufour (1841) and Kraepelin (1873) thought of a smearing function. Forel (1878), Trojan (1930), and Schlusche (1936) proposed a relation to the sexual organs because in ant, bee, and wasp queens, the glands are bigger than in workers. Moreover Schlusche observed that this gland in wasp females becomes larger at the time of ovary activation. On the other hand, Carlet (1890) contested the sexual as well as the smearing function, because the unfertile workers of honeybees, wasps, and ants possess the same gland as queens, and because the secretion is not sebaceous but liquid. Based on very untrustworthy mixture and injection experiments with the secretions of the "acid" venom gland and the "alkaline" Dufour gland, he concluded that the neutralized product of the two substances would be toxic, and that the venom gland secretion alone would paralyze. Because of this, the sphecids would have only one gland. The latter assertion was already refuted by Bordas (1894), who found Dufour glands in all hymenopterans studied, including the sphecids. Moreover in Vespidae both glands end separate from one another. It is only the secretion of the venom gland which is toxic. Janet (1898) supposed that the pretended alkaline secretion of the Dufour gland would neutralize the "acid" residues of venom at the sting which would otherwise be toxic for the venom producer itself. From this and the theories of Carlet, the trivial name alkaline gland resulted.

Thus the general function of the gland remains obscure. New chemical investigations by Cavill and Williams (1967) of the gland secretion in the ant *Myrmecia* indicate at best a lubricating function. The authors found the long chain hydrocarbons *cis*-heptadec-8-ene, pentadecane, and heptadecane. These are not released in the stinging act. Special functions may be expected in aculeates with greatly enlarged Dufour glands. Heselhaus supposed that bees with such glands (*Anthophora*) would use their secretions for nest building. Evidence, however, for this function is lacking. The toxic function of the venom gland in these bees remains. In the myrmicine, *Cremastogaster scutellaris*, the secretion of the large gland has a defensive function and substitutes the venom gland. The pheromone function of Dufour glands will be discussed later.

5. The Koshewnikow Gland

Koshewnikow (1899) found in workers and queens of *Apis* a gland which develops in pairs (Fig. 10, K gl). An oval mass of glandular cells lies upon the intersegmental membrane between the quadrate and spiracular plate. This membrane forms a reservoirlike structure under the gland. It consists of many short ducts in which the chitinous canaliculi of the single gland cells end in groups. They correspond to the Leydig gland cell type (Lukoschus, 1962b). Their fine structure has been studied in *Apis* by Hemstedt (1969). Other aculeates possess the same gland. Altenkirch (1962) found it in most Apidae genera which she studied. They are found more apically in the region of the bulb and are not so well organized as in the honeybee. Like Koshewnikow, Altenkirch observed this gland in *Vespa*, also. Here cells are rather isolated and end scattered over the entire membrane. Robertson (1968) found this gland also in the stings of primitive Formicidae (Ponerinae, Myrmeciinae), Thynnidae, and Scoliidae in the region of the triangular plate. They are lacking in Myrmicinae. The function of the Koshewnikow gland is still unknown. Koshewnikow thought it was a smearing gland. This function is contested by Zander (1922) and Snodgrass (1956). Altenkirch agrees with the opinion of Koshewnikow, the more so after she found in the sting chamber further gland cell associations of similar structure when the Koshewnikow gland was lacking. Ghent and Gary (1962) believe that the Koshewnikow gland secretion in *Apis* is the alarm substance. This could not be confirmed by Maschwitz (1964) (see Section II,C,3). In the honeybee queen the gland produces a pheromone (Butler and Simpson, 1965).

6. The Bordas Gland

Another accessory gland of the sting apparatus was described by Bordas (1895) in Terebrantia (Ichneumoninae, Cryptinae, Tryphoninae) and different sphecids. It is composed of multiple, densely lying cells whose canaliculi end within a gathering duct. The gland epithelium is covered on its outside by a basal membrane and has a cuticular intima. According to Bordas, it ends in the aculeus. This couldn't be confirmed by Rathmayer (1962). He studied this gland in sphecids. According to this author, the gland orifice lies at the point where the two sting sheaths are inserted at the oblong plates. While in *Tachysphex* this gland is still paired, in *Philanthus* and *Ammophila* it is fused into one. Its function is unknown.

7. The Sting Sheath Gland

In the sting sheath valves of various bees a gland can be found which here shall be called the sheath gland (Altenkirch, 1962; Maschwitz, 1964). In the honeybee, under the cuticle a high palisade gland epithelium is situated which resembles that of the wax glands and that of the superficial postgenal glands of the worker bee (Lukoschus, 1962a). Especially highly developed, it is beneath a strongly sclerotized strip on the outer sides of the sheaths. The gland is not developed in freshly emerged bees. The caste dimorphism is remarkable (see Table II).

The sting sheaths of the worker bee have lost their sensory function which yet can be found in the queen and have become glandular organs. In several Apidae genera (*Andrena, Anthophora, Colletes*) the gland epithelium is not found at the sting sheath cuticle, but at a membrane which runs between the two sheath edges. The function of this gland is unknown.

8. Further Sting Glands

Pawlowsky (1927) described a fascicular gland at the sting of *Sphex flavipennis* (Sphecidae) which may end in the bulb. Altenkirch (1962) found palisade gland epithelium in different Apidae on different skeletal plates of the sting apparatus. Furthermore, in different bees this author could demonstrate small ampule gland cells in the sting chamber, especially in those forms lacking the Koshewnikow gland.

9. The Anal Gland

In the ant subfamily of Dolichoderinae, whose sting apparatus is highly reduced, a new organ, the venom producing anal gland, can be found. It doesn't belong to the sting apparatus and shall be treated only because it is an abdominal pugnatory gland which has taken over the function of the sting. It was first described by Forel (1878). Dorsal to the rectum lie two large chitinous sacs which lead into one smaller ampule. The latter ends with a short duct in a cross cleft. The gland cells are found singly or clustered like grapes dorsally on the sacs. From the gland cells emerge canaliculi. They end singly but in the same region (*Tapinoma, Liometopum, Dolichoderus*; Forel, 1878) or fused in a grapelike fashion to paired ducts (*Bothriomyrmex*) Forel, 1878; *Iridomyrmex*, Pavan and Ronchetti, 1955) dorsally in the reservoir. The reservoir and the gland cells are covered by slender muscles. These glands produce odorless (*Iridomyrmex humilis, Bothriomyrmex meridionalis*) or strong-smelling (*Tapinoma, Liometopum*) secretions whose compounds are well known (see Section II,C,2).

C. Function

1. Stinging Process and Venom Emission

When an aculeate hymenopteran stings, the pointed aculeus by which the venom is injected must first be protruded out of the sting chamber and jabbed into the surface skin of the opponent. Then by boring movements of the lancets it penetrates deeper into the interior of the victim's body. At this point the venom secreted from the venom gland can be injected efficiently into the victim.

The movements of the aculeus are mainly performed by the described motion apparatus with its different plates and its muscular system, which is connected with the aculeus by the rami. The stinging process can be subdivided into several phases (Snodgrass, 1935; Rietschel, 1938); the protruding of the sting, the sawing movements of the lancets, and the withdrawing and retraction of the sting.

a. The Protruding of the Sting (*Fig. 14*). Previous authors supposed

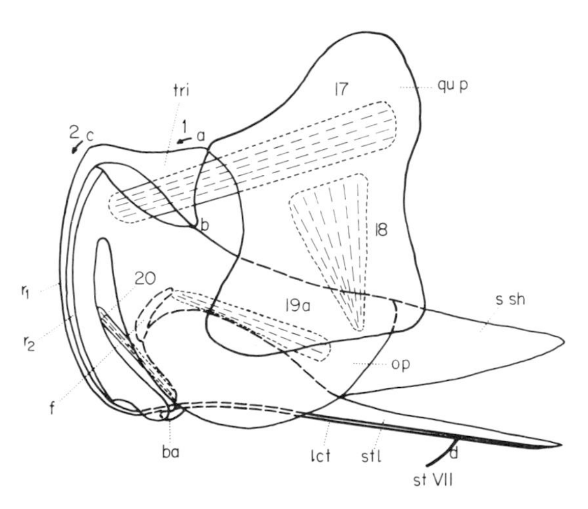

FIG. 14. Schematic figure of the most important skeletal parts and muscles of the sting apparatus of an aculeate. ba, basal articulation; f, furcula; lct, lancet; op, oblong plate; qu p, quadrate plate; r1, ramus of the lancet; r2, ramus of the stylet; stl, stylet; tri, triangular plate; 17, 18, 19a, muscles no. 17, 18, 19a; st VII, sternum No. VII. For further details see text.

that the blood pressure was responsible for the protruding of the sting, because direct muscles for this purpose are lacking (Kraepelin, 1873). Rietschel (1938), however, demonstrated that in bees and wasps the protruding is effected by indirect muscle movements. The sting is drawn downward, but the margin of sternite VII (st VII) directs this movement backward. As muscle 19a contracts, it draws the furcula (f) and the aculeus clockwise around ba. At d the sting lies on sternite VII. Here the down-movement becomes a backward gliding movement. This results in the elastic rami (r1, r2) being stretched. Their natural curves are flattened. The resulting tension is of importance for the return movement. The sting is now protruding. The first intrusion of the sting into the victim is effected by the whole strength of the body.

b. The Protrusion of the Lancets. This is the result of the contraction of muscle 17a + b. This effects the forward movement of the quadrate plate. At this point b remains almost immobile, a goes in the direction of arrow 1, the triangular plate turns counterclockwise. Hence c is going in the direction of arrow 2 and the lancet glides caudally.

c. The Retraction of the Lancets. By contraction of muscle 18 the lancet can be drawn back again and the quadrate plate brought back to its resting position. In this movement b remains still while a moves in the opposite direction of arrow 1, and thus the lancet is retracted. The pushing forward and the drawing back of the lancets is repeated several times. Because the right and left lancet are not connected, they can be moved independently of each other. As one lancet moves, the other rests. Subsequently the second lancet performs the same movement while the first is resting, and so on. In this way the lancets intrude deeper and deeper into the enemy's tissue and thereby pull the stylet (stl) with them. The venom now can be injected through the aculeus into the interior of the victim.

d. The Retraction of the Sting Apparatus. The retraction of this apparatus into the sting chamber is effected by the elasticity of the rami when muscle 19a relaxes, and additionally by contraction of muscle 20. In the honeybee worker the sting can be moved only in the median line. Vertical movements are prevented by a notch in sternum VII—a special structure in the honeybee—in which the sting glides. Furthermore the sting is not stiff enough for side movement because of the lack of connection of the spiracular and quadrate plates. These are still connected in *Vespa* and therefore, corresponding with the powerful intersegmental muscles of segments VII/VIII (Nos. 1, 2, and 3, Table I) and VIII/IX (Nos. 10 and 11, Table I), great sting mobility is possible.

The principles of venom ejection are different in several groups, varying according to the presence or absence of lancet valves in the aculeus and the structure of the venom reservoir. The most common procedure is found in all aculeates with the exception of the Vespidae, Pompilidae, Meliponinae, and the Camponotinae. The emptying of the reservoir is connected with the boring action of the sting. It is effected by the reciprocal movements of the valves which can be found in the position of rest on the same level at the lancets within the bulb. At the protrusion of one lancet the valve sucks out venom from the reservoir like a piston when the opening muscle is contracted (see Section II,B,2). This flows through a cleft which is opened between the two valves by the movement of the lancet. The venom which has reached the caudal side of the valves is driven to the tip of the sting; when these move forward again fresh venom is sucked out at the same time. During the caudally directed protrusion of the lancets, the two saillike appendices (Fig. 6, m) of the valves are spread like wings and their pouchlike structure (Fig. 6, p) is unfolded; these (m and p) fold together at the retraction of the lancets. The valves are thus able to glide through the venom-filled interior of the aculeus without much resistance (Trojan, 1922). Possibly at the lancet retraction the venom duct is also closed. Because of the small diameter of the venom duct, capillary action possibly aids the venom flow. This is extruded through a cleft between the lancets at the caudal end of the aculeus. Because of the locking apparatus (*Apis*, Myrmicinae), venom must not be pumped out with every movement of the lancets. The insect is enabled to eject the venom at will according to the deepness of the sting or its own excitement. This type of venom emptying may be called "pump sting type" in contrast to the "spray sting type" (Olberg, 1959). The latter is found in vespids and pompilids whose lancets have no pumping valves. By contraction of the venom reservoir muscles the secretion is injected into the victim. Schlusche (1936) could not find any locking apparatus in the wasp sting. Such a formation seems unnecessary, because in contrast to the pump sting type, stinging movements and venom emptying are separate.

We should also mention how the venom is given off in the camponotines, which have a well-devèloped venom gland but no functioning sting (Forel, 1878). The very weak wall muscles of the reservoir are not able to eject the venom. Therefore the latter is given off by the "abdominal press" that is, by contraction of the intersegmental muscles which increase the inner pressure of the gaster. Generally the formic acid produced by these ants is delivered in droplets. The *Formica rufa* group, however, is able to spray off the venom in this way up to a distance of 30 cm. Possibly the secretion of the Dolichoderinae anal glands is emptied in a similar way.

2. Venoms and Their Effects

The sting apparatus of the Hymenoptera accomplishes its original function of an ovipositor in the symphytes, terebrantes, and the Dryinidae and Sapygidae (Aculeata). A lot of zooparasitoid terebrantes paralyze their victims with the secretion of an ovipositor venom gland (Hase, 1924; Beard, 1952). The same is true for the bethyloid families Bethylidae, Cleptidae, Dryinidae, and a few species of the Chrysididae (Berland and Bernard, 1951). In some terebrantes the ovipositor also serves for defense (*Pimpla*, *Ichneumon*, *Ophion*; Pawlowsky, 1927). In the highly evolved *Aculeata s. str.* the sting became a true fighting instrument. With the exception of some ants, it serves mainly for venom production and transfer.

a. Venom Composition. In Table III a summary of the chemical components of venoms is given. As the Table shows, the more primitive Aculeata possess, as far as is known today, chemically complex venoms which are described mostly as "proteinlike." They contain enzymes, peptides, and amines. More advanced groups such as the ant subfamilies Camponotinae and Dolichoderinae possess other venoms, such as formic acid, various ketones, and lactones. The formic acid of the Camponotinae has been investigated by a number of famous chemists such as Marggraf, Doebereiner, Berzelius, and Liebig (Pawlowsky, 1927). Quantitative analyses were performed on several Camponotinae by Donisthorpe (1901) and Stumper (1922, 1960). The last author observed in the *Formica rufa* group and *Camponotus ligniperda* acid concentrations of about 50%; Osman and Brander (1961) observed in the same objects 61–65%. Furthermore several components were found and identified (ca. 1% of the crude venom weight; venom being isolated from the reservoir) as proteins, amino acids, ammonia, and metal ions. The relative weight of the venom differs from 0.5 to 20% of the body weight. *Polyergus rufescens* with 0.5% is an exception which may be explained by the biology of this ant. It is a slave-raiding social parasite which fights exclusively with its sabrelike mandibles.

Freshly emerged ants possess only very small amounts of venom. The full quantity is not produced until their first weeks. The same is observed in wasps and honeybee workers. In the bee, the venom production is correlated with the social function of the worker, which alters during its lifetime. The full amount of venom may be found when the worker bee becomes a guardian (Autrum and Kneitz, 1959; Lindauer, 1961).

The toxic ketones and lactones of the dolichoderines' anal glands are mixed with nontoxic substances (iridodial, dolichodial). These substances act as fixatives and solution mediums for the volatile venoms (Pavan, 1959).

b. Venom Effects against Insects. The venoms of the aculeates are directed mainly against insects. Originally these substances have served mainly for predation. Most of the aculeates, with the exception of the bees, several vespids and formicides, hunt for their brood, paralyzing or killing animals with their sting. In solitary aculeates the sting is mainly used for defense when the animals are molested themselves. In the social insects the defense of the nest plays a very important role.* The formic acid of the camponotines injures the opponent as much by its acid as by its reducing effects. It shows strong insecticidal effects as a liquid or a gas (Otto, 1960; Osman and Kloft, 1961). The volatile hydrocarbons, ketones, and esters of the Dufour gland (Regnier and Wilson, 1968, 1969; Bergström and Löfquist, 1968), which are delivered in addition to the venom, are efficient spreading agents enhancing the penetration abilities of formic acid. They may also have additional venom effects. The dolichoderine venoms such as iridomyrmecin (Pavan, 1955) and the amine components of the myrmicines

* The fact that only female Hymenoptera possess a venom apparatus might explain why, in contrast to termites, the males never took part in the formation of communities. Because of the lack of a sting they never were able to capture prey or to defend the colony. They remained in the brood care society of the developing colonies as uneconomical feeders, which if possible, were nourished only till mating.

TABLE III

COMPOSITION OF HYMENOPTERAN VENOMS

Suborder and superfamily	Family and species studied	Substances produced	References
Terebrantia			
Ichneumonoidea	Braconidae		
	Microbracon hebetor	Protein	Beard (1963)
Aculeata			
Scolicoidea	—	—	—
Bethyloidea	—	—	—
Pompiloidea	—		
Apoidea, Vespoidea	Apidae, Vespidae		
	Apis mellifica *Vespa vulgaris* *V. crabro*	Protein (hyaluronidase, phospholipase), peptides Histamine Serotonin Acetylcholine	Habermann (1968); see also contributions of Habermann in this volume
Sphecoidea	Sphecidae		
	Philanthus triangulum	Protein	Rathmayer (1962)
	Sceliphron caementarium	Protein, hydroxy acids, lecithin	Rosenbrook and O'Connor (1964)
Formicoidea	Formicidae		
	(Ponerinae)		
	Paraponera clavata	Protein	Hermann and Blum (1966)
	(Myrmeciinae)		
	Myrmecia gulosa *M. pyriformis*	Protein (hyaluronidase), histamine	De la Lande *et al.* (1963) Cavill *et al.* (1964)
	(Pseudomyrmicinae)		
	Pseudomyrmex pallidus	Protein	Blum and Callahan (1963)

(Myrmicinae)		
Solenopsis saevissima	Amines (alcaline two-phase system)	Adrouny *et al.* (1959), Blum *et al.* (1958), Blum *et al.* (1961)
Solenopsis xyloni		
Myrmicaria natalensis	*d*-Limonene	Quilico *et al.* (1961)
Myrmica ruginodis	Protein (hyaluronidase), histamine	Jentsch (1969)
(Camponotinae)	Formic acid	Donisthorpe (1901), Stumper (1922, 1960)
All species		
(Dolichoderinae)		
Iridomyrmex humilis	Iridomyrmecin ($C_{10}H_{16}O_2$)	Pavan (1955), Cavill *et al.* (1956)
Iridomyrmex nitidus	Isoiridomyrmecin	Cavill and Locksley, (1957)
Dolichoderus scabridus	($C_{10}H_{16}O_2$)	Cavill and Hinterberger (1961)
Iridomyrmex detectus		Cavill *et al.* (1956), Cavill and Ford (1960),
Iridomyrmex conifer	Iridodial ($C_{10}H_{16}O_2$)	Cavill *et al.* (1956),
Iridomyrmex rufoniger		Cavill and Hinterberger (1961),
Iridomyrmex nitidiceps		Cavill and Hinterberger (1961),
Tapinoma nigerrimum		Trave and Pavan (1956)
Iridomyrmex rufoniger		Cavill and Hinterberger (1961),
Iridomyrmex myrmecodiae		Cavill and Hinterberger (1961),
Dolichoderus clarki	Dolichodial ($C_{10}H_{14}O_2$)	Cavill and Hinterberger (1961),
Dolichoderus scabridus		Cavill and Hinterberger (1961),
Dolichoderus dentata		Cavill and Hinterberger (1961),
Iridomyrmex rufoniger		Cavill and Hinterberger (1961),
Iridomymex nitidiceps	2-Methylhept-2-en-6-one ($C_8H_{14}O$)	Cavill and Hinterberger (1961),
Dolichoderus scabridus		Cavill and Hinterberger (1961),
Tapinoma nigerrimum		Trave and Pavan (1956)
Dolichoderus clarki	4-Methylhexan-2-one ($C_7H_{14}O$)	Cavill and Hinterberger (1961)
Tapinoma nigerrimum	Propylisobutylketone ($C_8H_{16}O$)	Trave and Pavan (1956)
Iridomyrmex pruinosus	Methyl-*n*-amylketone ($C_7H_{14}O$)	Blum *et al.* (1963)

Solenopsis saevissima and *Solenopis xyloni* (Blum *et al.*, 1958) are strongly insecticidal.

In many aculeate families (and in terebrantes) specific effects of the venom against insects are known (review article: Piek and Simon Thomas, 1969; see also Clausen, 1940; and Evans, 1966). In Bethyloidea, Sphecoidea, Pompiloidea, Scolicoidea, and partly in Vespoidea, the victims are not killed but paralyzed. The paralysis may be transient, e.g. *Pompilus biguttatus* (Pomp.), *Larra* (Sphec.), *Thiphia* (Tiph.), or permanent: *Philanthus, Ammophila* (*Sphec.*), *Pterombrus* (*Tiph.*). In *Philanthus* this may be determinated by the venom dose (Rathmayer, 1962). Only rarely are the victims immediately killed by the sting (social wasps, ants, some sphecids). In Sphecidae, Pompilidae, and Vespidae the paralyzed victims are transported to the nest where eggs are laid on them. Scolicoidea and Bethyloidea build no nests. The viable state of the prey after it has been stung prevents it from decaying and becoming unsuitable for brood nutrition. The site of venom action in most cases is still unknown (Piek and Simon Thomas, 1969). Many aculeates sting their prey in the region of ganglia (e.g., *Ammophila, Liris, Megascolia*). Thus Fabre in 1879, discoverer of the paralyzing phenomenon, supposed that the sting hit the CNS and thus effected the paralysis. Other aculeates however sting their prey at random or in such a way that the CNS cannot be affected (*Synagris, Tiphia, Cerceris*). The investigations of Rathmayer (1962) on *Philanthus triangulum* (Sphecidae) and Beard (1952) on *Microbracon hebetor* (Terebrantes) demonstrate that in these animals the venom has a peripheral action. Though *Philanthus* stings the honeybee in the thorax ganglion region, the venom is distributed by the hemolymph and causes peripheral muscle paralysis. Possibly this holds true also for other hymenopterans. Piek (1969) and Piek and Engels (1969) demonstrated that the venom of *Philanthus* and *Microbracon* blocks the neuromuscular synapse most likely at the presynaptic site. Heart and intestinal muscle activity is not impaired. The cause for this may be that here synapses with other transmitter substances are present. The chemical nature of the paralyzing substances is unknown. The widespread supposition that the venom preserves the victim seems to be rather improbable, as paralyzed victims of *Microbracon hebetor* are more susceptible to the insect pathogen *Bacillus thuringensis* than unparalyzed ones (Tamashiro, 1960).

According to Rathmayer the venom of the sphecid *Philanthus* is not prey-specific. *Philanthus*, itself, however is immune against its own venom. The often expounded supposition that prey specifity of Sphecidae and Pompilidae is a result of a physiological correlation between venom and venom sensitivity to prey tissue therefore does not seem to be true. It is possible that if a species' prey is limited to a few animals the predator can acquire an inherited knowledge of their morphology and habits of life, e.g., time of flight and feeding plants. In this way the ecological niche which is represented by one or few similar prey objects can be optimally used.

Here may be mentioned that the sting apparatus of some sphecids has a special function. Various *Oxybelus* species and *Crossocerus elongatus* (Olberg 1959; Evans, 1962) carry prey to their nests by spitting it on the sting.

c. Venom Effects against Vertebrates. With exception of bee, wasp, and hornet venoms, which will not be treated in this contribution, we know very little of the effects of aculeate venoms against vertebrates. Moreover our knowledge is restricted mostly to humans, while effects in the natural vertebrate enemies are mostly unknown.

For many of the solitary apids it is difficult to penetrate the human skin. Their venoms moreover have only weak effects. The same is true in Sphecidae and even partly in the giant Scoliidae. Some bethylids (*Scleroderma*) are known to sting painfully. The stings of Pompilidae are painful, but the effects short-lived. Very frightening, however, are Mutillidae stings, which cause inflammation, fever, even delirious conditions. In some primitive societies they are therefore used to test the courage of young men in initiation ceremonies. According to Baer (Pawlowsky, 1927) several stings of a Peruvian species may be lethal. The stings of a series of primitive tropical ants of the subfamilies Ponerinae (*Dinoponera, Paraponera, Paltothyreus*) and Myrmeciinae are known to be painful and dangerous and may surpass hornets in their venom effects (Pawlowsky, 1927; Stumper, 1960; Hermann and Blum, 1966). They cause inflammation, high fever, or even paralysis. *Myrmecia* stings are as potent as bee stings (De la Lande *et al.*, 1963). The toxic effects of *Solenopsis* venom have been reported by Caro *et al.*, (1957), Blum and Callahan (1961), and Blum *et al.*, (1958, 1961). The effects of *Pseudomyrmex* venom have been described by Blum and Callahan (1963). Of the European sting-bearing ants, the venoms of the *Myrmica* species and *Neomyrma rubida* are most painful to human beings. These ants are, however, eaten in large quantities without damage by toads.

d. Fungicidal and Bactericidal Venom Effects. In different hymenopteran venoms, bactericidal and fungicidal effects could be demonstrated (*Solenopsis*, Blum *et al.*, 1958; Dolichoderinae; Pavan, 1958; *Formica*; Sauerländer, 1961). The biological significance of these effects is doubtful; they might be rather secondary effects of a general toxic action of the venom against living cells (see also Maschwitz *et al.*, 1970).

3. Communication

The sting apparatus of many social Hymenoptera is used partly for communication. Its glands produce alarm and trail pheromones.

a. Alarm. Many social hymenopterans are able to alarm their nest mates by excreting odorous mandibular and abdominal gland substances when they are attacked. In this article only the latter are treated (for mandibular alarm substances see review articles by Wilson, 1965; Maschwitz, 1966; and Blum, 1969). This function has developed independently in the three families where social forms can be found: in Apidae, Vespidae, and Formicidae. Chemical alarm developed only in higher organized communities. In more primitive (*Polistes, Bombus*) colonies and those with only a few inhabitants (*Ponera, Myrmecina,*) this ability has not yet been developed. An important gland producing alarm substances is the venom gland (*Vespa,* Myrmicinae, *Formica*). But in the Dufour gland also, alarm substances can be produced (Camponotinae, Myrmicinae). In the honey bee worker the gland producing the alarm substance has not yet been localized; the secretion is stored in the bristled membrane between the oblong plates. In the Dolichoderinae the anal defense gland has taken over the alarm function. The abdominal alarm substances are either produced separately from the venom (*Apis,* Camponotinae) and mixed with it when being released, or they are produced in the venom gland or in the anal gland itself. In this case the alarm secretion and the venom are identical (Dolichoderinae, *Formica*) or different from each other (most of Myrmicinae, *Vespa*). Some of the investigated forms possess several glands with alarm substances, e.g., *Formica* or *Lasius*. The alarm substances are not species specific as a rule; some even act intergenerically, e.g., Camponotinae. The alarm secretions generally are released at the attack; however, they can be released independently by a true alarm reaction. Mostly the alarmed colony members, especially in the nest attack flee; workers flee less often. This latter reaction can be observed especially at the feeding place. Aggressive ants, however can be stimulated to attack at the feeding place also. A larger insect which is smeared with the alarming venom attracts additional ants which help to overwhelm it. Honeybees and wasps, (*Vespa*) when stinging mark the enemy with alarm substance; thus the enemy is attacked more often and more intensively.

b. Trail Marking. According to Wilson (1959a), the alarm effect of the Dufour gland secretion in *Solenopsis* is only a secondary effect of a general attractivity. Mainly it serves for marking trails to feeding places and new nests. The ant trails the secretion on the ground with the protruded sting. The nest mates are attracted by the odor and follow it out of the nest. The same could be observed in *Pheidole fallax* (Wilson, 1963); four attine genera (Moser and Blum, 1963; Blum, Moser, and Cordero, 1964); *Tetramorium* (Blum and Ross, 1965); *Monomorium* and *Huberia* (Blum, 1966). With the exception of *Solenopsis* and *Pheidole*, the substance, however, is produced in the venom gland. In *Atta* the sting apparatus is reduced and serves only for trail marking. Possibly gastral alarm substances of other Myrmicinae are used as trail-marking substances at the same time (Maschwitz, 1966). In Ponerinae, Dorylinae, and Camponotinae trail-marking substances are not produced in sting glands, but in the hindgut. Dolichoderinae and *Cremastogaster* possess special trail pheromone glands (Carthy, 1951; Wilson and Pavan, 1959; Blum and Wilson, 1964; Hangartner and Bernstein, 1965; Blum and Portocarrero, 1964; Blum, 1966; Fletscher and Brand, 1968; Leuthold, 1968a).

4. Repellent Effects

Liepelt (1963) discovered a further effect of hymenopteran venoms. They can act as repellents. He observed that bee and wasp venoms have a disgusting taste to some birds who, therefore, will not eat these insects. The release of venom in these aculeates which could be observed when they were molested is thus of further importance. Odor-repellent effects against insects exist in Dolichoderinae and Camponotinae venoms (review article, Cavill and Robertson, 1965).

D. Evolutionary Peculiarities

1. The Sting Loss of Honeybee Workers

If a honeybee worker attacks a mammal, the sting apparatus is caught in the skin and lost. By this act the bee is wounded so badly that it dies soon after. For some observers this seemed to be a biological defect detrimental to the survival of the species (Darwin, 1859). Darwin attributed this to strong lancet barbs supposedly inherited from wood-boring ancestors. However, it was generally supposed that the loss of the sting apparatus in a mammal skin was an unimportant accident. Normally only insects would be stung and the sting would not be caught. Weinert (1920) even thought that because of the sting loss the venom apparatus could not have an offensive function. The formic acid of the venom (which does not really exist) would serve for preservation of the honey and disinfection of the hive air.

Rietschel (1938) discovered from his studies rather evident explanations for the loss of the sting that have recently been confirmed (Maschwitz, 1964). The sting loss is no accident but a natural reaction developed by selection, an autotomy. It is directed against the largest enemies of the honeybee, against honey- and brood-robbing mammals and birds. The following facts demonstrate this. As already mentioned (see Table II), the sting apparatus of the honeybee worker possesses a series of peculiarities in comparison to the more primitive queen bee sting and to those of solitary bees. The sting possesses a preformed breaking point: the spiracular plates and muscles which in the queen bee prevent the loss of the sting are reduced strongly in the worker. The lancets possess many strong barbs in contrast to the queen and to solitary bees. In this way the anchorage of the sting in the skin of the victim is stronger than its breaking strength. The numerous sensory bristles of the sting sheaths have almost fully disappeared and its nerve is reduced. In the queen, which stings only strongly sclerotized rival queens, they are well-developed. In the worker the last ganglion also is ripped out with the sting. Thus the sting apparatus is able to continue working and can inject the whole venom into the enemy.

Several modes of attack behavior also imply a specialization against mammals. The sting spot is chosen by the worker primarily by optical stimuli and only secondarily by tactile ones. Worker bees will attack dark spots of a much larger size than insects; the velvetlike structure of the mammal coat effects greater intensity and permanence of stinging. Furthermore, there are two stimuli found only in warm-blooded animals that provoke attack and stinging: warm, moist breath (Maschwitz, 1964) and possibly the odor of mammal sweat (Free, 1961; Lecomte, 1961). Moreover, the worker bee has reflective movements (torsions, flying away) which help to rip out the sting.

These adaptations of morphology and behavior, which effect the fixation of the sting in the coat of vertebrates, bring to the honeybee colony several advantages. The small, well-fixed sting apparatus, which the enemy is not able to strip off as it could do with a whole bee, injects the total venom amount of the reservoir into the opponent who is unable to defend himself against this. Furthermore, the stung spot is marked with the total amount of the odorous alarm substance stored in the setose membrane (see Section II,C,3). This increases the number and the aggressiveness of the attackers and makes possible for them to locate the sting spot. These facts might probably have been of selective advantage for the honeybee. The attack of a large vertebrate endangers the survival of the whole colony. Therefore the loss of a few bees in warding off enemy attack is not a significant loss for this society.

2. Sting Reductions in Ants

Sting reductions can be found in many Formicidae. The evolutionary reasons for this are not yet clear. Stumper (1960) and Cavill and Robertson (1965) attempted to correlate feeding habits with sting reduction, stating that species with intact stings are carnivores and those with reduced ones are herbivores. This does not correspond to the facts. For example, the predatory Dorylinae have a reduced sting apparatus which is no longer suitable for killing prey.

The following reflections should serve to clear up this question. The most important enemies and prey objects of the Formicidae are not vertebrates but arthropods, among those the most important adversaries are the numerous ants themselves. The hymenopteran sting is because of its position on the end of the body, a poorly movable instrument when small and fast insects are to be hit. The characteristic of all Apocritia, the narrowing of the body middle and the acquisition of a jointed petiolus in many Aculeata, leads to an increase of the mobility of the abdomen and concomitantly of the sting apparatus. All ant families possess a petiolus, and some ant subfamilies have even a further narrowed segment before the gaster, the postpetiolus. Nevertheless if one observes a stinging ant, it is very difficult for a larger species to hit other ants or small prey objects with their sting. It can only penetrate the intersegmental membranes since many ants and other insects possess a hard cuticle. If such objects must be hit quickly, a high degree of specialization as is known in sphecids seems necessary. The specialization for one or a few objects probably made it possible for these wasps to acquire in their evolution an innate knowledge of the prey morphology and of the optimal stinging spot (Rathmayer, *Philanthus*, 1962; see Section II, G.2). Highly differentiated sensory organs for finding the stinging spot have developed in these insects. Because of the numerous enemies and prey objects with different fighting methods, such specializations are impossible for the ants. If an ant wishes to sting, a very time consuming search of a sting spot begins after it has fixed itself with the mandibles on the opponent. Consequently the ant sting apparatus is a relatively inefficient instrument for fighting against many arthropods. Only the more primitive forms have retained its full use (e.g., Ponerinae and Myrmeciinae). Foerster (1912) found in Myrmicinae, the largest ant subfamily, among 11 genera 5 with more or less reduced sting apparatus: *Atta, Messor, Pheidole, Aphaenogaster,* and *Cremastogaster.* Dorylinae, Dolichoderinae, and Camponotinae also have reduced stings. Over 50% of all ant species probably possess atrophied stings. Sting-bearing ants fight often with their mandibles. Apparently the Dorylinae are specialized for this manner of fighting. The same is true in most Myrmicinae with reduced stings. In fight experiments it could be demonstrated that mandibular fighters such as *Pheidole* or *Atta* easily overwhelm the sting-fighting *Myrmica.* In some parasitic species which are known as mandibular fighters, e.g. *Strongylognathus* and *Harpagoxenus*, sting reduction could be demonstrated (Maschwitz, unpublished). Only very small species seem to be able to use their stings successfully during fights with other ants, for example, the tiny *Solenopsis fugax* which feeds upon the brood of other ants (Hölldobler, 1965).

Ants with "new" chemical weapons, which no longer possess the ponderous sting, have developed independently several times. Whole subfamilies are characterized by the development of modified defense glands, the toxic secretions of which can be delivered and applied rapidly to the enemy. These organs are the anal glands of the Dolichoderinae and the cushion-formed venom glands of the Camponotinae (Section II,B,3).

The ketones and lactones produced in the anal glands act in the same way as the formic acid of the Camponotinae as contact (Pavan, 1958; Otto, 1960) and respiratory (Osman and Kloft, 1961) venoms. Because of their lipophilic properties they are able to penetrate the cuticle of the opponent. With the help of such secretions each insect enemy and prey object can be attacked immediately with a minimum of body contact. The specific activity of some venoms also indicates that these weapons are directed especially for use against insects. The lactones of the Dolichoderinae are specific insecticides which partly surpass the killing power of DDT.

The new chemical fighting methods have apparently influenced the morphological evolution of these ants. Both subfamilies have developed no vulnerable superthin and supermovable petiolus and postpetiolus, which would improve the movability and target exactness of the sting. Also the thinness of the cuticle which is especially seen in some Dolichoderinae, may be a consequence of the new chemical weapons. Since ants with spray venoms are only a very short time in contact with the opponent, a strong sclerotization is not necessary for their protection. Thus thin-skinned ants with a great movability could be selected.

In ants other glands also can take over the fighting function of the original venom gland with its complicated ejection mechanism, the mandibular gland (Blum *et al.*, 1969, *Veromessor*) or the Dufour gland (*Cremastogaster scutellaris*, Maschwitz, unpublished). As mentioned, in *Cremastogaster* the Dufour gland is strongly enlarged and its volume is many times that of the original venom gland. Upon attack the secretion can be instantly expelled.

Possibly the sting reduction of the Meliponinae resulted from similar evolutionary causes. Strikingly, we can find in the most primitive stingless bees a remarkable analogy with the ants in the development of new chemical weapons. Here also the chitinous sting parts are reduced. The venom gland is modified and strongly enlarged. The new secretions are lipophilic. We do not know at the present time, however, whether they are indeed used as defense secretions.

III. THE URTICATING HAIRS OF LEPIDOPTEROUS LARVAE

A. Scope of the Problem

Several families of the order Lepidoptera (butterflies and moths) are known to have larvae (caterpillars) equipped with poisonous hairs. As will be pointed out, the poisonous effects are due to several types of hairs or spines. Normally a caterpillar is indeed the urticating stage of Lepidoptera whether the hairs or spines are acting directly or indirectly. But in some cases the other stages have urticating effects too, as Scheme 1 demonstrates.

Scheme 1 shows which possibilities exist for direct (whole line) and indirect (broken line) urticating effects of the different stages. Most effective are larval urticating hairs which act directly or indirectly via exuviae, larval

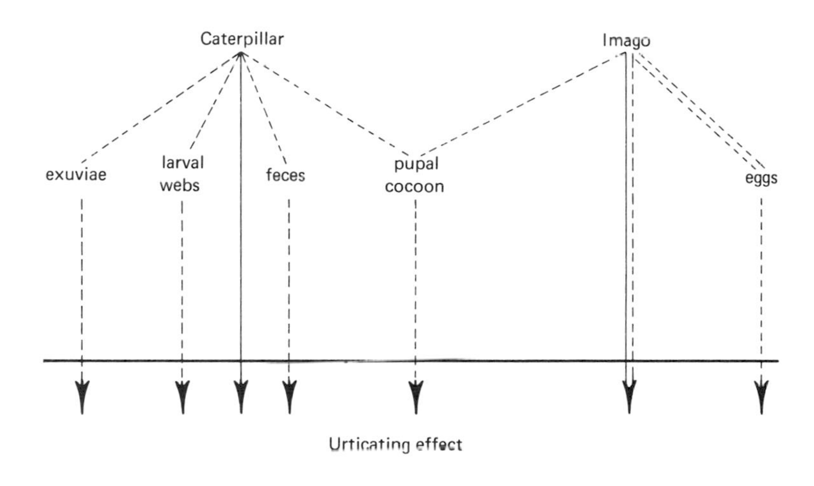

Scheme 1

webs or feces, to all of which they adhere. Often these larval hairs are woven into the pupal cocoon which itself becomes urticating or allows a transfer to the imago. As first demonstrated by Anderson (1885) in the case of imagines of the "brown-tail" ("Goldafter") *Euproctis chrysorrhoea* L. (indirect), urtication is due to larval hairs spun into the cocoon, and removed therefrom by the anal tuft under abdominal torsion (*Eltringham,* 1912) of the adult as it emerges through the cocoon wall. In some cases, however, real imaginal hairs of the anal tuft might cause urticaria on the skin of sensitive persons, as was shown by Weidner (1936) for females of *Lymantria dispar* L., by Kemper (1956) for the "brown-tail" moth and by Leger and Mouzels (1918) for *Hylesia.*

As for eggs, the double broken line between imago and eggs indicates in our scheme both possibilities of covering the egg layers with imaginal anal hairs as well as with larval hairs, transferred via cocoon to the anal tufts of the emerging females.

Caterpillars or other stages of nine families are reported to have urticating properties. Only one of these belongs to the suborder Rhopalocera; all others belong to the suborder Heterocera. A synopsis of the Lepidoptera families with urticating stages and the type of their irritating hairs is given in Table IV.

Since there exists no generally accepted "natural system" of the Lepidoptera, it is difficult to find out the valid names in higher as well as in lower

TABLE IV

SYNOPSIS OF LEPIDOPTERAN FAMILIES WITH URTICATING HAIRS

Superfamily	Family	As urticating well-known genera (incomplete list)	Stage, type of irritating hairs	Zoogeographical area
I. Rhopalocera Papilionidea	Nymphalidae	*Vanessa*, *Euvanessa*	Caterpillars with normal hairs, having the form of stiff pointed, ramified spines	Palaearctic, Nearctic
II. Heterocera	Cochlidiidae (=Limacodidae)	*Sibine*, *Natada*, *Parasa*	Caterpillars with spines, containing a poisonous liquid. Additional *Spiegelhaare* and small, starlike prickles	Indoaustralian, Ethiopic, Neotropic, some Palaearctic
Zygaenoidea	Megalopygidae	*Megalopyge*, *Carama*, *Lagoa*	Caterpillars with spines, containing a poisonous liquid. Additional *Spiegelhaare* and small starlike prickles	Most neotropic, singles are Nearctic, Palaearctic, Ethiopic
Noctuoidea	Arctiidae	*Lithosia*, *Arctia*, *Parasemia*, *Adolia*, *Callimorpha*	Caterpillars extremely hairy, with soft long normal, sometimes stiff pointed hairs	All regions, most species Neotropic and Orientalic
Noctuoidea	Lymantriidae (=Liparidae)	*Lymantria*, *Euproctis*, *Porthesia*, *Hemerocampa*, *Ocneria*, *Orgyia*, *Dasychira*	Caterpillars with cuticular zones bearing very short, heavy urticating hairs, so-called *Spiegelhaare* (Type α). Larval urticating hairs in cocoons too; or caterpillars with long urticating hairs, forming brushes (*Bürstenhaare*) in cocoons too. Imagines of some species ("brown-tail") with urticating hairs in anal tufts (esp. females), which can contain larval *Spiegelhaare* too. Eggs sometimes protected with larval or imaginal urticating hairs	Most Indoaustralian and Ethiopic, fewer Neotropic and Palaearctic, very few Nearctic

Noctuoidea	Noctuidae	*Acronicta, Apateta, Catocala*	Caterpillars with normal hairs, sometimes having the form of stiff pointed, ramified spines. Caterpillars of some species with longer urticating hairs, forming brushes ("*Bürstenhaare*") some tropical forms are said to have poisonous spines	All regions, most Neotropic
Bombycoidea	Thaumetopoeidae	*Thaumetopoea* (= *Cnethocampa*) *Anaphe*	Caterpillars with cuticular zones bearing very short, heavy urticating hairs (*Spiegelhaare*, Type β). Urticating larval hairs in webs and cocoons	Palaearctic, orientalic, Ethiopic
Bombycoidea	Lasiocampidae	*Macrothylacia, Cosmotriche, Dendrolimus, Lasiocampa, Malacosoma, Taragama*	Caterpillars mostly with clinging (sometimes sticking out) hairs. These seem to be smooth, but have at the distal tip microscopic small barbed hooks, directed to the base	Most forms tropic (New and Old World), less species Palaearctic, Nearctic and Australian
Bombycoidea	Saturniidae	*Automeris, Cricula, Dirphia, Hemileuca, Hylesia, Samia,*	Caterpillars often with spines, containing a poisonous liquid. Some species (for example, *Hylesia* sp.) with spine-like poisonous hairs as adults	Most forms Neotropic and Ethiopic, less Indoaustralian, very few Palaearctic and Nearctic

taxa. The same holds true for the numerous synonyms of genera and species. Therefore we use in this contribution the commonly used names and refer to Schröder (1925) Weber (1949), Grandi (1951), and the other authors, cited in references to Picarelli and Valle, Rotberg, and Pesce and Delgado in this volume.

It is impossible to give figures for the number of Lepidoptera with poisonous stages. All the examined Megalopygidae were found to be poisonous and this character is probably a universal one in the family, which covers about 200 species. The same might be true for the family Limacodidae with about 800 species and all the processionaries (Thaumetopoeidae, ca. 250 species). All Lymantriidae (Liparidae) examined were poisonous, a condition probably common to all members of the family having about 1200 species. Depending on individual differences in human sensitivity to substances from strange organisms, we have to expect reactions (burning, itching, urticaria, dermatitis) after direct or indirect contacts with numerous Lepidoptera having only unspecific "normal" hairs and spines. These effects can cause severe reactions, such as anaphylactic shock, by repeated contact especially in the case of gardeners, farmers, foresters, or preparators in entomological museums.

Urticating caterpillars or other stages of lepidopterans are subjects of entomological as well as of medical interest. This is the reason why the literature is spread over so many different medical (especially dermatology, opthalmology, and hygiene) and entomological journals. It is impossible to give a complete reference of all the literature. For literature not cited in this article see the excellent compilation of Gilmer (1925) and Weidner (1936).

B. Types of Urticating Hairs and Spines

Urticating effects of lepidopterous larvae are caused by hairs or spines. Spines are normally understood to be well sclerotized, stiff, pointed hairs. In case of the Cochlidiidae, Megalopygidae, and Saturniidae, the poisonous spines are a different formation. Real urticating hairs have some morphological and possibly also chemical characteristics, if we compare them with normal hairs. Several authors assumed glandular cells with poisonous secretions at the base of nettling hairs to be the origin of urtication. Histological studies often demonstrated the existence of glandular cells. But as was clearly shown by Weidner (1936), the hypodermis forms specialized, glandlike cells at the base of each hair. Therefore we should outline here some of the morphology of insect hairs in general. Real hairs of insects (macrotrichia,* setae) articulate with the cuticle. They are hollow and connected

* Microtrichia are thin protrusions of the upper cuticular layer (exocuticle).

with a hypodermic cell by a pore going through the cuticle. The hair-forming (trichogen) cell is always much larger than the surrounding epidermal cells. Generally there is found with the trichogen cell another rather large one, the so-called membrane cell. It forms the articular membrane and the basal ring of the hair. The latter is sometimes strengthened by special enlarged epidermal cells. The trichogen cell is enclosed from all sides by the membrane cell. If there is no membrane cell, the hair has no basal ring. In this case it stands in a small depression of the cuticle. The existence of the enlarged membrane cells, trichogen cell, and basal wings suggested the speculation that urticating hairs were equipped with poisonous glands. But the poisonous nature of the secretion has not yet been definitely established.

In the following section we give a detailed description of the different hairs and spines which have been reported to cause urticating effects.

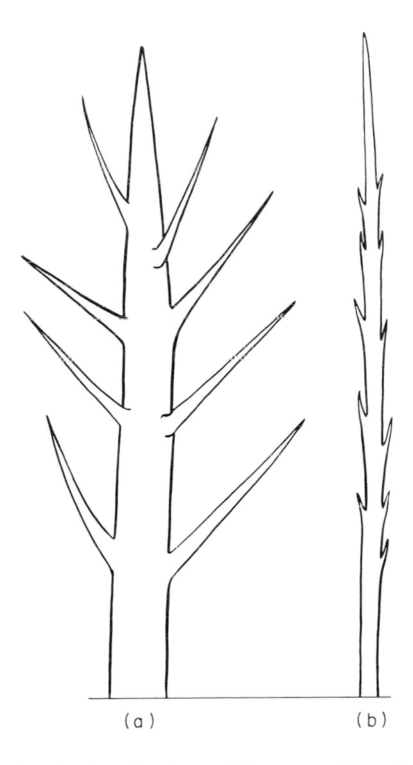

FIG. 15. (a) Stiff pointed, dorsal spine of the caterpillar *Vanessa urticae* L. (Nymphalidae). (b) Stiff spine hair of the caterpillar of *Arctia caja* L. (Arctiidae). These hairs stand in bunches from 20–40 setae on dorsal and lateral verrucae. They have up to 10 mm length and are often colored.

1. Normal Hairs

As we pointed out above, each insect hair can cause reactions in persons with a strong tendency to allergies. But the normal soft and long hairs of several caterpillars, especially of the Noctuidae, are also reported to sting. We do not know how these act. In the case of stiff, ramified and pointed hairs (Fig. 15) of Nymphalidae, Arctiidae, or Noctuidae we know that reactions are due to broken tips or side branches of the spines which penetrate into the human or animal skin after contact. These hairs have a blunt base, are conic on their distal end, and have often a long but blunt tip. The side spines are directed toward the outside and break off easily, especially if the hairs are old and dry.

2. Urticating Hairs with Special Structures at the Distal End

Caterpillars of the family Lasiocampidae are reported to have urticating hairs, though these seem to be normal ones. The hairs mostly cling to the insect although sometimes they stick out. As described by Kemper (1958), the hairs are rather thick, very different in length and have a blunt base which is sometimes thickened. Under normal magnification they seem to be totally smooth. Higher magnification with a phase-contrast microscope, however, shows behind the sharp distal tip very small barbed hooks (Fig. 16). These

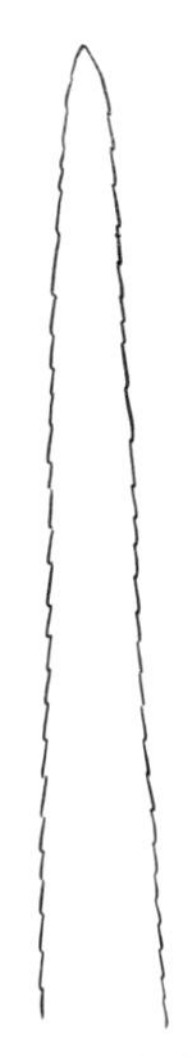

FIG. 16. Distal end of the nettling hair of the caterpillar of *Lasiocampa quercus* L. (Drawing from a photograph of Kemper, 1958.)

look like roofing slates or scales of fishes and are directed with their free end toward the base. On penetration into vertebrate skin, the barbed hooks take care of fixation of the hairs or their broken tips and can in this way cause irritations. Kemper found these hairs in caterpillars of *Macrothylacia rubi* L., *Lasiocampa quercus* L., *L. trifolii Esp.*, and *Dendrolimus pini* L., but it can be assumed that they are characteristic of the whole family.

3. Urticating Hairs with Pointed Bases

The urticating hairs described here are different in size and arrangement and are larval or imaginal formations. All these different types, however, have one common character: they have sharp pointed bases and mostly barbed hooks directed distally against the free end of the seta. All these hairs are more or less easily freed from the caterpillar's or moth's body and penetrate with the sharp base into the skin of enemies. Since the barbed hooks reach nearly down to the base, they can prevent the hairs from being withdrawn.

a. "Spiegelhaare." Very short, thin hairs which sit closely together on small round spots of certain abdominal tergites are known as Spiegelhaare. The hairs are very dense and have a uniform length, forming an even surface. We distinguish two types of such urticating hairs, both with strong urticating properties.

i. "Spiegelhaare" of the Lymantriidae. A considerable number of the caterpillars of the Lymantriidae (=Liparidae) and, according to Gilmer (1925), some Notodontidae possess typical short urticating hairs. The form of larvae of the "brown-tail (*Euproctis chrysorrhea*) is typical for this group [Fig. 17 (a)]. The caterpillars have urticating hairs on the first subdorsal tubercles of the first abdominal segment; in later stages they appear on subdorsal and supraspiracular tubercles of the first eight abdominal segments. This means there are to be found four "mirrors"on (Spiegelhaare, mirror-hair) on each segment [Fig. 17 (a)]. Upon these tubercles are minute papillae or cuplike structures closely crowded together. As is shown in a microscopic section [Fig. 17 (a) 2], these cups contain from three to a dozen short spiculelike hairs of the described structure [Fig. 17 (a)3]. They are from 0.07 to 0.1 mm long and from 4 to 5 μ in diameter at the larger end. According to Weidner (interpreting the drawings of Tonkes, 1933), each hair has its own trichogen cell (the lower-lying cells) but several hairs together have one common membrane cell, forming the cup. The space between the cups is usually less than the diameter of the cups themselves, the cells lying closely packed in the hypodermis. The trichogen cells correspond in number with the hairs and each cell is connected by a strand of cytoplasm with the hair. The hairs have in their interior a pore canal, but Gilmer (1925) has been unable to trace the cytoplasmatic strands to this pore canal. The existence of a pore canal gives no indication whether the *Spiegelhaare* have poisonous secretions in their interior since it might be a functionless structure left over from the hair formation.

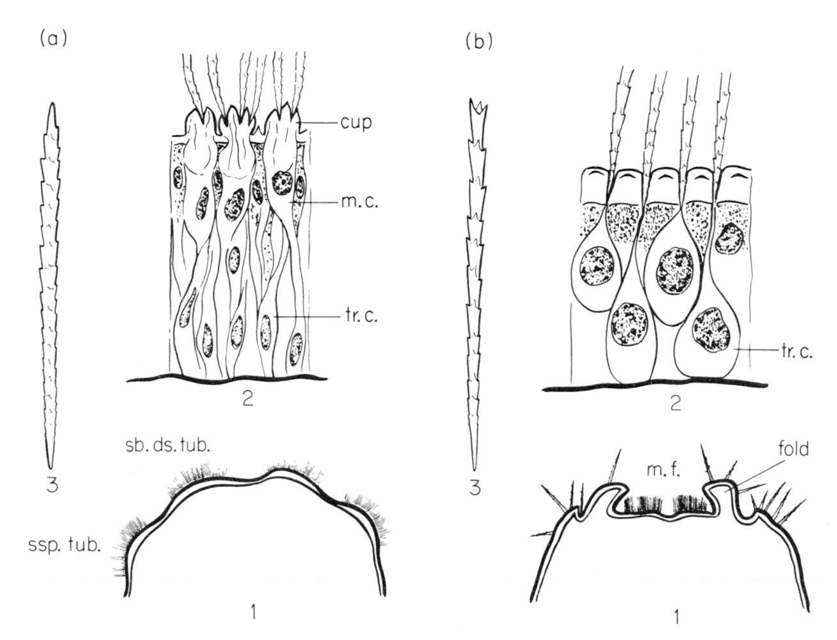

FIG. 17. (a) *Spiegelhaare* of a caterpillar of the brown-tail (*Euproctis chrysorrhoea*). 1, schematic cross section through the dorsal cuticle of an abdominal segment demonstrating the four tubercles with mirror-hairs (sb. ds. tub., subdorsal tubercle; ssp. tub., supra spiracular tubercle). 2, section through the hypodermal and cuticular area of a *Spiegelhaar* field (m.c., membrane cell; tr.c., trichogen cell). 3, a single *Spiegelhaar*. Drawings are modifications of Kephart (1915) and Tonkes (1933). (b) *Spiegelhaare* of the processionaries type. 1, schematic cross section through a mirror (m.f., mirror field) surrounded by folds with long normal setae. 2, schematic section through cuticle and hypodermal region of a mirror, following Weidner (1936) (tr.c., trichogen cell). 3, a single *Spiegelhaar*.

ii. "Spiegelhaare" of the Processionaries (Thaumetopoeidae). The processionaries *Thaumetopoea processionea* L., *Th. pinivora Tr.*, and *Th. pityocampa* Schiff are the best-known caterpillars with urticating hairs. According to Scheidter (1934), *Spiegel* with urticating hairs are to be found on the abdominal tergites (segments 4–12). Originally each segment had six cuticular plates, but by fusion the area now has four parts. These plates lie in fields in small depressions of the dorsal cuticle which are formed by surrounding folds. By approaching these folds—posterior and anterior folds can be approached by making the backside concave—the field of the hairs can be totally covered. In this way they can be turned in and out like pockets. On the margins of the fold stand long more or less spiny setae which might protect the fields and (or), as we presume, lead to the action of the urticating hairs. If the marginal setae or *Spiegelhaare* are heavily bumped by an enemy—you can make this experiment with a needle or a forceps—the fields open and bunches with thousands and thousands of the small urticating hairs gush out. It looks like an eruption. We do not know the mechanism by which the hairs are shot out—it could be due to local hemolymph pressure or to contractions of intersegmental muscles. There is no doubt that this mechanism is effective since the pointed bases [Fig. 17 (b)3] are freed and the hairs can penetrate in quite large numbers into the enemy's skin, mucous membranes of mouth and nose, or into his eyes. Scheidter (1934) estimated a total number of 630,000 *Spiegelhaare* for one caterpillar in the last stage of the *Th. processionea.*

The urticating hairs appear in all three above-mentioned processionaries for the first time in the third larval stage. In *Th. processionea* the first *Spiegelhaare* appear on segment 11, in *Th. pityocampa* they appear also on segments 4, 5, and 10. In the following stages urticating hairs appear progressively thicker from the anterior to the posterior segments. According to Weidner (1936), the urticating hairs have no membrane cells and corresponding cup-like structures. They stand singly in the cuticle; each has its trichogen cell (Fig. 17 (b)2).

The length of the urticating hairs increases with the development of the caterpillar. Commonly in *Th. processionea* they measure 0.106 mm in the third stage, 0.135 mm in the fourth stage, 0.165 in the fifth stage, and 0.213 mm in the sixth stage. Not all hairs in the same stage are exactly the same size (Weidner). As mentioned in the synopsis (Table IV), the *Spiegelhaare* are woven into the cocoon of the processionaries.

Kemper (1958) demonstrated with the phase-contrast microscope that these hairs (and any insect hairs) have in their interior a narrow channel which seems to contain air or some other gas. Studies with the help of an electron microscope* have disclosed the hollow interior (Fig. 18), but do not provide clues to content.

b. Brush-Hairs ("Bürstenhaare"). Caterpillars of the families Lymantriidae and Noctuidae have hairs spread over the lateral and dorsal parts of their whole body. Certain circumscribed areas are equipped with specialized urticating hairs; the short ones—*Spiegelhaare*—have been treated in the previous section. Several species have longer urticating hairs which form thick bunches looking like brushes. Therefore we call this type of hair *Bürstenhaare* (brush-hairs). Very typical are the brushes of *Dasychira pudibunda* L. (Lymantriidae), which are on the tergites of segments 4–7 (Kemper, 1956). Each of these four brushes contains about 2,000 hairs of bright yellow or yellow-gray color. Since these hairs have about the same length (about 4 mm), the brushes look like shaving brushes (Fig. 19,1). These brush-hairs

* For kind help with ultramicrotome sections and all preparations and photographs, we thank our colleague Mrs. A. Gluud-Becker.

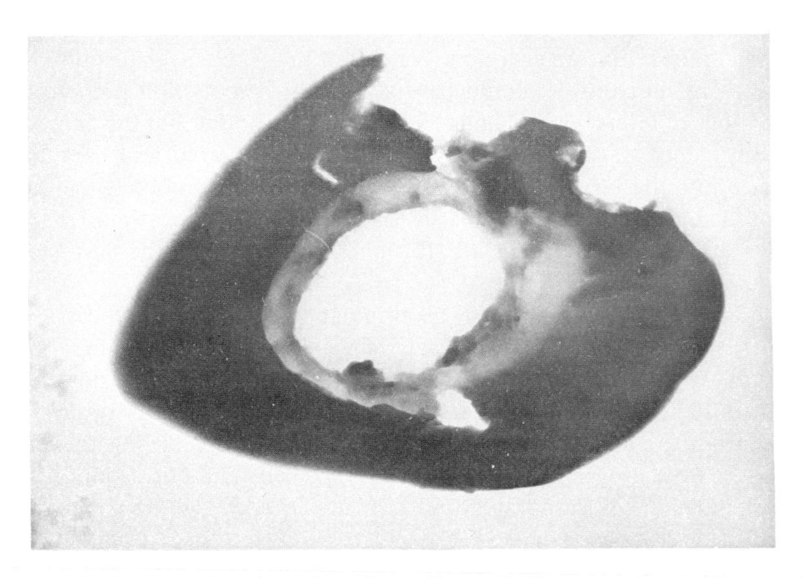

FIG. 18. Cross section through a *Spiegelhaar* of *Th. processionea*. Photographed with an electron microscope, magnification × 20,000.

have, in contrast to all other hairs, a sharp pointed base and side-spines reaching up to the free tip. The side spines become smaller and smaller against the base and have the efficacy of barbed hooks (Fig. 19,2). Experiments of Kemper (1956,1958) demonstrated that these long urticating hairs could penetrate into the skin as well as the short ones. The effect on human skin is said to correspond with the stinging of blood-sucking arthropods. The brush-hairs are also spun into the pupal cocoons.

c. Hairs of Imaginal Anal Tufts. Imaginal moths in the family Lymantriidae have within some species special hair-formations at the posterior end. Such anal tufts can be found in both sexes; nevertheless the females generally have better developed tufts since the tuft hairs are often used to protect the egg layers. Anal tuft hairs have urticating effects on human skin which could also be due to the transfer of larval *Spiegelhaare*. Kemper (1955) was able to demonstrate with anal tufts in the "brown-tail," that there can be found two types of hairs; the majority are smooth, about 1.8 mm long and tapered against both ends. The proximal end has a small ball, the distal end is slightly bent and appears oblique. Hairs of a second type have about the same length but are somewhat thinner. Several hairs (2–6) originate from a common base. The base is sharply pointed and has microscopic barbed hooks (Fig. 20). The upper part of this hair has small side spines, directed against the base. There is no doubt that these hairs might be highly efficient as urticating hairs whether they penetrate with their base or with the distal end. We find again

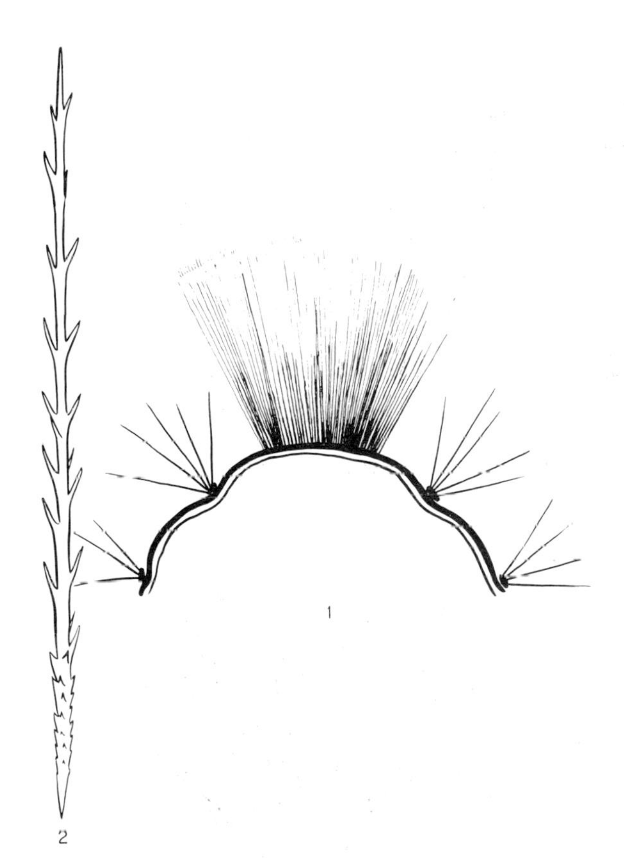

FIG. 19. *Bürstenhaare* of a caterpillar of *Dasychira pudibunda* L. (Lymantriidae). 1, schematic cross section through abdominal tergite with a brush in the middle. Besides the brush papillae with normal hairs. 2, a single *Bürstenhaar*.

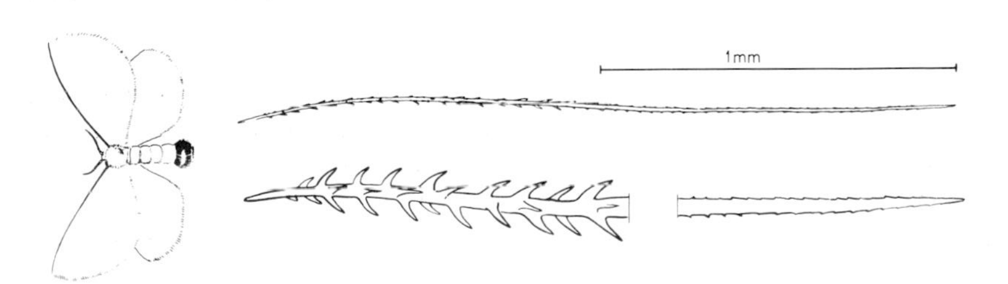

FIG. 20. Schematic drawing of a female of the "brown-tail" with its golden-brown anal tuft. Under that an urticating hair is shown at two different magnifications.

a common morphological sign of most urticating hairs—a sharp pointed base.

4. Poisonous Spines

The poisonous apparatus of the caterpillars of the families Cochlidiidae (Limacodidae), Megalopygidae, and Saturniidae is quite different from the hairs described above. According to Gilmer (1925), a similar type of poisonous spines occurs in several tropical forms of the family Noctuidae also. Certain well-marked morphological structures, differentiating it from the simple hair type, characterize the spine type. The primary difference is the fact that it is not the product of a single enlarged hypodermal cell, but is an evagination of the body wall, lined throughout its whole length except the tip, by hypodermis but little differentiated from ordinary body wall hypodermis. The spines vary widely, both in length and diameter. They are generally associated in some manner with a tubercule or a verruca. Often they are arranged in a rosette form, or may branch off rather irregularly from a main axis. (Fig. 21)

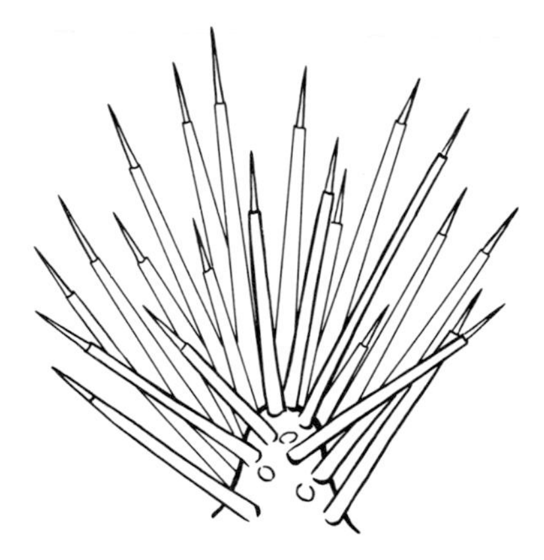

FIG. 21. Bunch of poisonous spines on a dorsal tubercle of a medically important South African caterpillar (*Parasa vivida*), Family Cochlidiidae. (Modified from Zumpt, 1956.)

They have never been found indiscriminately over the body but are confined usually to the dorsal, subdorsal, or lateral tubercles. The saddle-back, *Sibine stimulea*, known as one of the most venomous American forms, has the following weapons. Practically every segment carries two or more spine-bearing tubercles. Upon the anterior thoracic segments can be found a pair of fairly large, erect, hornlike tubercles, which are the point of heaviest armature. They are thickly covered with quite long, stout spines. The spines of the posterior segments, however, are rather short and stout.

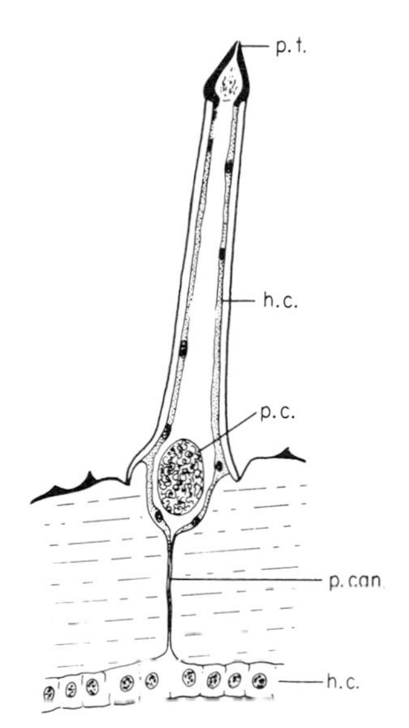

FIG. 22. Schematic section through a poisonous spine of a caterpillar of the family Cochlidiidae (modified from Weidner, 1936) h.c., hypodermal cells; p.can., pore canal; p.c., poison cell; p.t., pointed tip with a fine channel.

The spines have an acutely pointed tip, the point being longer and tapering (in case of *Parasa*, Fig. 21) or shorter (Fig. 22), but sharply differentiated from the spine proper. The spine is lined throughout its length by a hypodermis of flattened cells. The point is without hypodermis, but is opened just at its sharp pointed tip by a narrow channel, looking like the tube of a syringe. At the point of union of the spine and the body wall is a bulblike cavity composed of the widened proximal portion of the spine and the much widened outer end of the pore canal. This cavity is egg-shaped and is more or less buried in the cuticle of the body wall. In the Megalopygidae (Fig. 23), the point of the spine seems not to be as sharply pointed as are the spines in Figs. 21 and 22. It seems to be prolonged and has in its interior a very flat hypodermal layer nearly up to the distal end. The tip is held against the bulb by a chitinous diaphragm. These tips break off easily if they penetrate into the enemy's skin, but the broken end is still strong enough for the spines to penetrate the skin further. In case of the "puss caterpillar," *Megalopyge opercularis*, there is no connection between the body wall hypodermis and the epithelium of the spine interior. Only a trace looking like a radix (Fig. 23) gives an idea that the spine was "tied off" after its evagination. In Cochlidiidae (Fig. 22) there is a rather narrow pore canal leading from the base of the bulb through the cuticule to the hypodermis. The hypodermis may be seen entering the pore canal, passing through and emerging at the base of

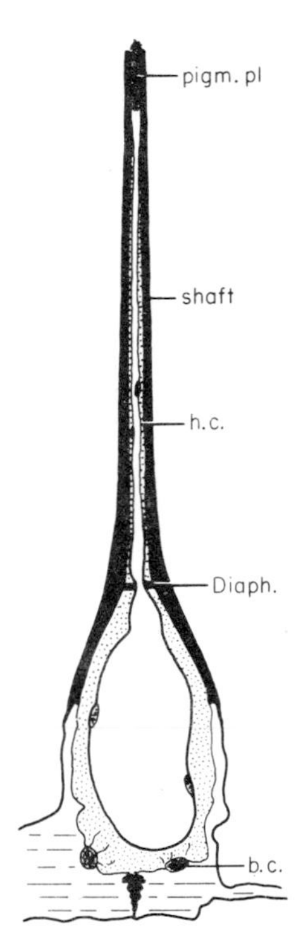

FIG. 23. Schematic section of a spine of *Megalopyge opercularis* (modified from Foot, 1922). pigm. pl., pigmented plug at the tip of the seta; shaft, shaft of the prolonged tip; Diaph., chitinous diaphragm between bulb and tip; b.c., bulb sac cells with large nuclei (poison gland?).

the bulb to spread out and line both the bulb and spine except for the penetrating tip. It forms a sac filled with a poisonous liquid. Lying within the sac is a large cell with a lobate nucleus which might be polyploid. In *M. opercularis* (Fig. 23) there is no separated glandular cell, but some of the epithelial cells of the sac have large nuclei.

Similar spines are to be found in caterpillars of Saturniidae. Valle *et al.*, (1954) found the mean weight of each poisonous spine of *Dirphia* (Saturniidae) to be 0.86 mg. The poisonous spines work like the tubes of a syringe; they inject their poisonous liquid into organisms which attack them or disturb them through contact. The ejection of the secretion is effected by the pressure which occurs when the spine is touched. More of the secretion can be squirted out if to the external pressure is added an internal one. This can be effected through contractions of the entire caterpillar's body which increase the hemolymph pressure. Local contractions of intersegmental muscles lead to similar effects. Often the caterpillars use internal pressure to erect the spines and direct them against an aggressor or they bristle all the spines at the approach for an animal. We do not know how the caterpillars sense an approaching body. It might be that certain sensory setae standing around or between the spines have a sensory function. Mostly the spines are located on tubercles or verrucae as mentioned above. Caterpillars of Cochlidiidae have a lateral poison apparatus, described by Weidner (1936). Poisonous spines (Fig. 24a) with the typical injection tip stand around a central plug. On the outside they are accompanied by a second type of hollow spine with a long tip looking like a whip (Fig. 24b). This whip is very long and flexible.

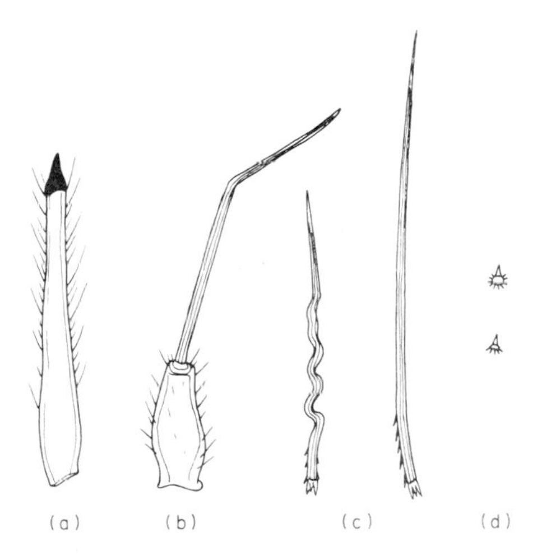

FIG. 24. Different types of spines and urticating hairs of a caterpillar of the family Cochlidiidae (modified from Weidner, 1936). a, normal poisonous spine; b, poisonous spine with long and flexible, whiplike tip; c, urticating hairs with pointed bases; d, small, starlike urticating prickles.

Therefore it is impossible for it to penetrate into skin, though those spines contain liquid which can be pressed out at the tip. Weidner supposes them to be sensory setae since they extend above the normal spines. Independent of these spines with whips there might be other sensory hairs of the normal type. It should be mentioned that the poison organs of Chochlidiidae are often equipped with both spines and urticating hairs. Weidner demonstrated *Spiegelhaare* (Fig. 24c) with pointed bases and barbed hooks. A further very small type of urticating structures are starlike spine clusters (Fig. 24d). These consist of a central spine on a hemisphere from which smaller spines project radially. These starlike prickles are concentrated on the central cone of the lateral poisonous apparatus (Fig. 25). They break off very easily and have urticating effects after penetration into skin.

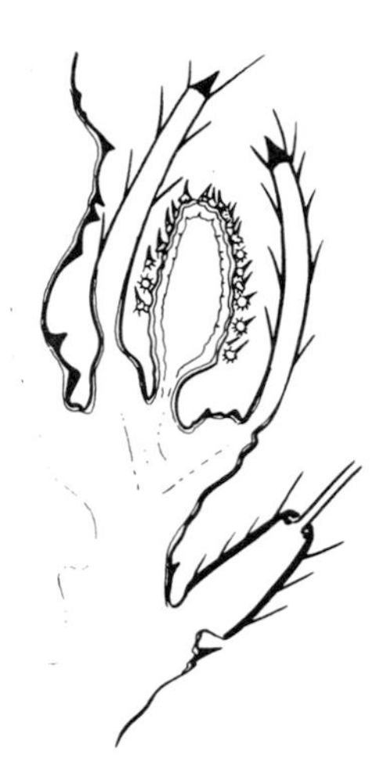

FIG. 25. Longitudinal section through the lateral poisonous apparatus of a caterpillar of Cochlidiidae (following Weidner). Round a central plug with starlike prickles stand spines, outside (right-hand) sits a spine with whiplike tip.

We have to mention another type of poisonous secretion extruded from the spines. The lower part of the bulb's cuticle is soft whereas the upper part and the prolonged tip are well sclerotized (Fig. 23). This forms a poorly defined articulation. The tip of the spine is bent off, by external pressure the liquid in its poison sac becomes squeezed and flows out under pressure. The amounts of liquid squeezed out can be relatively high. Caterpillars of *Automeris* are reported to roll in upon themselves after heavy contact, bristling all their spines like a hedgehog. Since they eject liquid from most of the spines, the surroundings of the caterpillar become wet.

According to Weidner, the liquid of the spines of *Automeris illustris* is clear as water. It contains microscopic small green droplets. In coming to the surface of the spine the secretion acquires a greasy consistency of blue/green colour. Loss of water by evaporation and oxidation might lead to this effect which sometimes causes plugs to form in the distal end of the tip's channel (Fig. 22, 23).

The contents of the poisonous liquid have been studied with pharmacological methods by Valle *et al.* (1954). The authors extracted whole spines of *Dirphia* and *Megalopyge* in a porcelain mortar with different solvents. On the base of pharmacological tests it could be calculated that *Dirphia's* spines contain 0.1 to 0.2 μg of histamine per milligram of wet tissue. As the setae average 0.86 mg., each one may have from 0.086 to 0.172 μg of histamine. The average seta's weight after exposure at 100°C for 24 hours is 0.43 mg, which corresponds to a water loss of 430 cubic microns. If this is the fluid volume of each seta, then the histamine concentration within the *Dirphia's* spines would correspond to a 0.02–0.04% solution. In *Megalopyge's* spines the histamine concentration was found to be very low. Acetylcholine could not be detected in either cases.

Toxicity on dogs and guinea pigs, hypotensive effect on dogs and cats, and cutaneous reactions on human beings suggest the presence, besides histamine, of other pharmacologically active substances, probably of a protein nature. According to Esable *et al.*, (1945), cited by Valle *et al.*, 1954) the venom of *Megalopyge urens* is of a globulin nature with necrotic, hemolytic, and immunological properties.

C. Ecological Importance of Urticating Hairs and Spines

The lack of reproducible experiments on the correlation between feeding habits of insect-feeding vertebrates (amphibians, reptiles, birds, mammals) and equipment of caterpillars with urticating hairs and spines allows no final interpretation of their ecological importance. Generally they are believed to be defense mechanisms. Several observers report that hairy caterpillars, especially forms with urticating hairs, are not eaten by all species of insectivorous birds. Only the cuckoo feeds on all hairy caterpillars; he even seems to have a preference for them. But some other birds also feed on hairy caterpillars. The smaller song birds often refuse large caterpillars whether they are hairy or not, especially the processionarie which have the highest development of mirrors with urticating hairs in their last stage. Weidner pointed out this would argue against the protection by mirror-hairs. The smaller caterpillars, which would need more protection since they are attacked by more bird species, have only a few urticating hairs or none. Nevertheless it can be supposed that the urticating hairs might have a certain effectiveness against birds, reptiles, and amphibians. They should be most effective against mammals since these show an olfactoric orientation and can inhale the urticating hairs into the sensitive mucous membranes of mouth and nose. The high rate of effectiveness against human beings supports the theory of special protection against mammals.

The behavior of caterpillars of the "white-marked Tussock moth" (*Hemerocampa leucostigma* Smith and Abbot) seems to indicate a connection with the function of urticating hairs: It was noted that when disturbed the larvae had a habit of elevating the last few segments and wagging them back and forth. Since these segments bear the urticating hairs, this habit is possibly associated with the presence of these hairs and used as a means of protection. Under such circumstances the urticating hairs would have a direct protective function. The poisonous spines might be much more effective as direct weapons, but mostly the urticating hairs are loosely attached so that they drop out without difficulty. In this way they might act indirectly as well as directly, via exuviae, larval webs, feces, pupal cocoons (compare the scheme in Section III, A). In addition the hairs are spread out and widely distributed by air currents. Especially after moltings, which occur more or less synchronously in populations of caterpillars, large amounts of irritating hairs can be spread out. There is no doubt that the population of the caterpillars might gain a certain ecological advantage. The repellent effects against some vertebrates could increase slightly the chance of survival and lead to a selective advantage for the species. On the other hand, there is no effective protection against insect enemies, whether they are predators or parasites. We know a lot of entomophagous species of arthropods feed on caterpillars with urticating hairs and spines. According to Weidner, several authors supposed that the total contamination of the biotope with urticating hairs could suppress other phytophagous insects, possibly by external and internal (midgut) penetration into their organisms. In this case competition for food would be eliminated or reduced; this could be presumed to be of real ecological importance for a population equipped with urticating hairs.

In caterpillars and in adult lepidopterans several poisonous chemical defense mechanisms against predation exist. Since the glandular and non-glandular defense apparatus as well as the chemistry and effectiveness of defensive substances have been comprehensively reviewed by Th. Eisner ("Chemical Ecology," Academic Press, 1970, E Sondheimer and J. B. Simeone, eds.), we refer the reader to this book.

REFERENCES

Adrouny, G. A., Derbes, V. J., and Jung, R. C. (1959). *Science* **130**, 449.
Altenkirch, G. (1962). *Zool. Beitr.* **7**, 161–238.
Anderson, J. (1885). *Entomology* **18**, 43–45.
Autrum, H.-J., and Kneitz, H. (1959). *Biol. Zentrbl.* **78**, 598–602.
Beard, R. L. (1952). *Conn. Agr. Expt. Sta., New Haven, Bull.* **562**.
Beard, R. L. (1963). *Ann. Rev. Entomol.* **8**, 1–18.
Bergström, G., and Löfquist, J. (1968). *J. Insect Physiol.* **14**, 995–1011.
Berland, L. and Bernard, F. (1951). *In* "Traité de Zoologie" (P.-P. Grassé, ed.), Vol. X, p. 771–1276. Masson, Paris.
Betts, A. (1922). *Bee World* **3**.
Beyer, O. W. (1891). *Jena. Z. Naturw.* **25**, 26–112.
Bischoff, H. (1927). "Biologie der Hymenopteren," Springer, Berlin.
Blum, M. S. (1966). *Proc. R. entomol. Soc. Lond.* (A) **41**, 155–160.
Blum, M. S. (1969). *Ann. Rev. Entomol.* **14**, 57–80.
Blum, M. S., and Callahan, P. S. (1961). *Proc. 11th Intern. Congr. Entomol., Vienna, 1960,* Vol. 3, 290-293.
Blum, M. S., and Callahan, P. S. (1963). *Psyche* **70**, 69–74.
Blum, M. S., Moser, J. C., Cordero, A. D. (1964). *Psyche* **71**, 1–7.
Blum, M. S., Padovani, F., Curley, A., Hawk, R. E. (1969). *Comp. Biochem. Physiol.* **29**, 461–465.

Blum, M. S., Padovani, F., Hermann, H. R., Kannowski, P. B. (1968). *Ann. Ent. Soc. Amer.* **61**, 1354–1359.

Blum, M. S., and Portocarrero, C. A. (1964). *Ann. Entomol. Soc. Am.* **57**, 793–794.

Blum, M. S., Roberts, J. E., and Novak, A. F. (1961). *Psyche* **68**, 73–74.

Blum, M. S., and Ross, G. N. (1965). *J. Insect Physiol.* **11**, 857–868.

Blum, M. S., Walker, I. R., Callahan, P. S., and Novak, A. F. (1958). *Science* **128**, 306–307.

Blum, M. S., Warter, S. L., Monroe, R. S., and Chidester, J. C. (1963). *J. Insect Physiol.* **9**, 881–885.

Blum, M. S., and Wilson, E. O. (1964). *Psyche* **71**, 28–31.

Bordas, L. (1894). *Zool Anz.* **17**, 385–387.

Bordas, L. (1895). *Ann. Sci. Nat. Zool.* [7] **19**, 289–344.

Bordas, L. (1897). "Description anatomique et étude histologique des glandes de venin des insectes hyménoptères." Paris.

Brown, W. L., Jr. (1954). *Insectes Sociaux*, 21–31.

Bucher, G. E. (1948). *Can. J. Res.* **D 26**, 230–281.

Butlcr, C. G., and Simpson, Y. (1965). *Ustav védeckotechnichých informací* **MZ** LVII,33–36.

Callahan, P. S., Blum, M. S., and Walter, J. R. (1959). *Ann. Entomol. Soc. Am.* **52**, 573–590.

Carlet, G. (1890). *Ann. Sci. Nat. Zool.* [7] **9**, 1–17.

Caro, M. R., Derbes, V. J., and Jung, R. C. (1957). *A.M.A. Arch. Dermatol.* **75**, 475–488.

Carthy, J. P. (1951). *Behaviour* **3**, 304–318.

Cavill, G. W. K., and Ford, D. L. (1960). *Australian J. Chem.* **13**, 296–310.

Cavill, G. W. K., and Hinterberger, H. (1961). *Proc. 11th Intern. Congr. Entomol., Vienna, 1960.* Vol. **3**, pp. 53–59.

Cavill, G. W. K., and Locksley, H. D. (1957). *Australian J. Chem.* **10**, 352–358.

Cavill, G. W. K., Ford, D. L., and Locksley, H. D. (1956). *Australian J. Chem.* **9**, 288–293.

Cavill, G. W. K., Robertson, P. L., and Whitfield, F. B. (1964). *Science* **164**, 79–80.

Cavill, G. W. K., and Robertson, P. L. (1965). *Science* **149**, 1337–1345.

Cavill, G. W. K., and Williams, P. J. (1967). *J. Insect Physiol.* **13**, 1097–1103.

Clausen, C. P. (1940). Entomophagous Inscets. McGraw-Hill, New York.

Cohic, F. (1948). *Rev. Franc. Entomol.* **14**, 229–276.

Darwin, C. (1859). "On the Origin of Species by Means of Natural Selection." Murray, London.

De la Lande, I. S., Thomas, D. W., and Tyler, M. (1963). *Second International Pharmacological meeting* **9**, 71–76.

Dewitz, H. (1875). *Z. Wiss. Zool.* **25**, 174–200.

Dewitz, H. (1877). *Z. Wiss. Zool.* **28**, 527–536.

Donisthorpe, H. (1901). *Trans. Entomol. Roy. Soc. London, Proc. S. 13.*

Dufour, L. (1841). *Mem. Acad. Roy. Sci.* **7**, 265–647.

Eltringham, H. (1912). *Proc. Entomol. Soc. London* pp. 78–81.

Evans, H. E. (1962). *Evolution* **16**, 468–483.

Evans, H. E. (1966). *Ann. Rev. Ent.* **11**, 123–154.

Fabre, J. H. (1879). "Souvenirs Entomologiques." Délagrave, Paris.

Flemming, H. (1957). *Z. Morphol. Oekol. Tiere* **46**, 321–341.

Fletscher, D. J., and Brand, I. M. (1968). *J. Insect Physiol.* **14**, 783–788.

Foerster, E. (1912). *Zool. Jahrb., Abt. Anat. Ontog. Tiere* **34**, 347–380.

Foot, N. C. (1922). *J. Exptl. Med.* **35**, 737–753.

Forel, A. (1878). *Z. Wiss. Zool.* **30**, Suppl., 28–68.

Free, J. (1961). *Animal Behav.* **9**, 193–196.

Fröhlich, K. O., and Kürschner, I. (1964). *Beitr. Entomol.* **14**, 507–524.

Gerstaecker, A. (1867). *Arch. Naturg.* **30**, 1–95.
Ghent, L. R. (1961). Doctoral thesis, Cornell Univ., Ithaca. N.Y.
Ghent, L. R., and Gary, N. E. (1962). *Psyche* **69**, 1–6.
Gilmer. P. M. (1923). *J. Parasitol.* **10**, 80–86.
Gilmer, P. M. (1925). *Ann. Entomol. Soc. Am.* **18**, 203.
Goldmann, L. *et al.* (1960). *J. Invest. Dermatol.* **34**, 67–78.
Grandi, G. (1951). "Introduzione allo Studio dell 'Entomologia," Vol. II, pp. 68–275. Edizione Agricola, Bologna.
Habermann, E. (1968). *Ergebnisse der Physiologie* **60**, 220–325.
Hangartner, W., and Bernstein, St. (1964). *Experientia* **20**, 392–393.
Hanna, A. D. (1934). *Trans. Roy. Entomol. Soc. London* **82**, 107–136.
Hase, A. (1924). *Biol. Zentrbl.* **44**, 209–243.
Hase, A. (1939). *Anz. Schädlingskunde* **15**, 133–142.
Hemstedt, H. (1969). *Z. Morph. Tiere* **66**, 51–72.
Hermann, H. R., and Blum, M. S. (1966). *Ann. Entomol. Soc. Am.* **59**, 397–409
Heselhaus, F. (1922). *Zool. Jahrb., Abt. Anat. Ontog. Tiere* **43**, 369–464.
Hölldobler, K. (1965). *Mitt. Schweiz. Entomol. Ges.* **38**, 71–81.
Holmgren, E. (1896). *Entomol. Tidskr.* **17**, 81–85.
James, H. C. (1926). *Proc. Zool. Soc. London* pp. 75–182.
Janet, C. (1898). "Études sur les Fourmis, les Guêpes et les Abeilles," Notes 17 and 18. Georges Carré et C. Naud, Paris.
Jentsch, J. (1969). Proc. VI. Congr. IUSSI, 69–75, Bern.
Jones, D. L., and Miller, J. H. (1959). *Arch. Dermatol.* **79**, 81–85.
Imms, A. D. (1919). *Quart. J. Microscop. Sci.* [N.S.] **63**, 293–374.
Imms, A. D. (1957). "A General Textbook of Entomology," 9th rev. ed. Methuen, London.
Kahlenberg, H. (1895). Thesis, Univ. Erlangen.
Kemper, H. (1955). *Z. Angew. Zool.* **42**, 37–59.
Kemper, H. (1956). *Z. Angew. Zool.* **43**, 103–128.
Kemper, H. (1958). *Proc. 10th Intern. Congr. Entomol., Vienna,* **1956**, pp. 719–723.
Kephart, C. F. (1915). *J. Parasitol.* **1**, 95–103.
Kerr, W. E., and De Lello, E. (1962). *J. N.Y. Ent. Soc.* **70**, 190–214.
Koshewnikow, G. A. (1899). *Anat. Anz.* **15**, 519–528.
Kraepelin, R. (1873). *Z. Wiss. Zool.* **23**, 289–330.
Lacaze-Duthiers, H. (1849). *Ann. Sci. Nat. Zool.* [3] **12**, 353.
Lacaze-Duthiers, H. (1850). *Ann. Sci. Nat. Zool.* [3] **14**, 21.
Lalesque, F., and Marder, C. (1909). *Bull. Stat. Biol. d'Arcachon* **12**, 61.
Lecomte, J. (1961). *Ann. Abeille* **4**, 165–270.
Leger, M., and Mouzels, P. (1918). *Bull. Soc. Pathol. Exotique* **11**, 104.
Leuenberger, F. (1954). "Die Biene." Sauerländer, Aarau.
Leuthold, R. H. (1968*a*). *Psyche* **75**, 233–248.
Leydig, F. (1859). *Arch. Anat. Physiol.* pp. 33–183.
Liepelt, W. (1963). *Zool. Jahrb. Abt. Physiol. Allgem. Zool. Tiere* **70**, 167–176.
Lindauer, M. (1957). *Ber. Wanderversammlung Deut. Entomol.*, Berlin, **8**, 71-78.
Lindauer, M. (1961). "Communication among Social Bees," Harvard Univ. Press, Cambridge, Massachusetts.
Lukoschus, F. (1962*a*). *Z. Bienenforsch.* **6**, 72–76.
Lukoschus, F. (1962*b*). *Z. Morphol. Oekol. Tiere* **51**, 261–270.
Maidl, F. (1934). "Lebensgewohnheiten und Instinkte der staatenbildenden Insekten," Wagner, Wien.
Martini, E. (1952). "Lehrbuch der Medizinischen Entomologie," Fischer, Jena.

Maschwitz, U. (1964). *Z. Vergleich. Physiol.* **47**, 596–655.
Maschwitz, U. (1966). *Vitamins and Hormones* **24**, 267–290.
Maschwitz, U., Koob, K., and Schildknecht, H. (1970). *J. Insect Physiol.* **16**, 387–404.
Meinert, E. (1860). *Kgl. Danske videnskab. Selskab Skrifter, Raekke, Nat. Mat. Afd.* **5**.
Morison, G. D. (1928). *Quart. J. Microscop. Sci.* **71**, 563–651.
Moser, J. C., and Blum, M. S. (1963). *Science* **140**, 1228.
Oeser, R. (1961). *Mitt. Zool. Museum Berlin* **37**, 1–119.
Olberg, G. (1959). "Das Verhalten der solitären Wespen Mitteleuropas," VEB Deutscher Verlag der Wissenschaften, Berlin.
Osman, M. F. H., and Brander, J. (1961). *Z. Naturforsch.* **11**b, 749–753.
Osman, M. F. H., and Kloft, W. (1961). *Insectes Sociaux* **8**, 383–395.
Otto, D. (1960). *Zool. Anz.* **164**, 42–57.
Ouljanin, (1872). *Z. Wiss. Zool.* **22**, 289.
Pavan, M. (1955). *Chim. Ind.* (*Milano*) **37**, 625–627.
Pavan, M. (1956). *Ricera Scient.* **26**, 144–150.
Pavan, M. (1958). "Significato chimico e biologico di alcuni veleni di insetti," Tipografia Artigianelli, pp. 1–75. Pavia.
Pavan, M. (1959). *Proc. 4th Intern. Congr. Biochem., Vienna, 1958* Vol. 12, p. 15–36. Pergamon Press, London.
Pavan, M., and Ronchetti, G. (1955). *Atti. Soc. Ital. Sci. Nat. Milano* **94**, 379–447.
Pawlowsky, E. N. (1914). *C.R. Soc. Biol.* **76**, 351–354.
Pawlowsky, E. N. (1927). "Gifttiere und ihre Giftigkeit," Fischer, Jena.
Pawlowsky, E. N., and Skin, A. K. (1927). *Z. Morphol. Oekol. Tiere* **9**, 615–637.
Piek, T. (1969). *Acta Physiol. Pharmac. Néerl.* **15**, 104–105.
Piek, T., and Engels, E. (1969). *Comp. Biochem. Physiol.* **28**, 603–618.
Piek, T., and Simon Thomas, R. T. (1969). *Comp. Biochem. Physiol.* **30**, 13–31.
Quilico, A., Grünanger, P., and Pavan, M. (1961). *Proc. 11th Intern. Congr. Entomol., Vienna, 1960*, pp. 66–68.
Ratcliffe, N. A., and King, P. E. (1967). Proc. R. ent. Soc. Lond. (A) **42**, 49–61.
Rathmayer, W. (1962). *Z. Vergleich. Physiol.* **45**, 413–462.
Réaumur, R. A. F. de (1740). "Mémoires pour servir à l'histoire des Insectes V." Paris.
Regnier, F. E., and Wilson, E. O. (1968). *J. Insect Physiol.* **14** 955–970.
Regnier, F. E. and Wilson, E. O. (1969). *J. Insect Physiol.* **15**, 893–898.
Rehm, E. (1940). *Z. Morphol. Oekol. Tiere* **36**, 89–122.
Remane, A. (1952). "Die Grundlagen des natürlichen System der vergleichenden Anatomie und der Phylogenetik," Geest u. Portig, Leipzig.
Rietschel, P. (1938). *Z. Morphol. Oekol. Tiere* **33**, 313–357.
Robertson, P. L. (1968). *Aust. J. Zool.* **16**, 133–166.
Rosenbrook, W., and O'Connor, R. (1964b). Can. J. Biochem. **42**, 1005–1010.
Ross, H. H. (1937). *Illinois Biol. Monographs* **15**, No. 2.
Ruttner, F. (1961). *Z. Bienenforsch.* **5**, 253–266.
Sauerländer, S. (1961). *Naturwissenschaften* **48**, 629.
Scheidter, F. (1934). *Pflanzenkrankh. Pflanzenschutz* **44**, 223, 362, 385, and 497.
Schlusche, M. (1936). *Zool. Jahrb., Abt. Anat. Ontog. Tiere* **61**, 77–99.
Schröder, C. (1925). "Handbuch der Entomologie," Vol. III, pp. 852–941. Fischer, Jena.
Snodgrass, R. E. (1925). "Anatomy and Physiology of the Honey Bee," McGraw-Hill, New York.
Snodgrass, R. E. (1931). *Smithsonian Inst. Misc. Collections* **85**, No 6.
Snodgrass, R. E. (1933). *Smithsonian Inst. Misc. Collections* **89**, No 8.
Snodgrass, R. E. (1935). "Principles of Insect Morphology," McGraw-Hill, New York.

Snodgrass, R. E. (1942). *Smithsonian Inst. Misc. Collections* **103**, No 2.
Snodgrass, R. E. (1956). "Anatomy of the Honeybee," Cornell Univ. Press, Ithaca, New York.
Soliman, H. S. (1941). *Bull. Soc. Fouad Entomol.* **25**, 1–96.
Sollmann, A. (1863). *Z. Wiss. Zool.* **13**, 528–540.
Stumper, R. (1922). *Compt. Rend.* **174**, 66–67.
Stumper, R. (1960). *Naturwissenschaften* **47**, 457–463.
Swammerdam, J. (1752). "Bibel der Natur," Leipzig.
Tamashiro, M. (1960). *Berlin. J. Insect. Pathol.* **2**, 209–219.
Tonkes, P. R. (1933). *Bull. Biol.* **67**, 44–135.
Trave, R., and Pavan, M. (1956). *Chim. Ind.* (*Milano*) **38**, 1015–1019.
Trojan, E. (1922). *Arch. Mikroskop. Anat. Entwicklungsmech.* **96**, 340–353.
Trojan, E. (1930). *Z. Morphol. Oekol. Tiere* **19**, 678–685.
Trojan, E. (1936). *Z. Morphol. Oekol. Tiere* **30**, 597–628.
Valle, J. R., Picarelli, Z. P., and Prado, J. L. (1954). *Arch. Intern. Pharmacodyn.* **98**, 324–334.
von Ihering, H. (1886). *Entomol. Nachr.* **12**, 177.
Weber, H. (1933). "Lehrbuch der Entomologie," Fischer, Jena.
Weber, H. (1954). "Grundriss der Insektenkunde," Fischer, Stuttgart.
Weidner, H. (1936). *Z. Angew. Entomol.* **23**, 432–484.
Weinert, H. (1920). *Naturwiss. Wochschr.* **19**, 225–236.
Whelden, R. M. (1960). *Ann. Entomol. Soc. Am.* **53**, 793–808.
Wilson, E. O. (1959a). *Science* **129**, 643–644.
Wilson, E. O. (1963). *Ann. Rev. Entomol.* **8**, 345–368.
Wilson, E. O. (1965). *Science* **149**, 1064–1071.
Wilson, E. O., and Pavan, M. (1959). *Psyche* **66**, 70–76.
Zander, E. (1899). *Z. Wiss. Zool.* **66**, 289–333.
Zander, E. (1922). "Handbuch der Bienenkunde III. Der Bau der Biene." Ulmer, Stuttgart.
Zumpt, F. (1956). "Insekten als Krankheitsüberträger." Kosmos, Stuttgart.

Chapter 45

Chemistry, Pharmacology, and Toxicology of Bee, Wasp, and Hornet Venoms

*E. HABERMANN**

INSTITUT FÜR PHARMAKOLOGIE UND TOXIKOLOGIE DER UNIVERSITÄT WÜRZBURG, WÜRZBURG, GERMANY

I. Introductory Remarks 61
II. Biochemistry and Pharmacology of Single Venom Constituents 63
A. Biogenic Amines 63
B. Peptides and Small Proteins 64
C. Hyaluronidase (Bee and Wasp Venoms) 73
D. Phospholipase A (Bee, Wasp and Hornet Venoms) 74
E. Phospholipase B (Wasp, and Hornet Venoms) 84
III. Possible Hazards for Man and Their Prevention 85
A. Envenomation of Man 85
B. Antibody Production and Desensitization to Bee Venom Components 86
C. Allergic Reactions 87
References 89

I. INTRODUCTORY REMARKS

Besides their zoological classification as products of hymenopterans, bee, wasp, and hornet venoms share many properties. A highly specialized apparatus serves for their secretion, storage, and ejection. This is in contrast to the conditions under which the venoms of snakes are produced and applied to the victims: in biting animals, parts of the digestive system, including salivary glands and teeth, are modified as part of the envenomating tool. This difference is also reflected in the different composition of hymenopteran and certain snake venoms. The former nearly exclusively contain basic substances

* *Present address:* Department of Pharmacology, University of Giessen, Giessen, Germany.

of high pharmacological activity but only a few enzymes splitting phospholipids and mucopolysaccharides. The latter are composed in a more complex manner; except for elapid venoms, they are enriched with proteolytic and nucleotide-splitting enzymes, various esterases, and amino acid oxidase, which can be understood to be digestive factors that have undergone biochemical evolution into dangerous agents. The frontier between "simple" body constituent and "venom" is not sharply marked. Bee venom, for example, contains histamine; wasp venom contains histamine and serotonin; and hornet venom, in addition to both, contains acetylcholine. These biogenic amines are normally present in tissues; only the relative concentration in an envenomating apparatus reveals their toxic potencies. The same is true for phospholipase A and hyaluronidase, which are constituents of some organs but become extremely concentrated in venoms. Although not identical with plasma kinins of warm-blooded animals, wasp and hornet kinins resemble them in some respects. Finally, certain basic peptides and proteins of mammalian tissues share some (not all) pharmacological properties with similar substances from bee venom, e.g., melittin.

Like snake venoms, hymenopteran venoms are concentrated mixtures of water-soluble, nitrogen-containing substances. The relative emphasis given to single compounds has changed with time. The detection of phospholipase A in bee venom, the failure of attempts to separate toxic and phospholipid-splitting potencies of *Crotalus d. terrificus* venom (Slotta and Fraenkel-Conrat, 1938), and the identification of many enzymes in snake venoms (see Zeller, 1951) led to the working hypothesis that most of the pharmacological effects of bee and snake venoms might be due to their enzyme content. This view was strengthened by the characterization of the α-toxin of *Clostridium perfringens* as phospholipase C (MacFarlane and Knight, 1941). Fractionation of bee

TABLE I

COMPOSITION OF HYMENOPTERAN VENOMS

	Bee	Wasp	Hornet
Biogenic amines	Histamine	Histamine Serotonin	Histamine Serotonin Acetylcholine
Peptides and small proteins	Apamin Melittin Mast cell degranulating peptide	Wasp kinin	Hornet kinein
Enzymes	Phospholipase A Hyaluronidase	Phospholipase A Phospholipase B Hyaluronidase	Phospholipase A Phospholipase B Hyaluronidase?

venom, however, allowed for the first time a clear distinction of toxic peptides and proteins without enzymic properties from enzymes which are less active on classic pharmacological objects and also less toxic (Neumann *et al.*, 1952, 1953). In this manner, bee venom, because of its relatively simple composition, has been used successfully as a model, not only for the composition and mode of action of the hymenopteran venoms, but also for other venomous secretions, e.g., that of snakes. The present status (Habermann, 1965, 1968) is given in Table I.

II. BIOCHEMISTRY AND PHARMACOLOGY OF SINGLE VENOM CONSTITUENTS

A. Biogenic Amines

These substances play a double role in envenomation by hymenopteran venoms: in part they are contained in them, and in part their release from tissue stores is induced by bee venom and probably also by wasp venom (see Sections II,B,3,*c* and II,D,2,*b*).

The existence of histamine in bee venom has been shown by various means: precipitation as the picrate (Tetsch and Wolff, 1936; Reinert, 1936), paper chromatography (Schachter and Thain, 1954), paper electrophoresis (Neumann and Habermann, 1954a; see Fig. 1), gel filtration (Habermann and Reiz, 1964, 1965). The amount of histamine present is still a matter of controversy, since the substance has been determined by biological methods which are sensitive also to other venom constituents, e.g., the isolated guinea pig ileum (Ackermann and Mauer, 1944) is also sensitive to melittin and phospholipase A (Habermann, 1957a). Using cat's blood pressure as a measure, only 0.1% of dry substance as histamine was found (Salzmann, 1953; Neumann and Habermann, 1960); others report 1–1.5% (Reinert, 1936; Schachter and Thain, 1954). Histamine is the only biogenic amine of pharmacological interest found in bee venom. Other low molecular basic compounds (Neumann and Habermann, 1954a; Habermann and Reiz, 1964) have no definite action.

Wasp venom contains, when investigated by paper chromatography and subsequent bioassay, 2% of the dry substance as histamine; application of Code's procedure yields 1.6% (Jaques and Schachter, 1954).

Hornet venom contains 3–30 mg histamine/gm of dry venom sac, as has been shown by bioassay with or without previous paper chromatography (Neumann and Habermann, 1956; Albl, 1956; Bhoola *et al.*, 1961).

Serotonin has been detected in wasp venom by bioassay after previous chromatography in an amount of about 0.32 mg/gm (Jaques and Schachter, 1954), in hornet venom (Neumann and Habermann, 1956; Albl, 1956) of 7–19 mg/gm (Bhoola *et al.*, 1961).

Very interestingly, hornet venom contains, in addition to histamine and serotonin, remarkable amounts of acetylcholine. This has been identified by parallel assays on various isolated organs, by paper chromatography and by sensitivity to cholinesterase and inhibitors. Up to 5% was found in the dry substance (Neumann and Habermann, 1956; Albl, 1956; Abrahams, 1955; Bhoola *et al.*, 1961).

The pharmacology of biogenic amines can be discussed here only with respect to the role they play in the venoms. Since histamine, serotonin, and acetylcholine can be classified among pain-producing substances, they may be of some importance in the initial (not the prolonged) pain following hymenopteran stings. Exact data are, however, not available. The general toxicity of the venoms is certainly connected with other compounds, but the acute circulatory effects following intravenous injection are caused by biogenic amines to a considerable degree in wasp and hornet venoms, but only slightly in bee venom. Their effects should be taken into consideration when testing the venom on smooth muscular organs such as guinea pig ileum, or when testing for local increase of capillary permeability. The activity of hornet venom on heart and skeletal muscle is practically exclusively due to the acetylcholine present in it.

B. Peptides and Small Proteins

I. Wasp and Hornet Kinins

The term "kinin" [as part of the name "bradykinin" (Rocha e Silva *et al.*, 1949)] has been applied to peptides originating from plasma proteins which lower the blood pressure and contract smooth muscular organs. The term has been extended to other peptides with corresponding pharmacological properties (Schachter and Thain, 1954). Rapid reversibility of effects and lack of tachyphylaxis should be added to the above definition; otherwise a lot of venom components, such as melittin from bee venom, would have to be included in this group. With these restrictions in mind, bee venom is free from kininlike activity which can, on the other hand, be found in wasp (Jaques and Schachter, 1954) and hornet venom (Bhoola *et al.*, 1961). Wasp venom kinin is destroyed by trypsin and chymotrypsin; hence it is a peptide. It is slowly dialyzable and very stable against heating in the neutral or slightly acidic range. It causes slow contractions of various isolated smooth muscular organs, e.g., guinea pig ileum or rabbit jejunum. When tested on the arterial blood pressure of rabbits and cats, it is an extremely hypotensive agent (Schachter and Thain, 1954) and raises, like "classical" kinins, the vascular permeability of the guinea pig and rabbit skin. Furthermore, it produces pain when applied to the base of a blister (Schachter, 1960). It can be distinguished from the plasma kinins, bradykinin and kallidin, by paper chromatography and by sensitivity to trypsin (Schachter, 1963). It seems to be homogeneous (Schachter and Thain, 1954). A survey of nonmammalian kinins and wasp kinins has been made recently. The structure of the main kinin from Polistes has been elucidated and confirmed by synthesis to be Pyr-Thr-Asn-Lys-Lys-Lys-Leu-Arg- Gly- [Bradykinin] (Piano, 1970).

Hornet kinin is similar to, but not identical with, wasp kinin. Whereas the pharmacological actions of both substances are qualitatively indistinguishable, quantitative comparison showed comparatively less activity of hornet kinin on the guinea pig ileum. It differs from bradykinin for the same reason. It can be distinguished from wasp kinin by its resistance to trypsin and by its migration in butanol–acetic acid–water, whereas wasp kinin remains at the starting point (Bhoola *et al.*, 1961; Schachter, 1963).

The kinins, like the biogenic amines, are very active on blood pressure, on vascular permeability, and on sensory nerve endings. They participate in the effects of whole venoms on various pharmacological objects. Their relative importance, however, cannot yet be estimated because of lack of knowledge concerning their concentration in the venoms.

2. Apamin (from Bee Venom)

Apamin has been found by screening bee venom fractions obtained by gel filtration (Fig. 2). It is eluted between the melittin and the histamine peak,

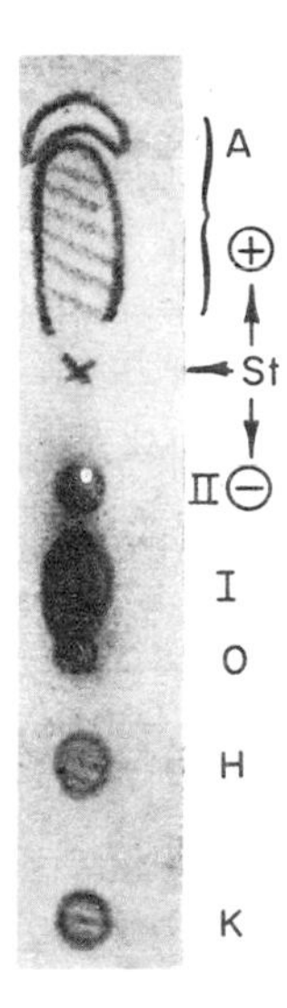

FIG. 1. Paper electrophoresis of a single bee sting. 0.1 *M* phosphate buffer, pH 7.0, 4 hours at 4 V/cm and 20°C on Whatman I paper. Ninhydrin staining. St, area of stinging; K, unknown cathode component; H, histamine; O, I, II, fraction O, I, II, respectively; A, unknown anodic components.

which argues for a relatively low molecular weight. It accounts for only about 2% of the dry venom so that extensive analysis was hampered by scarcity of substance. Further purification can be achieved by chromatography on carboxymethyl cellulose. It is a basic peptide consisting of 18 amino acids and can be stained with amido black and ninhydrin. The peptide is destroyed by performic acid, trypsin, and chymotrypsin, but not by pepsin or carboxypeptidase A. Disulfide connect bridges pos. 1 with 11 and pos. 3 with 15 (Callewaert *et al.*, 1968).

TABLE II

AMINO ACID SEQUENCES OF THE BEE VENOM PEPTIDES: MELITTIN, APAMIN, AND MAST CELL DEGRANULATING PEPTIDE (MCD PEPTIDE)

1. Melittin[a]
   Gly-Ile-Gly-Ala-Val-Leu-Lys-Val-Leu-Thr-
   Thr-Gly-Leu-Pro-Ala-Leu-Ile-Ser-Trp-
   Ile-Lys-Arg-Lys-Arg-Gln-$GluNH_2$
2. Apamin[b]
   Cys-Asn-Cys-Lys-Ala-Pro-Glu-Thr-Ala-Leu-Cys-
   Ala-Arg-Arg-Cys-Gln-Gln-$HisNH_2$
3. Mast Cell Degranulating Peptide[c]
   Ile-Lys-Cys-Asn-Cys-Lys-Arg-His-Val-Ile-
   Lys-Pro-His-Ile-Cys-Arg-Lys-Ile-Cys-Gly-Lys-
   $AsnNH_2$

[a] Habermann and Jentsch, 1967
[b] Haux *et al.*, 1967; Shipolini *et al.*, 1967
[c] Haux, 1969

The outstanding pharmacological action of apamin is on the central nervous system, whereas no remarkable influences have been found so far on isolated organs or rabbit's blood pressure. The vascular permeability of the skin is locally increased by apamin applied intracutaneously to the rabbit's back. Sublethal doses administered intravenously, intramuscularly, or subcutaneously render mice unusually quiet for 5–15 minutes; then, extreme hypersensitivity and badly coordinated hypermotility are impressive. Death (LD_{50}, mouse i.v. 4 mg/kg) occurs with signs of respiratory distress as a result of hypermotility. Epileptiform convulsions are absent. When the mice survive, abnormal excitability and motility last, depending on dosage, for up to 3 days. The reaction is stopped completely by transsection of peripheral nerves, and only partially by transsection of the spinal cord. Therefore, the site of action is in the brain and spinal cord. Rats and rabbits react in a similar manner; in these animals, a stuporous component of the envenomation is obvious even in stages of hyperexcitability (Habermann, 1965; Habermann and Reiz, 1964,1965).

Following application of whole venom, the apamin effect is prominent only with certain venom charges and only if not injected intravenously. The animal would be killed by the more concentrated melittin and phospholipase before apamin intoxication developed fully. It was observed 25 years ago, however, that a relatively quickly dialyzable part of whole bee venom causes central excitation (Hahn and Leditschke, 1937), a further argument for the low-molecular character of apamin.

3. Melittin (from Bee Venom)

a. Biochemistry. This substance is by weight the main constituent of bee venom; about 50% of the Folin-positive material consists of it. In venom charges prepurified by precipitation with picric acid, it can be found enriched up to 73% since it is more easily precipitated than are phospholipase A and hyaluronidase. Melittin has been purified first by paper electrophoresis as a so-called "Fraction I" from bee venom in contrast to "Fraction II" containing the enzymes mentioned (Fig. 1); but, according to our present knowledge, apamin migrates, under the conditions employed, in Fraction I also. Nevertheless, the fundamental analysis of pharmacological and biochemical properties was successful using that simple separation procedure (Neumann *et al.*, 1952; Neumann and Habermann, 1954a) because of the low apamin content of whole venom when contrasted to melittin. A second purification procedure on columns of partially oxidized cellulose (Fischer and Neumann, 1953) was worked out later and the amino acid composition determined by quantitative paper chromatography (Fischer and Dörfel, 1953). Recently, as for other venom constituents, gel filtration on Sephadex G-50, followed by chromatography on carboxymethyl cellulose, proved successful for isolating melittin (Fig. 2). Determination by an amino acid analyzer yielded 26 acids; neither the sulfur-containing acids nor phenylalanine, tyrosine, or histidine were among them. The only N-terminal residue was glycine; the C-terminal was resistant to carboxypeptidase A. Melittin is sensitive to trypsin, chymotrypsin, and pepsin (Habermann and Reiz, 1965). Besides its basicity (isoelectric point above pH 10), its ability to lower the surface tension of water—to a much stronger degree than does any other protein—deserves special mention. Since all strong surface-active agents promote hemolysis, the so-called "direct" hemolysis (see below) can be correlated immediately with this unique physico-chemical property of the causative substance. In addition to basicity and surface activity, the adsorbability of the polypeptide to organic constituents might be of importance for its pharmacological effects: melittin causes turbidity of suspensions of cellular particles (Habermann, 1958b) and is bound by frog skeletal and heart muscle (Andrysek, 1952; Späth, 1952). Altogether, these unusual properties make melittin a "structural" poison. Amino acid sequence (Table II) shows an invert soap-like structure for melittin. Positions 1–20 are occupied mainly by amino acids with their

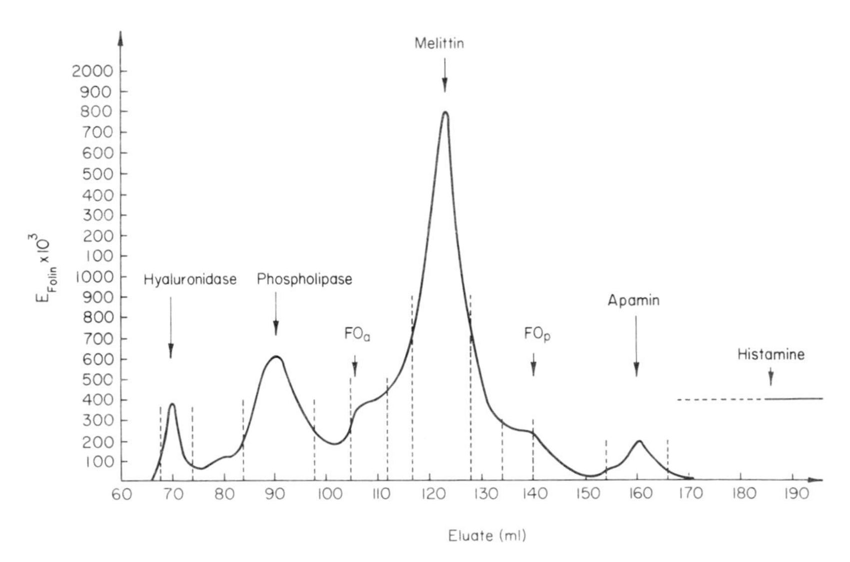

FIG. 2. Gel filtration of dried bee venom on Sephadex G-50. 0.1 *M* ammonium formate buffer, pH 4.5, column dimensions 300 × 1 cm, 294 mg bee venom applied. Room temperature. Dotted lines represent the cuttings for further purification.

residues neutral and/or hydrophobic, whereas the C-terminal remainder is composed of basic constituents, with one exception. This might explain the pharmacological properties of melittin in general (Habermann and Jentsch, 1967). However, the effects of modification reactions (Habermann and Kowallek, 1970) caution against an oversimplifying view. Melittin and some congeners have been synthesized recently (Schröder *et al.*, 1971) and were compared with native melittin (Habermann and Zeiner,1971).

Melittin has—because of its general attack on structures—a very broad spectrum of pharmacological activities. Their simplest model is the melittin-erythrocyte interaction.

b. Direct Hemolysis. In the action of whole bee venom on erythrocytes two components are involved: melittin and phospholipase A. Only melittin attacks "washed" erythrocytes directly. Phospholipase A, in contrast to phospholipase C from *C. perfringens,* acts only indirectly hemolytically by splitting extracellular lecithin to lysolecithin (Neumann *et al.*, 1953). Erythrocyte stromata can serve as a source of lecithin, so it is obvious that melittin hemolysis can be intensified by adding phospholipase A to the reaction mixture (Habermann, 1958a). Therefore, hemolysis by whole bee venom is always a combination of "direct" and "indirect" processes.

The melittin hemolysis itself is, with respect to the behavior of unhemolyzed, melittin-treated erythrocytes, of the nonosmotic type. Velocity and degree increase with temperature and pH value. In hypertonic medium, the degree of hemolysis is diminished. Most of the hemoglobin is released during the first minutes of incubation. Its release is preceded by a prolytic loss of potassium. Blood plasma, lecithin, polysaccharide sulfates, and, in higher concentrations, also citrate, diminish the hemolytic potency of melittin, probably by acid-base interaction with it. During lysis melittin is used up; erythrocyte ghosts bind it too. The fixation by intact erythrocytes is, however, not very strong: if the erythrocyte number in the incubation medium is increased, more hemoglobin per weight of melittin is liberated (Habermann, 1958a). Melittin and similarly bee venom produce characteristic changes of the erythrocyte membrane, as can be visualized by electron or dark field microscopy. In melittin concentrations of around 0.1 mg/ml, the erythrocyte stromata shrink and develop a netlike structure, which does not extend through their interior. This precipitating process on the stromata is in sharp contrast to the lytic effect of lysolecithin (Habermann and Mölbert, 1954).

In vivo, sublethal or lethal doses of melittin, applied intravenously, are not very harmful to circulating erythrocytes of rabbits. In mice, there was a considerable rise of the hematocrit value. In both animals, the cause of death by melittin is not hemolysis (Stockebrand, 1965).

c. Liberation of Pharmacologically Active Substances. Release of potassium before and during hemolysis (see above) is only a special case of appearance of a pharmacologically active substance following cellular damage by melittin. Potassium loss occurs also in frog skeletal muscle bathed in melittin solution (Habermann, unpublished). Damage to other potassium barriers could be a cofactor in producing pain or contraction of various organs possessing smooth and striated musculature, but further experimental evidence should be accumulated.

Like erythrocytes, tissue mast cells which contain histamine and serotonin are disrupted by melittin. This can be made visible by staining rat mesenterial pieces following incubation with melittin (Breithaupt and Habermann, 1968; Kachler, 1958). The release of histamine is directly measurable in the incubation fluid of the isolated rat diaphragm (Striebeck, 1958). It is still unknown whether this occurs also *in vivo*; furthermore, the question is open whether the liberation of histamine (Feldberg and Kellaway, 1937) and adrenaline (Feldberg, 1940) from isolated perfused organs can be achieved by melittin also. Serotonin appears in free form after destruction of thrombocytes by melittin (Habermann and Springer, 1958). Its release from mast cells concomitant with that of histamine appears to be highly probable because of the nonspecific mode of action of melittin.

d. Effects on Smooth Musculature. The above-mentioned liberation of pharmacologically active substances offers an explanation for the smooth muscular action of melittin; this is, however, certainly not the only cause of melittin contracture since it is resistant to antihistaminics. It is true that low melittin doses are inhibited by atropine; the higher ones are antagonized by the nonspecific papaverine (Albl, 1956). Melittin contraction differs from that following histamine by its slower onset and longer duration; it differs from the bradykinin contraction by its tendency to tachyphylaxis, and after higher doses, destruction of the ability of the organs to contract at all. Sensitivity to lower doses of histamine, tachyphylaxis, and damage of the contractile mechanism were described by Feldberg and Kellaway (1937) for whole venom; these effects are at least partially due to melittin. Histamine and phospholipase A (see Section II,D,2,*c*) are also active on smooth musculature, but in a different manner.

e. Skeletal Musculature and Its Innervation. Melittin imitates all the effects of whole bee venom on skeletal musculature hitherto investigated, whereas the other constituents, especially the enzyme fraction (Neumann and Habermann, 1954a; Habermann, 1957a; Röthel, 1953) and apamin (Habermann, unpublished) are inactive when applied alone. The possibility of potentiation—like that observed in hemolysis—has not yet been ruled out.

The isolated frog musculature reacts to melittin with an often biphasic contracture: the immediate phase is followed by a slow shortening, lasting up to 1 hour. Tachyphylaxis is very pronounced. The direct electrical excitability is also diminished, although to a lesser degree, perhaps because melittin—in contrast to electrical stimulation—may not reach all the fibers of the exposed muscle. The muscle becomes depolarized (Heydenreich, 1957) and loses potassium (Habermann, unpublished); the normal length cannot be restored by anodic repolarization or prolonged washing (Hofmann, 1952a).

Damage of neuromuscular junctions can be demonstrated best on the phrenicus-diaphragm preparation of the rat; under certain conditions, it occurs on frog's musculature too. The block is more of the decamethonium than of the curare type since it is resistant to physostigmine and accompanied by contracture. It differs from the decamethonium block by lack of reversibility. The difference between doses affecting direct and indirect excitability is small (Hofmann, 1952b; Röthel, 1953). Even strongly contracting doses of whole bee venom do not affect conduction in the phrenic nerve of the rat (Hofmann, 1952b) or the *N. ischiadicus* of the frog (Röthel, 1953). Under the conditions used, the sequence of sensitivity is: neuromuscular junction $>$ muscle fiber $\gg$ nerve. On the other hand, axons of the squid prepared free are very sensitive to whole bee venom (Rosenberg and Podleski, 1963); so the surrounding tissue determines at least partially the relative sensitivity to bee venom of conducting structures.

A role of such effects in general intoxication by bee venom or melittin is improbable; the indirect excitability of the *m. tibialis anterior* and the diaphragm of cats is not diminished by amounts causing respiratory arrest (von Bruchhausen, 1955).

f. Ganglionary and Central Synapses. The universal cellular damage by melittin manifests itself also on ganglionary and central synapses. When applied to the isolated perfused ganglion cervicale superius of the cat, low doses produce increased sensitivity of the nictitating membrane to preganglionary stimulation and often a long-lasting contraction. Higher melittin doses first stimulate, then paralyze the ganglion to preganglionary as well as to acetylcholine challenge. In this arrangement, apamin is without effect; lysolecithin (see Section II,D,2,*b*) exerts another mode of ganglionary paralysis (Habermann, 1954a; Seifert, 1958).

In vivo, neither ganglionary excitation nor paralysis followed the injection of even lethal doses of bee venom. By such treatment, the nictitating membrane itself becomes sensitized to electrical or histamine stimulation (von Bruchhausen, 1955).

The mixture of excitation and paralysis and the consecutive variability of results obtained even on the same animal makes analysis of the central nervous effects of melittin very difficult. Intracarotid and intraventricular (IVth ventricle) injections in cats result in increase or decrease of the systemic blood pressure and irregularities or depression of the respiration as well as occasional depression of the homolateral flexor reflex (Rossbach, 1955). Melittin probably does not act directly upon the centers but on certain receptors because the reaction follows the application immediately and often leads to tachyphylaxis. Reflectory (von Bruchhausen, 1955) and excitability changes (Gerlich, 1950) observed following systemic application of whole venom are ambiguous, in view of the detection of apamin (see Section II,B,2).

g. Circulatory Effects. Melittin excites and paralyzes functions of the heart as it does those of skeletal muscle. On the isolated frog's heart, low doses of melittin act positively inotropic; high doses lead to contracture which cannot be reversed by prolonged washings. Arrhythmias and changes of frequency are indicative of a simultaneous influence on the conducting system (Neumann and Habermann, 1954a).

The vessels of the hindquarters of the frog, perfused according to Laewen-Trendelenburg, react to melittin with an often biphasic constriction. The flow of Tyrode's solution through the vessels of isolated rabbit ears can be either increased or diminished (Silber, 1953; Habermann, 1954a).

On the narcotized but otherwise intact cat, a few milligrams per kilogram of melittin—and similarly whole bee venom—produce for several minutes a picture similar to the Bezold-Jarisch reflex: The blood pressure decreases to very low values; extreme bradycardia and single extrasystoles establish that the heart is involved; simultaneously, respiration is severely depressed up to temporary arrest; vagal impulses are of importance, since cutting of the nerve normalizes the circulation. Tachyphylaxis is very pronounced. Repeated application leads only to a very transient fall of blood pressure, followed by a longer lasting increase. Finally, the rise of blood pressure becomes more and more prominent. Under artificial respiration, extreme hypertension is reached accompanied by irreversible cardiac damage with extrasystoles, blocks of various localization, and ventricular flutter quite distinct from the reversible changes following initial high bee venom doses. Temporary fall and prolonged rise of blood pressure are probably due to direct effects of melittin; this again contrasts with the reflex-like action of initial high doses. As shown with whole bee venom, the rise of blood pressure is unaffected by pretreatment with ergotamine or adrenalectomy; the fall is not prevented by atropine or antihistaminics. The finer mechanism of the circulatory effects is not yet clear. Besides the direct action on heart and vessels demonstrated on isolated systems, it is still unknown if potassium liberated from erythrocytes or tissues is of importance. Further investigations are needed also for elucidating the mechanism of desensitization to acetylcholine and to vagal stimulation which has been observed on blood pressure of cats pretreated with higher doses of bee venom (Habermann, 1954a; Salzmann, 1953; Hackstein, 1953). Whole bee venom produces hemoconcentration (Feldberg and Kellaway, 1937); perhaps this indication of general capillary damage is also an effect of melittin.

h. Local and General Toxicity. All pharmacologically active constituents of bee venom are involved in its local toxicity: histamine by damaging small vessels and nerve endings (see Section II,A), hyaluronidase by increasing permeability of interstitial ground substance (see Section II,C), phospholipase A by raising capillary permeability (see Section II,D,2,*d*), MCD peptide by releasing histamine (see Section II,B,4), and, finally, melittin.

When injected into the skin of the human forearm, melittin produces pain, erythema, and edema. The same is true when it is applied into the conjunctival sac of the rabbit. Higher doses induce cutaneous necrosis at the application site in the rabbit's back; lower doses increase the permeability of cutaneous vessels for circulating dye. It can be assumed that melittin alters all the cells of connective tissue which it contacts. Besides these direct effects, the reaction to melittin may involve the liberation of various pharmacologically active substances such as potassium, histamine, and serotonin (see Section II,B,3,*c*). Metabolism may be disturbed, especially by loss of inorganic and organic phosphates, as has been demonstrated with frog musculature (Heydenreich, 1957). Cells once permeable may undergo further damage by interaction of melittin with their enzyme systems, e.g., inhibition of oxidative processes, uncoupling phosphorylation from oxidation (Habermann, 1954b), and activation or inhibition of the succinic dehydrogenase system (Habermann, 1955a). Melittin is, therefore, directed toward the same structurally bound enzyme systems as is phospholipase A (see Section II,D,4).

The mode of general toxicity of melittin is not yet understood. There is a wide gap between the i.v. LD_{50} (mice = 3.5 mg/kg) and the subcutaneous LD_{50} which, as deduced from experiments with whole venom, is at least 10 to 20 times higher. This might be explained either by adsorptive binding of melittin at the site of injection or by quick destruction in the body or by need of a short-lasting high melittin level in blood for inducing fatal damages. The symptoms of death are not very characteristic: the animals are quiet, dyspneic; antemortem excitation often occurs. Death occurs usually during the first hours (Wloszyk, 1954). It is unknown whether permeability changes of small vessels, e.g., of the lung, are terminating factors; in rabbits, hemolysis is not.

4. Mast Cell Degranulating Peptide

Compared with apamin and melittin, the other basic components are of lesser importance with respect to concentration and pharmacological activity. One of them, a so-called "Fraction 0," characterized by high electrophoretic migration velocity, has been described, but no pharmacological activity could be ascribed to it (Neumann and Habermann, 1954a). Meanwhile, by application of gel filtration on Sephadex G-50 followed by chromatography on carboxymethyl cellulose, a series of such basic peptides that migrate on electrophoresis with the velocity of melittin or somewhat faster has been differentiated. FO_a appears on Sephadex ahead of melittin, FO_{pI} with it, FO_{pII} behind it. As far as investigated, no hemolytic activity comparable with that of melittin has been found. However, there was increase of vascular permeability if the venom fraction was applied intracutaneously (Habermann and Reiz, 1965). A few years later, Fredholm (1966) found evidence for a mast cell degranulating factor in bee venom which was different from melittin. Breithaupt and Habermann (1968) purified that peptide and identified it as one of the FO-components. Like the classic releaser of histamine, compound 48/80, it acts mainly in rats where it is much more effective than melittin. MCD peptide and compound 48/80 depress the blood pressure of rats in a quantitatively and qualitatively similar manner with crossed tachyphylaxis. The amino acid sequence is given in Table II.

C. Hyaluronidase (Bee and Wasp Venoms)

The existence of hyaluronidase in bee venom has been known since the work of Chain and Duthie (1940). On electrophoresis, it migrates with phospholipase A in the so-called Fraction II (Neumann and Habermann, 1954a). Separation from this enzyme can be achieved by chromatography on Amberlite IRC-50 with continuously increasing ionic strength (Habermann, 1957c) or, with better recovery, by gel filtration on Sephadex G-50. As the next step of the latter procedure, impurities must be adsorbed by treatment with Amberlite IRC-50 (Habermann and Reiz, 1965). Hyaluronidase is more labile than phospholipase, so it is nearly completely inactivated by the treatment with acid necessary for preparing commercial, picrate-precipitated venom. In solution, partial reactivation occurs. Its pH optimum is sharper than that of testis hyaluronidase. Depending on buffer and testing procedures used, it has been located between pH 4.0 and 5.0. Like hyaluronidases of other origin, it is inhibited by plasma constituents and polysaccharide sulfates like heparin, and is strongly dependent on the ionic composition of the incubation medium. A 20- to 30-fold increase in purity has been attained (Habermann, 1957c).

The bee venom enzyme splits hyaluronate to the same end products as does testis hyaluronidase. The bulk of substrate is hydrolyzed to a tetra- and a hexasaccharide containing equimolar amounts of *N*-acetylglucosamine and glucuronic acid with the sequence

$$\text{GpA } 1\xrightarrow{\beta}3\text{GNAc } 1\xrightarrow{\beta}4\text{GpA } 1\xrightarrow{\beta}3\text{GNAc}(1\xrightarrow{\beta}4\text{GpA } 1\xrightarrow{\beta}3\text{GNAc})$$

There are indications that bee venom also splits chondroitin sulfates A and C, heparin, and blood group-specific substances A and B (Barker *et al.*, 1963).

Bee venom hyaluronidase develops, besides its characteristic effects on permeability of connective tissue, no other pharmacological potencies (Habermann, 1957c). No better information is available for wasp venom hyaluronidase (Jaques, 1955). Proof for a corresponding enzyme in hornet venom is still lacking.

D. Phospholipase A (Bee, Wasp, and Hornet Venoms)

I. Biochemistry of Phospholipase A and Lysophosphatides

a. General Remarks. Phospholipase A is an ingredient of many animal venoms; it is contained in all snake venoms hitherto investigated and in bee, wasp, and hornet venom. Furthermore, it is a normal constituent of some organs of warm-blooded animals, e.g., pancreas (see Ercoli, 1940), intestinal mucosa (Epstein and Shapiro, 1959), brain (Gallai-Hatchard *et al.*, 1962), duodenal content (Vogel and Zieve, 1960), and blood (Zieve and Vogel, 1960; Habermann, 1963). Earlier contributions are summarized in several monographs (see Ercoli, 1940; Zeller, 1951; Slotta, 1960). During the past 10 years, however, some important discoveries have been made concerning mode of action and purification of phospholipase A. Only part of these investigations have been done with hymenopteran venoms; therefore this survey will cover to some extent phospholipases of other origins also.

The phospholipase A content of bee venom is relatively high; at least 12% of dry venom consists of this enzyme (Habermann and Reiz, 1965). Commercial, prepurified venom contains relatively less, since the enzyme is less easily precipitated by picric acid than are the concomitant basic peptides.

The content in wasp and hornet venom is still unknown; comparisons with bee venom are difficult to establish because the phospholipase B activity of these venoms would interfere with biological (e.g., hemolysis by lysolecithin) or titrimetric estimations. There exists no experimental proof that the phospholipase A or B action in wasp or hornet venom is due, as is commonly believed, to two distinct enzymes and not to a single protein.

Phospholipase A splits one fatty acid from phosphatidyl compounds, thus leaving a monoacyl phosphatide as a lyso compound. According to recent findings, this action is conditioned by the chemical structure of the substrate as well as by physicochemical parameters given by the medium.

b. Structural Conditions for the Action of the Enzyme

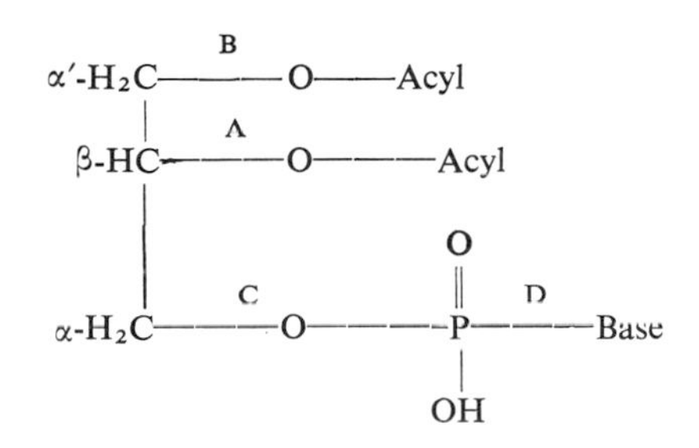

Sites of attack of phospholipases A–D

As generally accepted, natural phosphatidyl compounds have their phosphoryl residue attached to the α-position. For a long time there existed considerable doubt whether phospholipase A splits the fatty acid from the α'- or the β-linkage. Not too long ago, one seemed to have conclusively demonstrated that the lysolecithin formed from natural lecithins remains acylated at β. Tattrie (1959) and Hanahan *et al.* (1960), however, brought evidence by application of structure-specific enzymes that lysolecithin is an α'-acyl phosphatide. Direct proof has been possible by subjecting mixed-acid lecithins to hydrolysis by *Crotalus adamanteus* (de Haas *et al.*, 1960) and human pancreatic (van Deenen *et al.*, 1963) phospholipases. Irrespective of a saturated or unsaturated character of acyl residues, that in the β-position is always liberated. Long *et al.* (1963) attached themselves to this view. When using mixed natural substrates, e.g., egg lecithin, *C. adamanteus* venom splits most quickly the β-saturated and α-unsaturated, followed by the β-unsaturated and α-saturated, and finally the α,β-saturated lecithin molecules (Moore and Williams, 1963). There seems to exist a preference for the liberation of longer (C_{20}, C_{22}) fatty acids (Marinetti *et al.*, 1960) although shorter ones, e.g., that from dihexanoyllecithin (Roholt and Schlamowitz, 1961), can be set free by *Crotalus d. terrificus* venom. Choline plasmalogens are more resistant to snake venom enzymes than are lecithins (Marinetti *et al.*, 1959). This difference is stronger with *Crotalus atrox* venom than with cobra venom (Gottfried and Rapport, 1962). The position of the aldehydogenic group in plasmalogens was doubtful for a long time, but the positional specificity of phospholipase A detected with ester phosphatides as well as by comparative experiments between the products of alkaline hydrolysis and incubation with *Agkistrodon piscivorus* venom make the attachment of this group to the α'-position highly probable (Pietruszko and Gray, 1962).

The basic component attached to the phosphatidic acid is no determinant for the substrate specificity; it influences, however, micelle formation and solubility and in this way the accessibility of the substrate. The mode of action of *C. adamanteus* phospholipase, splitting off the acid in β-position, is the same on synthetic mixed acid, L-α-phosphatidylethanolamines, as on choline compounds (de Haas and van Deenen, 1961a). Phosphatidylserine from brain can be transformed by *A. piscivorus* venom into the corresponding lyso compound (Rathbone *et al.*, 1962); the position of residual fatty acid is—as found with synthetic mixed-acid phosphatidylserines—always in α' (de Haas and van Deenen, 1961b). It seems to be still unknown whether ethanolamine and serine plasmalogens can serve as substrates for phospholipase A as does choline plasmalogen. For historical remarks on structure and breakdown of phosphatidylethanolamine, see Robins (1963). Degradation of phosphatidic acid to lysophosphatidic acid by pancreatic phospholipase (Rimon and Shapiro, 1959) occurs probably by unmasking the β-hydroxy group (van Deenen *et al.*, 1963). Inositol phosphatides are not substrates for *A. piscivorus* venom (Long and Penny, 1957). Apparently, there exist specific phosphoinositidases in the liver which split either two fatty acids or phosphoinositide from the glycerol share (Kemp *et al.*, 1961).

β-Lecithins, e.g. (α, γ-distearoyl)-β-glycerylphosphorylcholine are hydrolyzed by *Crotalus adamanteus* venom, although more slowly, to α-stearoyl-β-glycerylphosphorylcholine. α-Stearoyl-γ-margaroyl-β-glycerylphosphorylcholine also loses only the γ substituent. Therefore, phospholipase A cannot be used to distinguish between α- and β-lecithins (de Haas and van Deenen, 1963). D-Lecithins, however, are resistant to the actions of *C. adamanteus* (de Haas and van Deenen, 1963) or *A. piscivorus* (Long and Penny, 1957) venoms.

c. Environmental Conditions for the Action of the Enzyme. Neither substrate of phospholipase A delivers ideal aqueous solutions. The enzymic breakdown is limited therefore by size and shape of the micelles and the arrangement of the substrate and enzyme molecules at the micellar surface. By adding certain organic solvents or emulgators, the turnover of the substrate by the same enzyme concentration can be raised manyfold. The optimum mixture of the ingredients depends on the enzymes and substrates used; comparisons between the results of different experimental procedures are difficult. Moreover, further micelle-forming, surface-active agents appear as products of enzymic action: these are lyso compounds and fatty acids. The composition of reaction mixtures for phospholipases A follows several empirical formulas, one or the other being preferable for certain purposes.

Diluted, buffered egg yolk is an excellent substrate for testing various phospholipases A against a standard; it forms a relatively stable, very enzyme-sensitive emulsion. If purified substrates are needed, activators must be incorporated. The action of bee venom phospholipase on purified egg lecithin is greatly enhanced by deoxycholate or Tween 20, or to a lesser degree, by serum albumin (Habermann, 1957b). The activation by deoxycholate has also been used for controlling the purification of human pancreatic phospholipase A (Magee *et al.*, 1962). Phosphatidylethanolamines, however, in the absence of deoxycholate are more susceptible to this enzyme (van Deenen *et al.*, 1963). Whereas bee venom phospholipase (Habermann, 1957b) is inhibited by oleic acid, addition of fatty acids to the particulate preparation from rat mucosa containing phospholipase A and B activity accelerates the lecithin breakdown (Epstein and Shapiro, 1959). Roholt and Schlamowitz (1961), using (as micelles) the water-soluble dihexanoyllecithin as the substrate and *Crotalus terrificus* venom as the enzyme, found activation by ovolysolecithin. Habermann (1957b) had the same results with bee venom phospholipase and egg lecithin as the substrate.

Other systems make use of solubilization by organic solvents. Neumann and Habermann (1954b) first solubilized egg yolk lecithin by adding 40% methanol for reaction with bee venom phospholipase. More recently, a system containing about 9% diethyl ether has been applied successfully for assay of *A. piscivorus* phospholipase (Magee and Thompson, 1960) but not for human pancreatic phospholipase A (Magee *et al.*, 1962). In this system and probably in others, the ζ-potentials of the substrate micelles are of importance for their reaction with cobra venom phospholipase A. Differences in ζ-potentials explain at least partially discrepancies found between the pH optima when splitting phosphatidylcholine or ethanolamine. Enrichment in fatty acids leads to decreased negativity of ζ-potential and inhibition of enzyme activity. Addition of ether restores both. Dicetyl phosphate diminishes the ζ-potential and inhibits enzymic breakdown; restoration of the potential by cetyltrimethylammonium hydroxide is, however, not followed by normalization of the enzymic activity. Other factors are, therefore, involved (Dawson, 1963).

An extreme reaction mixture with respect to organic solvents has been worked out by Hanahan (1952): together with its substrate, phospholipase A dissolves in wet diethyl ether. The lysolecithin formed is ether insoluble and thus can be separated easily from the soluble fatty acids. Although not very flexible, the system is very useful for preparative purposes.

The progress of the enzymic reaction can be followed by determination of the loss of acyl bonds (Stern and Shapiro, 1953; Magee and Thompson, 1960), by titration (Fairbairn, 1945), or manometric estimation (Bovet-Nitti, 1947; Habermann, 1957b) of the acids liberated. Lysolecithin release can be measured by hemolytic potency (see Neumann and Habermann, 1954b). The heat coagulation of egg yolk is prevented by previous incubation with hymenopteran venoms. In the absence of proteolytic enzymes, this is a measure of phospholipase A, as has been shown by parallel assay of coagulation time and hemolytic potency under various conditions (Habermann and Neumann, 1954a).

d. Structure-Activity Relationship of Lysophosphatides. From the foregoing, their general structure is as follows:

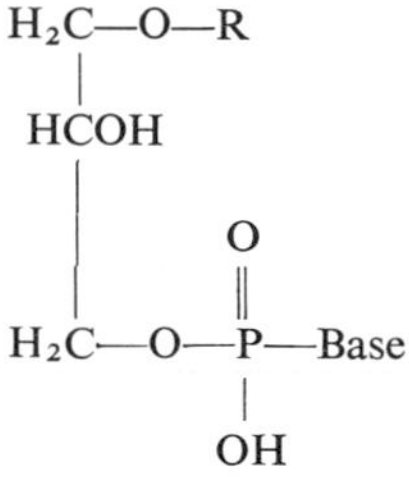

$$\begin{array}{l} H_2C\text{—O—R} \\ \quad | \\ HCOH \qquad\qquad O \\ \quad | \qquad\qquad\quad \| \\ H_2C\text{—O—P—Base} \\ \qquad\qquad\qquad | \\ \qquad\qquad\quad OH \end{array}$$

R can be an acyl or an aldehydogenic group. Choline, ethanolamine, and serine are possible bases. The pharmacological effects seem to be more related to the amphipathic character of such molecules than to the peculiarities in the R or base group. Lysophosphatidylserine (Rathbone *et al.*, 1962), lysophosphatidalcholine, and hydrogenated lysophosphatidalcholine (Hartree and Mann, 1960) possess a hemolytic value comparable with that of lysolecithin. No exact values are available for pure lysophosphatidylethanolamine, but the hemolytic potency of the older "lysocephalin" preparations was comparable with that of lysolecithin (Levene *et al.*, 1924).

The pharmacological and biochemical activities of the lysophosphatides are closely related to their solubilizing potency. The clearing of brain homogenates (Grosse and Tauböck, 1942; Webster, 1957), liver particulate preparations (Habermann, 1958b), and egg yolk (Habermann and Neumann, 1954a), the solubilization of lecithin, cholesterol, and oleic acid (Habermann, 1955a, 1958b), and the "dissolution" of erythrocyte stromata (Habermann and Mölbert, 1954) all might be due to the surface activity of lysolecithin (Habermann, 1955b; see Robinson, 1961, for review). Similar processes may take place on more complicated pharmacological systems; it has been emphasized that any lysolecithin present or formed in the body could be important for regulating permeability and solubilizing processes of many kinds (Neumann and Habermann, 1957).

e. Purification and Some Chemical Properties of Bee Venom Phospholipase. Separation from nonenzymic constituents was achieved first by paper electrophoresis (Neumann *et al.*, 1952). Isolation of larger amounts became possible by a combination of chromatography on Amberlite IRC-50 (retaining basic peptides) and alumina oxide (retaining hyaluronidase) (Habermann and Neumann, 1957). Gel filtration on Sephadex G-50 with subsequent chromatography on Amberlite IRC-50 (Habermann and Reiz, 1965; Stockebrand, 1965) proved very useful. The enzyme is easily soluble in water, precipitable with picric acid without loss of activity, homogeneous on paper electrophoresis and various chromatographic procedures. As N-terminal amino acid, it contains leucine (or isoleucine) only. It is slowly dialyzable; this plus its retardation on Sephadex G-50 argues for a relatively low molecular weight of about 19,000. It shares its reaction conditions (see Neumann and Habermann, 1954b; Habermann, 1957b) with other phospholipases A.

2. Pharmacological Effects of Bee Venom Phospholipase and Lysolecithin

a. Hemolysis. Phospholipase A only indirectly leads to hemolysis by producing lysophosphatides from accessible phosphatides. The phospholipids in intact, washed erythrocytes are resistant against that enzyme. This was found first with bee venom (Neumann and Habermann, 1954a) and then corroborated with cobra venom (Habermann and Neumann, 1954b) and various Australian snake venoms (Doery and Pearson, 1961). Phospholipase C, however, attacks the phospholipids of the intact erythrocyte even in very low concentrations. Since the phospholipids of lysed cells serve as substrates for phospholipase A, this enzyme is an autocatalytic factor involved in the "direct" hemolysis which finds its manifestation in the dose-response curve of whole venom hemolysis when compared with that of the "direct" hemolysin melittin (Habermann, 1958a).

The fundamental mechanism of lysolecithin hemolysis is based on the amphipathic character of the molecular structure, which is responsible for its surface activity and solubilizing potency against lecithin and cholesterol (Habermann, 1955a,b); the same may be true for other lysophospholipids. Lipids are relevant factors in the erythrocyte structure. Their disaggregation should provoke increased membrane permeability. The inactivation of lysolecithin by cholesterol (Delezenne and Fourneau, 1914) and lecithin (Habermann, 1958a) favors this view. The initially raised permeability finds its expression in the marked spherocytosis observed *in vitro* by Bergenhem and Fahraeus (1936) and in the higher osmotic susceptibility of lysolecithin-treated erythrocytes and their prolytic potassium loss (Habermann, 1958a). The binding of lysolecithin on intact erythrocytes is evident from the inverse relationship between initial erythrocyte count and hemoglobin liberated (Habermann, 1958a). The importance of adsorptive phenomena can be deduced further from Collier's finding (1952) confirmed by Habermann

(1958a) that lysolecithin hemolysis proceeds with lower temperature to higher values. For the velocity of hemolysis, the reverse is true (Habermann, 1958a; Jung, 1959). The controversy as to whether a monolayer of lysolecithin on the cell surface is necessary for hemolysis (Gorter and Hermans, 1943; Jung, 1959) is without basis; a true monolayer is impossible because lysolecithin forms mixed aggregates with lecithin (see Robinson, 1961). The relation between lysolecithin load of erythrocytes and morphological changes has been determined by aid of quantitative paper chromatography. If up to 0.6×10^{-11} μmoles/cell are bound, the cell shape remains normal; up to 1.3×10^{-11} μmoles/cell results in crenation; 2×10^{-11} μmoles/cell produce spherocytosis, whereas about 2×10^{-10} μmoles/cell would be hemolyzing. Changes in shape (and this means permeability too) and lysolecithin content can be reversed by washing with albumin solution (Klibansky and de Vries, 1963). Loading rabbit erythrocytes with lysolecithin decreases their cholesterol content (Klibansky *et al.*, 1962).

In vivo, phospholipase A and lysolecithin affect erythrocytes in a similar manner as *in vitro*. Injection of partially purified phospholipase A from bee venom (Habermann and Krusche, 1962) or *Vipera palestinae* venom (Klibansky *et al.*, 1962) into rabbits abruptly decreases the plasma lecithin. The concomitant rise of plasma lysolecithin is far from equivalent, so that most of the lysolecithin must have been removed from the plasma in some way. With snake venom phospholipase, part of the lysolecithin lost has been found reversibly attached to the erythrocytes. It is not enough to produce significant hemolysis but is sufficient for raising the hematocrit value and sphering.

The i.v. LD_{50} of bee venom phospholipase differs widely in mice (7 mg/kg) and rabbits (below 0.5 mg/kg). Both animals die from lung edema which develops while hemolysis is beyond its peak. In rabbits, even smaller enzyme doses precipitate sudden death if injected rapidly (Stockebrand, 1965). De Vries (1961) also produced spherocytosis in rabbits by injecting plasma lysophosphatides. The survival time of red blood cells of rabbits injected with *V. palestinae* enzyme is shortened (de Vries *et al.*, 1962). The strong effect of phospholipase A on plasma lipids followed by nonlethal erythrocyte changes *in vivo* is in contrast to that of phospholipase C which damages the erythrocytes, thus leading to death by hemolysis, whereas the plasma lecithin level is only slightly lowered (Habermann and Krusche, 1962).

b. Liberation of Pharmacologically Active Substances. Lysolecithin increases not only the permeability of erythrocytes but also that of many other cells. Muscle cells lose part of their potassium (Habermann, unpublished) with concomitant decrease of membrane potential (Heydenreich, 1957). Mast cells of the isolated mesenterial tissue of the rat are destroyed (Högberg and Uvnäs, 1957), which is an indication not only of the release of histamine but also of serotonin and heparin. The action of phospholipase A on mast cells is probably an indirect one. Whereas Högberg and Uvnäs (1957) reported disruption of mast cells in the mesenterial pieces by very low doses of bee venom enzyme, Rothschild (1965) found it effective on isolated, washed cells only when phospholipid was added. The liberation of histamine (Feldberg *et al.,* 1938) and adenylic compounds (Kellaway and Trethewie, 1940) from isolated perfused organs, the release of suprarenin from adrenal tissue *in vitro* and *in vivo* (Feldberg 1940), the mobilization of acetylcholine from guinea pig brain *in vitro* (Gautrelet and Corteggiani, 1939), and of serotonin from rabbit thrombocytes (Habermann and Springer, 1958) point to a common, relatively nonspecific membrane damage by lysolecithin. The serotonin stores of thrombocytes, like erythrocytes and probably mast cells, are highly resistant to any direct action of bee venom phospholipase.

Lysolecithin itself is only one pharmacologically active substance formed by phospholipase A. Fatty acids are another product, and some of them may also be of pharmacological interest. Egg yolk incubated with venoms containing phospholipase A contains certain acids which induce a slow contraction of smooth musculature (Feldberg *et al.*, 1938; Vogt, 1957). Substances behaving similarly have been detected in the perfusion fluid of the guinea pig's lung after application of bee venom phospholipase (Schütz and Vogt, 1961). It is, however, still open to question whether these smooth muscular effects are the result of the genuine fatty acids or of autooxidation products which are formed easily from unsaturated acids and are highly active (Dakhil and Vogt, 1962).

c. Effects on Smooth and Striated Muscles and on Synapses. On the isolated guinea pig ileum, bee venom phospholipase A generates a slow contraction with consecutive strong tachyphylaxis (Habermann, 1957a). Lysolecithin, on the other hand, acts like a spasmolytic on the ileum, rendering it less sensitive to histamine and acetylcholine (Feldberg *et al.*, 1938). Neither bee venom phospholipase A nor lysolecithin has a significant effect on the phrenicus-diaphragm preparation of the rat (Habermann, unpublished). Whereas the enzyme is ineffective, lysolecithin diminishes the sensitivity of the isolated ganglion cervicale superius of the rat for acetylcholine and, to a lesser degree, electrical excitation. The inhibition is reversible (Habermann, 1954a; Seifert, 1958). As can be deduced from experiments with whole bee venom, an effect of phospholipase A on ganglionary or neuromuscular synapses is certainly not predominant *in vivo* (von Bruchhausen, 1955). Whereas the intact nerve is relatively resistant against the enzyme, crudely prepared squid axons are sensitized by snake venoms for subsequent application of various pharmacologically active substances, among them curare. There is some evidence that phospholipase A is the responsible factor (Rosenberg and Podleski, 1963; Rosenberg and Ng, 1963).

d. Effects on Circulation. Purified bee venom phospholipase strongly lowers the blood pressure of cats and rabbits, where it is highly tachyphylactogenic (Habermann, 1957a). It may be of interest to remember that with comparable enzyme doses (see Habermann and Krusche, 1962) nearly the whole blood lecithin of rabbits is transformed into lysolecithin and fatty acids. Lysolecithin itself lowers the systemic blood pressure and raises the pulmonary artery pressure of the cat and the portal vein pressure of the dog (Feldberg *et al.*, 1938).

The isolated frog heart is not markedly influenced by bee venom phospholipase A; there is only a slight positively inotropic and contractile effect with doses as high as 1:5000. Lysolecithin 1:1000 produces a gradual contracture too (Habermann, 1957a). The isolated rabbit ear responds to bee venom phospholipase A by vasoconstriction and consecutive tachyphylaxis (Stockebrand, 1965).

3. Toxicity

a. Local Toxicity. In contrast to melittin, bee venom phospholipase A is a very weak irritant of the conjunctiva of the rabbit. When injected intracutaneously in that animal, it increases the vascular permeability (Habermann, 1957a). Intramuscular injection produces swelling of thighs of mice; this may be a manifestation of local myolysis found with a phospholipase fraction from habu venom (Maeno *et al.*, 1963).

b. General Toxicity. Purified bee venom enzyme is less toxic in mice (about 7.5 mg/kg, i.v.) than other venom constituents, e.g., melittin or apamin. Since maximally about one fifth of the native venom consists of phospholipase A, it contributes only to a small extent to whole venom toxicity. The causes of death are not yet completely understood. There are signs of intravital hemolysis, but, more important, also of lung edema, probably as a result of generally raised permeability. In rabbits, phospholipase A is about tenfold more toxic than is melittin (LD_{50} about 0.5 mg/kg). If injected quickly, even lower doses precipitate sudden death of unknown mechanism. The LD_{50} of lysolecithin (150 mg/kg i.v., mouse) is much higher; lung and erythrocyte damage has been observed also (Habermann, 1957a). Nevertheless, it is still controversial whether death by phospholipase A is exclusively mediated by lysolecithin formation or is due to loss of intact phospholipids. Lack of correlation between phospholipase content and toxicity has been demonstrated for electrophoretic fractions of *Naia haje* venom also (Radomski and Deichmann, 1958).

4. Phospholipase A and Enzymic Processes

a. Blood Coagulation. Bee venom retards blood coagulation *in vitro* (Dyckerhoff and Marx, 1944). Tissue thromboplastin (brain, lung) is inactivated mainly by bee phospholipase A, as can be deduced from activation and inhibition procedures, distribution of inhibitory potency between electrophoretic venom fractions, ability of lysolecithin to inactivate thromboplastin, and, finally, by lysolecithin formation in thromboplastin preparations (Habermann, 1954c). Whether loss of lipid activators (for structure, see Hecht and Slotta, 1962) or change of surface charge (see Papahadjopoulos *et al.*, 1962) is more important must still remain a matter of speculation. Snake venoms contain a principle acting similar to bee venom phospholipase, but its effect is often obscured by other factors affecting blood coagulation. The melittin effect on tissue thromboplastin is relatively weak when compared with that of phospholipase A (Habermann, 1954c).

Intravenous injection of bee venom phospholipase in mice does not change the bleeding time (Redelberger, 1958). The coagulation time of whole blood of rabbits is not markedly changed by intravenous injection of sublethal or lethal doses of bee venom phospholipase A or melittin (Stockebrand, 1965).

b. Respiratory Enzymes. The effect of bee and snake venoms on respiratory enzymes has been found independently in various laboratories (Ghosh and Bhattacharya, 1952; Fleckenstein *et al.*, 1950; Chatterjee, 1949). The relationship between inactivation of such systems by heated snake venoms (Braganca and Quastel, 1953) or bee venom fractions (Neumann *et al.*, 1952; Neumann and Habermann, 1954a) and phospholipase A content was of paramount importance for understanding these and other metabolic venom effects. Structural changes are important since only desmoenzymes are inhibited (Braganca and Quastel, 1953). It is known that lipids are integral parts of mitochondria. The problem of whether destruction of phospholipids or formation of the solubilizing lysolecithin is the more important factor in inactivation of respiratory enzymes is the same as with tissue thromboplastin. Some points, however, argue for a relation between loss of mitochondrial function and the former: mitochondria treated with snake venom phospholipase differ in their electron microscopic picture from lysolecithin-treated samples (Nygaard *et al.*, 1954); the amount of lysophospholipids formed is too small to induce damage in normal mitochondria. The degree of susceptibility varies from member to member in the respiratory chain, the link between cytochrome b and c being especially sensitive (Nygaard and Sumner, 1953; Nygaard, 1953a). Lysolecithin inhibits the oxidation of glutamate with O_2 (Habermann, 1954b), of succinate and β-hydroxybutyrate with ferricyanide, but not of succinate with phenazine methosulfate (Witter *et al.*, 1957). Low doses of phospholipase can increase respiration, perhaps by raising permeability (Nygaard and Sumner, 1953). Inhibition of succinic acid oxidase and formation of lysolecithin can be prevented temporarily by addition of succinate and/or ATP (Nygaard, 1953b). This can be explained by prevention of mitochondrial swelling caused by lysolecithin when ATP is added (Witter and Cottone, 1956).

c. Inhibition of Oxidative Phosphorylation. This system is closely linked spatially and functionally with the respiratory chain. The connection is especially sensitive to bee venom phospholipase, to lysolecithin (Habermann, 1954b), and to oleate. The effect of snake venoms is also due to their phospholipase A content (Habermann, 1955a). The uncoupling by lysolecithin takes place at the steps related to electron transfers to cytochrome c and to oxygen (Witter *et al.*, 1957). Coupling in the range of cytochrome c oxidase is also preferentially interrupted in brain and liver mitochondria of mice treated with cobra venom *in vitro* or *in vivo* (Aravindakshan and Braganca, 1961a). Intactness of particles is an important prerequisite for sensitivity against lysolecithin or cobra venom; oxygen consumption and ATP production of fragmented mitochondria decrease simultaneously. Mitochondrial swelling precedes inhibition of oxidative phosphorylation by lysolecithin or snake venom phospholipase. As with respiratory inhibition, lysolecithin formed in mitochondria by cobra venom is too small in amount to produce uncoupling of phosphorylation; loss of phospholipid is, therefore, more important (Aravindakshan and Braganca, 1961b).

Recent investigations argue for mitochondrial damage that also results from *in vivo* application of venoms containing phospholipase. Aravindakshan and Braganca (1959) reported decrease of the P/O quotient from 2.6 to 1.5 by injecting heated cobra venom in mice. Untreated venom was much more active. Mitochondria from such animals swelled *in vitro* faster than did normal controls. It could, however, be argued that these changes are possibly produced by contacting the mitochondria during preparation with phospholipase or lysophospholipids distributed in the extracellular space of the living animal.

d. Solubilization of Enzymes. The clearing effect of lysolecithin on tissue suspensions (see Section II,D,1,*d*) raises the question of the solubility behavior and activity of desmoenzymes in such samples. Lysolecithin and heated venom from *Vipera ammodytes* solubilize the ATPase and the pyrophosphatase of rat liver mitochondria. Simultaneously, pyrophosphatase has been partially inhibited (Habermann, 1955a), and the Mg-dependent ATPase activated (Witter *et al.*, 1957). Mitochondria from rat liver swell when in contact with lysolecithin; this can be prevented partially by addition of ATP (Witter and Cottone, 1956). Human erythrocytes lose about 44% of their true cholinesterase when incubated with lysolecithin (Greig and Gibbons, 1956), brain slices lose true and pseudocholinesterase (Marples *et al.*, 1959). Such spatial derangement of enzymes is certainly fundamental for the impairment of respiratory and phosphorylating chains, as mentioned above. It may be of importance in the local damage following venom application.

E. Phospholipase B (Wasp and Hornet Venoms)

The lysolecithin formed by phospholipase A can enter two metabolic pathways: it can be again acylated to lecithin (Lands, 1960; Webster, 1962) or broken down to glycerylphosphorylcholine by phospholipase B which is present not only in wasp and hornet venoms (see Ercoli, 1940) but also in various mammalian tissues (see Marples and Thompson, 1960). Liver (Dawson, 1956) and pancreatic (Shapiro, 1953) enzymes are unable to hydrolyze lecithin whereas in an enzyme preparation from *Penicillium notatum*, phospholipase A and B activities were firmly associated (Bangham and Dawson, 1959). The question whether venom phospholipase B hydrolyzes electively the α'-acyl bond of lysolecithin or the β-bond of lecithin also has not yet been settled. Nothing is known concerning the structural requirements of phospholipase B or its pharmacological significance.

III. POSSIBLE HAZARDS FOR MAN AND THEIR PREVENTION

A. Envenomation of Man

On local application, the well-known exudative reactions occur. In rabbits (Langer, 1897) and men (for review, see Jensen, 1962) exudation, emigration of blood cells, and enrichment of mobile tissue cells around a circumscribed necrosis are histological sequelae of bee stings. The concentration of every active constituent in the venom droplet is far above their thresholds. It is indeed higher than any concentration that could be applied, at this time, in the form of solutions of isolated, pure constituents of the native venom. Supported by hyaluronidase, all venom parts play roles which overlap each other. Local reactions may be evoked: in bee venom by melittin, MCD peptide, phospholipase A, and histamine, and to a lesser extent by apamin and FO-compounds; in wasp venom by wasp kinin, histamine, serotonin, and very probably also by phospholipase A; in hornet venom by hornet kinin, histamine, serotonin, acetylcholine, and very probably by phospholipase A.

The therapy of toxic local reactions can be restricted to nonspecific alleviation of the pain connected with the acute exudation and the direct venom effects on nerve endings. We have found a topical antihistamine useful, although the histamine present in or liberated by the venoms is certainly only of moderate importance in local reaction. Intravenous injection of an antihistamine before stinging (Neoantergan) lowers the cutaneous temperature in the area of sting but not the other local symptoms (Geske and Jung, 1950).

The degree and mode of general toxicity of wasp and hornet venom is still unknown. For bee venom, one can assume a mean toxicity in mice of 6 mg/kg i.v. for various batches of crude dry venom. If man were in the same range of susceptibility as is the mouse and if we assume 0.1 mg of dry substance per sting, then some thousands of bees would have to sting man (intravenously!) to reach a lethal dose. On subcutaneous injection, at least a tenfold amount of venom is needed in animals (see Section II,B,3,*h*). This explains the lack of clinical cases of lethal poisoning of adults by any hymenopteran venom. Among beekeepers it is well known that hundreds of bee stings can be tolerated with only moderate ill feeling. This systemic reaction need not be due to resorbed venom but can be explained simply by the involvement of large cutaneous areas damaged. Intravasal hemolysis with the corresponding urine, as a sure indication for systemic envenomation, is extremely rare. One such reaction followed thousands of bee stings but the patient survived (Koszalka, 1949). Treatment, e.g., exchange transfusion, is the same as in intravascular hemolysis of other origins. It must be kept in mind, however, that cases of death caused by hymenopteran venoms do not resemble the picture of envenomation seen in animals neither with respect to symptomatology nor to dose-activity relationships. All these observations argue for an allergic pathogenesis.

B. Antibody Production and Desensitization to Bee Venom Components

The antigenic properties of bee venom are not only of theoretical but also of practical interest, e.g., with respect to the resistance of beekeepers and to anaphylactic reactions in hypersensitive patients. Hitherto only some pharmacologically active components have been tested for their ability to induce circulating antibodies. Injection of whole venom in rabbits or guinea pigs raised the antiphospholipase and antihyaluronidase titer of blood serum. Venom effects due to melittin, however, were not more influenced by previous mixing with antiserum than with normal serum. In this category fell the cardiac effects *in vitro* as well as the general and local toxicity; melittin hemolysis took place in the presence of the antiserum (Habermann and El Karemi, 1956). Apamin has not yet been subjected to immunological investigation. Antigenicity of phospholipase A and hyaluronidase was to be expected since they are typical proteins. The resistance of beekeepers perhaps may be partially due to antibody production against the former enzyme. Mohammed and El Karemi (1961) gave evidence for phospholipase inhibition by the blood of such persons, although the evidence was meager; Gastpar *et al.* (1956) observed a raised antihemolytic value of beekeepers' blood which might be explained as inhibition of the action of phospholipase A in direct hemolysis (see Section II,B,3,*b*). The antihemolytic potency of blood plasma is, however, the result of many variables, since both melittin and lysolecithin (as the active product of phospholipase A) are antagonized by normal serum constituents; therefore, the findings with beekeepers' serum are still questionable.

The lack or at least very low antigenicity of melittin is in accord with chemical considerations and with animal experiments. Melittin hardly can be called a protein since it consists of only 26 amino acids; the absence of tyrosine, phenylalanine, histidine, aspartic acid, and of the acids containing sulfur (see Table II) may reduce antigenicity. Since melittin is responsible for many pharmacological venom effects, the failure to demonstrate active local or general immunization in rabbits is easily explained by nonreactivity of the antibody-producing system to this substance. Retrospectively, the negative results of Dold's experiments (1917) and of Anton's very thorough investigations (1946) are clearly understood. But these findings, although unequivocal, cannot be applied to human nonreactivity because time may be an important factor. The animal experiments mentioned lasted at most a few months whereas humans may be in contact with bee venom their whole lives, which could suffice for a reaction against very weak antigens.

It should be stressed here that diminished reactivity of beekeepers against the sequelae of stings is generally assumed according to experience but has not been established by comparative experiments. Most of 40 beekeepers interviewed reported diminished swelling of the stung area. All still felt pain which lasted longer (up to 20 minutes) than pain induced by histamine application (Habermann, unpublished) and it might therefore be due to melittin. Some observations seem to indicate a restriction of the "immunity" to areas which generally receive stings while the beekeepers are working with bees. Since the antigenic enzymes hyaluronidase and phospholipase A are without doubt involved in the development of local reactions, additional antibody production against melittin is not needed to explain the statements of the beekeepers.

C. Allergic Reactions

In European countries, fatal accidents caused by hymenopteran venoms are more frequent than such accidents following other animal poisoning. They cannot be ascribed to true toxicity of bee, wasp, or hornet venoms, but must be catalogued among anaphylactic reactions. This view was substantiated by the following points which were discussed in detail, together with case reports, by Jensen (1962).

There is no relation between severity of reactions and number of stings. If beekeepers are stung by a lot of bees, there is, besides local pain, only moderate discomfort. Koszalka's patient who was badly stung, but survived with unconsciousness and hemoglobinuria, has already been mentioned. From quantitative considerations of animal toxicity, it seems to be highly improbable that an adult human will ever receive sufficient venom to be stung to death (see Section III,A). The often-cited "intravenous" bee sting never has been proved. Stings into the pharynx also rarely occur in the statistics of fatal events.

On the contrary, practically all fatal reactions develop the clinical picture of anaphylactic shock. In most cases, previous sensitization is demonstrable.

TABLE III

SYNOPSIS OF EFFECTS OF BEE VENOM COMPONENTS AND LYSOLECITHIN[a]

Effects	Histamine[b] 0.1–1%	Melittin[b] 50%	Apamin[b] 2%	MCD peptide[b] 2%	Phospholipase A[b] 12%	Lysolecithin[b] 0%	Hyaluronidase[b] <3%
General toxicity[c] (mg/kg)	250–300[d]	4	4	>40	7.5	150	Low or 0
Local toxicity							
Pain production	++	++	?	?	?	?	0
Increase of capillary permeability	++	++	+	+	+	+	Indirectly
Cellular damage	0	++	?	mast cells	+	++	0
Systemic neurotoxicity	0	Low or 0	++	+	Low or 0	Low or 0	Low or 0
Direct hemolysis	0	++	0	0	0	++	0
Circulatory effects	++	++	0	++	+	+	0
Neuromuscular effects	0	++	0	?	0	?	0
Smooth muscular effects	++	++	0	?	+	+	0
Ganglionary blockade	0	++	0	?	0	++	0
Histamine release	0	++	0	++	Indirectly	++	?
Indirect hemolysis	0	0	0	0	++	0	0
Inhibition of blood coagulation	0	+	?	?	++	+	0
Inhibition of electron transport	0	+	?	?	++	+	0
Inhibition of oxidative phosphorylation	0	+	?	?	++	+	0
Spreading	0	0	0	0	0	0	++
Surface activity	0	++	?	?	0	++	?
Antigenicity[e]	0	0	?	?	++	0	++

[a] ++ strong, + demonstrable, 0 not demonstrable, ? not investigated. Emphasis is given on the basis of absolute activity of the components and not on their relative content in a typical venom; [b] approximate, in a typical dry venom; [c] mg/kg, mouse, i.v.; [d] from Guggenheim (1951); [e] in rabbits and guinea pigs.

Usually one single sting challenges the anaphylaxis. Persons suffering from venom allergy are particularly endangered. The symptoms belong to the acute allergic phenomena (itching, urticaria, Quincke edema, allergic asthma, in severe cases circulatory collapse) which appear within a few minutes and fade within a few hours.

Unless it is given immediately, therapy will always be too late. Therefore for humans of known high sensitivity an "emergency kit" is recommended, consisting mainly of isoproterenol tablets, epinephrine inhaler, and tourniquet (Shaffer, 1961). There exists neither a specific local nor general measure against the hymenopteran venoms. However, the usual antiallergic therapy, consisting of corticosteroids, Ca^{++}, adrenaline, perhaps antihistamines, is indicated. The local reactions in unsensitized patients are reported to be mitigated by an antihistamine ointment (Strauss, 1949). Other measures, even if catalogued in widely read textbooks (Moeschlin, 1964), are mainly of folkloric interest.

Most of the human population has occasion to become sensitive to hymenopteran venoms. The role of venom as antigen source can be deduced from the successful desensitization achieved by injection of increasing doses of purified venom and also from observations in our laboratory where—during some years of continuous occupation with bee venom—about half of the personnel became sensitised by inhaling dry venom dust. The nature of the antigens responsible is still unknown. Besides the true antigens, hyaluronidase and phospholipase A, other venom constituents could function as haptens. Furthermore, body constituents and pollen adhering to the stinging apparatus have been discussed. Body extracts of some Hymenoptera share some antigenic compounds (Foubert and Stier, 1958); on the other hand, 85% of positive reactants on intracutaneous injection of body extracts are hypersensitive against more than one species (Mueller, 1959).

Table III summarizes our overall knowledge concerning the biochemical, pharmacological, and toxicological properties of the bee venom components. A comparable systematic survey of wasp and hornet venom would be premature.

REFERENCES

Abrahams, G. (1955). Inaugural Dissertation, University of Würzburg.
Ackermann, D., and Mauer, H. (1944). *Arch. Ges. Physiol.* **247**, 623.
Albl, F. (1956). Inaugural Dissertation, University of Würzburg.
Andrysek, K. (1952). Inaugural Dissertation, University of Würzburg.
Anton, H. (1946). *Z. Immunitaets Forsch.* **105**, 241.
Aravindakshan, I., and Braganca, B. M. (1959). *Biochim. Biophys. Acta* **31**, 463.
Aravindakshan, I., and Braganca, B. M. (1961a). *Biochem. J.* **79**, 80.
Aravindakshan, I., and Braganca, B. M. (1961b). *Biochem. J.* **79**, 84.

Bangham, A. D., and Dawson, R. M. (1959). *Biochem. J.* **72**, 486.
Barker, S. A., Bayyuk, S. I., Brimacombe, J. S., and Palmer, D. J. (1963). *Nature* **199**, 693.
Bergenhem, B., and Fahraeus, R. (1936). *Z. Exptl. Med.* **97**, 555.
Bhoola, K. D., Calle, J., and Schachter, M. (1961). *J. Physiol.* (*London*) **159**, 167.
Bovet-Nitti, F. (1947). *Experientia* **3**, 283.
Braganca, B. M., and Quastel, I. H. (1953). *Biochem. J.* **53**, 88.
Breithaupt, H., and Habermann, E. (1968). *Naunyn-Schmiedebergs Arch. Pharmakol. Exp. Pathol.* **261**, 252.
Callewaert, G. L., Shipolini, R., and Vernon, C. A., *FEBS Letters* **1**, 111–113 (1968).
Chain, E., and Duthie, E. S. (1940). *Brit. J. Exptl. Pathol.* **21**, 324.
Chatterjee, A. K. (1949). *Indian J. Med. Res.* **37**, 241.
Collier, H. (1952). *J. Gen. Physiol.* **35**, 617.
Dakhil, T., and Vogt, W. (1962). *Arch. Exptl. Pathol. Pharmakol.* **243**, 174.
Dawson, R. M. C. (1956). *Biochem. J.* **64**, 192.
Dawson, R. M. C. (1963). *Biochem. J.* **88**, 414.
de Haas, G. H., and van Deenen, L. L. M. (1961a). *Biochim. Biophys. Acta* **48**, 215.
de Haas, G. H., and van Deenen, L. L. M. (1961b). *Biochem. J.* **81**, 34P.
de Haas, G. H., and van Deenen, L. L. M. (1963). *Biochem. J.* **88**, 40P.
de Haas, G. H., Mulder, I., and van Deenen, L. L. (1960). *Biochem. Biophys. Res. Commun.* **3**, 287.
Delezenne, C., and Fourneau, E. (1914). *Bull. Soc. Chim. France* p. 421.
de Vries, A. (1961). *Rev. Hematol.* **6**, 25.
de Vries, A., Kirschmann, C., Klibansky, C., Condrea, E., and Gitter, S. (1962). *Toxicon* **1**, 19.
Doery, H. M., and Pearson, J. E. (1961). *Biochem. J.* **78**, 820.
Dold, H. (1917). *Z. Immunitaets Forsch.* **26**, 284.
Dyckerhoff, H., and Marx, R. (1944). *Z. exptl. Med.* **113**, 194.
Epstein, B., and Shapiro, B. (1959). *Biochem. J.* **71**, 615.
Ercoli, A. (1940). *In* "Handbuch der Enzymologie" (F. F. Nord and R. Weidenhagen, eds.), Vol. I, pp. 480–494. Akad. Verlagsges., Leipzig.
Fairbairn, D. (1945). *J. Biol. Chem.* **157**, 633.
Feldberg, W. (1940). *J. Physiol.* (*London*) **99**, 104.
Feldberg, W., and Kellaway, C. H. (1937). *Australian J. Biol. Med. Sci.* **15**, 461.
Feldberg, W., Holden, H. F., and Kellaway, C. H. (1938). *J. Physiol.* (*London*) **94**, 232.
Fischer, F. G., and Dörfel, H. (1953). *Biochem. Z.* **324**, 465.
Fischer, F. G., and Neumann, W. P. (1953). *Biochem. Z.* **324**, 447.
Fleckenstein, A., Tippelt, H., and Kroner, H. (1950). *Arch. Exptl. Pathol. Pharmakol.* **210**, 380.
Foubert, E. L., and Stier, R. A. (1958). *J. Allergy* **29**, 13.
Gallai-Hatchard, J., Magee, W. L., Thompson, R. H. S., and Webster, G. R. (1962). *J. Neurochem.* **9**, 545.
Gastpar, H., Jager, J., and Seitz, W. (1956). *Klin. Wochschr.* **34**, 729.
Gautrelet, J., and Corteggiani, E. (1939). *Compt. Rend. Soc. Biol.* **131**, 951.
Gerlich, N. (1950). *Arch. Exptl. Pathol. Pharmakol.* **211**, 97.
Geske, H., and Jung, F. (1950). *Klin. Wochschr.* **28**, 477.
Ghosh, B. N., and Bhattacharya, K. L. (1952). *Sci. Cult.* (*Calcutta*) **18**, 253.
Gorter, E., and Hermans, J. (1943). *Rec. Trav. Chim.* **62**, 681.
Gottfried, E. L., and Rapport, M. M. (1962). *J. Biol. Chem.* **237**, 329.
Greig, M. E., and Gibbons, A. J. (1956). *Arch. Biochem. Biophys.* **61**, 335.
Grosse, A., and Tauböck, K. (1942). *Z. Rheumaforsch.* **5**, 429.

Guggenheim, M. (1951). "Die Biogenen Amine," p. 481. Karger, Basel.
Habermann, E. (1954a). *Arch. Exptl. Pathol. Pharmakol.* **222**, 173.
Habermann, E. (1954b). *Naturwissenschaften* **41**, 429.
Habermann, E. (1954c). *Arch. Exptl. Pathol. Pharmakol.* **223**, 182.
Habermann, E. (1955a). Thesis, Würzburg.
Habermann, E. (1955b). *Arch. Exptl. Pathol. Pharmakol.* **225**, 158.
Habermann, E. (1957a). *Arch. Exptl. Pathol. Pharmakol.* **230**, 538.
Habermann, E. (1957b). *Biochem. Z.* **328**, 474.
Habermann, E. (1957c). *Biochem. Z.* **329**, 1.
Habermann, E. (1958a). *Z. Ges. Exptl. Med.* **129**, 436.
Habermann, E. (1958b). *Z. Ges. Exptl. Med.* **130**, 19.
Habermann, E. (1963). *Proc. 1st Intern. Pharmacol. Meeting, Stockholm,* 1961 Vol. 2, pp. 171–173. Pergamon Press, Oxford.
Habermann, E. (1965). *Proc. Int. Pharmacol, Meet. 2nd, 1963,* **9**, 53–62.
Habermann, E. (1968). *Ergeb. Physiol.* **60**, 220.
Habermann, E., and El Karemi, M. M. A. (1956). *Nature* **178**, 1349.
Habermann, E., and Jentsch, J. (1967). *Z. Physiol. Chem.* **348**, 37.
Habermann, E., and Kowaleck, H. (1970). *Z. Physiol. Chem.* **351**, 884.
Habermann, E., and Krusche, B. (1962). *Biochem. Pharmacol.* **11**, 400.
Habermann, E., and Mölbert, E. (1954). *Arch. Exptl. Pathol. Pharmakol.* **223**, 203.
Habermann, E., and Neumann, W. (1954a). *Z. Physiol. Chem.* **297**, 179.
Habermann, E., and Neumann, W. (1954b). *Arch. Exptl. Pathol. Pharmakol.* **223**, 388.
Habermann, E., and Neumann, W. P. (1957). *Biochem. Z.* **328**, 465.
Habermann, E., and Reiz, K.-G. (1964). *Naturwissenschaften* **51**, 61.
Habermann, E., and Reiz, K.-G. (1965). *Biochem. Z.* **341**, 451.
Habermann, E., and Springer, H. (1958). *Naturwissenschaften* **45**, 133.
Habermann, E., and Zeüner, G. (1971). *Naünyn-Schmiedebergs Arch. Pharmak.* **270**, 1.
Hackstein, F. G. (1953). Inaugural Dissertation, University of Würzburg.
Hahn, G., and Leditschke, H. (1937). *Ber. Deut. Chem. Ges.* **70**, 1637.
Hanahan, D. J. (1952). *J. Biol. Chem.* **195**, 199.
Hanahan, D. J., Brockerhoff, H., and Barron, E. (1960). *J. Biol. Chem.* **235**, 1917.
Hartree, E. F., and Mann, T. (1960). *Biochem. J.* **75**, 251.
Haux, P. (1969). *Z. Physiol. Chem.* **350**, 536.
Haux, P., Sawerthal, H., and Habermann, E. (1967). *Z. Physiol. Chem.* **348**, 737.
Hecht, E., and Slotta, K. H. (1962). *Acta Physiol. Pharmacol. Neer.* **10**, 278.
Heydenreich, H. (1957). Inaugural Dissertation, University of Würzburg.
Högberg, B., and Uvnäs, B. (1957). *Acta Physiol. Scand.* **41**, 345.
Hofmann, H. T. (1952a). *Arch. Exptl. Pathol. Pharmakol.* **214**, 523.
Hofmann, H. T. (1952b). *Arch. Exptl. Pathol. Pharmakol.* **216**, 250.
Jaques, R. (1955). *Helv. Physiol. Acta* **13**, 113.
Jaques, R., and Schachter, M. (1954). *Brit. J. Pharmacol.* **9**, 53.
Jensen, O. M. (1962). *Acta Pathol. Microbiol. Scand.* **54**, 9.
Jung, F. (1959). *Acta Biol. Med. Ger.* **2**, 481.
Kachler, H. (1958). Inaugural Dissertation, University of Würzburg.
Kellaway, C. H., and Trethewie, E. R. (1940). *Australian J. Exptl. Biol. Med. Sci.* **18**, 63.
Kemp, P., Hübscher, G., and Hawthorne, J. N. (1961). *Biochem. J.* **79**, 193.
Klibansky, C., and de Vries, A. (1963). *Biochim. Biophys. Acta* **70**, 176.
Klibansky, C., Condrea, E., and de Vries, A. (1962). *Am. J. Physiol.* **203**, 114.
Koszalka, M. F. (1949). *U.S. Army Med. Dept., Bull.* **9**, 212.
Lands, W. E. M. (1960). *J. Biol. Chem.* **235**, 2233.

Langer, J. (1897). *Arch. Exptl. Pathol. Pharmakol.* **38**, 381.
Levene, P. A., Rolf, I. P., and Simms, H. S. (1924). *J. Biol. Chem.* **58**, 859.
Long, C., and Penny, I. F. (1957). *Biochem. J.* **65**, 382.
Long, C., Odavič, R., and Sargent, E. J. (1963). *Biochem. J.* **87**, 13P.
MacFarlane, M. G., and Knight, B. C. J. G. (1941). *Biochem. J.* **35**, 884.
Maeno, H., Mitsuhashi, S., Okonogi, T., Hoshi, S., and Homma, M., (1963). *Japan J. Exptl. Med.* **32**, 55.
Magee, W. L., and Thompson, R. H. S. (1960). *Biochem. J.* **77**, 526.
Magee, W. L., Gallai-Hatchard, J., Sanders, H., and Thompson, R. H. S. (1962). *Biochem. J.* **83**, 17.
Marinetti, G. V., Erbland, J., and Stotz, E. (1959). *Biochim. Biophys. Acta* **33**, 403.
Marinetti, G. V., Erbland, J., and Stotz, E. (1960). *Biochim. Biophys. Acta* **38**, 534.
Marples, E. A., and Thompson, R. H. (1960). *Biochem. J.* **74**, 123.
Marples, E. A., Thompson, R. H., and Webster, G. R. (1959). *J. Neurochem.* **4**, 62.
Mohammed, A. H., and El Karemi, M. M. A. (1961). *Nature* **189**, 837.
Moeschlin, S. (1964). "Klinik und Therapie der Vergiftungen," pp. 666–668. Thieme, Stuttgart.
Moore, J. H., and Williams, D. L. (1963). *Biochim. Biophys. Acta* **70**, 348.
Mueller, H. L. (1959). *J. Allergy* **30**, 123.
Neumann, W., and Habermann, E. (1954a). *Arch. Exptl. Pathol. Pharmakol.* **222**, 367.
Neumann, W., and Habermann, E. (1954b). *Z. Physiol. Chem.* **296**, 166.
Neumann, W., and Habermann, E. (1956). *In* "Venoms," Publ. No. 44, pp. 171–174. Am. Assoc. Advance. Sci., Washington, D.C.
Neumann, W., and Habermann, E. (1957). *Nature* **180**, 1284.
Neumann, W., and Habermann, E. (1960). *In* "Handbuch der physiologisch- und pathologisch-chemischen Analyse" (K. Lang and E. Lehnartz, eds.), Vol. IV, Part 1, pp. 801–844. Springer, Berlin.
Neumann, W., Habermann, E., and Amend, G. (1952). *Naturwissenschaften* **39**, 286.
Neumann, W., Habermann, E., and Hansen, H. (1953). *Arch. Exptl. Pathol. Pharmakol.* **217**, 130.
Nygaard, A. P. (1953a). *J. Biol. Chem.* **204**, 655.
Nygaard, A. P. (1953b). *Arch. Biochem.* **43**, 493.
Nygaard, A. P., and Sumner, J. B. (1953). *J. Biol. Chem.* **200**, 723.
Nygaard, A. P., Dianzani, M. U., and Bahr, G. F. (1954). *Exptl. Cell Res.* **6**, 453.
Papahadjopoulos, D., Hougie, C., and Hanahan, D. J. (1962). *Proc. Soc. Exptl. Biol. Med.* **111**, 412.
Pietruszko, R., and Gray, G. M. (1962). *Biochim. Biophys. Acta* **56**, 232.
Piano, J. J. (1970). *In* "Handbook of Experimental Pharmacology" Volume 25 (E. G. Erdös, ed.), p. 589, Springer, Berlin.
Radomski, J. L., and Deichmann, W. B. (1958). *Biochem. J.* **70**, 293.
Rathbone, L., Magee, W. L., and Thompson, R. H. S. (1962). *Biochem. J.* **83**, 498.
Redelberger, W. (1958). Inaugural Dissertation, University of Würzburg.
Reinert, M. (1936). *Festschr. Emil Barell* pp. 407–421.
Rimon, A., and Shapiro, B. (1959). *Biochem. J.* **71**, 620.
Robins, D. C. (1963). *J. Pharm. Pharmacol.* **15**, 701.
Robinson, N. (1961). *J. Pharm. Pharmacol.* **13**, 321.
Rocha e Silva, M., Beraldo, W. T., and Rosenfeld, G. (1949). *Am. J. Physiol.* **156**, 261.
Roholt, O. A., and Schlamowitz, M. (1961). *Arch. Biochem. Biophys.* **94**, 364.
Rosenberg, P., and Ng, K. Y. (1963). *Biochim. Biophys. Acta* **75**, 116.
Rosenberg, P., and Podleski, T. R. (1963). *Biochim. Biophys. Acta* **75**, 104.

Rossbach, F. (1955). Inaugural Dissertation, University of Würzburg.
Röthel, M. (1953). Inaugural Dissertation, University of Würzburg.
Rothschild, A. M. (1965). *Brit. J. Pharmacol.* **25**, 59.
Salzmann, G. (1953). Inaugural Dissertation, University of Würzburg.
Schachter, M. (1960). *In* "Polypeptides" (M. Schachter, ed.), pp. 232–246. Pergamon Press, Oxford.
Schachter, M. (1963). *Ann. N. Y. Acad. Sci.* **104**, 108.
Schachter, M., and Thain, E. M. (1954). *Brit. J. Pharmacol.* **9**, 352.
Schröder, E., Lübke, K., Lehmann, M., and Beetz, I. (1971). *Experientia* (in press).
Schütz, R. M., and Vogt, W. (1961). *Arch. Exptl. Pathol. Pharmakol.* **240**, 504.
Seifert, E. (1958). Inaugural Dissertation, University of Würzburg.
Shaffer, J. H. (1961). *J. Am. Med. Assoc.* **177**, 473.
Shapiro, B. (1953). *Biochem. J.* **53**, 663.
Shipolini, R., Bradbury, A. F., Callewaert, G. L., and Vernon, C. A. (1967). *Chem. Commun.* p. 679.
Silber, B. (1953). Inaugural Dissertation, University of Würzburg.
Slotta, K. H. (1960). *In* "The Enzymes" (P. D. Boyer, H. Lardy, and K. Myrbäck, eds.), 2nd rev. ed., Vol. 4, p. 551. Academic Press, New York.
Slotta, K. H., and Fraenkel-Conrat, H. L. (1938). *Ber. Deut. Chem. Ges.* **71**, 1076.
Späth, W. (1952). Inaugural Dissertation, University of Würzburg.
Stern, I., and Shapiro, B. (1953). *Brit. J. Clin. Pathol.* **6**, 158.
Stockebrand, P. (1965). Inaugural Dissertation, University of Würzburg.
Strauss, W. T. (1949). *J. Am. Med. Assoc.* **140**, 603.
Striebeck, C. (1958). Inaugural Dissertation, University of Würzburg.
Tattrie, N. H. (1959). *J. Lipid Res.* **1**, 60.
Tetsch, C., and Wolff, K. (1936). *Biochem. Z.* **288**, 126.
van Deenen, L. L. M., de Haas, G. H., and Heemskerk, C. T. T. (1963). *Biochim. Biophys. Acta* **67**, 295.
Vogel, W. C., and Zieve, L. (1960). *J. Clin. Invest.* **39**, 1295.
Vogt, W. (1957). *J. Physiol.* **136**, 131.
von Bruchhausen, F. (1955). Inaugural Dissertation, University of Würzburg.
Webster, G. R. (1957). *Nature* **180**, 660.
Webster, G. R. (1962). *Biochim. Biophys. Acta* **64**, 573.
Witter, R. F., and Cottone, M. A. (1956). *Biochim. Biophys. Acta* **22**, 372.
Witter, R. F., Morrison, A., and Shepardson, G. R. (1957). *Biochim. Biophys. Acta* **26**, 120.
Wloszyk, L. (1954). Inaugural Dissertation, University of Würzburg.
Zeller, E. A. (1951). *In* "The Enzymes" (J. B. Sumner and K. Myrbäck, eds.), Vol. 1, Part 2, pp. 986–1013. Academic Press, New York.
Zieve, L., and Vogel, W. C. (1960). *J. Lab. Clin. Med.* **56**, 959.

Chapter 46

The Venomous Ants of the Genus Solenopsis

*PABLO RUBENS SAN MARTIN**

MUSEO DE HISTÓRIA NATURAL, MONTEVIDEO, URUGUAY

I. Introduction 95
II. Systematics 98
III. Biological Cycle of *Solenopsis saevissima richteri* 100
References 101

I. INTRODUCTION

There are many species of Formicidae that have a venom-inoculating apparatus of different degrees of complexity and toxicity. The sting of these species may cause very painful reactions in human beings and animals, but only seldom do deadly accidents occur.

In worker ants, the sting is used both defensively for the protection of the nest, and offensively to kill enemies and insects used as larval food. In America, chiefly in the Amazon region, the ant presents a more or less serious threat to the human economy as a consequence of the distribution of venomous species over a wide geographical area.

The venom apparatus of several species has been intensively studied. It consists of a pair of gland filaments lying free in the posterior body cavity, a spheroidal venom reservoir, from which a duct supplies the bulb of the sting, an accessory gland, also supplying the sting bulb, and the sting mechanism itself, which may be supported on each side by interconnected chitinous plates formed by the last three abdominal segments.

Quilico *et al.* (1961) have been successful in identifying the ant venom as an aromatic substance consisting of 80% *d*-limonene and 20% *l*-limonene. An ether extract of the whole body of *Myrmicaria natalensis* from the African

* Deceased.

FIG. 1. The nest of *Solenopsis s. saevissima.*

Congo contains 35% acetic acid, 31% isovalerianic acid, and 22% propionic acid, with traces of isobutyric acid.

Weckering (1961) reported that in the venoms of *Formica rufa* the concentration of formic acid is 21–71%, and that the small tropical ants of *Iridomyrmex humilis* secrete in their anal glands a venomous mixture, mortal for several insects but inoffensive for man. This kind of venom was purified and crystallized by Pavan in 1956 (see Pavan, 1959) under the name "iridomyrmecine." The structural formula has been established by Fusco *et al.* (1955).

In the same year, Cavill, Ford, and Locksley (1956) extracted from *Iridomyrmex nitidus* of Australia a very similar substance, named "iso-iridomyrmecine."

The venom compound of *Myrmecia gulosa* (Myrmeciinae) has been extensively studied by Cavill, Robertson, and Whitfield (1964). It is proteinaceous and may be separated by electrophoresis into eight components. The venom contains histamine, hyaluronidase, and a direct hemolytic factor. It shows also kininlike action.

Pavan, in 1956, isolated an aromatic oil of the mandibular glands of *Dendrolasius fuliginosus* and named this "dendrolasine." It seems that all these substances may have insecticide properties.

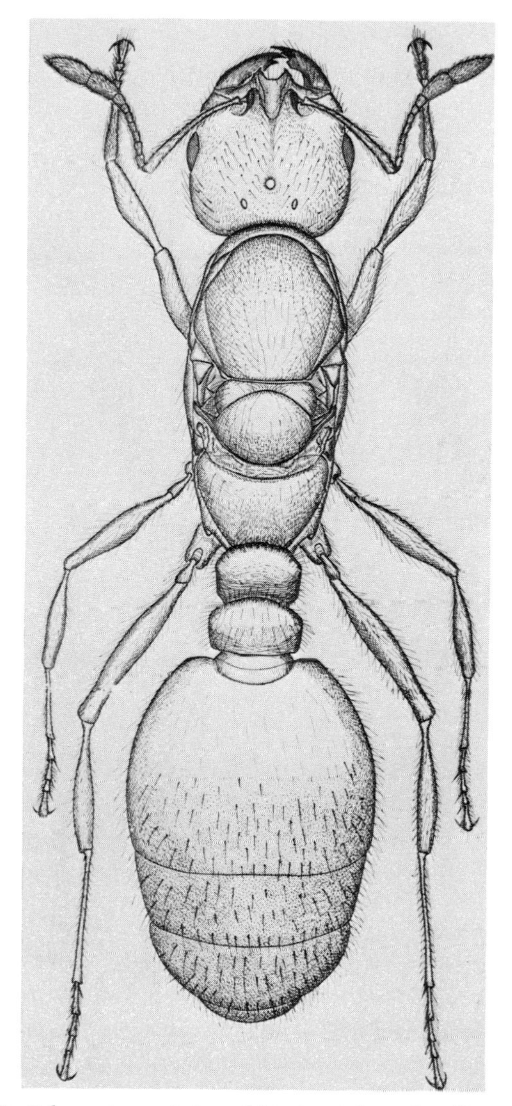

FIG. 2. *Solenopsis saevissima richteri*—adult worker (4–6 mm).

In 1960, Cavill and Hinterberger detailed, from a wide range of dolichodernic ants, the isolation of the following unusual terpenoid extractives: iridomyrmecin, iridolactone, iridodial, and dolichodial, together with the ketones: methyl heptenone, methyl hexanone, and propyl isobutyl ketone.

Büchel and Korte (1961) succeeded in isolating the crystalline *d,l*-iridomyrmecine from the Argentine ant, *Iridomyrmex humilis*, which acts not

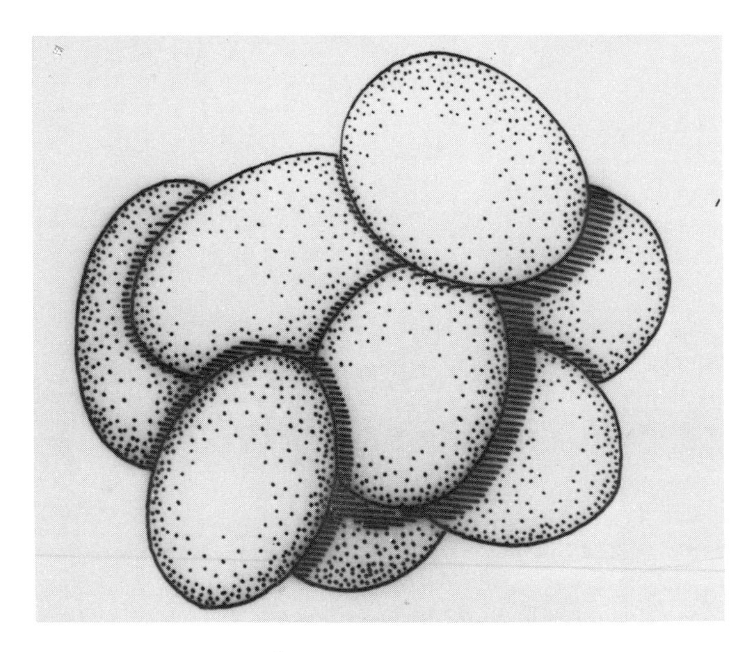

FIG. 3. *Solenopsis saevissima richteri*—eggs.

only as an insecticide but also as an antibiotic against typhus, paratyphus, and cholera. This venom is produced by special anal glands.

We must distinguish in ants three parts of the venom apparatus: mandibular glands, anal glands, and the venom apparatus proper with its sting. The collection of sting-bearing ants in large quantities and the extraction of their venom present many difficulties; 0.3 mg per venom reservoir or 0.35% of the body weight of *Myrmecia gulosa* was the quantity of venom obtained after reservoir dissection by Cavill *et al.* (1964).

One of the most important venomous ants is found in the genus *Solenopsis* (Solenopsidini-Myrmecinae). This genus, with a worldwide geographical distribution, its capacity for rapid reproduction, and adaptation to different ecological niches, presents serious problems in some countries with large agricultural regions because of its irritant venom to man and animals. This is true in Central America and in some southern parts of the United States, where the subspecies *Solenopsis saevissima richteri* exists in several areas.

II. SYSTEMATICS

The species of *Solenopsis saevissima* with the popular names of "Hormiga de fuego," "Formiga do fôgo," "Lava-pé," and "Imported fire-ant" are

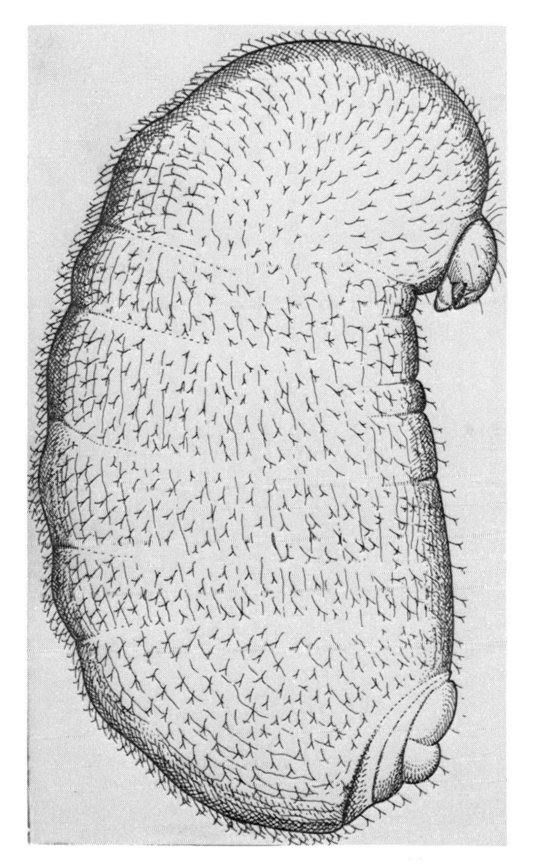

FIG. 4. *Solenopsis saevissima richteri*—larva (2–4 mm).

distributed in South America with several subspecies or races; the most important are: *S. saevissima saevissima* and *S. s. richteri.* This last subspecies may be the most important and its frequency center may be the large area of the La Plata around Buenos Aires.

Unfortunately, the taxonomic question of the subspecies and the local races of *Solenopsis saevissima*, proposed by Forel, by Forel and Smith, and by Forel-Wheeler, Forel-Santschi, and Forel-Creighton, in the years 1915, 1916, and 1950, still is confused.

S. s. saevissima is distributed over a large part of South America and may have developed several local races, while *S. s. richteri* has only a limited spreading area. No morphological characters exist to distinguish both subspecies. Perhaps only *Solenopsis saevissima* may exist as a true species and all the other related species, subspecies, and local or ecological races have only "populational character."

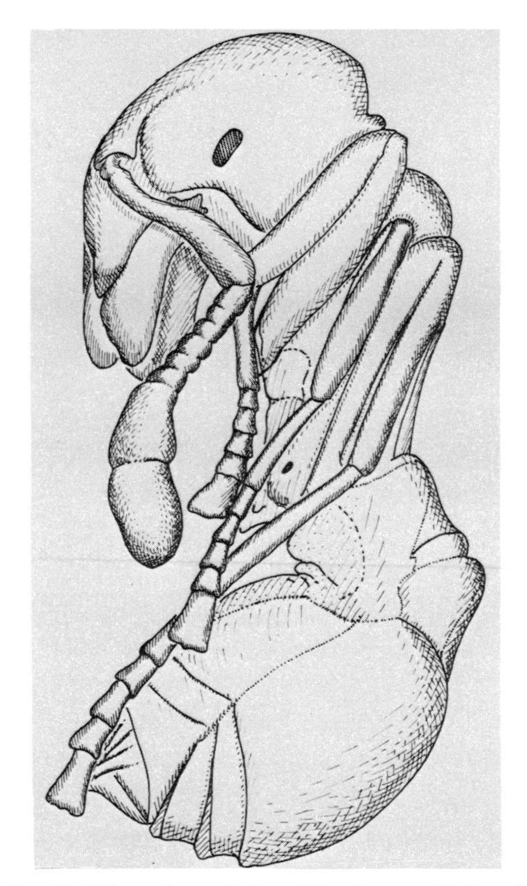

FIG. 5. *Solenopsis saevissima richteri*—pupa (2–3 mm).

III. BIOLOGICAL CYCLE OF *SOLENOPSIS SAEVISSIMA RICHTERI*

(*a*) *Eggs*. In Uruguay there are two periods of oviposition during the yearly cycle: from the middle or the end of January until March—called "the summer cycle"—and from the middle of July to the middle of September —"the winter cycle." Only ant workers are hatched from the summer cycle since in the succeeding months no larvae and pupae of sexed forms were noted. In the winter oviposition, which is much larger than the summer oviposition, workers as well as alated sexed individuals are born.

(*b*) *Larvae of Workers*. The larvae of workers originating in the winter generation exist from July to November, and decline slightly without disappearing in the middle of January. In March the larvae of the summer generation are present and disappear completely in the middle of June. In November the worker larvae are most abundant in the ant nest (winter generation).

(*c*) *Sexed Larvae.* The sexed larvae start in the beginning of August. They can be distinguished easily from the workers by their larger size. In the middle of November they reach the largest numbers and then begin to decline until they disappear in the middle of February.

(*d*) *Pupae of Workers.* The pupation of the workers also has two periods: one at the end of August (winter generation), attaining its climax at the end of November or at the beginning of December, declining afterward until it disappears in February. At that time the worker pupae of the summer generation begin, attain their climax at the end of March, and disappear at the end of June.

(*e*) *Pupae of Sexed Forms.* These appear at the end of August, climax at the end of November, and disappear at the end of February.

(*f*) *Alate Sexed Males and Females.* In the ant colonies males and females of the alate forms may be found during the whole year with a maximum in the first fortnight of December and a minimum during the months of May and June. The decline of sexed forms in January is due to individual flights which begin at this period and last until the middle of April.

REFERENCES

Büchel, K. H., and Korte, F. (1961). *Proc. 11th Intern. Congr. Entomol. Vienna, 1960,* Vol. III, Symp. 3, pp. 60–65.

Callaham, P. S., Blum, M. S., and Walter, J. R. (1959). *Ann. Entomol. Soc. Am.* **52,** 573–590.

Cavill, G. W. K., and Hinterberger, H. (1960). *Australian J. Chem.* **13,** 45.

Cavill, G. W. K., Ford, D. L., and Locksley, H. D. (1956). *Australian J. Chem.* **9,** 288.

Cavill, G. W. K., Robertson, P. L., and Whitfield, F. B. (1964). *Science* **146,** 79–80.

Creighton, W. S. (1930). *Proc. Am. Acad. Arts. Sci.* **66,** No. 2, 39–151.

Creighton, W. S. (1950). *Bull. Museum Comp. Zool. Harvard* **104,** 1–585.

Fusco, R., Trave, R., and Vercellone, A. (1955). *Chim. Ind.* (*Milan*) **37,** 251.

Herman, H. R., and Blum, M. S. (1965). *Ann. Entomol. Soc. Am.* **58,** 81–89.

Pavan, M. (1959). *Proc. 4th Intern. Congr. Biochem., Vienna, 1948,* Vol. 12, pp. 15–36. Pergamon Press.

Quilico, A., Grünanger, P., and Pavan, M. (1961). *Proc. 11th Intern. Congr. Entomol. Vienna, 1960,* Vol. III, Symp. 3, pp. 66–68.

Walker, J. R., and Clower, D. F. (1961). *Ann. Entomol. Soc. Am.* **54,** 92–99.

Weckering, R. (1961). *Proc. 11th Intern. Congr. Entomol. Vienna, 1960,* Vol. III, Symp. 3, 102.

Wilson, E. O. (1951). *Evolution* **5,** No. 1, 68–79.

Wilson, E. O. (1952). *Mem. Inst. Oswaldo Cruz* **50,** 49.

Wilson, E. O. (1953). *Evolution* **7,** No. 1, 262–263.

Chapter 47

Pharmacological Studies on Caterpillar Venoms

ZULEIKA P. PICARELLI AND J. R. VALLE

DEPARTMENT OF BIOCHEMISTRY AND PHARMACOLOGY, ESCOLA PAULISTA DE MEDICINA, SÃO PAULO, BRAZIL

I. Historical Data on Lepidopterism 103
II. Urticating Caterpillars 104
III. Venom Apparatus 105
IV. Pharmacology of Extracts from Caterpillar Setae 108
A. Extract from *Dirphia* Setae 109
B. Extract from *Megalopyge* Setae 112
References 117

I. HISTORICAL DATA ON LEPIDOPTERISM

It has been known for many years that certain lepidopterous larvae possess urticating hairs which cause cutaneous reactions in man and animals. This lepidopterism has been described by many authors since Pliny and Dioscorides mentioned *Cnethocampa pityocampa* Den. & Schiff for its vesicating properties.

The first publications on the subject were merely descriptions of symptoms that appeared following contact with various kinds of larvae, but in some of them could already be found data of value for explaining the mechanism producing such phenomena. Among the early significant works about lepidopterism were those of Will (1848), Morren (1848), and Goossens (1881, 1886), followed by those of Anderson (1885), Keller (1883), Laudon (1891), Packard (1894), and Beille (1896), and the excellent histological study of Holmgren (1896). After a pause of some years, lepidopterism was again taken up by vòn Gorka (1907), Tyzzer (1907), Bleyer (1909), von Ihering (1914), Chalmers and Marshall (1918), Bishopp (1923), Lapie (1923), Mills (1923, 1925), Dallas (1928, 1936), Tisseuil (1935), Jörg (1935), Cheverton (1936), Lucas (1942), and Bercowitz (1945).

The various authors devoted much discussion to whether the effect of these urticating hairs was purely mechanical or whether they were rather poisonous. After the experiments of Tyzzer (1907) there was strong evidence that a specific poison was present in the hairs. In spite of that, Fabre (1919) attributed the dermatitis produced by European larvae to contamination of their hairs with excretion products. He disregarded data from some earlier authors that showed that lepidopterous larvae have a special venom apparatus. Actually the poison glands of the larvae of the brown-tail moth, *Euproctis chrysorrhoea*, had already been described by Kephart (1914), and the dermatitis caused by the larval stage of the moth *Megalopyge opercularis*, also known as the puss caterpillar, had been attributed by Foot (1922) to a venom introduced in the skin by the hollow poisonous spines of the larvae. Descriptions of the various stages of development of the stinging hollow hair from *Euproctis chrysorrhoea* were subsequently published by Tonkes (1933), and the role played by the hairy caterpillar of the pale tussock moth in hop dermatitis was suggested by Whitwell (1943).

Today it is agreed that lepidopterous larvae cause cutaneous reactions through a special venom apparatus but there is still little knowledge about the nature of the poison in the different urticating species.

II. URTICATING CATERPILLARS

The following are the principal families of Lepidoptera that include urticating caterpillars (Craig and Faust, 1951):

a. Nymphalidae. These have long and branched spines; they include *Euvanessa antiopa* and *Vanessa io*.

b. Megalopygidae. Short spines are radially arranged on elevated ridges: *Megalopyge crispata*, *M. urens*, *M. opercularis*, *M. lanata*, etc.

c. Eucleidae. In these larvae, which are the spine type, the shafts of the spines are lined with hypodermal cells: *Sibine stimulea*, *Parasa chloris*, *P. latistriga*, etc.

d. Thaumetopoeidae or Notondontidae. These are processionary caterpillars with modified barbed hairs: *Thaumetopoea processionea* and *T. wilkinsoni*.

e. Limantriidae or Liparidae. The larvae have a modified single hair type, with setose hairs in groups arising from cuticular cups or with hairs borne in dense groups on tussocks or tubercles: *Euproctis chrysorrhoea*, *Porthesia similis*, etc.

f. Arctiidae. Hairs are borne on dorsal tufts; an example is *Euchaetis egle*.

g. Noctuidae. Slender, sharp spines break off into the skin: *Apatela americana*, *Catocala* sp., etc.

h. Saturnidae and Hemileucidae. Spines are distributed on dorsal and lateral tubercles: *Automeris io*, *Hemileuca oliviae*, *H. maia*, *Dirphia sabina*, and *D. multicolor*.

Urticating caterpillars are worldwide in distribution. Lists of the European specimens will be found in Tyzzer's article (1907) and in Railliet's textbook (1895). In Ceylon and India, caterpillars from Bombycidae have been known to cause trouble (Castellani and Chalmers, 1913) and *Taragania igniflua*, a Philippine caterpillar, was mentioned by Tyzzer (1907). Castellani and Chalmers (1913) mentioned four families of caterpillars living in Africa that are capable of producing urtication: a member each of the Arctiidae, Eucleidae, Liparidae, and Bombycidae.

In North America the brown-tail moth, *Euproctis chrysorrhoea*, is the best known caterpillar; four more commonly encountered are *Lagoa crispata*, *Hemileuca maia*, *Sibine stimulea*, and *Automeris io*.

South America is the habitat of many noxious caterpillars. Five groups in that region must be mentioned: Megalopygidae, represented by the common group of *tataranas*, from *tatá-raná*, a Tupy-Guarany word meaning "like fire" (von Ihering, 1914; Bleyer, 1909); Eucleidae; Arctiidae; Hemileucidae, and Saturnidae. In Brazil specimens of each group may be found. The reader should refer to the article by von Ihering (1914), by far the best paper on urticating caterpillars of Brazil, for further particulars concerning them (da Costa Lima, 1945; da Fonseca, 1949; Barth, 1954).

III. VENOM APPARATUS

According to Packard (1894) the poisonous spines of *Lagoa crispata* are secreted by large trichogen cells lying under the rest of the hypodermis and connected with the spines through pore canals in the cuticle. In addition, there are other smaller cells which are supposed to secrete the poison. Ingenitsky (1897) clearly demonstrated the existence of two cells connected with the hair, the larger being the trichogen cell and the smaller the poison-secreting gland.

In all the examples studied by the different authors, the hairs were said to belong to one of two general types. The first type is a simple hair tapering gradually from the base to the tip and containing the poisonous material (Fig. 1); the second type is practically the same except that the tip of the hair is a heavily chitinized cone, readily detachable. When a hair of the first type comes in contact with the skin the point breaks off and its content is liberated; with the second type, a spine hair, the tip breaks off and enters the skin.

FIG. 1. Urticating caterpillar hairs (*Megalopyge* sp.) tapering gradually from the base to the top and containing the urticating material.

Kephart (1915) described a very different type of poison hair in the brown-tail caterpillar; they are pointed at the base, enlarging gradually toward the tip, with three rows of barbs covering their entire length. However, they may be considered just a modification of the first hair type, as stated by Gilmer (1925) in his excellent article concerning a comparative study of the poison apparatus of certain lepidopterous larvae. It is also possible to find hairs belonging to a modified spine type: they are branched like small pine trees (Fig. 2).

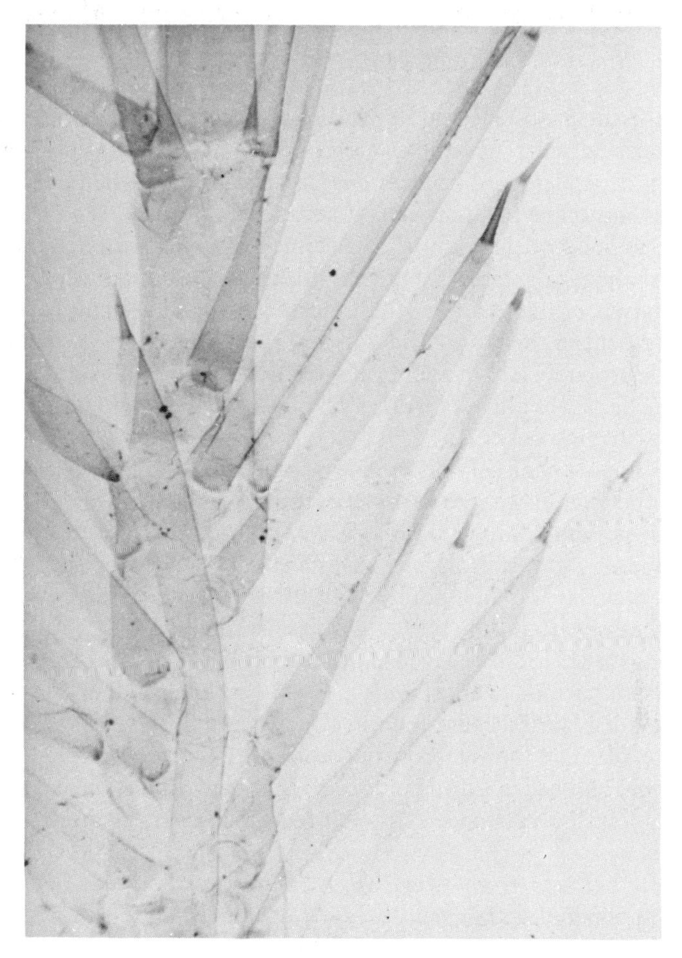

FIG. 2. Modified spine type of caterpillar hair (*Dirphia* sp.) branched like a small pine tree.

Each of the two hair types is supplied with a unicellular poison gland in immediate communication with its lumen. The simple spine type may exhibit an hypodermal gland cell or a gland cell placed well within the lumen of the spine itself, completely out of contact with the body hypodermis. In the branching spine type, the gland cells are completely within the lumen of the spine, the greater part of it being filled with ordinary body fluids.

The urticating hairs may be irregularly distributed over the body of the larvae among the ornamental hairs or may be grouped on tussocks.

IV. PHARMACOLOGY OF EXTRACTS FROM CATERPILLAR SETAE

The experiments of Tyzzer (1907) were the first to indicate that the effect of the urticating hairs of lepidopterous larvae was not simply mechanical but mainly dependent on a poisonous secretion inside them. However, no definite conclusion as to the chemical composition of the venom came from this work, in spite of the author's experiments on its solubility, its reaction to heat and chemicals, and its tendency to break up human erythrocytes.

The analysis of the nature of the venom was made difficult by the minute amounts present in the hairs. The poison is the product of a special unicellular gland, probably a sister cell of the trichogen cell, and is not available as a separate chemical entity since the cytoplasm of the gland cell functions as the venom-carrying agent.

A very good review of the opinions that had been published about the nature of the venom was given by Pawlowsky and Stein (1927). However, very little is presently known on the subject; there is no agreement even as to whether one or more poisonous substances are present in the hairs of the different larvae. As was well pointed out by Tonkes (1933), it is very difficult to accept the view that only one kind of poison is involved.

Goosens (1881, 1886) believed that the hairs of processionary caterpillars contained cantharidin, while Laudon (1891) supposed that they contained formic acid, disregarding that the introduction of this acid alone in the skin could not explain all the symptoms produced by the sting of the caterpillars. Bleyer (1909) noticed a strong odor of this acid in the broken spines of caterpillars, but the substance extracted from them was alkaline when tested with litmus.

Aqueous extracts from *Megalopyge urens* were studied by Gaminara (1928), who concluded that the observed effects of these extracts could not be due to formic acid or cantharidin but to complex proteinaceous substances, which would make them very similar to animal venoms.

According to Estable *et al.* (1946) the venom of *Megalopyge urens* hairs is of a globulin nature with necrotic, hemolytic, and immunological properties. Mazzella and Patetta (1946), studying the same species, claimed to have prepared a heat-sensitive extract and a powder that kept its activity for months. Recently Ardao *et al.* (1966) separated this venom into four fractions by paper electrophoresis: one fraction presented hemolytic activity only; a second, hyaluronidase activity; a third, proteolytic activity and a very low hemolytic effect; and the action of the fourth fraction was undetermined.

In 1947, Emmelin and Feldberg attributed the triple response, and the itching and burning sensation evoked by the sting of the common nettle, *Urtica urens*, to the concomitance of histamine and acetylcholine in the fluid

FIG. 3. Drawing of specimen of *Dirphia* sp. caterpillar. The green, hard spines like pine trees were distributed on dorsal and lateral tubercles.

of the nettle hair. On the basis of this work, we decided to study some pharmacological properties of crude extracts prepared from hairs of urticating caterpillars found in Brazil, since the cutaneous reactions following contact with them resembled those produced by the nettle *Urtica urens*.

We studied a species of *Dirphia* and a species of *Megalopyge*. What follows is a brief report of such observations, part of which have already been published (Valle *et al.*, 1954; Picarelli *et al.*, 1966).

A. Extract from *Dirphia* Setae

The *Dirphia* sp. specimens presented green spines like small pine trees distributed on dorsal and lateral tubercles (Fig. 3). They were cut and kept dried over $CaCl_2$. From this material, extracts were prepared with distilled water or saline brought to pH 5.6 by addition of 0.1 *N* HCl and filtered on paper. When not in use these filtrates were kept in the freezer. The fresh spines averaged 0.86 mg but after exposure to 100°C for 24 hours their average weight was 0.43 mg. This corresponded to a water loss of 0.43 μl, which may be assumed to be the fluid volume of each spine.

By assaying the extracts on isolated guinea pig ileum, it has been possible to demonstrate the presence of 0.086 to 0.172 μg of histamine on each spine, which leads to the assumption that they contained a 0.02–0.04% solution of histamine. The contractile effect produced by the extracts on the gut was only due to histamine, since it was completely abolished by pyrilamine (Fig. 4) but not by atropine or lysergic acid diethylamide. No acetylcholine was detected in the extracts even with more sensitive preparations such as toad rectus abdominis muscle (Fig. 5) or by effect on cat and dog blood pressure. In the cat, a fall followed by a rise in blood pressure was obtained either before or after atropinization of the animal (Fig. 6). In the dog, a slight increase of carotid blood pressure was followed by a fall persisting for several minutes or by death if the dose was sufficiently high (Fig. 7); atropinization did not affect this biphasic response.

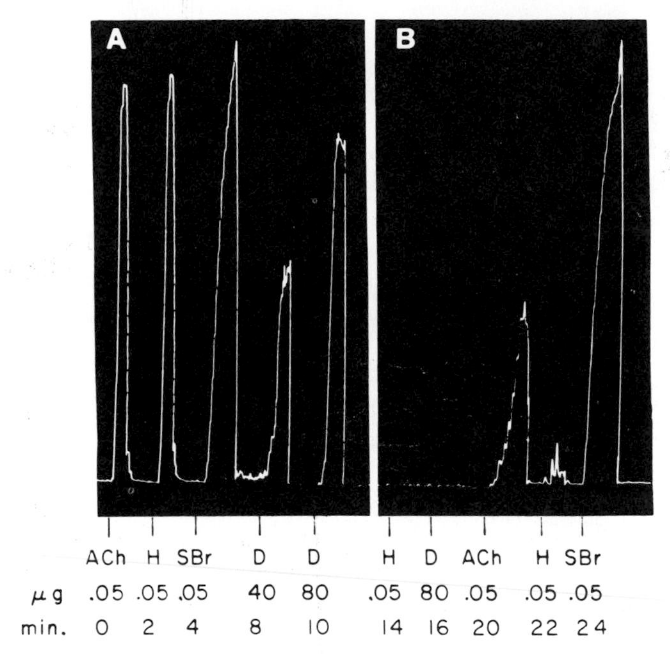

FIG. 4. Effect of pyrilamine on the response of isolated guinea pig ileum to *Dirphia* spine extract. Organ suspended in 10 ml of aerated Tyrode's solution maintained at 35°C. ACh, acetylcholine; H, histamine; SBr, synthetic bradykinin; D, *Dirphia* spine extract. Between A and B, 1 μg of pyrilamine was added to the organ bath. The effects of both histamine and *Dirphia* spine extract were completely abolished by the drug while bradykinin was not affected and acetylcholine was partially inhibited.

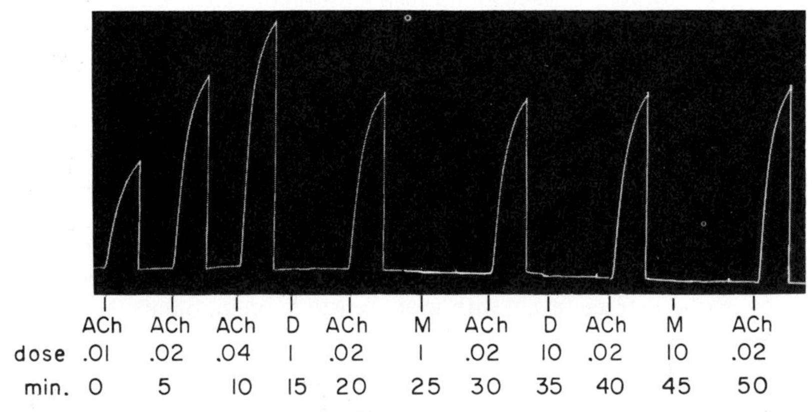

FIG. 5. Absence of acetylcholine in caterpillar setae extracts. Isolated toad rectus abdominis muscle in 10 ml of aerated Ringer's solution at room temperature. Symbols as in Fig. 4. M, *Megalopyge* hair extract. Doses of acetylcholine in micrograms and of extracts in milligrams of dried material. Even 10 mg of *Dirphia* spines or *Megalopyge* hairs did not elicit any effect on this preparation.

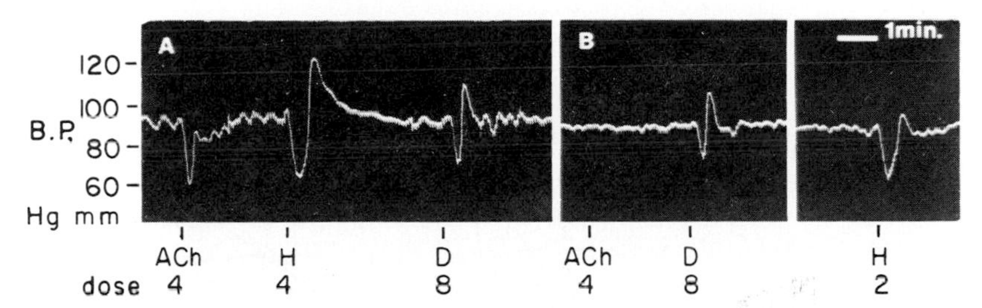

FIG. 6. Action of *Dirphia* spine extract on cat carotid blood pressure. A 1.8-kg male cat was anesthetized with pentobarbital and ether. Symbols as in Fig. 4. Doses of histamine and acetylcholine in micrograms and of the extract in milligrams of dried material. Between A and B, the cat was injected with 0.5 mg of atropine. This treatment was enough to block the effect of acetylcholine but did not affect either the dual response of the extract or that of histamine.

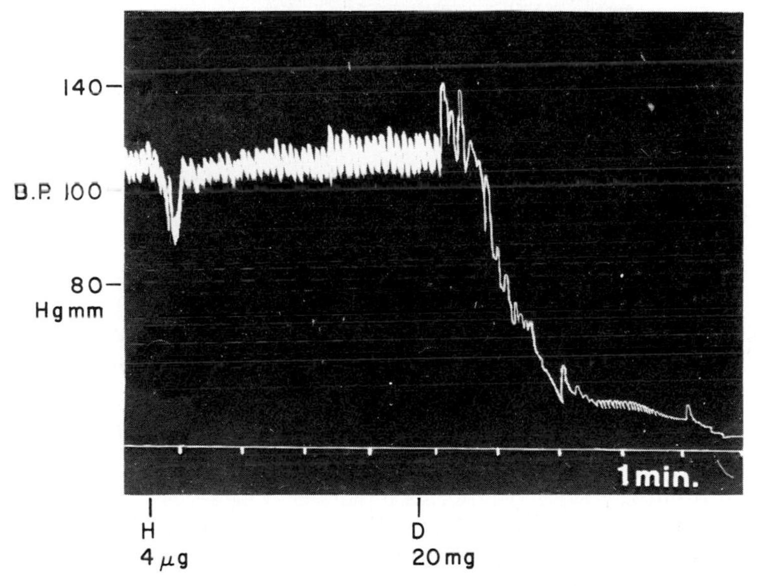

FIG. 7. Effect of *Dirphia* spine extract on dog carotid blood pressure. A 5.5-kg male dog was anesthetized with morphine and barbituric derivative (Somnifène, Roche). Symbols as in the preceding figures. Injection of 20 mg of *Dirphia* spine extract, which ought to have had the same effect as 4 μg of histamine, caused death of the animal within 3 minutes.

These effects were not parallel to those elicited by equivalent amounts of histamine (Fig. 7), indicating that they would depend on either the presence or the liberation of other pharmacological substances besides histamine.

Released histamine was suggested by Broadbent (1953) as the probable mediator of the itching sensation. A single *Dirphia* spine rubbed over the flexor surface of the human forearm produced an immediate and sharp

FIG. 8. Drawing of specimen of *Megalopyge* sp. caterpillar. Hidden among the great number of brownish-red ornamental hairs on these caterpillars were short, hard, hollow setae which were the urticating ones.

pricking, followed 5 minutes later by the appearance of a circumscribed wheal and spreading congestion; later on, a burning sensation became evident but no definite itching was felt on any occasion, even on the next day when the local edema and the spots were still persistent. Moreover, histamine assays by the method of Code (1937) on dog blood collected before the injection of spines extract or during the hypotension caused by it gave no different results.

Feldberg and Kellaway (1937), studying bee venom, and Kellaway *et al.* (1938), studying bacterial toxins, had already suggested the eventual release of histamine and other active pharmacological principles as adenyl compounds by lepidopterous larvae. The observed effects of *Dirphia* spine extracts on blood pressure could indeed be explained by the liberation of such adenyl compounds. However, experiments performed to study this aspect of the problem have not been conclusive.

B. Extract from *Megalopyge* Setae

These caterpillars (Fig. 8) presented short, hard, hollow setae hidden among a great number of long ornamental hairs (Fig. 1). The short setae were cut and dried in vacuum over $CaCl_2$. Distributed in amber vials closed

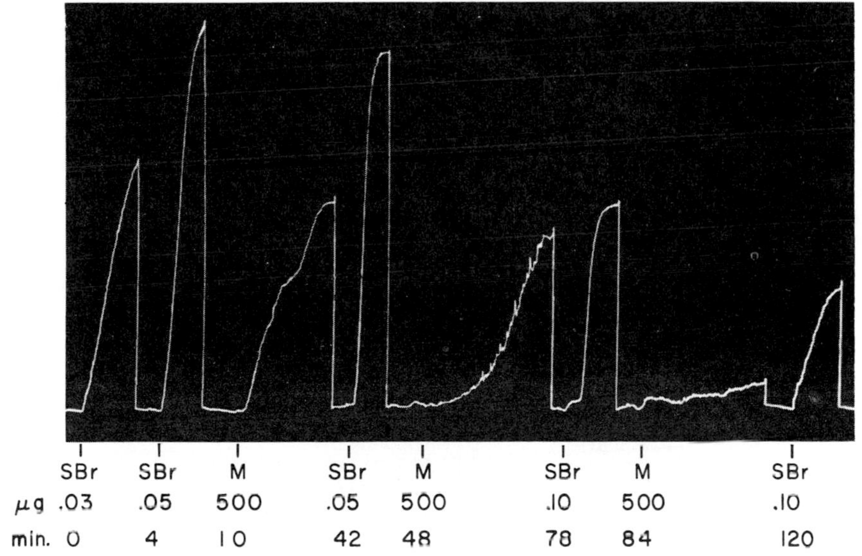

FIG. 9. Effect of *Megalopyge* hair extract on isolated guinea pig ileum. Same conditions as in Fig. 4 except that Tyrode's solution contained atropine (1 μg/ml) and pyrilamine (1 μg/ml). The gut, after an initial slow type of contraction in response to the extract, became less and less sensitive.

under nitrogen atmosphere this material was kept active in the freezer for a long period of time. After drying, 67% of its original weight was lost. When needed, a sample of these dried setae was crushed in a mortar and extracted several times with phosphate buffer, pH 6; the extracts were filtered on paper. Only 10% of the dried material was extracted by this procedure.

The extracts contained neither histamine nor acetylcholine or 5-hydroxytryptamine, since they produced a slow type of contraction of isolated guinea pig ileum (Fig. 9) and rabbit duodenum, not abolished by antihistaminic drugs and not affected by atropine or lysergic acid diethylamide. After a first response the pharmacological preparations became less sensitive or even insensitive to further additions of the extracts (Fig. 9). No detectable amounts of acetylcholine were found by assay of these extracts on toad rectus abdominis or on the cat blood pressure. On rats or dogs, the injection of such extracts gave rise to an elevation of the carotid blood pressure followed by a fall persisting for several minutes (Fig. 10). Atropine, antihistamines, and lysergic acid diethylamide did not interfere with these results but a decrease in response to further injections was observed. This seemed to indicate that the effects were due to the release of substances other than acetylcholine, histamine, or 5-hydroxytryptamine.

Injected intradermally in rats that had received a previous Evan's blue

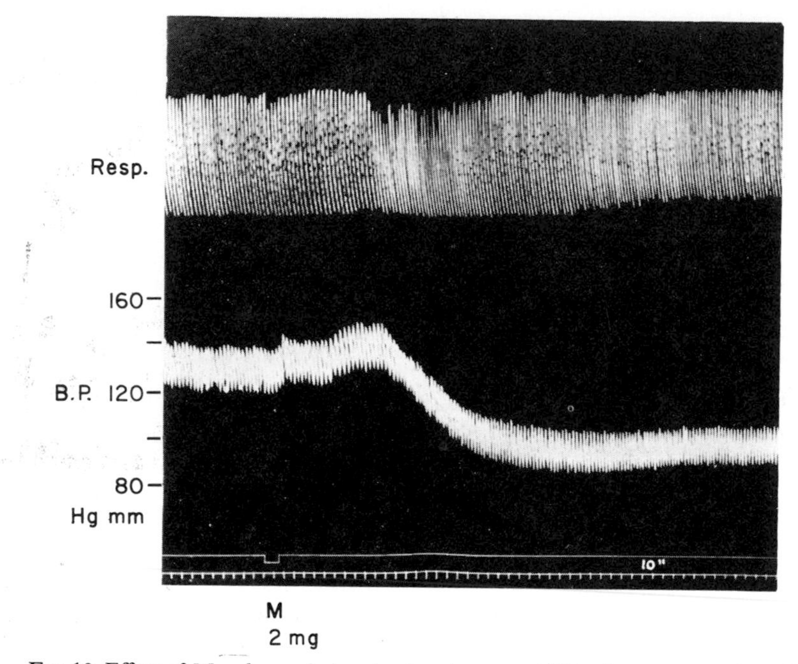

FIG. 10. Effect of *Megalopyge* hair extract on dog carotid blood pressure and respiration in 10.3-kg male dog. Same conditions as in Fig. 7. The effect of 2 mg of *Megalopyge* hair extract on the respiration was discrete, while the slow and persistent fall in blood pressure was very marked.

intravenous injection these extracts produced extravasation of the dye in the rat skin. Figure 11 shows the responses obtained with 10 or 20 μg of dried setae injected 5 minutes after 0.5 ml of 0.5% Evan's blue. Previous treatment of the rats with antihistaminic drugs or lysergic acid diethylamide did not modify these effects.

Attempts were made to verify the presence of kinins, kininogenases, or kininogens in these extracts of *Megalopyge* setae, since such a system of substances shares some properties with the extracts; these investigations did not lead to any conclusion.

In searching for other properties of the *Megalopyge* setae extracts that would facilitate isolation and identification of the substance, or substances, responsible for their effects, it was observed that when assayed on a gelatin film, the extracts were able to digest the protein, as evidenced by the clear spots which appeared on the methylene blue-treated film exactly where samples of the extracts were applied. However, this proteolytic activity was not confirmed by the classical Anson's method (1938) of determining proteolytic activity using globin as substrate.

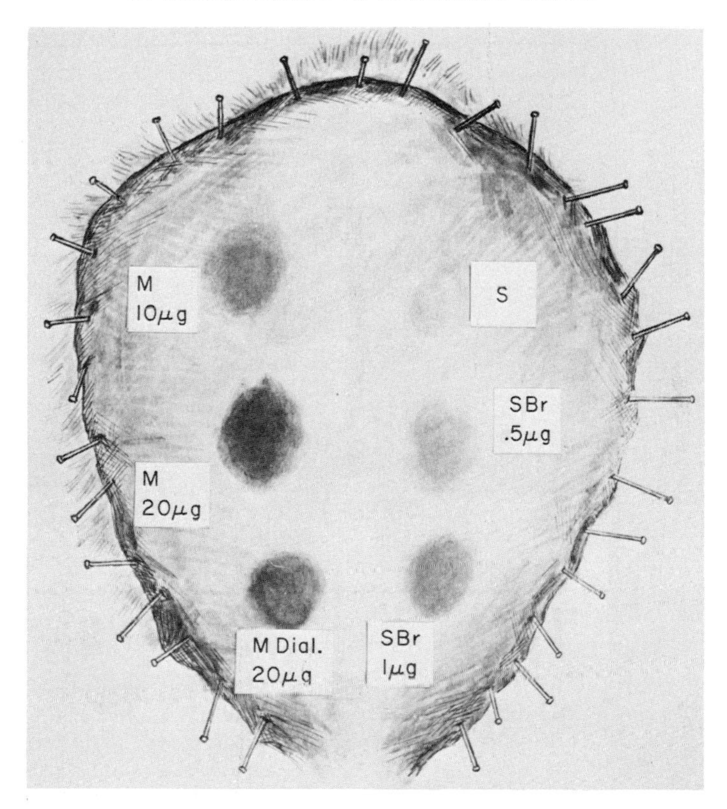

FIG. 11. Effect of *Megalopyge* hair extracts on extravasation of Evan's blue in rat skin. S, saline; M, *Megalopyge* hair extract; M Dial., same extract after 24 hours of dialysis against phosphate buffer, pH 6, in the refrigerator. The effect of the extract increases with the dose; the material is not dialyzable.

The *Megalopyge* setae extracts showed hemolytic activity on dog, cat, and human erythrocytes. This was determined by a modification of the method of de Hurtado and Layrisse (1964) in which the dose of substance producing 50% hemolysis was obtained in a diagram where logarithmic concentrations of the hemolytic substance were plotted against the probits corresponding to percentage of hemolysis produced. For *Megalopyge* setae this dose was found to be 30 μg/ml, whether the erythrocytes used in the assay were washed or not.

Considering that only 10% of the dried material was soluble, these extracts exhibited a very high degree of activity as either vascular permeability-increasing or hemolytic agents.

With the aid of these two properties exhibited by the extracts it has been possible to verify that the substance (or substances) responsible for them was

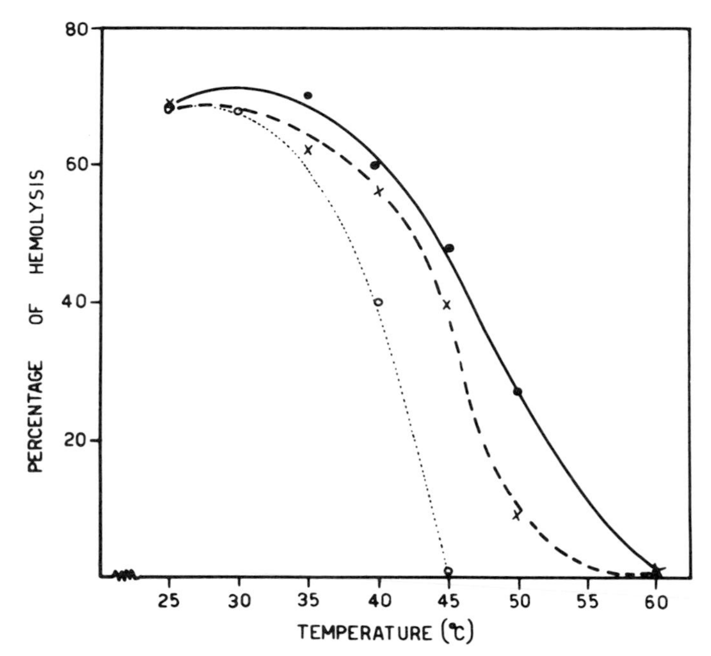

FIG. 12. Heat lability of the hemolytic agent (or agents) present in *Megalopyge* hair extracts. Diagram relating percentage of hemolysis of dog erythrocytes and temperature (in °C) of preincubation. Samples of *Megalopyge* hair extracts (2 mg/ml) were preincubated for 2 (●——●), 10 (×- - -×), or 30 (○ · · · ○) minutes at various temperatures. Hemoglobin determinations at 540 mμ, in a Coleman Jr. spectrophotometer. After 30 minutes preincubation of the extracts at 45°C, they did not hemolyse any more dog erythrocytes. If the temperature of preincubation was 60°C, only 2 minutes were necessary for their complete inactivation.

not dialyzable, was not soluble in ethyl ether, acetone, or ethanol, and was very heat labile, being inactivated by 30-minute warming at 45°C or even 2-minute warming at 60°C (Fig. 12). Maximum activity was detected at pH 6; no activity appeared at pH 2 or 10. It was precipitated by 75% saturation of the extract with ammonium sulfate and was digested by incubation with trypsin, pepsin, or chymotrypsin.

These data seem to indicate that the agent (or agents) responsible for the effects described here is proteinaceous in nature.

Taking into account the high activity of these extracts as agents for direct hemolysis or for increasing vascular permeability, it seems worthwhile to extend these preliminary observations to the isolation and identification of the substance or substances responsible. Actually, the component which seems to damage the pharmacological preparations could be eliminated during the purification procedures, thus allowing a better study of other pharmacological properties of the extracts.

ACKNOWLEDGMENTS

We are much indebted to Dr. Lauro Travassos, Dr. Rudolf Barth, and Dr. H. R. Pearson for classification of the specimens studied and to Dr. Lygia C. Abreu for assays on vascular permeability changes. We also wish to express our appreciation for the technical assistance we received from Miss Vera Lucia Silveira, Mr. Benedito Vieira Dias, and Mr. Luiz Francisco Ribeiro. Synthetic bradykinin was obtained through the courtesy of Dr. E. Nicolaides from Parke Davis & Co. Drawings were made by Jesuino Ribeiro.

REFERENCES

Anderson, J., Jr. (1885). *Entomologist* **18**, 43.
Anson, M. L. (1938). *J. Gen. Physiol.* **22**, 79.
Ardao, M. I., Perdomo, C. S., and Pellaton, M. G. (1966). *Nature* **209**, 1139.
Barth, R. (1954). *Mem. Inst. Oswaldo Cruz* **52**, 125.
Bercowitz, S. (1945). *U.S. Army Med. Dept., Bull.* **4**, 464.
Broadbent, J. L. (1953). *Brit. J. Pharmacol.* **8**, 263.
Castellani, A., and Chalmers, A. J. (1913). "Manual of Tropical Medicine." Baillière, Tindall and Cox, London.
Chalmers, A. J., and Marshall, A. (1918). *J. Trop. Med. Hyg.* **21**, 197.
Cheverton, R. L. (1936). *Trans. Roy. Soc. Trop. Med. Hyg.* **29**, 555.
Code, C. F. (1937). *J. Physiol. (London)* **89**, 257.
Craig, C. F., and Faust, B. C. (1951). "Clinical Parasitology." Lea & Febiger, Philadelphia, Pennsylvania.
da Costa Lima, A. (1945). "Insetos do Brasil." Escola Nacional de Agronomia, Rio de Janeiro, Brasil.
da Fonseca, F. (1949). "Animais Peçonhentos." Inst. Butantan, São Paulo, Brasil.
Dallas, E. D. (1928). *4th Reunion Soc. Arg. Patol. Reg. Norte* p. 691.
Dallas, E. D. (1936). *8th Reunion Soc. Arg. Patol. Reg. Norte* p. 469.
de Hurtado, I., and Layrisse, M. (1964). *Toxicon* **2**, 43.
Emmelin, N., and Feldberg, W. (1947). *J. Physiol. (London)* **106**, 440.
Estable, C., Ferreira-Berruti, P., and Ardao, M. I. (1946). *Arch. Soc. Biol. Montevideo* **12**, 186.
Fabre, J. H. (1919). "Souvenirs entomologiques," 6th Ser. Librairie Delagrave, Paris.
Feldberg, W., and Kellaway, C. H. (1937). *Australian J. Exptl. Biol. Med. Sci.* **15**, 461.
Gaminara, A. (1928). *Bull. Soc. Pathol. Exotique* **21**, 656.
Gilmer, P. M. (1925). *Ann. Entomol. Soc. Am.* **18**, 203.
Goossens, T. (1881). *Ann. Soc. Entomol. France* **1**, 231.
Goossens, T. (1886). *Ann. Soc. Entomol. France* **6**, 461.
Holmgren, E. (1896). *Entomol. Tidskr.* **17**, 81.
Ingenitzky, J. (1897). *Horae Soc. Entomol. Ross* **30**, 129.
Jörg, M. E. (1935). *9th Reunion Soc. Arg. Patol. Reg. Norte* **3**, 1617.
Kellaway, C. H., Trethewie, E. R., and Turner, A. W. (1938). *Australian J. Exptl. Biol. Med. Sci.* **16**, 253.
Keller, C. (1883). *Z. Entwl. u. einh. Weltanschauung* **13**, 302.
Lucas, T. A. (1942). *J. Am. Med. Assoc.* **119**, 877.
Mazzella, H., and Patetta, M. A. (1946). *Arch. Soc. Biol. Montevideo* **13**, 131.
Morren, C. F. A. (1848). *Bull. Acad. Roy. Sci. Bruxelles* **2**, 132.
Packard, A. S. (1894). *Proc. Am. Phil. Soc.* **32**, 275.
Pawlowsky, E. N., and Stein, A. K. (1927). *Z. Morphol. Oekol. Tiere* **9**, 615.

Picarelli, Z. P., Abreu, L. C., and Valle, J. R. (1966). *Abstr. 3rd Intern. Pharmacol. Congr., São Paulo, 1966,* p. 262.
Railliet, A. (1895). "Traité de Zoologie Médicale et Agricole." Asselin et Houzeau, Paris.
Tisseuil, J. (1935). *Bull. Soc. Pathol. Exotique* **28**, 719.
Tonkes, P. R. (1933). *Bull. Biol. France Belg.* **67**, 44.
Valle, J. R., Picarelli, Z. P., and Prado, J. L. (1954). *Arch. Intern. Pharmacodyn.* **98**, 324.
von Ihering, R. (1914). *Anais Paulistas Med. Cir.* **3**, 129.
Whitwell, G. P. B. (1943). *Lancet*, 245, 305.
Will, F. (1848). *Froriep's Not. Geb. Natur-u. Heilk.* **7**, 145.

Chapter 48

Poisoning from Adult Moths and Caterpillars

HUGO PESCE AND ALVARO DELGADO*

FACULTAD DE MEDICINA, INSTITUTO DE MEDICINA TROPICAL "DANIEL A. CARRION," UNIVERSIDAD NACIONAL DE SAN MARCOS, LIMA, PERU

I. Introduction 120
II. Lepidopterism 120
III. The Poisonous Lepidoptera 120
A. Classification 120
B. Distribution 121
C. Morphology 121
D. Toxicology 124
IV. Ecology 125
A. Synecology 125
B. Etiopathogenesis 126
V. Symptomatology of Lepidopterism 129
A. Toxic Syndrome 129
B. Allergic Syndrome 130
VI. Prevention and Treatment of Lepidopterism 131
A. Prevention 131
B. Treatment 132
VII. Erucism 133
VIII. Poisonous Caterpillars 133
A. Classification 133
B. Distribution 134
C. Somatic Morphology 135
D. Morphology of the Venom Apparatus 138
E. Toxicology 143
IX. Ecology 145
A. Synecology 145
B. Etiopathogenesis 147
X. Symptomatology of Erucism 149
A. Histopathology 149
B. Circumscribed Toxic Syndromes 150

* Deceased.

C. General Toxic Syndromes 152
D. Allergic Syndrome 152
XI. Prevention and Treatment of Erucism 153
A. Prevention 153
B. Treatment. 154
References 155

I. INTRODUCTION

Poisonous accidents, both local and general, caused in man by adult lepidopterans have been conveniently distinguished as lepidopterism, while those caused by caterpillars are known as erucism (from the Latin "*eruca*," caterpillar).

In this chapter, we summarize the available data about both syndromes, and present the results of observations carried out recently in Peru.

II. LEPIDOPTERISM

While erucism was known as early as Graeco-Roman times (Phisalix, 1922) and has, in modern times, been the subject of numerous studies, lepidopterism was first observed only at the end of the last century (Laudon, 1891) and important findings were not made until recent decades.

This syndrome seems to be limited to South America. In addition to individual cases from French Guiana, Uruguay, and Argentina, several small epidemic outbreaks in Argentina and the Caribbean region, outbreaks involving several hundred cases in Brazil and several thousand cases in Peru are noted in the literature.

The published information on the epidemiology, clinical treatment, morphology, and poisonous apparatus of Lepidoptera is satisfactory; yet, our knowledge of the histology of the lesion, insect histology, and toxicology, remains inadequate.

The geographic distribution of this syndrome, the magnitude of its epidemic outbreaks, as well as the constant contributions made in this field have made it advisable to review available reports on this type of envenomation and its place within the pathological patterns of the tropical region and the adjacent areas.

III. THE POISONOUS LEPIDOPTERA

A. Classification

The only adult lepidopteran definitely known to be capable of causing anthropotoxic effects is a genus whose taxonomic position (Brues *et al.*, 1954) is as follows: class Insecta, subclass Pterygota, order Lepidoptera (Glossata), suborder Heteroneura, division Ditrycia, superfamily Bombycoidea, family Saturniidae, subfamily Hemileucinae, genus *Hylesia* Hubner.

The only adult lepidopterans capable of producing zootoxic effects belong to the superfamily Zygaenaidea, family Zygaenidae, with only one venomous genus, *Zygaena*.

B. Distribution

The genus *Hylesia* is neotropical. More than 300 species are found in America (Bouvier *apud* Boyé, 1932). The anthropotoxic ones have been observed only in South America, and have the following distribution: in French Guiana: *H. urticans* Floch and Abonnenc (Floch and Abonnenc, 1944); in Argentina: *H. nigricans* Berg (Dallas, 1928) and *H. fulviventris* (Jörg, 1935); in Venezuela: *Hylesia* sp. (P. Anduze *apud* Floch and Constant, 1950); in Peru: *H. valvex* Dyar (Allard and Allard, 1958); in Brazil: *Hylesia* sp. (Gusmão *et al.*, 1961); in Mexico: *H. linda* (*alinda* Druce) is very common, also occurs in Arizona, U.S. However, toxic accidents are not reported (Riley and Johannsen, 1938); cases of poisoning from the Mexican *H. continua* (Bouvier *apud* Boyé, 1932) are unknown. Furthermore, in Japan, a species of Lymantriidae, *Euproctis flava* (Mirisita *et al.*, 1955; Higuchi and Urabe, 1959), is regarded as venomous.

The zootoxic genus *Zygaena* has been observed in Liguria, Italy (Rocci *apud* Phisalix, 1922).

C. Morphology

1. General Description

The adults of the genus *Hylesia* are medium-sized, and have a wing span of 3–5 cm and a body length of 1.5–3 cm. The female is usually larger than the male and has a bent abdomen. The body is covered with hairs, bristles, and scales. The rear wings are smaller than the front ones, having a different shape but similar coloration. The outer edges of the wings are slightly notched. The discal cells are closed in both pairs of wings. In the rear wings only one vein is distinct; the M2 vein is in front of the middle of the apical cell. The frenulum is absent, as is the tibial spur. The frontal process is flattened laterally and the antenna arises from the frontal protuberance. The antennae are pectinated in their distal portion in both sexes (twice bi-pectinated in the males of *H. urticans*). The anepisternum is small; the labial palpi are long.

They are dull-colored moths with rather uniform coloration, which varies with the species. In *H. nigricans* from Argentina (Fig. 1), the female has

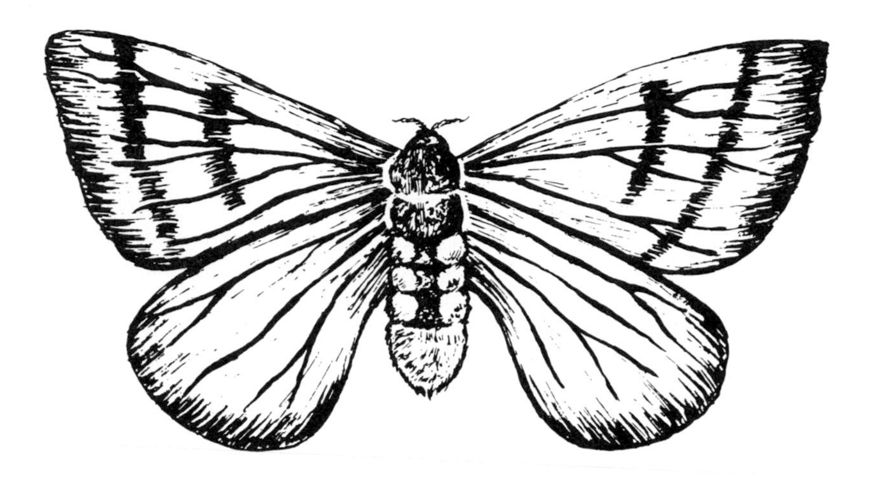

FIG. 1. *Hylesia nigricans*, Argentina.

blackish wings, and the abdomen is whitish with a medial dorsal, dark band, except in the last segment; the male has yellow earth-colored wings. *H. fulviventris* has a yellowish-brown abdomen. *H. linda* from Mexico (Fig. 2) has pale marble-gray wings with transversal clear brown bands; the body is dark brown. *Hylesia* sp. from Amapá, Brazil, is of a uniform darkish gray color, while the end of the abdomen is yellowish pink. *H. valvex* from Peru has dusty black wings.

The complete metamorphosis of this genus lasts 3 months, and has been experimentally studied by Boyé (1932) who gives an extensive description of all metamorphic changes; they have been omitted here since we are particularly interested in the full-grown moth.

2. The Sting Weapon

The phanerotoxic animals possess various poisonous apparatuses composed of unicellular or pluricellular poisonous secreting glands and a sting weapon, features serving to distinguish them from the cryptotoxic animals whose tissues and humors have special primary functions and acquire a venomous function only secondarily.

Although studies on the histology of the lepidopteran poison glands are not available, the presence of sting weapons filled with poison in the genus *Hylesia* can be definitely affirmed; for this reason this genus belongs to the great group of phanerotoxic animals.

FIG. 2. *Hylesia linda*, Mexico.

Léger and Mouzels (1918) in Cayenne found, on the tegument of adult *Hylesia* sp. (*urticans*, 1944), "tiny spines barbed on the tip" of 150 μ in length, 3 μ in diameter, which bore the venomous function. Later, Boyé (1932) called them "flechettes" and demonstrated that they cover the abdomen except along a ventral dorsal band. He also observed that the fine burrs or "barbelures" located on the distal third result from the disarrangement of the distal orifice of telescoped chitinous cylinders (Fig. 3), and are hollow and grayish; they are arranged around the base of curved golden-pink setae 10 times longer and 25 times less in number. He made a detailed description of the tegument, but not being able to study the histological structure, he failed to demonstrate the connection of the flechettes with the poison gland cell. Tisseuil (1935) established that only the adult female exhibits these flechettes. Jörg (1933), in Argentina, found, on imagos of *H. nigricans*, dense hairs clothing the ventrolateral surface of the abdomen as well as the dorsal surface of the last abdominal segments. These consist of uniform arch-shaped, flattened hairs 1 mm in length that carry in the free apex a reverse hook like a single arrow, besides shorter, rigid spines carrying on the tip two or three hooks like compound arrows. Both types exhibit three layers: pneumatic or medular, a middle layer of cubic cells, and a cuticle of chitinized lining cells. He believes that the numerous glands of the coelomatic cavity manufacture the poisonous secretion found in the urticating hairs. Recently, in Brazil Forattini (see Gusmão *et al.*, 1961) found, in female specimens of *Hylesia* sp., that the flechettes are distributed predominantly

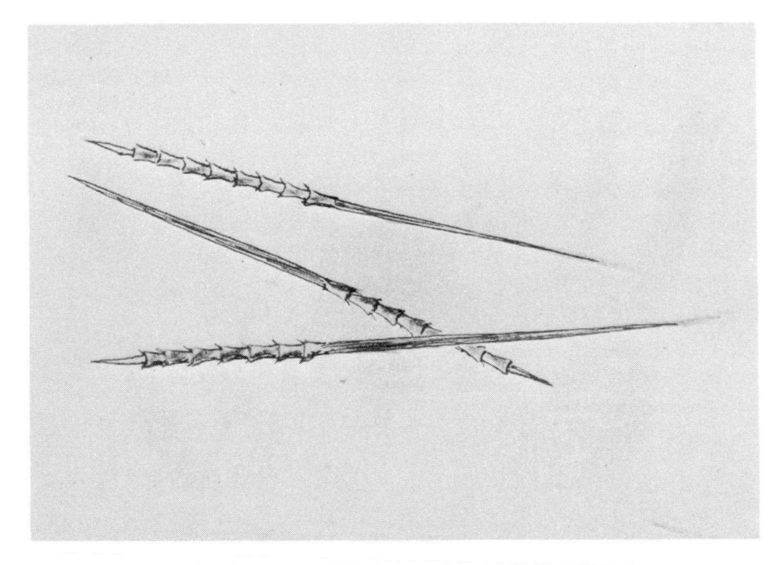

FIG. 3. Venomous "flechettes" from *Hylesia* female.

on the central and lateral surface of the last segments: these are of dark color, 170 μ in average length and 4–5 μ in diameter, barbed all along their entire length but more intensely in the distal fourth part.

Most of the authors cited have obtained experimental dermatitis of the same type as the accidental, using material taken from the abdominal surface of the adult insect containing the flechettes. We can be certain, then, that the sting weapon just described is responsible for toxic accidents.

D. Toxicology

It seems reasonable to assume, therefore, that the moth is not a diffuse cryptotoxic source but actually has phanerotoxic sources in the flechettes which the adult female carries on her abdomen. Although the presence of a cell gland associated with a seta, as is the case in urticating caterpillars, has not yet been demonstrated histologically, it would appear that the setae are filled or daubed with poison whose properties are retained long after they are shed by the moth.

Léger and Mouzels (1918) working with *H. urticans* demonstrated that its venom is rather soluble in water and insoluble in alcohol since the centrifugate of the watery macerate possesses the same venomous properties as the setae, a finding that has been confirmed by Jörg (1933) in *H. nigricans.* Rotberg (see Gusmão *et al.*, 1961) using insect dust composed of scales and setae obtained positive epidermal reactions; in addition, he found that the venom is insoluble in ether.

Jörg (1963) using chromatographic analysis of *H. fulviventris* found a substance that he believes is an allergine. Much remains to be done in the field of toxicology of *Hylesia* as it relates to histology, biochemistry, and pharmacodynamics before we can gain a thorough understanding of the poisonous secretions involved.

According to Phisalix (1922), Rossi (1917) treated the golden zootoxic secretion of *Zygaena* with hydrolyzing products, obtaining a fixed atoxic fraction and a toxic volatile fraction whose nature is ketonic or aldehydic. Inoculation in mice and frogs produces paralysis and death.

IV. ECOLOGY

A. Synecology

1. Relation to Moths

In a given region the adults appear in a brood briefly of 3 to 6 days (Floch, 1952) followed or imbricated by other overlapping broods for 2 to 4 weeks (Tisseuil, 1935).

Their annual appearance coincides with the end of the rainy season (Argentina, Peru, Brazil). Nevertheless, Boyé (1932) in the constant warm climate of Guiana observed a trimestral generation; such case probably represents the most favorable ecological conditions. Tisseuil (1935) confirmed the same periodicity but with some interruptions. Floch (1952, 1954) for the same country reported several irregular broods a year. In Argentina where there is a cold winter, the brood is annual (Dallas, 1933) and occurs during the first three warmest months of the year (Jörg, 1933).

The caterpillars inhabit many kinds of host plants, and larvae of the same genus have been found feeding on different species. The adults commonly inhabit only rural or forested regions and are rarely found in towns; for this reason, few observations have been made. The male flights precede bisexual flight by several days (Tisseuil, 1935). They prefer to fly in the evening, not at night; it is during these few hours that the peasants turn on their house lights; thus these evening fliers are guided by a positive phototropism, as are most of the Heterocera.

2. Relation to Humans

The epidemiologic studies show that the "direct" erucic accidents are individual events caused through direct contact with urticating caterpillars; otherwise, both "indirect" erucic accidents and "indirect" lepidopterism are predominantly collective phenomena caused by the atmospheric diffusion of the urticating setae, either from the larvae or from the adult moths.

Laudon (1891) in Europe, was the first to call attention to lepidopterism although his work appears to have gone substantially unnoticed at the time. During his studies of several species of *Thaumetopoea* (Notodontidae), already known to cause massive indirect erucism by their processionary larvae, he observed that the "epidemics of urticaria" with conjunctivitis pharyngitis, and sometimes laryngitis previously reported by other observers were closely associated with the first flights of emerged adults; he assumed that the urticating hairs shed in the air by the moths were the etiological agents.

Many years later, Léger and Mouzels (1918) observed, among the children of Cayenne, collective accidents caused by *Hylesia* sp. (*urticans* 1944). Jörg (1933) observed them among the workmen of a sawmill; Floch (1952, 1954), in people from main walks of life; P. Anduze (Floch and Constant, 1950) in large groups of oil workers; various observers in the Caribbean region noted its presence in Italian sailors of a ship docked in Caripito (May 1952) and in a group of American sailors on a ship in the Gulf of Paria (June 1952) according to Allard and Allard (1958); Jörg (1935) reported an outbreak in a small village; Gusmão *et al.* (1961) in a crowded Brazilian town with 707 cases, the equivalent of 40% of the inhabitants; and Allard and Allard (1958) observed 3000 cases, equivalent to 70% of the population in a Peruvian city.

Beginning with Boyé (1932) and continuing with the episodes cited by various authors, the appearance of the broods and the emergence of adults of genus *Hylesia* has been proved to have a close relationship with the duration of an epidemic outbreak of dermatitis. The flight of the male adults does not produce toxic accidents (Tisseuil, 1935); besides, a close relationship between the number of human cases and the proportion of infestation by adult females (3% to 33%) has been noted (Floch, 1952, 1954). Statistical studies made by Gusmão *et al.* (1961), in Brazil, during the outbreak of Amapá have revealed that the distribution of the 707 cases according to age, sex, and color coincides with the corresponding composition of the population, demonstrating in this way an indiscriminate occurrence.

B. Etiopathogenesis

Direct lepidopterism occurs only in exceptional conditions, when the adult insect is handled or is rubbed against human skin, or comes into contact with the eye.

By far the most common accident is indirect lepidopterism, different types of which we have classified, defined, and described as follows.

1. Anthropotoxic, True Lepidopterism, Due to Casual Injurious Action

Each adult female is a carrier of several tens of thousands of urticating flechettes which are constantly shed, together with other setae and scales;

during flight they frequently rub against leaves, flowers, and window screens, resulting in clouds of easily visible dusty particles. When man approaches an area infested with these moths, or when the latter overrun a town in great numbers, the venomous flechettes inevitably fall on the skin or come into contact with the upper respiratory duct, producing dermatitis or other toxic disturbances. Furthermore, the dusty particles retained on the bedclothes serve as a reservoir of injurious weapons.

The active offensive mechanism of the moth does not intervene in the process, and the injurious action occurs casually.

Massive epidemic events are predominant, while individual accidents are infrequent.

2. Zootoxic and Anthropotoxic, Meta-lepidopterism, Due to Passive Defense

Pawlowsky (1912), in Russia, was the first to observe that in the "nest" of *Euproctis chrysorroea* the eggs are laid in clusters that the female covers with her abdominal urticating setae. This was confirmed by rubbing the human skin with this "nest," causing rash and papules identical to those seen in erucism. Thus, Schade, some time later (*apud* Jörg, 1935) found the same phenomena working with "nests" of Megalopyge. Jörg believed the "egg-clusters" or "nest" (Fig. 4) of *H. nigricans* (1933) and those of *H. fulviventris* (1935) to be toxic. Boyé (1932), after these initial observations in *Hylesia* sp, (*urticans*) had been fully confirmed, was able to carry out

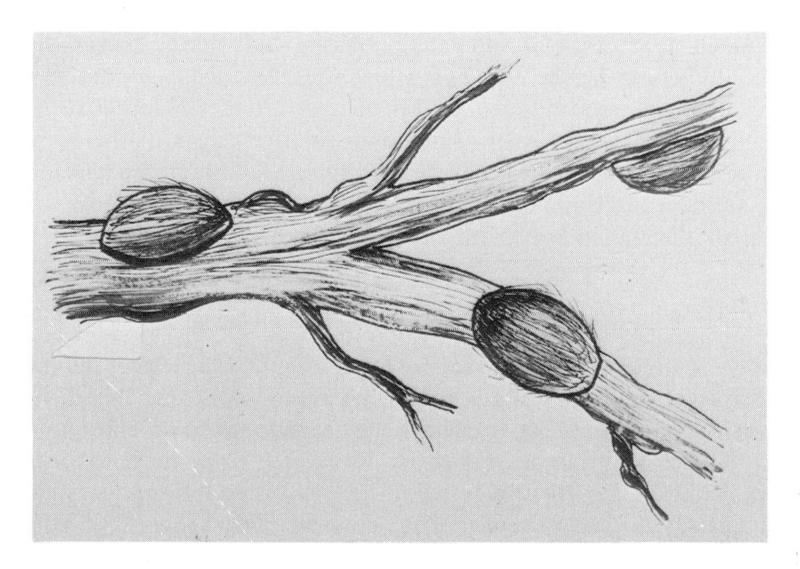

FIG. 4. "Nests" of *Hylesia* (mass of eggs, covered with setae of adult female).

further investigations. He observed that the larvae just emerged from the egg cluster are contaminated with maternal urticating hairs till the first molting; he was therefore able to produce experimental dermatitis with very young caterpillars.

Bearing in mind that there are some predators that attack the larvae in their egg state, while others attack the newly hatched naked caterpillars, we can readily understand that this process constitutes a defensive survival trait, which is a primary attribute of the venomous function in animals, achieved by means of a mechanism of passive defense. Obviously the primary victims are other arthropods and only accidentally humans.

Another important observation, made by Tisseuil (1935), was the fact that rubbing adult male specimens of *H. urticans* against the skin produced (as proved by Dallas, 1933, with *H. nigricans*) precocious rash and urticarial elements of short duration. Since the males are harmless, and have only cylindrical blunt hairs, there is reason to suspect that during mating they acquire numerous poison flechettes that remain attached to their bodies. It should be pointed out that, in this case, the existence of an organic protective mechanism performed by venomous function cannot be admitted; this event is merely an aberrant expression of passive defense. Some arthropoda, and occasionally humans, may suffer the effects of this action.

In all human cases already discussed, the immediate venomous source proved to be the egg-cluster coverlet that would constitute the primary cause of an apparent ovism or tecto-ovism; the very young caterpillars would be incriminated as a cause of simulated erucism; and finally the adult male occasionally would be the cause of a lepidopterism. Considerable evidence indicates that these are different forms of inadvertent lepidopterism caused by usually poisonless specimens since the primitive origin of the poisonous flechettes is the adult female. Because the decisive factors in the development of this accident are the mechanical vectors of poisoning weapons, we have designated it meta-lepidopterism or "consigned" lepidopterism.

3. Zootoxic, Para-lepidopterism, Due to Active Defense

Many adult lepidopterans do not exhibit urticating setae but possess special stink glands producing a golden nauseous secretion; this distinctive odor even at a distance has repellent and sometimes toxic effect on small dipterans. Such is *Zygaena*, studied by Rocci (1917), in Italy, according to Phisalix (1922). This repellent function against other arthropods and small birds is usually not casual, but one of active defense.

The individual quantity of venom is to a great extent harmless for man and consequently there are no accidents of this nature.

4. Anthropotoxic, Pseudo-lepidopterism, Due to Casual Injurious Action

In some lepidopterans, specimens of the genus *Euproctis*, for instance, which are urticating in the larval stage, pupation occurs in a cocoon built of silk and poisonous larval hairs, and when the adult imago emerges from the cocoon, it is contaminated with a considerable number of these hairs. This is true of *E. chrysorrhoea* from the U.S.A., according to Riley and Johannsen (1938) and probable with *E. flava* from Korea, according to Mills (1923). Adults of both species have been known to produce toxic dermatitis.

Evidently the urticating hairs originate with the caterpillar, and the adult is only a carrier of a larval weapon that casually produces in man pseudo-lepidopterism or "consigned" erucism.

In this case there is no active intervention by the moth and the casual accident is always individual.

V. SYMPTOMATOLOGY OF LEPIDOPTERISM

A. Toxic Syndrome

1. Dermatitis

Various investigators have described lepidopteran dermatitis, sometimes showing differences of opinion among themselves: Léger and Mouzells (1918), Dallas (1926, 1927, 1928, 1933), Boyé (1932), Jörg (1933, 1935), Tisseuil (1935), Floch and Abonnenc (1944), Floch and Constant (1950, 1954), Floch (1954), Allard and Allard (1958), and Gusmão *et al.* (1961). The main features and symptoms are as follows.

Dermatitis usually occurs on the limbs and neck and more rarely on the abdomen. It appears a few minutes after the skin has come in contact with urticating material and the patient immediately experiences intense itching.

According to the majority of authors, the first objective signs are urticarial papules of various sizes, isolated, or, at times, crowded together. These are soon followed by erythematous patches, swelling, indurate areas and by burning pain; less frequently by monomorphic eruptions, single or associated with stiff micropapules, only slightly erythematous, crowned at times by little vesicles. In case of intimate contact with the moth, this gives rise to an extensive urticariform eruption with production of bullae.

According to another authority (Gusmão *et al.*, 1961), the lesion consists of "erythematous patches with diffuse micropapular reaction, often associated with vesicles and exulcerations."

Dermatitis lasts from 6 to 8 days, but cases lasting 14 days are not infrequent.

The primary toxic nature of the urticating syndrome, suggested by its immediate appearance and by the identical symptoms in all the patients during an outbreak, has been experimentally confirmed by Allard and Allard (1958) who rubbed the urticating material obtained from Peruvian moths on the skin of persons not previously affected in zones both near and distant from the outbreak. Similar experiments were carried on by Rotberg (Gusmão *et al.*, 1961) who also succeeded in producing epidermal reactions.

2. Localizations on External Mucous Membranes

Boyé (1932) pointed out the absence of localizations on external mucous membranes, except in the case of the severe conjuntivitis attributable to a bombicid, which he observed in Madagascar.

It appears that the only two localizations encountered were: the ulcerous lepidopteran keratitis reported by the ophthalmologist Goerger (1940) according to Floch and Constant (1950), and the intense stomatitis suffered by a cat that had eaten an *Hylesia* (Floch and Abonnec, 1944).

3. Localizations in the Upper Respiratory Duct

Floch and Constant (1950) described rhinopharyngitis and trachaeitis in persons suffering from lepidopteran dermatitis or in their relatives, whose signs were whooping cough, bronchitis, fever, and other general disturbances not being in evidence.

4. General Symptoms

It is essential to bear in mind that dermatitis is of slow onset; thus, there is an initial period of irruption which progresses to an acme in 2 days and then regression occurs in about 6 to 8 days; for this reason Tisseuil (1935) pointed out that the component of general intoxication is constant.

In cases of diffuse urticaria of more than 2 weeks duration, dizziness, oppression, headache, nausea, dyspnea, and malaise, have been frequently observed (Higuchi and Urabe, 1959).

B. Allergic Syndrome

The epidemic outbreak of the syndrome and the early occurrence of symptoms make it difficult to attribute all the cases observed to an allergic mechanism.

Jörg (1933) prepared an aqueous extract from the imago setae of *H. nigricans* which he mildly rubbed on the skin of two persons. One of them previously suffered lepidopterism, the other did not. There were nongeneral signs and the cutaneous reactions obtained in both were practically the same from the standpoint of morphology and duration. It would seem that this experiment has shown a primarily toxic reaction in both cases, using an extract that may possibly carry most of the venom-containing setae. The same author (Jörg, 1935) observed that the people affected by the 1932 outbreak of epidemic lepidopterism had become immune or had at least acquired resistance when reexamined 3 years later. Over a period of 10 years, only one person displayed sensitization reactions each time he entered the zone of endemic lepidopterism; the author desensitized this patient using an extract of the tiny spines of *Hylesia*. These observations would demonstrate that in all the cases the lesions are toxic in nature, but there are persons with a congenital hyperergia or who seem to have acquired it.

Jörg (1963) announced having found an allergen in *H. fulviventris*; it has also been pointed out in individual cases of probable allergic syndrome, such as rhinopharyngitis, comparable to the asthmatic condition in a man who handled adult *H. urticans* (Floch and Constant, 1950). We believe, with Gusmão *et al.* (1961), that one can accept the statement that a few affected individuals "become sensitive and later become victims of allergic conditions, which succeed or coexist with the primary irritative lesions."

To define the existence and importance of an allergic component it is necessary first of all to take into account the statistically predominant incidence of the toxic type of clinical picture with early manifestations, or the late clinical pictures of allergic type among those who have suffered the first accident, as well as the establishment of resistance and immunity or the appearance of hypersensitivity among those who suffer successive accidents. It is, furthermore, convenient to make experimental tests in humans living in the endemic and undamaged zones.

It is also necessary to try to demonstrate in each species the existence of venoms of direct action or the possible existence of allergens. The investigation must concentrate on a given species and demonstrate the quantity, constancy, and variation of the poisonous substances encountered, as well as the level at which they induce clinical effects.

VI. PREVENTION AND TREATMENT OF LEPIDOPTERISM

A. Prevention

1. Sanitary Measures

Among the rural population, as well as in public parks, only the knowledge of the plant hosts containing large numbers of egg clusters, cocoons, larvae, and adults will permit the application of chemical and biological control measures. In such cases slow-acting insecticides are useful.

In the affected town, measures should be taken against the moths at the beginning of the invading brood. Fast-acting insecticide must be applied to the surface of dwellings, by means of aerosol bombs. These insects must be killed immediately to avoid their filling the air with urticating setae. A useful formula of adequate concentration for this purpose is the following: in an ordinary tank of 52 gallons 1,200 gm of lead arsenate are dissolved in 200 liters of water, 600 gm of dehydrated lime are then added. The men employed in spraying the solution must wear safety goggles, besides the usual fumigator's outfit. In this way, the temporal and spatial invasion is effectively limited and the number of accidents is therefore greatly reduced, saving many hours and even days of working time.

2. Individual Measures

Since this moth is an evening flyer and visits illuminated dwellings in search of light, it is convenient to use the same easy measure adopted in Cayenne (Boyé 1932) of putting out the lights for 1 to 3 hours each evening. A preventive measure observed in Serra do Navio, Brazil (Gusmão *et al.*, 1961) consists of keeping the bedclothes safe from contamination until the very moment of going to sleep. Daily washing of contaminated linen is advisable. During and after infestation, lamps and light features should not be handled, because their surfaces are covered with lepidopteran dust (Boyé, 1932).

3. Humoral Modifications

Individuals showing special symptoms from the first contact, and those who in successive attacks have become sensitive, must be protected against probable new attacks during the seasonal appearance of the moths if they are to remain in or visit the endemic zone.

It is sometimes useful to immunize the sensitive individuals with non-specific foreign protein. If the causal agent is known and the antigen available, it is worthwhile to make attempts to use a minimal progressive dose for a sufficient time. This treatment was used in Argentina by Jörg (1935) several years ago, with good results against *H. nigricans*. If several species are causal agents, it is advisable to use polyvalent antigens.

B. Treatment

1. Local

The early application, during the first half hour, of 50% sodium hyposulfite lotion is regularly followed by the abolishment of itching and rapid disappearance of the elementary cutaneous lesions. This therapeutic response may be due to a reduction of the poisonous substance, according to Floch and Constant (1950). This treatment is operative during the early hours after infection, but when cellular and tissue reactions have developed, it is no longer effective.

2. General

The same authors recommend the oral administration of synthetic antihistaminics. The effect of an early antihistaminic administration consists of a complete remission of pruritus and a prompt relief of the cutaneous eruption. When the administration is delayed, it produces only neutralization of pruritus and therefore remission of sleeplessness, which may last for several days. Higuchi and Urabe (1959) reported disappearance of the exanthema in 2–3 days in four patients with intramuscular administration of 30 mg ACTH gel.

VII. ERUCISM

Caterpillars, strictly speaking, or "*erucae sensu strictu*," are abundantly equipped with setae; among those which possess poisonous glands, porous setae play the role of offensive weapons. Such are the "phanerotoxic" caterpillars, capable of introducing poisons into the human skin.

There are also other caterpillars that have glands capable of secreting toxic substances, but they do not have defensive setae: their poison acts by direct contact, through volatile emanations, or is projected from a certain distance by means of an emission tube. These are the "cryptotoxic" caterpillars, and their action is generally directed against other arthropods, small birds, and rodents.

In the present work, we have concerned ourselves principally with the first group, and their characteristic "anthropotoxic" action, with special emphasis on the defensive weapons of Peruvian caterpillars, and on two series involving 621 cases of erucism observed in the Peruvian jungle. We shall also have occasion to refer to erucism in Meso-America and the Caribbean region.

VIII. POISONOUS CATERPILLARS

A. Classification

The caterpillars causing that form of poisoning known as "erucism" belong to the phylum Arthropoda, class Insecta (Hexapoda), order Lepidoptera (Glossata), suborder *Heteroneura* (Frenatae).

1. Families Including Phanerotoxic Species

The main families of phanerotoxic species are: Arctiidae, Bombycidae, Eucleidae (Limacodidae, Cochlidiidae), Lasiocampidae, Lithosiidae, Lymantriidae (Liparidae), Megalopygidae, Morphoidae, Noctuidae, Notodontidae, Nymphalidae, Saturniidae, and Sphingidae.

2. Families Including Cryptotoxic Species

The main families of cryptotoxic species are: Cossidae, Lycaenidae, Notodontidae, Papilionidae, Pieridae.

B. Distribution

The principal genera of this group, whose species possess caterpillars bearing offensive setae with anthropotoxic action, are distributed as follows, by region and according to families:

(*a*). *Neoarctic Region.* Arctiidae: *Euchaetia, Halisidota, Spilosoma.* Eucleidae: *Adoneta, Empretia, Parasa, Phobetron, Sibine, Sisyrosea.* Lasiocampidae: *Lasiocampa, Malacosoma.* Lymantriidae: *Euproctis, Hemerocampa, Porthetria, Stilpnotia.* Megalopygidae: *Lagoa, Megalopyge, Norape.* Noctuidae: *Apatela, Catocala.* Nymphalidae: *Euptoieta, Vanessa.* Saturniidae: *Automeris, Coloradia, Hemileuca, Pseudohazis,* Sphingidae: *Sphinx.*

(*b*). *Neotropic Region.* Eucleidae: *Sibine.* Lasiocampidae: *Tolype.* Megalopygidae: *Megalopyge, Podalia.* Morphoidae: *Morpho.* Saturniidae: *Automeris, Dirphia, Eacles, Rotschildia.*

(*c*). *Paleoarctic Region* (*Europe*). Bombycidae: *Bombyx.* Lasiocampidae: *Tolype.* Lithosiidae: *Lithosia.* Lymantriidae: *Euproctis, Liparis, Orgyia, Porthesia, Stilpnotia.* Notodontidae: *Anaphe, Thaumetopoea.* Nymphalidae: *Vanessa.*

(*d*). *Paleoarctic Region* (*Asia*). Eucleidae: *Parasa.* Lymantriidae: *Euproctis.* Notodontidae: *Thaumetopoea.*

(*e*). *Ethiopic Region.* Eucleidae: *Parasa.* Notodontidae: *Thaumetopoea.* Saturniidae: *Ludia.*

More than 100 species with known poisonous caterpillars are included among the 41 different genera cited.

Genera with Cryptotoxic Species

The principal genera of this group, whose species possess caterpillars without offensive setae and with a characteristically zootoxic function, are distributed as follows, by region and families:

(*a*). *Neoarctic Region.* Cossidae: *Cossus, Prionoxystus.* Lycaenidae: *Lycaena.* Notodontidae: *Cerura, Schizura.* Papilionidae: *Papilio.* Pieridae: *Pieris.*

(*b*). *Neotropic Region.* Pieridae: *Pieris.*

(*c*). *Paleoarctic Region.* Cossidae: *Cossus.* Lycaenidae: *Lycaena.* Notodontidae: *Cerura.* Papilionidae. *Thais.* Pieridae: *Pieris.*

(*d*). *Oriental and Australian Regions.* Notodontidae. *Cerura.*

C. Somatic Morphology

1. General Descriptive Summary of Poisonous Caterpillars

The great majority of poisonous caterpillars belong to the larval state of heterocerous Lepidoptera. The hairy ones are responsible for erucism.

The phytophagous hunger of the larvae stimulates rapid growth, which, because of the unexpandable exoskeleton, forces repeated changes, usually five within the first 2 or 3 months.

The caterpillar larva is composed of a head and 14 segments or somites, forming a soft, cylindrical body.

The head is usually retractable, with a strong bucal masticating apparatus, rudimentary antennae, and punctiform ocelli.

The parts of the body are: a thorax of three somites, each one of which has a pair of true feet, and an abdomen of 11 somites, having several pairs of prolegs, or false feet, which are unsegmented and fleshy.

The integument of the urticating caterpillars possesses tubercles and setiferous papillae located and arranged symmetrically, features that are used to elaborate quetotoxic maps. Among the setae may be found flexible hairs varying between a few centimeters in length and microscopic size; rigid spiniform bristles; and spinules of varied morphology. The poisonous setae in communication with the poisonous secretory glands are tubular or porous, and are filled with poison.

2. Brief Note about Some Tropical Caterpillars of Peru

Among the Peruvian caterpillars that we have studied, there is a special prevalence of species belonging to the family Megalopygidae. We believe that the following are deserving of special attention:

(*a*). *Megalopyge* sp. ("*cuy dorado*" or "golden guinea pig" caterpillar, Satipo, Junín, Peru), which is described in detail in the following paragraphs. Frequency of observation: 28.9%.

(*b*). *Megalopyge* sp. ("*cuy rojizo*" or "reddish guinea pig" caterpillar, Satipo); body, 2.5 × 0.4 cm; including hairs 5 × 1.5 cm. The long, straight hairs are 12 mm long and have a reddish color; the short hairs are about 4 mm long, are more rigid than the long hairs, and are pale pink in color. Frequency 6.2% (Fig. 5).

(*c*) *Megalopyge* sp. ("*utcu bayuca*" or "cottony acorn" caterpillar, Rioja, San Martín, Peru). Body size, 2.5 × 0.5 cm; body size including hairs, 9 × 4 cm. The three types of setae that we are going to describe in connection with the offensive apparatus all have a cream color, while the tegument is a very pale pink. Frequency 16.8% (Fig. 6).

(*d*). *Podalia* spp. ("*sarohmé blanco amarillento*," Satipo, Peru) are

FIG. 5. Caterpillar of *Megalopyge* sp. "cuy rojizo" ("reddish guinea pig" caterpillar from Satipo, Peru). (a) Mature larva. (b) Cocoon of the caterpillar. (c) Chrysalid of the caterpillar.

morphologically similar to the "*cuy dorado*," but have a prismatic triangular shape. Frequency 10.2%.

We have also observed caterpillars belonging to the Saturniidae and Sphingidae families.

3. Description of a Typical Peruvian *Megalopyge* Caterpillar

We will describe a *Megalopyge* sp. known as "*cuy dorado*" in Satipo, Junín, Peru.

Soft, fleshy, eruciform larva, possessing the physical and color charac-

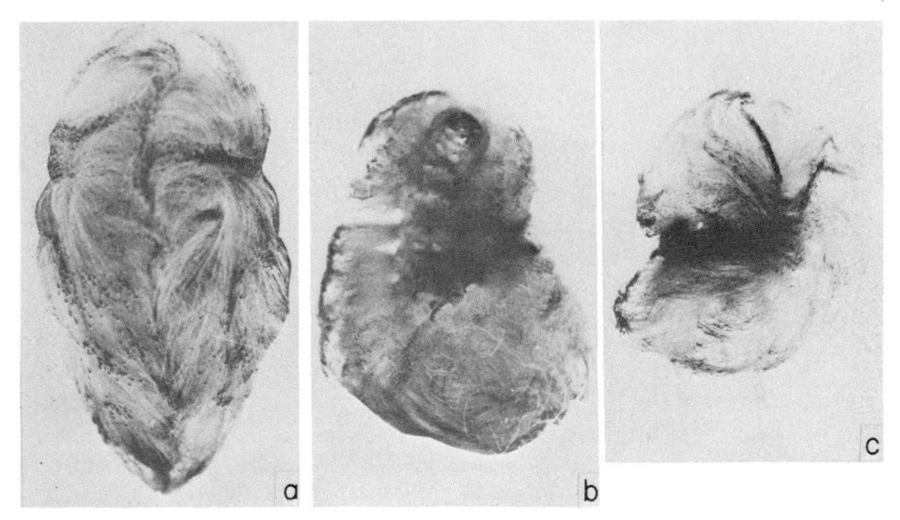

FIG. 6. Caterpillar of *Megalopyge* sp. "*utcu bayuca*" ("cottony acorn" caterpillar from Rioja, Peru). (a) Dorsal view. (b) Ventral view (c) Laterocephalocaudal view.

teristics of that state of development; an initial yellowish tinge gives way to a golden hue on the body itself, while the extremes take on a orange color [Fig. 7 (a) the three upper larvae; Fig. 7 (b)]; as the time of metamorphosis approaches, the larvae lose their long hairs and display only short, black hairs [Fig. 7 (a), the two lower larvae; Fig. 8 (a) and Fig. 8 (b)]. The body attains a length of 75 mm and a diameter of 10 mm.

The head [Fig. 7 (c)] is hemispherical, retractable, and protractable, bald, dark pink in color, and protected by three hoods. The chitinous masticatory apparatus [Fig. 7 (d)] is powerfully built and black in color; the jaws are notched. The phytophagia emit a gnawing sound. They possess rudimentary eyes and antennae.

The trisegmented thorax has an extensive prothorax. Each one of the thoracic segments has a pair of true feet, which are conical and segmented [Fig. 7 (d)]. The tarsae end in a robust chitinous claw, black in color.

The abdomen [Fig. 8 (b)] has eight distinct somites and three smaller ones in the caudal segment. The seven pairs of prolegs are distributed among the segments from 2 through 7, and 10, provided with fleshy suckers and armed with little hooks, except for those of somites 2 and 7. The prolegs are elastic, retractable, mamelonated, fleshy, and adorned with pubescent circular combs.

The ventral body surface is nearly bald, while the dorsal surface is shaggy with setae of varying colors, lengths, and consistencies.

The long, golden-yellow hairs taken together give the caterpillar its characteristic triangular anphioxiform shape. In general the longer hairs

FIG. 7. Caterpillars of *Megalopyge* sp. "*cuy dorado*" ("golden guinea pig" caterpillar from Satipo, Peru). (a) Three mature larvae and two larvae next to pupation. (b) Two mature larvae in their natural habitat. (c) Detail of the cephalic segment. (d) Detail of mouth parts and thoracic legs.

that give the caterpillar its color are neatly arranged vertically and horizontally, as if they had been combed. Next to the cephalic segment, there is an orange-yellow tuft that looks somewhat like a horn. Along the sides, there are eight pairs of projecting tufts of a lighter color, that the caterpillar moves as if they were ambulacral feet. The cephalothoracic pair causes the caterpillar to look as if it had wings.

Below the layer of yellowish hair, one encounters a superabundance of short black hairs.

The urticating spiniform bristles are hidden and dispersed.

Furthermore, there are very small urticating hairs which are diffusely distributed among the other types of hair.

D. Morphology of the Venom Apparatus

It is necessary to distinguish between two very different types of poisonous apparatus: those that operate on the basis of setae, and those that are not so provided.

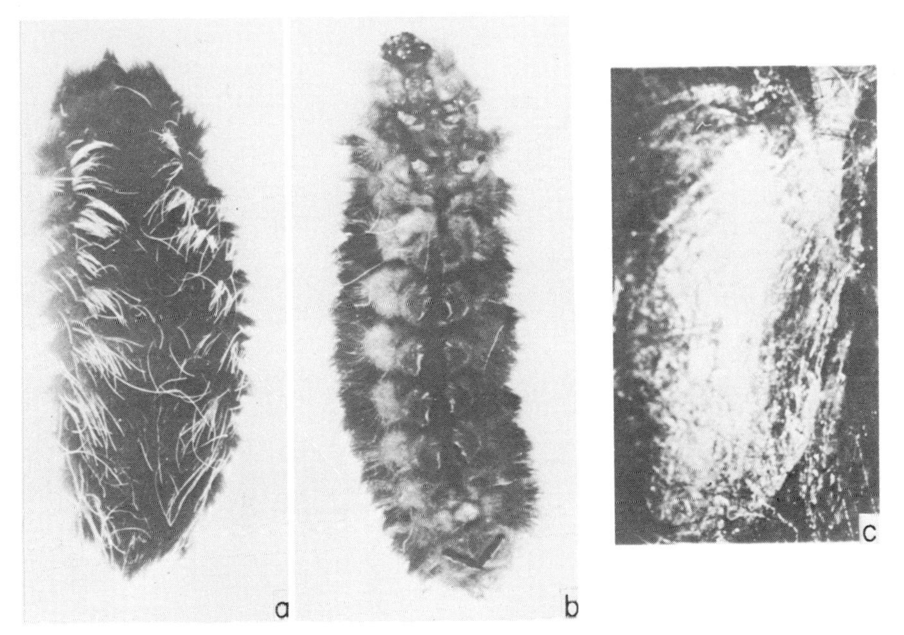

FIG. 8. Caterpillar and cocoon of *Megalopyge* sp. "*cuy dorado*" ("golden guinea pig" caterpillar from Satipo, Peru). (a) Late stage, dorsal view. (b) Late stage, ventral view. (c) Cocoon of the caterpillar.

I. Poisonous Setae

Among the "erucae" or hairy caterpillars, the toxic function is effected by means of special unicellular or multicellular toxic glands, of cuticular, hypodermic, or subhypodermic location, connected with differentiated grooved or hollow setae that play the role of offensive weapons.

We distinguish five principal types, some of which have been observed in Peruvian caterpillars.

(*a*). *Hairs of Gilmer's Primitive Type*. These are large spiculated, more or less flexuose, and brightly colored; several varieties are found on a single caterpillar, ranging in length from 1 to 3 cm, with cylindrical rounded, lanceolated, spatulated, or feathery ends (Fig. 9, 1).

Only in a few species are these connected with hypodermic or subhypodermic, unicellular or multicellular poisonous glands, as for example in the case of some caterpillars of the Lymantriidae, e.g., *Hemerocampa leucostigma* (H. H. Knight, 1922; Gilmer, 1923); some of the Arctiidae, e.g., *Euchaetis egle*; and some of the Noctuidae of the genus *Apatela* (Riley and Johanssen, 1938).

It has not been possible to establish a relationship between the long hairs and the poisonous glands in the majority of the Megalopygidae (Gaminara,

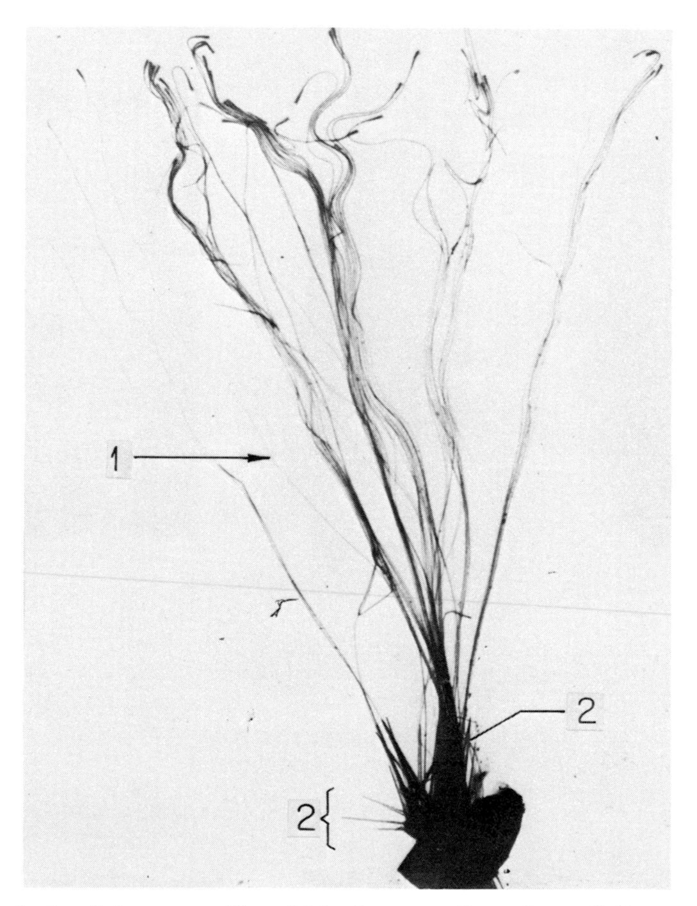

FIG. 9. A tuft from caterpillar of *Megalopyge* sp. "*utcu bayuca*" ("cottony acorn" caterpillar from Rioja, Peru). 1, Long hairs; 2, spiniform bristles.

1928; Jörg, 1935; Estable *et al.*, 1946; da Fonseca, 1949). We have observed that the irritating effects of these hairs are either null or minimal; for which reason we believe that only occasionally do they become contaminated by the poison emitted by the spiniform bristles.

(*b*). *Spiniform Bristles, of the Type Studied by Foot (1922) or the More Frequent Type Studied by Mills (1925)*. These are shorter, more rigid, round, sharply pointed, and with a length of 6 mm; they are connected with the intracuticular gland. In the type studied by Foot, the so-called sac is a multicellular poisonous gland lined by hypodermic cells (Gilmer, 1925); in the type studied by Mills, the intracuticular gland is unicellular, and communicates with the hypodermic by means of a pore canal. In both types, the modification of the seta consists in the fact that it is transformed into a

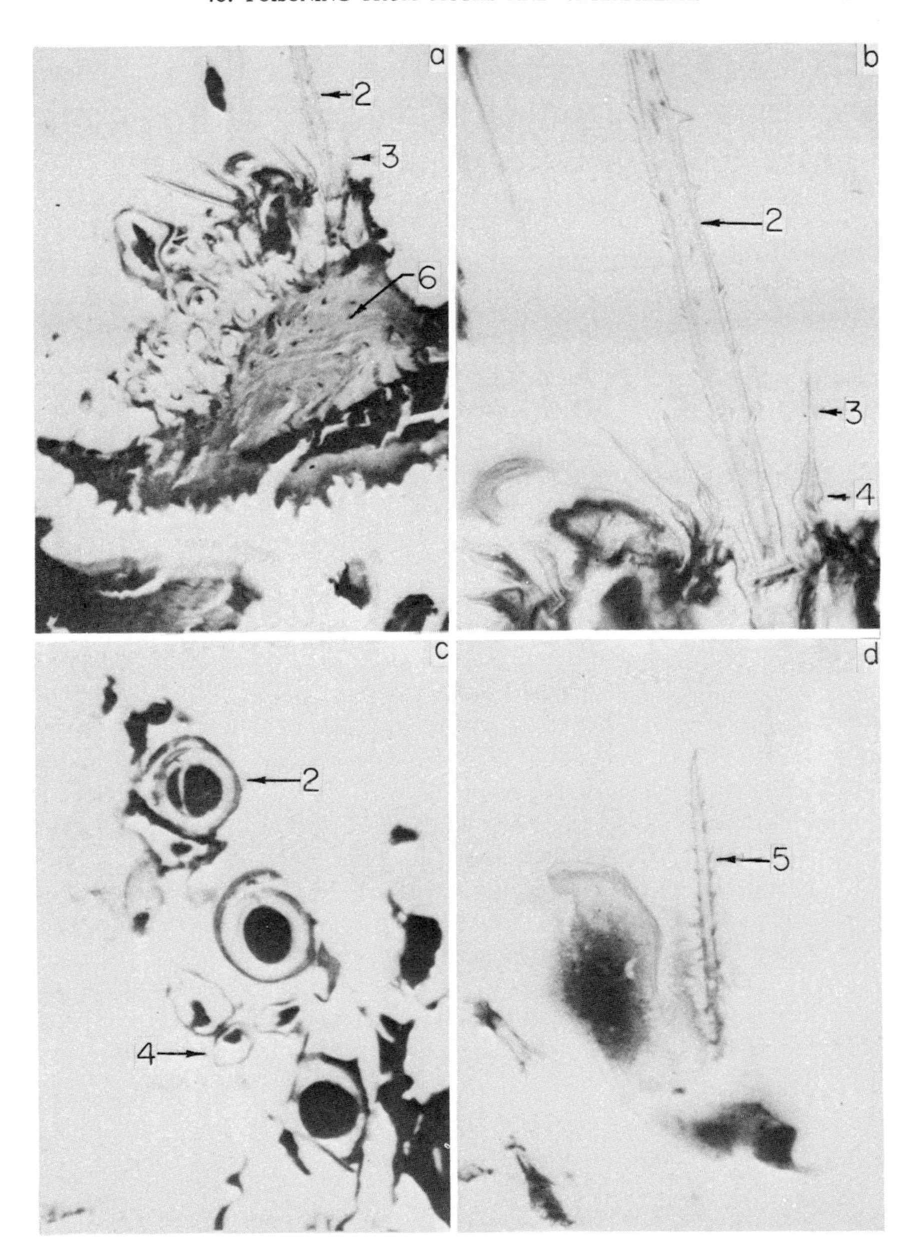

FIG. 10. Structure of venom apparatus of *Megalopyge* sp. "*utcu bayuca*" ("cottony acorn" caterpillar from Rioja, Peru). (a) Frontal cut of a tubercle. (b) The same magnified. (c) Transversal cuts of bristles and of smooth minuscle hairs just passing through the basal sheet. (d) Longitudinal cut of a single spiculated minuscle hair. 2, Spiniform bristle; 3, smooth minuscle hair; 4, basal sheet; 5, spiculate minuscle hair; 6, cuticle.

thick spiniform hollowed bristle [Fig. 10 (b) 2] lined all the way with hypodermic cells.

This is true of the great majority of Megalopygidae, e.g., *M. opercularis* (Foot, 1922; Gilmer, 1925), *M. urens* (Gaminara, 1928; Jörg, 1935), *Megalopyge* spp. (da Fonseca, 1949); in some Eucleidae, e.g., *Sibine stimulea* (Gilmer, 1925), *Parasa hilarata* (Mills, 1925); and one genus of Noctuidae, *Catocala* spp.; and in some Saturniidae, e.g., *Hemileuca* spp. (Caffrey, 1918), *Automeris io* (Gilmer, 1925), *Dirphia* sp. (Valle *et al.* 1954).

(*c*). *Microscopic Hairs, or Fuzz, to Which Little Attention Has Been Paid To in the Literature.* These are very short, rigid, smooth [Fig. 10 (b) 3] or spiculated [Fig. 10 (d) 5], sharp, and about 150 to 300 μ in length; they are connected with the unicellular intracuticular gland. This is true of the Megalopygidae, e.g., in *M. urens* of Uruguay (Gaminara, 1928) and in *Megalopyge* spp. studied in Peru, as we indicate below.

(*d*). *Aleiform Spinulae, or "Flechettes," of the Type Studied by Tyzzer (1907).* These are very short, being only 120 or 200 μ in length, have the shape of a dart, are porous, and are formed by four to six telescoped segments which are distally spiculed; from 3 to 12 are to be found on a single papilla, loosely bound to a secondary intracuticular poisonous cup, connecting the entire cluster by means of a common conduit, originating in the primary poisonous gland located in the hypoderma. This is true of Lymantriidae, e.g., *Euproctis phaerrhoea* (Tyzzer, 1907; Kephart, 1914; Gilmer, 1925). In some of the Notodontidae, e.g., *Thaumetopoea pityocampa,* there are thousands on the epicuticula growing out of every group of papillae or "mirror." These spinulae are 200 μ long, 3 μ wide, with an interior tube 1 μ wide, communicating with the unicellular pyriform hypodermic gland, which is 300 μ long and 40 μ wide (Beille, 1896). It is possible that the spinulae found by Ziprkowsky *et al.* (1959) in *Th. wilkinsoni* are of this type. As many as 630,000 have been counted on a single specimen.

(*e*). *Cardiform Spinules.* These are pyriform and are 300 μ long by 180 μ wide. They have a principal terminal spicule, and short lateral spicules, and their base is loosely connected to the epicuticula. This is true of some Eucleidae, e.g., *Adoneta, Empretia* (Packard, 1898).

2. Poisonous Setae of Caterpillars of Peru

Among the Peruvian caterpillars that we have studied, we have found in one *Megalopyge* sp. (so called "*utcu bayuca*" or "cottony acorn" caterpillar, Rioja, San Martin, Peru), the following setae. Hairs 5 cm long, which are flexible and have feathery ends (Fig. 9, 1), without evident glandular connection. Spiniform bristles, 5 or 6 mm long, rigid [Fig. 9, 2; Fig. 10 (a) 2, (b) 2, (c) 2,] with glandular connection. *Smooth microscopic* hairs, 300 μ long [Fig. 10 (a) 3; (b) 3] with glandular connection. Spiculated microscopic hairs [Fig. 10 (d) 5]. We have observed similar setae in *Megalopyge* sp.

("*cuy dorado*" from Satipo, Peru), and in *Megalopyge* sp. ("*cuy rojizo*" from Satipo, Peru).

3. Other Toxic Apparatuses

In hairless caterpillars, or caterpillars with very sparse hair, we find other apparatuses of active defense deprived of setae, composed of voluminous glands, sometimes eversible, which secrete odorous and repellent toxic substances which they sometimes project under pressure, while at other times they act only through emanations which often prove to be harmful to humans.

(*a*). *Eversible Gland without Emission Tube.* The most usual type is osmeterium. By means of a prothoracic dorsal notch, the caterpillar everts, when irritated, a tubular V-shaped fleshy process which contains secretive glands without an emission tube: from this it sprays an ill-smelling liquid, with a strong acidic reaction. This occurs in *Papilio asteria* (Burnett, 1854, *apud* Phisalix, 1922), *Orgya* (Packard, 1886, *apud* Phisalix, 1922), *Thais* (Packard, 1898), *Papilio cresphontes* (Chu, 1949). *Lycoena* has this organ in the tenth postcephalic segment.

(*b*). *Noneversible Gland with Slow Emission Tube.* Some caterpillars possess a pair of secretory tubes that measure each 9.0 × 4.0 mm with a storage sac of 2.5 × 3.5 mm and an excretional conduit measuring 1.2 mm that opens at the internal border of the jaw; this is the case in *Cossus ligniperda* of Europe, described by Bordas in 1902 (*apud* Phisalix, 1922), containing 0.4g of a neutral reactive liquid, and emanation that, besides producing a burning sensation when it comes into contact with the human skin, kills sparrows within 3 hours after the substance is injected, and a fly after 4 hours of exposure to the vapor. Analogous types are *Cerura* (Packard, 1898) and *Harpya* Oche. (Faust, 1924) with a highly irritating liquid acid. Similar types are *Cossus cossus* from Europe and *Prionoxystus robinae* from North America (Chu, 1949). Other caterpillars that possess a large subhypodermic gland secreting a fluid are *Pieris rapae*, cosmopolita (Matheson, 1950). The larvae of *Pieris brassicae* may be swallowed and survive in the human intestine, a phenomenon designated scoleciasis by Hope in 1837 (*apud* Riley and Johannsen, 1938) who reports six cases with toxic enteritis.

(*c*). *Noneversible Gland with Violent Emission Tube.* These exist in several Notodontidae (Poulton, 1886, *apud* Phisalix, 1922); *Macrurocampa* and *Schizura* are capable of emitting jets for distances of 2 or 3 cm (Packard, 1898).

E. Toxicology

The toxicological observations, both physicochemical and experimental, were carried out with different species, employing a wide range of scientific procedures.

The nature of the present work does not permit us to provide a comparative analysis of all the available literature, and accordingly we shall limit ourselves to a brief summary of results extracted from the following authors: for Lymantriidae: Tyzzer (1907) with *Euproctis phaerrhoea*; for Megalopygidae: Foot (1922) with *M. opercularis*, Gaminara (1928), Estable *et al.* (1946), Mazzela and Patetta (1946) with *M. urens*, Valle *et al.* (1954) with *Megalopyge* sp.; for Saturniidae: Valle *et al.* (1954) with *Dirphia* sp., Goldman *et al.* (1960) with *Automeris io*, Jörg (1964) with *A. coroesus*.

1. Extraction of the Poison

For the extraction of raw, total, natural toxin, the larvae were immersed in a 5-ml physiological solution for 15 minutes, stirring it slowly with a glass rod in order to stimulate secretion.

In order to obtain raw partial extract of toxins, setae of various kinds have been used, or the skin *in toto*, or dessicated and triturated glands, or the body *in toto*. The most powerful extracts have been of the hydric group, which were sometimes dried in vacuum.

In order to obtain pure partial toxin, hydric extract was treated with physicochemical methods having been obtained as a globulin. There appears to be no known test with ultracentrifugation. Some workers have used chromatographic methods.

One may assume that the various extracts obtained have a widely varying objective validity as representative of the actual *in toto* venom being studied.

2. Experimental Effects

Hemolysis was obtained *in vitro* with human and animal blood. In small mammals a constant inflammatory syndrome, sometimes with necrosis, neurotoxic action, varying arterial pressure, and histamine effects was observed; in dogs, an immunizing action was found that reached a tolerance to five times the lethal dose.

3. Crude Toxin: Characteristics and Properties

Characteristics of the crude toxin are an acid reaction, thermic lability between 55° and 60°C, atmospheric lability of the hydric extracts, good conservation of the dry extracts in cold, chemical lability with medium alkaline solutions.

4. Pure Toxin: Nature and Composition

Because the literature cited above does not record the successful isolation of pure toxin, assertions as to its nature and composition must rest on our knowledge of certain chemical reactions and on observed characteristic toxic effects.

(*a*). *Protein Toxin.* Gaminara (1928) studied *Megalopyge urens* and obtained typically albuminoid reactions, which are in agreement with experimental effects. Estable *et al.* (1945), working with *M. urens* found globulin present in raw toxin and were partially successful in isolating it. Valle *et al.* (1954) inferred the protein nature, close to that of adenyl, of an important toxic fraction of *Megalopyge* sp. by experimental effects; its content in this species may be higher than that found in the toxin of the Saturniidae studied.

(*b*). *Histamine, Acetylcholine, Serotonin.* Gaminara (1928) in his study of *M. urens* indicates experimental effects of a histamine type. Valle *et al.* (1954) have made an especially important contribution by demonstrating a histamine content varying between 0.02% and 0.04% in the setae of *Dirphia* sp.; the absence of acetylcholine in *Dirphia* sp. and *Megalopyge* sp. has also been proved. Goldman *et al.* (1960) did not find evidence of histamine, serotonin, or 5-hydroxyindoleacetic acid in the aqueous extract of *Automeris io*. Jörg (1964) found histamine, acetylcholine and plasmokinin in *A. coroesus.*

(*c*). *Other Bodies.* Goldman *et al.* (1960) found a chromatographic fraction of the proteolytic enzyme type in the aqueous extract of *Automeris io*. Jörg (1964) found a nonprotein factor containing peptide, dolorigen, and neuritogen in a chromatographic analysis of the toxin of *A. coroesus.*

In our study of the clinical symptomatology of Peruvian Megalopygidae and Saturnidae, we have found that effects of the histamine type are dominant while those of the proteic type are less important. We believe, nevertheless, that it is convenient to bear in mind the possible action, as in the case of ophidism, of some toxins consisting of prosthetic groups, capable of acting even in quantities of very small weight.

IX. ECOLOGY

A. Synecology

From the biological point of view, synecology is the study, in this case, of the correlation between the caterpillars' habitat and way of life and the habits of human beings living in the same region. From the medical point of view, synecology is the epidemiology of erucism.

1. Relation to Caterpillars

In Paleoarctic and Neoarctic regions the life cycle of Lepidoptera is usually annual; the season that offers the most favorable climatic conditions for caterpillars, the end of spring and the beginning of summer, is short. For this reason, oviposition is great, and the hordes of young caterpillars are forced to invade the surrounding plant life *en masse*; and, since most of the plants have only seasonal foliage, they are forced to feed intensively during the time available to them. During their final stages, they lose great quantities of setae that, owing to the frequent winds and to the spaces free of forests, are freely and broadly shed. In addition to this circumstance, we must not forget the fact that a high percentage of the soil in highly developed countries is used for agriculture and the population is relatively dense. This combination of circumstances is responsible for the fact that the wind-blown setae enter into frequent contact with human beings, thus causing collective epidemics of indirect erucism. This is true principally of the Arctiidae, Lymantriidae, Notodontidae, and some Megalopygidae.

In the Neotropic, Ethiopic, and Oriental regions, the life cycle of the Lepidoptera is often semiannual and sometimes trimestrial, as determined by the periods of drought between the primary and secondary rainy seasons. Their preferred habitat is the dense jungle, with few winds, and a very low population density. Any loss of setae is absorbed by the vegetation and they do not, therefore, come into contact with the small human population. Collective epidemics are rare; direct erucism is responsible for almost all the cases, and displays the symptoms we shall describe in the following paragraphs.

The species chiefly responsible belong to the Megalopygidae and the Saturniidae.

In the Peruvian jungle, the first series of 433 cases studied (1958–1962) in seven provinces in three departments of the Republic were distributed throughout the year with a monthly proportion varying between 4.4 and 9.4%, except for the trimester from March to May, following the principal rainy season, when it went up to 14.5%.

2. Relation to Human Beings

Collective accidents of indirect erucism, characteristic of Europe, Korea, Japan, and part of the United States, occur preferentially in sizable human populations; thus, for example, in the Bois de Boulogne in Paris, in the seaside resorts of the Baltic Sea, in schools and colleges in Texas, and in many important agricultural centers.

Cases of individual direct erucism, although they also exist in small numbers and during certain months in some of the countries mentioned above (Eucleidae), are most common in the forest villages of Neotropic America, where great numbers of cases are to be observed throughout the year. The hinterland of such villages is a beachhead carved from the virgin forest, in which the production of food crops continues throughout the year. Of the cases observed in the first series of 433 cases studied (1958–1962), 65.9% took place while the person was engaged in hoeing and harvesting, 15.7% in lumbering, 11.3% while carrying loads, and only 7.2% in no agricultural tasks. Of the plants that harbored the caterpillar, 46.6% were food plants, 43.1% were commercial crops, mainly coffee and bananas, while all other plants taken accounted for only 10.3% of the cases reported. The direct accident, in regions like this, is an inevitable by-product of individual agricultural activity; even though the caterpillars are greatly feared, the people cannot afford to refrain on that account from tending their crops. Such poisoning must be considered an occupational disease of the regions in question: 42.0% of the people studied suffered from 2 to 5 accidents and 19.9% had from 6 to 15.

B. Etiopathogenesis

In order to gain a full understanding of etiopathogenesis with an ecological basis, it is necessary to devise some means of classifying the genuine and the apparent causal agents, the various vehicles of venom, the various means by which it is transmitted to human beings and other animals, and the various exogenous expressions of the venom function, determining the correlation of these factors among themselves, and adopting an appropriate system of consistent nomenclature. The classification that we describe below would appear to be reasonable when judged by the previously considered criteria.

1. True "Direct" Erucism

True direct erucism is produced by hairy caterpillars, "*erucae sensu stricto*," which have poisonous setae, through direct contact with live caterpillars. This means that, although the overall number of cases in a given region may be impressive, all of them are of the individual type. They are caused by the active defense of the caterpillar when it has been irritated.

(*a*). *Contact with the Tegument.* This is the most common means of human accident among agricultural workers, where they come into direct contact with the caterpillar-infested plant. This occurs, e.g., in Brazil, Paraguay, Uruguay, and Argentina. The most common families involved are the Saturniidae, Megalopygidae, and Eucleidae. Toxic dermatitis and general envenomation have been found among 621 cases observed in Peru.

(*b*). *Ingestion.* This is the type of erucism produced exclusively by the live caterpillar, and occurs only among animals, as for example with the cases of severe enteritis found in European ducks which have ingested caterpillars (Zurn, *apud* Phisalix, 1922), the fatal enteritis of South African pigs which have ingested *Nudanriella cytherea* caterpillars (Neveu-Lemaire, 1938), and the death of cattle in the Bavenda region of Africa, as a result of having consumed grasses infested by the "Khohe" caterpillars (Phisalix, 1922).

2. True "Indirect" Erucism

Indirect erucism is also produced by hairy caterpillars, but through indirect contact with the shed poisonous setae. This means that the accidents are usually of the indiscriminate collective type, because they result from the casual abrasive action of the larva during the months in which they are most avidly in search of food.

(*a*) *Contact with the Tegument.* This is the form of the disease known in the Paleo- and Neoarctic regions since ancient times, and is usually caused by the Arctiidae, Lymantriidae, and Notodontidae. Ziprkowsky *et al.* (1959) have recently called attention to 600 cases of collective envenomation in Israel. We have observed several cases originating near an infested tree in Chanchamayo, department of Junín, Peru (Pesce *et al.*, 1957). Toxic dermitis is the chief symptom; considerable lesions in the nasopharyngeal mucus, nudous opthalmia, and general envenomation are infrequent. Some cases of severe allergy are cited in the literature.

(*b*). *Ingestion.* Where loose or wind-blown setae are concerned, man as well as animals may fall victim to this type of accident. Artault (1901, *apud* Phisalix, 1922) has described many cases of erucic stomatitis caused by the ingestion of setae of the *Liparis chrysorrhoea* on contaminated fruit. Dogs ingesting grasses infested with setae of *Thaumetopoea pinivora* have also been known to suffer from severe stomatitis (Mengin, 1855, *apud* Phisalix, 1922).

3. Para-Erucism

Those poisoning accidents caused by hairless caterpillars through a toxic secretion of specialized glands, and without the use of setae are called para-erucism; they are the result of the caterpillars' active defense and occur mainly in animals and only occasionally in man. The direct form results from contact with the highly irritating acid and caustic secretions: this is the case of the *Harpya* and some species of Notodontidae, as, for example, *Cerura* and *Schizura*. The indirect form is more frequent and results from volatile secretions acting at a distance, not only by repellance, as in the case of some Papilionidae, but also by means of intensely poisonous emanations which are capable of killing flies and causing a skin sensation similar to that of contact with fire, as is the case of *Cossus ligniperda* (Bordas, 1902, *apud* Phisalix, 1922).

4. Meta-Erucism

Meta-erucism is a form of indirect erucism resulting from passive defense, caused by larval setae originating in the cocoon, and possessing toxic properties often retained for long periods of time, as is true of *Euproctis phaerrhoea* (Tyzzer, 1907) and *Hemerocampa leucostigma* (H. H. Knight, 1922). What is more, the imago contaminated by the same kind of setae as it leaves the cocoon, and retains them even in adulthood; this is true of *Euproctis phaerrhoea* (Riley and Joannsen, 1938) and probably of *E. flava* (Mills, 1923; Illingworth, 1926; Morisita, 1955). In these latter cases, the meta-erucism, which is nothing more than a form of "consigned erucism," takes on a pseudo- form of lepidopterism, caused by the hereditary larval hairs. Dermatitis is known to have resulted from this circumstance.

5. Pseudo-Erucism

Pseudo-erucism represents a form of passive defense exactly opposite to meta-erucism. The toxic setae are of maternal origin; they are deposited as a covering on the eggs, and confer toxic properties on the nests thus constructed; this problem has been studied by Schade (1927) working with *Megalopyge* sp., and by Jörg (1935) working with *Hylesia nigricans* and *H. fulviventris*. What is more, the larva can pick up and retain these setae among their ornaments; (Boyé 1932) was able to produce dermatitis from similar larva. Pseudo-erucism is, as we have already noted, really a form of meta-lepidopterism or "consigned" lepidopterism.

X. SYMPTOMATOLOGY OF ERUCISM

A. Histopathology

The principal histopathological changes produced by caterpillars in human beings and animals, would appear to be the following.

In the skin, congestion, basal infiltration and edema: with Lasiocampidae (Jörg, 1935), with Megalopygidae (Foot, 1922; Estable *et al.*, 1945; Mazzella and Patetta, 1946), with Notodontidae (Ziprowsky *et al.*, 1959), with Saturniidae (Jones and Miller, 1959; Goldman *et al.*, 1960).

In the dermis, infiltration of eosinophilic cells: with Megalopygidae (Foot, 1922; Piaggio Blanco and Paseiro, 1946; Mazzella and Patetta, 1946), with Saturniidae (Goldman *et al.*, 1960).

In the dermis, hemorrhagia: with Megalopygidae (Estable *et al.*, 1935; Mazzella and Patetta, 1946).

In the epidermis, vesicles and necrosis: with Lasiocampidae (Jörg, 1935), with Megalopygidae (Foot, 1922; Piaggio Blanco and Paseiro, 1946).

In blood, hemolysis *in vitro*: with Lymantriidae (Tyzzer, 1907). In the ocular conjuctiva, granuloma of the foreign-body type: with Arctiidae (Schweinitz and Shumway, 1904).

1. Principal Features

The most common symptom is an acute predominantly vascular inflammation, frequently of the urticary type, and sometimes with vesicles.

2. Principal Families

The most important histopathological involvement was noted in the case of Megalopygidae, Saturniidae, and Lasiocampidae.

B. Circumscribed Toxic Syndromes

Toxic syndromes embraced the local and regional symptomatology.

1. Observations of Various Authors

(*a*). *Simple Inflammatory Dermitis.* This condition is common to the great majority of cases; the erythema is constant, of sudden onset and usually disappears within 24 hours.

(*b*). *Urticarian Dermitis.* The edematous component is usually associated with erythema, and consists of isolated hives and local or regional edema, almost always accompanied by itching. In most cases it lasts for 24 hours, while in others it takes several days.

(*c*). *Vesicles and Erosions.* These skin symptoms are less common and are followed by small superficial necrosis. Cheverton (1936) found them in the ocular conjunctiva.

(*d*). *Nodules.* Schweinitz and Shumway (1904) observed their presence in the cornea and episclera.

(*e*). *Regional Urent Pain.* This is a common symptom caused by certain families of caterpillars. The presence of acute neuritis in some patients has been noted by Jörg (1935); arthralgia sometimes occurs, as it is observed in Mexico (Garza de los Santos, 1958).

2. Our Observations (cf. Pesce and Delgad o, 1963)

The first series was carried out in 14 districts of the Peruvian jungle between 1958 and 1962, and involved 433 cases of individual accidents; the following symptomatological observations are extracted from our original work.

It is important to note the existence of dermic syndromes of the histaminic or histaminoid type (Fig. 11), involving erythema (97.5%), edema (75.5%), papule and hives (37.9%), and petechiae (28.6%); while itching occurs in few cases (10.4%), it is usually masked by diffuse urent pains (71.8%).

Other inflammatory symptoms of toxic origin that are also of importance are the following: ganglionic infarct (31.2%), ganglionic pains (28.4%).

Neural involvement is common: urent pains (71.8%) are almost always superficial and diffuse, sometimes accompanied by neuralgia or by articular pain (3.7%).

Superficial necrotic effects: phlyctenae (13.3%).

Sequelae are almost always present: residual hyperchromia (92.6%); and

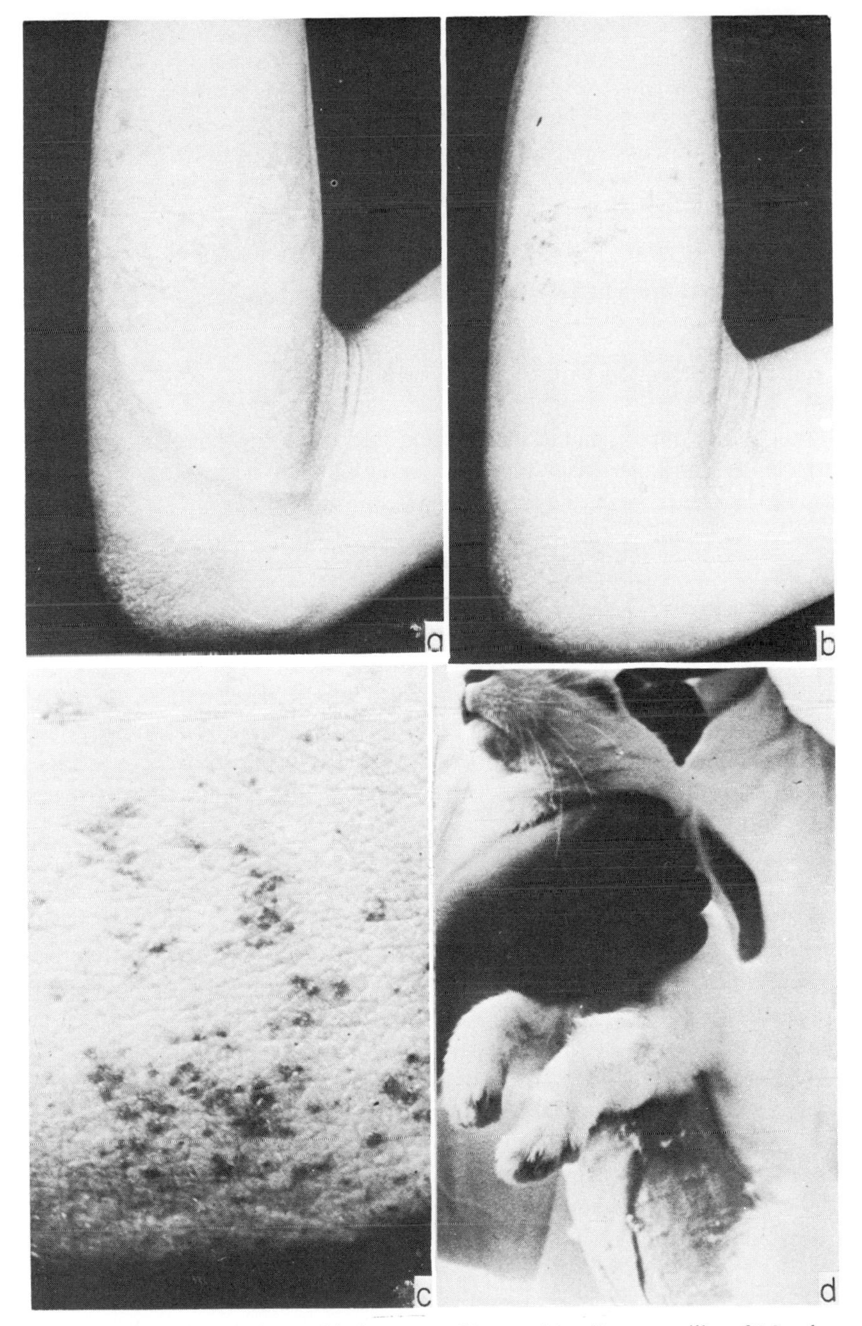

FIG. 11. Dermic experimental lesions caused by touching live caterpillar of *Megalopyge* sp. "*cuy dorado*" ("golden guinea pig" caterpillar from Satipo, Peru). (a) Hives at 7 minutes. (b) Hemorrhagic papules at 48 hours. (c) Detail of the papules. (d) In a rabbit, edematous giant papule at 3 hours.

others which occur only rarely: cicatricial traces (2.5%), which are left by some phlyctenae.

All this symptomatology is primarily due to Megalopygidae (61.1%) and Saturniidae (13.6%).

C. General Toxic Syndromes

1. Summary of Previous Research

A generalized eruption, often of the urticariform type, is worthy of note, most generally associated with the Notodontidae and only rarely with other families.

Numbness, cramps, headaches, nausea, and vomiting have been noted, although infrequently in cases brought on by contact with Megalopygidae (Lucas, 1942). Other general symptoms are considered rare.

2. Our Observations (cf. Pesce and Delgado, 1963)

Using the first series of 433 cases as the basis of our conclusions, we have developed the following systematic symptomatology (which is also extracted from our original report).

There is a significant incidence of general symptoms: malaise (26.8%), chills (14.3%), sensation of high temperature (29.1%) although this last figure may be subject to revision, for it could be partially due to a false interpretation and to an extension of the regional urent pain that at times becomes diffuse.

There are general toxic symptoms which are compatible with those of the histamine type: nausea (9.7%), vomiting (3.9%), diarrhea (2.5%), urticarial eruption (1.6%).

Some circulatory symptoms are attributable to the same cause: they are cerebral edema (20.1%) and anguish (7.4%).

Nervous symptoms are: confusion (2.8%), dromophilia (1.8%). In cases involving extreme pain, lethargy sometimes develops within 8 hours and lasts for about 10 hours thereafter.

Collapse occurs in its various stages: adynamia (26.4%), exhaustion (5.5%), and lypothymia (2.1%).

The systemic occurrences described are more frequent than one would expect on the basis of the available literature, especially in the case of the Megalopygidae.

D. Allergic Syndrome

We feel that it is impossible to individualize objectively allergic syndromes with their own exclusive symptomatology. The allergic nature of the symptoms can be ascertained only when there is evidence of previous sensitization, and when this fact is reinforced by subsequent aggravation.

Analyzing the clinical data of the 268 cases within our first group studied, which experienced more than one attack, we were able to determine that the most recent attack was of the same order of intensity as the earlier attacks in 82.5% of the same cases.

In 10.4% of these cases there was marked attenuation, while the symptoms became progressively more serious in only 7.1%. Nevertheless, the importance of both facts is somewhat diminished if we keep in mind the fact that one-third of the patients attribute the variation in intensity of the accidents to the fact that each attack was caused by a different species of caterpillar, and one-fifth of them felt that the variation was due to the fact that the intensity of each contact was different. Among those cases in which each new attack was of a different intensity than the previous ones, those who seem to have acquired a certain amount of resistance or immunity to erucism far outnumbered those who appear to have been suffering from an allergic sensitization.

These facts take on a new light when we remember that this group of 268 individuals suffered a total of 1,351 accidents, a number sufficient to permit the opportunity for allergic phenomena to become evident if they were of even median frequency.

We may add that the second series of 188 cases studied (1964) did not provide any evidence of allergic sensitization whatsoever.

It is necessary to bear in mind, furthermore, the possible existence of rare nonspecific general reactors. The so-called arthropod-sensitive patients so frequently reported in the United States may, in reality, be sensitive only to one group of insects, as, for example the Hymenoptera (Müller, 1956) and would show negative reactions if tested with caterpillars (Goldman *et al.*, 1960). Finally, there are some patients who display sensitivity to the toxic effects of only a given species of caterpillar. Schmitz (1917) has reported cases of severe hemorrhagic nephritis in Germany that coincided with an outbreak of erucic urticaria; this instance is in many ways reminiscent of the well-known universal capillaritis that produced fatal consequences as a result of bee stings.

It would appear that neotropical caterpillars afford few cases of allergic sensitization, probably, as Jörg has suggested (1964), because their toxin does not contain protein or other substance capable of acting as haptens.

XI. PREVENTION AND TREATMENT OF ERUCISM

A. Prevention

1. Sanitary Measures

An attack on the larvae in their habitat is the most radical measure, taking into account the plants especially preferred by the toxic species and the season favorable to their proliferation. If the invasion is of sufficient dimensions, the expense of spraying a solution of arsenate of lead and lime, or other substances that have also fast killing action, is justifiable.

Putting infested parks "off limits" can also be an effective measure, as we see in the Bois de Boulogne case in 1866 (A. Raillet *apud* Neveu-Lemaire, 1938). Bishopp (1923) reports that a caterpillar invasion forced Texas authorities to close the schools.

2. Individual Measures

In cases of indirect erucism caused by wind-blown setae, it is advisable to wash the floors and walls repeatedly with soap and water, using, for example, a 5% solution of saponate of cresol. Potter (*apud* Jörg, 1935) advises that underclothes and sheets be boiled and washed frequently. Protective goggles are also useful.

In cases of direct erucism, involving commercial agriculture, well-protected crews of exterminaters equipped with miniature flame-throwers can scour the field before the regular farmhands begin their workday.

3. Humoral Changes

In those persons subject to sensitization reactions to caterpillar toxin, progressive immunity may be achieved by slow prolonged treatment with very small doses of the specific antigen.

B. Treatment

1. External

External treatment consists of alkaline compresses, ammonia water, bicarbonate of soda; lotion of lime water containing 7% zinc oxide and 1.5% phenic acid; creams containing antihistaminics and novocaine. In cases involving diffuse intense urent pains, we obtained excellent results with the use of hot baths and hot packs.

2. Internal

In acute cases we obtained rapid correction by the use of orally administered antihistaminics; 10% calcium gluconate given intravenously also proved beneficial. Orally administered corticosterones are probably the best treatment for symptoms of more than 48 hours' duration. Hydration in conjunction with diuretics and cardiotonics, and venoclysis employing a physiological solution of periston-type polyvinyl antitoxic compounds are indicated for cases with severe general involvement. Specific antitoxic sera would, of course, be an ideal means of treating erucic accidents.

REFERENCES

Allard, H. F., and Allard, H. A. (1958). *J. Wash. Acad. Sci.* **48**, No. 1, 18–21.

Baliña, P. L. (1915). *Prensa Med. Arg.*

Berg, C. (1900). *Comun. Museo Nacl. Buenos Aires* [2] **6**, 206–208.

Boso, J. M. (1816). *In* "Viaje a las Montanas de Yucarhes," Chapter IV. Valdizan & Maldonado, *In* "La Medicina Popular Peruana," Vol. III, pp. 348–388, *vide* pp. 369–372. Lima, Peru, 1922.

Bourquin, F. (1936). *Rev. Soc. Entomol. Arg.* **8**, 125–132.

Bourquin, F. (1939). *Physis.* (*Paris*) **17**, 431–441.

Bourquin, F. (1941). *Rev. Soc. Entomol. Arg.* **11**, 22–30.

Bourquin, F. (1942). *Rev. Soc. Entomol. Arg.* **11**, 305–316.

Boyé, R. (1932). *Bull. Soc. Pathol. Exotique* **25**, No. 10, 1099–1107.

Breyer, A., and Orfila, R. N. (1945). *Rev. Soc. Entomol. Arg.* **12**, 299–304.

Brues, C. T., Melander, A. L., and Carpenter, F. (1954). *Bull. Museum Comp. Zool. Harvard* **108**, 226–304.

Cheverton, R. L. (1936). *Trans. Roy. Soc. Trop. Med. Hyg.* **29**, 555–557.

Chu, H. F. (1949). *In* "How to Know the Inmature Insects," pp. 149–189. W. C. Brown, Dubuque, Iowa.

da Costa Lima, A. (1945). *In* "Insetos do Brasil," Vol. V, pp. 164–180. Dept. Impr. Nacl., Rio de Janeiro.

da Costa Lima, A. (1950). *In* "Insetos do Brasil," Vol. VI, pp. 260–274. Dept. Impr. Nacl. Rio de Janeiro.

da Fonseca, F. (1949). "Animais Peconhentos," pp. 284–292. Inst. Butantan, São Paulo, Brazil.

Dallas, E. D. (1926). *Rev. Soc. Entomol. Arg.* **2**, 63–64.

Dallas, E. D. (1937a). *Semana Med.* (*Buenos Aires*) **12**, 760.

Dallas, E. D. (1937b). *Semana Med.* (*Buenos Aires*) **14**, 896.

Dallas, E. D. (1928). *4th Reunion Soc. Arg. Patol. Reg. Norte* **2**, 691–694.

Dallas, E. D. (1933). *8th Reunion Soc. Arg. Patol. Reg. Norte* **2**, 469–474.

da Matta, A. (1922). *Amazonas Med.* **4**, Nos. 13–16, 167–170.

Estable, C., Ferreira-Berruti, P., and Ardao, M. J. (1946). *Arch. Soc. Biol. Montevideo* **12**, No. 3, 186–198.

Faust, E. A. (1924). *In* "Tratado de Medicina Interna" (P. Mohr and J. Staehelin, eds.), pp. 328–330. Madrid.

Floch, H. (1952). *Arch. Inst. Pasteur Guy. Inini., Publ.* **262**, 297.

Floch, H. (1954). *Arch. Inst. Pasteur Guy Inini., Publ.* **326**, 105.

Floch, H., and Abonnenc, E. (1944). *Inst. Pasteur Guy., Publ.*, 89.

Floch, H., and Constant, Y. (1950). *Inst. Pasteur Guy. Inini, Publ.* **220**.

Floch, H., and Constant, Y. (1954). *Bol. Entomol. Venezolana* **9**, 9–12.

Gaminara, A. (1926). *4th Conf. Sudam. Hig. Microbiol. y Patol., Buenos Aires*, **19**, 26.

Gaminara, A. (1928a). *Arch. Trab. 3rd Congr. Nacl. Med. Arg.* **7**, 968–975.

Gaminara, A. (1928b). *Bull. Soc. Pathol. Exotique* **21**, No. 8, 656–662.

Garza de los Santos, A. (1958). *Medicina* (*Mex.*) **38**, 121.

Gilmer, P. M. (1925). *Ann. Entomol. Soc. Am.* **18**, 203–239.

Goeldi, E. A. (1913). "Die sanitärische-pathologische Bedeutung der Insekten und verwandten Gliedertiere." Friedländer & Sohn, Berlin.

Goldman, L., Sayer, F., Levine, A., Goldman, J., Goldman, S., and Spinanger, J. (1960). *J. Invest. Dermatol.* **34**, 67–79.

Gusmão, H. H., Forattini, O. P., and Rotberg, A. (1961). *Rev. Inst. Med. Trop. Sao Paulo* **3**, No. 3, 114–120.
Hase, A. (1926). *Z. Angewandten. Entomol.* 244–297.
Higuchi, K., and Urabe, H. (1959). *Hautarzt* **10**, 79–87.
Hope, H. (1837). *Trans. Roy. Entomol. Soc. London* **2**, 256–271.
Houssay, B. A. (1927). *Semana Med. (Buenos Aires)* **13**, 832.
Illingworth, J. F. (1926). *Proc. Hawaiian Entomol. Soc.* **6**, 267–270.
Jones, D. L., and Miller, J. H. (1959). *A.M.A. Arch. Dermatol.* **79**, 81–85.
Jörg, M. E. (1933). *8th Reunion Soc. Arg. Patol. Reg. Norte* **2**, 482–495.
Jörg, M. E. (1935). *9th Reunion Soc. Arg. Patol. Reg. Norte* **3**, 1617–1635.
Jörg, M. E. (1963). Personal communication.
Jörg, M. E. (1964). Personal communication.
Katzenellenbogen, I. (1955). *Dermatologica* **111**, 99.
Knight, E. (1908). *Hospital* [N.S.] **3**, No. 76, 545.
Koehler, P. (1931). *Rev. Soc. Entomol. Arg.* **6**, No. 3, 305–308.
Koehler, P. (1935). *Rev. Soc. Entomol. Arg.* **7**, 79–91.
Larson, O. A. (1927). *J. Econ. Entomol.* **20**, No. 4, 647.
Lucas, T. A. (1942). *J. Am. Med. Assoc.* **119**, 877–880.
Mazza, S. and Frias, D. (1926). *2nd Reunion Soc. Arg. Patol. Reg. Norte* 293–295.
Mazzella, H., and Patetta, M. A. (1946). *Arch. Soc. Biol. Montevideo* **13**, 131–136.
Morisita, T., *et al.* (1955). *Acta Scholae Med. in Gifu.* **2**, 347–354 and 471.
Neveu-Lemaire, M. (1938). *In* "Traité d'entomologie médicale et vétérinaire," pp. 776–782, Vigot, Paris.
Packard, A. S. (1898). "A Textbook of Entomology," pp. 187–201 and 375–396. Macmillan, New York.
Pawlowsky, E. N., and Stein, A. K. (1927). *Z. Morphol. Oekol. Tiere* **9**, 616–637.
Pesce, H., and Delgado, A. (1963). *Acta 8th Congr. Intern. Med. Trop., Rio de Janeiro,* **19**, 133.
Phisalix, M. (1922). *In* "Animaux venimeux et venins," Vol. I, pp. 343–356. Masson, Paris.
Piaggio Blanco, R., and Paseiro, P. (1946). *Arch. Uruguayos Med.* **29**, No. 1.
Reyes, H. (1963). Personal communication, Santiago de Chile.
Ribeiro, B. L. (1948). *Rev. Brasil. Biol.* **8**, 127–141.
Riley, W. A., and Johannsen, O. A. (1938). *In* "Medical Entomology," pp. 173–188. McGraw-Hill, New York.
Schade, F. (1927). *Entomol. Rundschau* **44**, No. 1, 4; No. 2, 7; No. 3, 12.
Tisseuil, J. (1935). *Bull. Soc. Pathol. Exotique* **28**, No. 8, 719–721.
Tonkes, P. R. (1933). *Bull. Biol. France Belg.* **67**, 44–99.
Valetta, G., and Huidobro, H. (1954). *Compt. Rend. Soc. Biol.* **148**, 1605.
Valle, J. R., Picarelli, Z. P., and Prado, J. L. (1954). *Arch. Intern. Pharmacodyn.* **98**, No. 3, 324–334.
Vellard, J. (1947). *In* "Los Animales Venenosos de Sud América," pp. 65–72. Tesis Bach. Cienc., Lima.
von Ihering, R. (1914). *Ann. Paulistas Med. Cir.* **3**, No. 6, 129–189.
Ziprkowsky, L., Hofshi, E., and Tahori, A. S. (1959). *Israel Med. J.* **18**, 26–31.
The Quotation "in extenso" will be found in: Schottler, W. H. A. (1954), *Mem. Inst. Butantan* **26**, 7–73.

Chapter 49

Lepidopterism in Brazil

A. ROTBERG

ESCOLA PAULISTA DE MEDICINA, SÃO PAULO, BRAZIL

I. History 157
II. Species 158
A. Caterpillars 158
B. Moths 159
III. Clinical Symptoms 159
A. Caterpillars 160
B. Moths 162
IV. Pathology 167
V. Treatment 167
References 167

I. HISTORY

Skin reactions to exposure to butterfly and moth larvae in Brazil stand among the oldest descriptions of human disease observed in the New World. In a "Letter from S. Vicente," Anchieta (1560), the famous priest and educator who accompanied the first colonizers, refers to the Indians' fear of the pain that followed skin contact with certain caterpillars; this fear, however, did not stop them from using the larvae for libidinous purposes, placing them on their genitalia and provoking thereby an intense and acute swelling which was often followed by necrosis and ulceration of the prepuce. "Tatáraná" (like fire) was the Tupy-Guarany name given to these larvae; this is the derivation of the present "tatorana" which was extended later to other caterpillars.

Marcgrave and Piso, the scientific advisors to the Prince of Nassau who governed Brazil under Dutch rule, stated in the 1648 and 1658 editions of the "Historia Naturalis Braziliae" (*apud* von Ihering, 1914) that *si vermis hic cutem humanam attingat, urit instar ignis* and *cutique deglubendae et vesicis excitandis* (*vim*) *aptam.*

II. SPECIES

Lepidopterism in Brazil results mainly from contact with larvae of nocturnal moths (5 families) and diurnal butterflies (1 family). Adult moths of a genus of one of these families are also known to cause dermatitis.

These Lepidoptera have been thoroughly studied by von Ihering (1914) and Almeida (1944), and they are mentioned in zoological textbooks by Monte (1934), da Fonseca (1949), and Lima (1945, 1950).

A. Caterpillars

1. Megalopygidae (tatorana properly called, sucuarana; lagarta de fogo = fire caterpillar; lagarta-cabeluda = hairy caterpillar).

These are the caterpillars that are most feared, particularly those of the *Megalopyge lanata* species, whose long-haired and white-spotted dark larvae are highly irritating. Other noxious caterpillars of this family include *Megalopyge superba*, *M. radiata*, *M. undulata*, *Podalia radiata*, *P. chryscc oma*, *P. albescens*, *P. orsilochus* and others (*Aidos*, *Trosia*).

Megalopygidae live on various, usually cultivated, plants. They occur most often on guava, avocado, nut, orange and cashew trees and on rose and coffee bushes.

2. Hemileucidae

The yellow and green caterpillars of this family have hair grouped in isolated tufts, which resemble pine needles. They are very noxious, particularly the various species of the genera *Automeris* and *Dirphia*. The most common varieties are *A. melanops*, *A. illustris*, *A. aurantiaca, A. viridescens*, *A. acuminata*, *A. incisa*, *A. complicata*, *D. sabina*, and *D. multicolor*.

The pharmacological properties of extracts from the urticating setae of *Dirphia* and *Megalopyge* have been thoroughly studied in Brazil by Valle *et al.* (1954). Avocado, tamarind, *Platanus*, *Ficus*, and several *Anonaceae* and *Ulmaceae* act as host plants.

As will be seen, *Hylesia*, whose adult phase is venomous, also belongs to this family.

3. Eucleidae

This family includes less noxious caterpillars, which show lateral structures and resemble spiders (lagarta aranha = spider caterpillar).

Euryda variolarus and *Phobetron hipparchia* are the most common species in Brazil. Orange, avocado, eucalyptus, and palm trees, as well as rose bushes, are some of the host plants.

4. Lasiocampidae

Hairy caterpillars of this family live gregariously under the same shelter or web, especially on ivy. *Artace cribraria, Macromphalia lignosa, Titya proxima, T. undulosa*, and *Euglyphes ornata* are the most common. *E. ornata* is habitually found on avocado trees.

5. Arctiidae

These dark or near-black caterpillars show hairy warty protuberances and are found on manioc, guava, cotton, castor bean, berries and fig, as well as certain uncultivated plants. Common Brazilian species include *Utheteisa ornatrix, U. pulchella, Antarctia fusca, Ecpantheria orsa, Eupseudosoma aberrans*, and *Eupseudosoma involutum*.

6. Morphidae

Of the diurnal Lepidoptera, Morphidae is the only family with urticating caterpillars. The iridescent gunmetal or lilac butterflies are highly appreciated by collectors and souvenir makers.

The species most often found are *Morpho hercules, M. achillaena achillaena, M. anaxibia, M. menelaus, M. rhetenor, M. laertes*, and *M. cypri*. Their dark-ribboned yellow hairy larvae usually live on the wild plants *Leguminosae, Musaceae*, and *Menisphermaceae* and are known to cause dermatitis.

B. Moths

A collective outbreak of dermatitis caused by adult moths of the genus *Hylesia*, family Hemileucidae, was reported in northern Brazil by Gusmão *et al.* (1961). The moth is dark gray and lives in the forests of the Amazon. The species could not be identified, but it does not seem to differ from other moths diversely classified as *Hylesia urticans, H. continua, H. canitia* (Guianas), *H. nigricans, H. fulviventris* (Argentina), and *H. volvex* (Peru).

III. CLINICAL SYMPTOMS

Clinical symptoms of lepidopterism in Brazil, as elsewhere, are usually of the dermatological type, with occasional general, ocular, and secondary reactions. Caterpillar dermatitis is more often an isolated inflammatory area at the site of contact, while the lesions of moth dermatitis are disseminated over large areas of the skin. It is still not sufficiently clear whether previous exposures make future contact more or less severe.

A. Caterpillars

Most cases of lepidopterism occur in summer and early autumn when larvae eclose from their eggs in central and southern Brazil, though cases have been noted in late fall and other seasons. In the north and northeast, the temperature is almost always high, and caterpillars are hatching most of the time, so that a seasonal influence is less marked.

Victims of lepidopterism are, as a rule, inexperienced children, and sometimes, other individuals who accidentally touch or press their skin against caterpillars in orchards and gardens.

The clinical symptoms vary markedly and depend on a variety of factors: the species and their diverse type of hairy coat and quality of venom; duration or amount of pressure of the caterpillar against the body, or both; and racial, individual or topographical characteristics of the individual affected. In spite of the diversity of manifestations, a general clinical picture of larvae lepidopterism may be presented.

a. Subjective Symptoms. These include mild burning sensations, with or without a slight pruritus, to violent stabbing and burning pains that may remain for hours and be accompanied or followed by more or less intense pruritus.

b. Objective Symptoms. A series of morphological changes may be noted whose succeeding and overlapping dermatological elements are more or less intense erythema, small whitish edematous papules, urticarial wheals, vesicles, bullae, petechiae, and dry reddish papules; these are followed by secondary erosion, excoriation, desquamation, and pigmentation.

c. Sequence of Phenomena. As in cases of dermatitis under medical observation, this begins with stinging immediately followed by an intense burning pain; a few minutes later a diffuse area of erythema develops. This area, within 2–15 minutes will show whitish edematous hemispheric papules, about 2–3 mm in diameter, which may be isolated but, more often, are grouped. They may coalesce and form a urticarial wheal, with a whitish rough surface, which itches violently for 1 or more hours. Within 24 hours the lesion will have developed into a congestive and slightly edematous cord showing numerous isolated or coalescent vesicles (Fig. 1), and, sometimes, blisters. These later break up, spontaneously or by trauma, and develop into erosions or excoriations that dry up, and, about 5–7 days after the sting, become covered with scabby small crusts. Thereafter the lesion subsides and passes through desquamation and residual pigmentation phases; the latter phase may continue for a few weeks.

This clinical picture may be complicated by secondary infection with suppurating pustules.

As considered above, considerable variations may occur in this sequence.

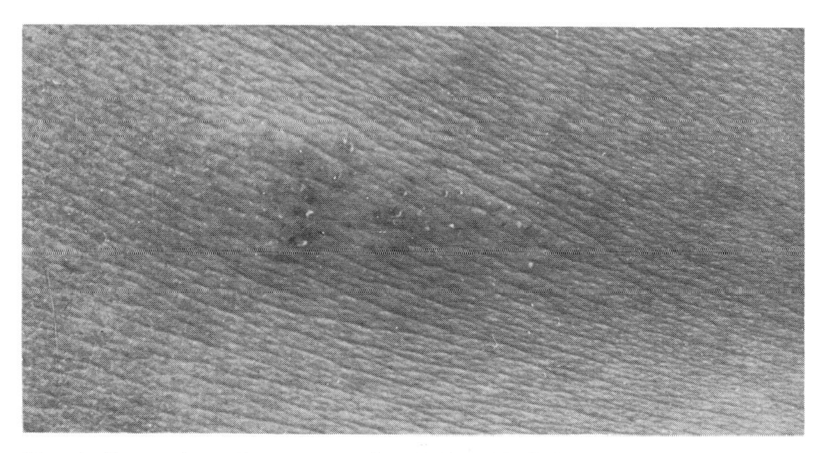

FIG. 1. Congestive, edematous, and vesicular skin lesion caused by contact with a caterpillar.

The urticarial wheal and the erosive excoriated lesions may be minimal or absent; hemorrhage may occur inside and/or around the lesions, with diffuse purpuric elements or small petechiae added to the clinical picture. Dry pink papules, lasting for some 3–4 days are sometimes seen at the end of the vesicular stage. Rarely, more lasting, minute nodules may appear as a foreign body reaction to the penetrating hairs and become ulcerated.

d. Regional and Generalized Symptoms. These result from more severe stings. A large area of edematous inflammation may be observed around the sting. When this occurs in a limb, the swelling may spread, deforming it, sometimes doubling its size; this is accompanied by lymphangitis, adenopathy and arthralgia which usually subside 24–48 hours after the sting.

At the height of these phenomena, it is not uncommon to observe general malaise, fever, insomnia, nausea, vomiting, and muscle spasms. Temporary local or regional neuritis, such as paraesthesia, anesthesia, paresis, and even paralysis, may occur.

Severe regional or generalized symptoms are more often noted after the stings of the Megalopygidae or after multiple contacts with less poisonous caterpillars.

A patient reported by Alvarenga (1912) rubbed his right hand against a group of caterpillars; his symptoms included edema of the hand, lymphangitis, and adenopathy up to the axilla the same day. The next day, the saliva and the urine became sanguinolent, and this condition lasted for 3 days.

Matta (1922) reports several cases of lepidopterism by larvae, among which was a man whose forehead was stung by a *Megalopyge*. Erythema, burning pain, and pruritus followed immediately; in a few minutes, a violent inflammatory reaction was seen on his forehead, eyelids, and nose, with lacrimation, photophobia, and cephalalgia. The next day vesicles appeared and involution began.

Other ocular reactions, characterized by photophobia, erythema, and edema of the eyelids, with conjunctivitis and phlyctenular keratitis, have been reported by Andrade (1940). No instances of ophthalmia nodosa by caterpillar hairs are found in the Brazilian literature.

References to lepidopterism are not infrequently seen in agricultural magazines. A reader's query in one of them (Editorial, 1952) concerned the case of a boy of 14 who had to be hospitalized for 30 days after having touched a group of caterpillars. Blisters developed locally and other symptoms included generalized pain, high fever, sleepiness, and gingival and suboral hemorrhages with considerable hematological alterations. In its answer the magazine classified the larva as a member of the Hemileucidae family, probably *Automeris*.

B. Moths

Skin eruptions caused by adult Lepidoptera were reported by Gusmão *et al.* (1961) in the small mining village of Serra do Navio (the Federal Territory of Amapá in the extreme north of Brazil, near French Guiana).

In May 1960, a cloud of moths of the genus *Hylesia*, from the Amazon, landed in the village. Attracted by the artificial light, they hit against window screens or fluttered around lamps, thereby releasing large amounts of dust that settled on furniture and bedding. Subsequent examination of this dust showed it to be a mixture of moth scales and barbed and pointed arrowlike setae about 170 μ long by 4–5 μ thick (Fig. 2). These setae are found abundantly in the covering of the terminal abdominal segments of the female *Hylesia* (Fig. 3).

Upon retiring, the inhabitants contacted that dust, which caused intense itching; scratching spread and aggravated the phenomena, causing insomnia and interfering seriously with the well-being and the capacity to work of those people affected.

A few minutes after the onset of the subjective symptomatology, a diffuse erythema was noted; this was aggravated by scratching. Within the next hours, and more distinctly, the next morning, numerous reddish-pink micropapules appeared which were disseminated or more or less densely grouped in plaques surrounded by isolated scattered lesions. The lesions were located preferentially on the trunk and upper limbs, but no skin area was exempt. Excoriated by scratching, they became topped by tiny scabs, while new papules kept appearing for about a week. Around the 15th day involution was complete, but only when precautions had been taken against further attacks, i.e., by storing the bedding until used to prevent further contamination.

FIG. 2. Poisonous setae of the covering of *Hylesia* sp. (Courtesy of Gusmão, Forattini, and Rotberg.)

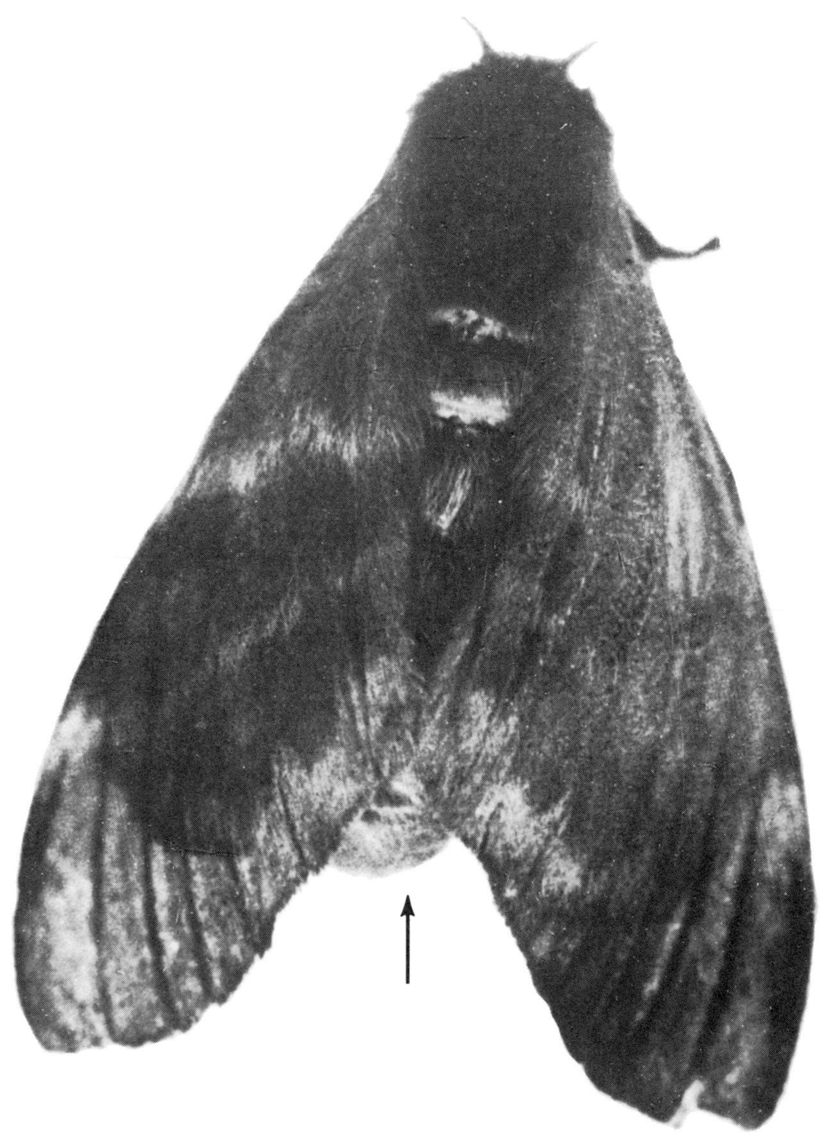

FIG. 3. *Hylesia* sp., female. The extreme abdominal (arrow) segments carry the poisonous setae. (Courtesy of Gusmão, Forattini, and Rotberg.)

Seven-hundred and seven individuals (39.8% of the 1777 inhabitants) of both sexes and all ages and races were affected in this outbreak, which lasted from May 7 to the beginning of June, i.e., the period of time following the eclosion of moths that took place in April.

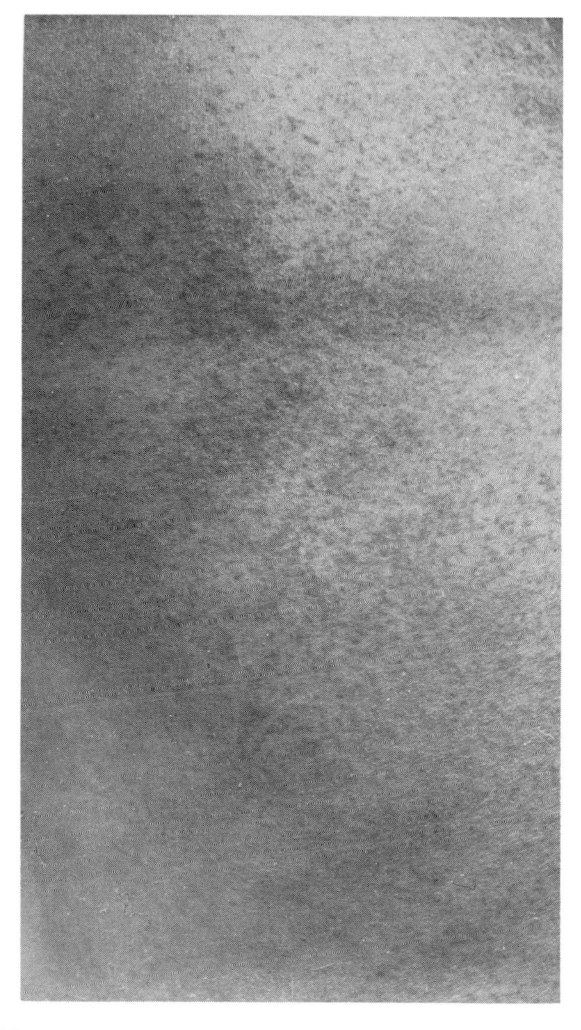

FIG. 4. Widespread, erythematous, papular reaction (spray) with 3-year-old *Hylesia* dust. (Courtesy of Rotberg and Boerner.)

It is likely that other collective outbreaks or isolated or unrecognized cases of lepidopterism by *Hylesia* have occurred in the Amazon region and along the whole Brazilian border, as other species of the genus have been known to cause such epidemics in South America (Leger and Mouzels, 1918; Boyé, 1932; Floch and Abonnens, 1944, in French Guiana; Dallas, 1933; Jörg, 1933, in Argentina; Allard and Allard, 1958, in Peru).

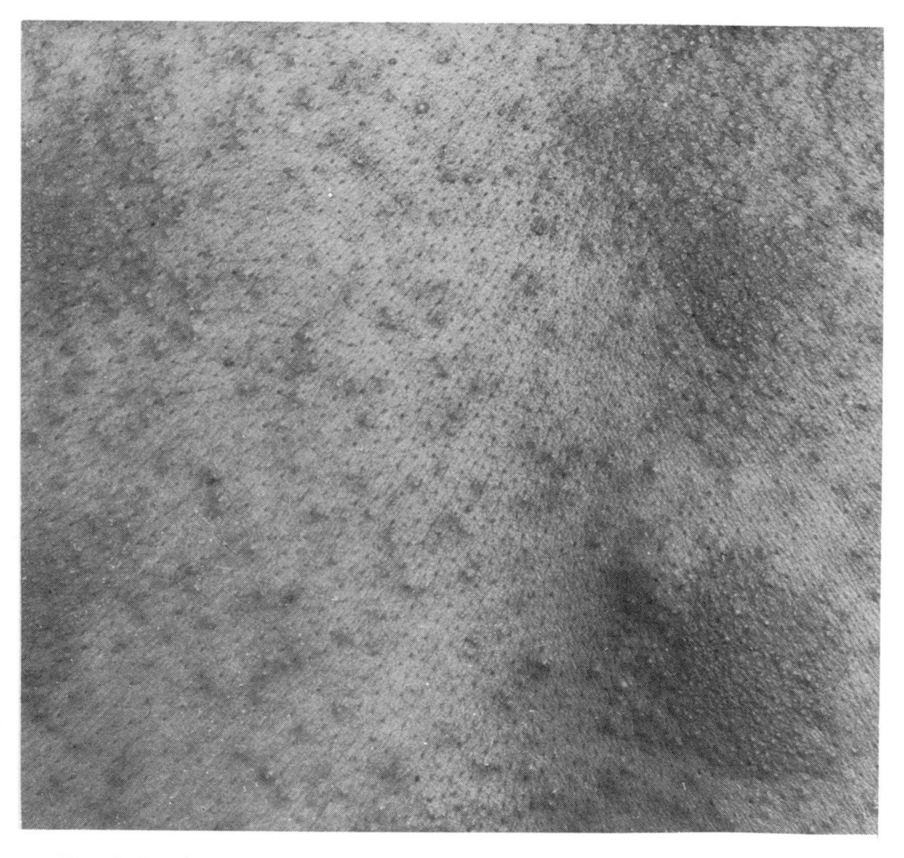

FIG. 5. Patch tests with different preparations from *Hylesia* abdominal covering. Isolated papules between test sites result from accidental contamination by setae while testing. (Courtesy of Rotberg and Boerner.)

Material from the abdomen of the moth, when applied to the skin of several volunteers immediately provoked a burning and pruritic erythematous patch; if kept in place for some hours, burning pain, edema, vesiculation, and superficial necrosis were observed. Rotberg and Boerner (1963) reported that after 3 years this material, kept without preservatives at room temperature, was still able to produce intense skin reactions on volunteers and on themselves when applied either by spraying (Fig. 4) or patch testing (Fig. 5).

Pradinaud (1969), who studied two outbreaks in 1968 at Cayenne, French Guiana, states that the toxic substance of the "arrows" of the *Hylesia urticans* is soluble in water and sweat, which would explain why there is spreading of the lesions after simple contact.

IV. PATHOLOGY

a. Caterpillar Dermatitis. The initial microscopic reactions are vasodilatation and edema corresponding to the immediate phase of erythema and whealing. Further changes occurring in the epidermis, such as intercellular edema and spongiosis, lead to vesicle formation. Vesicles measure about 1.5-2 mm; they contain some fibrin, a moderate number of lymphocytes, and, sometimes, fragments of caterpillar hair. A slight dermal congestion and edema are soon followed by a perivascular exudate of lymphocytes and some eosinophiles; polymorphonuclear leucocytes and histiocytes may also be seen. Lymphocytes and eosinophiles may be noted in lymphatic spaces and the lumen of capillaries.

b. Moth Dermatitis. The epidermis is practically normal initially, but may partially or totally erode away later. Edema, vasodilatation and a perivascular and periglandular infiltration of eosinophiles and histiocytes are the main dermal changes. Sometimes epithelioid and foreign-body types of giant cells can be seen; the latter occasionally contain fragments of setae.

V. TREATMENT

Uncomplicated dermatitis can be treated with slightly acid lotions and/or sedative powders, which usually are satisfactory, but in severe cases corticosteroids may be used. Pradinaud recommends local paintings with an alcoholic tannin solution.

REFERENCES

Allard, H. F., and Allard, H. A. (1958). *J. Wash. Acad. Sci.* **48**, 18.

Almeida, R. F. (1944). *Arquiv. Zool. Estado São Paulo* **4**, 33.

Alvarenga, Z. (1912). *Ann. 7th Congr. Brasil. Med. Cir. Belo Horizonte, Brasil, 1912,* p. 132.

Anchieta, J. (1560). "Carta fazendo a descripção das innumeras coisas naturaes que se encontram na provincia de S. Vicente, hoje S. Paulo." Translated from Latin by J. V. Almeida, Casa Eclectica, Sao Paulo, Brazil, 1900 (p. 26).

Andrade, C. (1940). "Oftalmologia tropical (sulamericana)." Rodrigues & Cia., Rio de Janeiro, Brazil.

Boyé, H. (1932). *Bull. Soc. Pathol. Exotique* **25**, 1099.

da Fonseca, F. (1949). "Animais Peçonhentos." Inst. Butantan, São Paulo, Brazil.

Dallas, E. D. (1933). *Reunion Soc. Arg. Patol. Reg. Norte* **8**, 469.

Editorial. (1952). *Chacaras Quintaes* **85**, 75.

Floch, H., and Abonnenc, E. (1944). *Inst. Pasteur Guy. & Terr. Inini*, Publ. No. 89.

Gusmao, H. H., Forattini, O. P., and Rotberg, A. (1961). *Rev. Inst. Med. Trop. Sao Paulo* **3**, 114.

Jörg, M. E. (1933). *Reunion Soc. Arg. Patol. Reg. Norte* **8**, 482.

Leger, M., and Mouzels, P. (1918). *Bull. Soc. Pathol. Exotique* **11**, 104.

Lima, A. C. (1945, 1950). "Insetos do Brasil," Vols. V and VI. Imprensa Nacional, Rio de Janeiro, Brazil.
Matta, A. (1922). *Amazonas Med.* **4**, 167.
Monte, O. (1934). "Borboletas que vivem em plantas cultivadas." Ofic. graf. Estat. Belo Horizonte, Brazil.
Pradinaud, R. (1969). *Rev. Med.* **6**, 319.
Rotberg, A., and Boerner, A. (1963). 20th Meeting Brazilian Dermatologists, Porto Alegre, Brazil, 1963.
Valle, J. R., Picarelli, Z. P., and Prado J., E. (1954) *Arch. Intern. Pharmacodyn.* **98**, 324.
Von Ihering, R. (1914). *Anais Paulistas Med. Cir.* **3**, 129.

VENOMOUS CENTIPEDES, SPIDERS, AND SCORPIONS

Chapter 50

Venomous Chilopods or Centipedes

WOLFGANG BÜCHERL

INSTITUTO BUTANTAN, SÃO PAULO, FELLOW OF THE CONSELHO NACIONAL DE PESQUISAS, RIO DE JANEIRO, BRAZIL

I. Introduction 169
History of Nomenclature of Chilopods 170
II. Description, Classification, Distribution, and Biology of *Scolopendromorpha* 172
A. Description 172
B. Classification and Distribution 180
C. Biology 189
III. Venom Apparatus and Toxicity of Scolopendromorph Venoms 190
A. Venom Apparatus 190
B. Toxicity of Scolopendromorph Venoms 193
C. Conclusions 195
References 195

I. INTRODUCTION

This essay will discuss only the larger representatives of venomous Chilopoda which can effect man. For the morphological description we select only those genera of the order Scolopendromorpha with large tropical and subtropical species. For classification we use short dichotomic keys only for the most important venomous species. Statements of distribution are based primarily upon localities of South and Central America which have been examined for specimens, and secondarily on literature references which we consider reliable. All biological data are derived from our experience; since 1940 we have kept different specimens alive in our laboratories, observing their life habits. The venom apparatus is studied in *Scolopendra viridicornis*, one of the largest and most aggressive tropical species of America. The toxicity of scolopendromorph venoms is studied in the genera

Scolopendra, Otostigmus, Cryptops, and *Scolopocryptops*, the most important representants of the order, in relation to white mice, guinea pigs, and pigeons.

History of Nomenclature of Chilopods

Linnaeus in 1758 placed under his "Insecta Aptera" the only two genera of myriapods recognized by him: *Julus* (milliped) and *Scolopendra* (chilopod). Latreille in 1802–1805 established the "Legion" Myriapoda with two orders, Chilognatha and Syngnatha. Leach in 1814 distinguished the Myriapoda as a separate class, coordinated with Crustacea, Arachnida, and Insecta, pointing out five genera, *Scutigera, Lithobius, Scolopendra, Cryptops,* and *Geophilus.* Newport in 1844 pointed out four families of the order Chilognatha: Scutigeridae, Lithobiidae, Scolopendridae, and Geophilidae. The subdivisions made by Meinert in 1868 between chilopods with 15 pairs of legs and chilopods with 21 or more pairs of legs have added nothing new. Haase, in 1881 distinguished as a suborder the Chilopoda Anamorpha, in which the larvae have only 7 pairs of legs and the adults 15 pairs, including the families Scutigeridae and Lithobiidae, and the second suborder Chilopoda Epimorpha, larvae and adults with the same number of legs, with the two families Scolopendridae and Geophilidae. The definition of the groups of chilopods made by Latzel are the same as those of Haase. The system of Bollmann (1893) is a regression to Brandt's older division (1840) and brought nothing new. Significant new progress was initiated by Pocock in 1902 with the elevation of the chilopoda to a separate class, subdivided as follows: subclass Pleurostigma, with four orders: Geophilomorpha, Scolopendromorpha, Craterostigmomorpha, and Lithobiomorpha, and subclass Notostigma. Chamberlin, in 1911 subdivided the class Chilopoda into three orders: Schizotarsia, Anamorpha, and Epimorpha, the last with two suborders: Scolopendroidea and Geophiloidea. Verhoeff in 1925 set up the following division:

Class Chilopoda
- Subclass Notostigmophora Verhoeff 1901
  - Order Scutigeromorpha Pocock 1902; family Scutigeridae (Gervais) 1837
- Subclass Pleurostigmomorpha Verhoeff 1901
  - Order Anamorpha Haase 1880
    - Suborder Craterostigmomorpha Pocock 1902
      - Family Craterostigmidae Pocock 1902
    - Suborder Lithobiomorpha Pocock 1902
      - Families Cermatobiidae Haase 1887 and Lithobiidae Newport 1844

Order Epimorpha Haase 1880
Suborder Scolopendromorpha Pocock 1896
Superfamily Cryptopina Verhoeff 1907
Families Cryptopidae Verhoeff 1907 and Newportiidae Pocock 1895
Superfamily Theatopsina Verhoeff 1907
Families Theatopsidae Verhoeff 1907 and Plutoniidae Bollman 1895
Superfamily Scolopendrina Verhoeff 1907
Families Scolopendridae Kräpelin 1903 and Scolopocryptidae Verhoeff 1907
Suborder Geophilomorpha Pocock 1896– Family Mecistocephalidae Verhoeff 1901
Superfamily Adesmata Verhoeff 1908
Families Geophilidae Cook 1895, Schendylidae Cook 1895, Brasilophilidae Verhoff 1925, Gonibregmatidae Cook 1895 and Himantariidae Cook 1895.

The suborder Scolopendromorpha was coordinated by Kohlrausch, in 1881 in Scolopendridae Heteropodes, Cribriferi, Morsicantes and Cryptopsii, and by Haase, in 1887, in Scolopendridae Holopneusticae, and Hemipneusticae. Pocock in 1895 elevated this suborder to the order Scolopendromorpha, with four families: Scolopendridae, Cryptopidae, Scolopocryptidae and Newportiidae. The most accepted system of the order Scolopendromorpha was set up by Attems in 1930, as follows:

Order Scolopendromorpha
1. Family Scolopendridae
   A. Subfamily Scolopendrinae
      1. Tribe Scolopendrini—genera: *Scolopendra, Trachycormocephalus, Arthrorhabdus, Cormocephalus, Campylostigmus, Rhoda, Scolopendropsis.*
      2. Tribe Asanadini—genera: *Asanada, Pseudocryptops*
   B. Subfamily Otostigminae
      1. Tribe Otostigmini—genera: *Otostigmus, Digitipes, Alipes, Ethmostigmus, Rhysida, Alluropus*
      2. Tribe Arrhabdotini—genus: *Arrhabdotus*
2. Family Cryptopidae
   A. Subfamily Cryptopinae—genera: *Cryptops, Paracryptops, Mimops, Anethops*
   B. Subfamily Theatopsinae—genus: *Theathops*
   C. Subfamily Scolopocryptopinae—genera: *Scolopocryptops, Otocryptops, Kethops, Kartops, Newportia.*

   In 1953 R. E. Crabill introduced the new genus, *Dinocryptops*, eliminating the genus *Otocryptops.*

II. DESCRIPTION, CLASSIFICATION, DISTRIBUTION, AND BIOLOGY OF SCOLOPENDROMORPHA

A. Description

The Scolopendromorphs are terrestrial animals in which the body has two main divisions: the head and the trunk or body. Both are enclosed in a chitinous exoskeleton. The body is slender, ribbonlike and a few times longer than broad and high, being depressed dorsoventrally; it is about 10 mm to 300 mm in length (Fig. 1). The color varies from light yellow to deep brown to blackish. There exist species of green, blue, or reddish-blue colors. The head and the last segment of the body are often differently colored than the other parts.

FIG. 1. *Scolopendra viridicornis.*

The head (Figs. 2 and 3) of Scolopendromorphs consists of the cephalic plate or dorsal plate with two antennae and(if present) four ocelli on each side, and of the ventral parts: clypeus, cephalic pleurites, labrum, one pair of mandibles, one pair of "first maxillae" and one pair of "second maxillae" (Fig. 4). The cephalic plate (Fig. 2) may be smooth or finely and sub-sparsely punctuated with microscopical short thin hairs. Two weak, longitudinal furrows along the middle line are common and may be complete or shortened at the anterior or posterior sides (Fig. 2); the furrows may be simple or branched posteriorly. In a few genera the cephalic plate shows only one median furrow; in others there may exist two basal plates at the posterior outer sides of the cephalic plate (Cormocephalus). The antennae (Figs. 2 and 3) generally bear 17 cylindrical articles, rarely as many as 34, of which the first three or six, may be smooth and shining, the others being covered with short light hairs, intermixed or not with longer setae. In a few genera the antennae are short, not much exceeding the length of

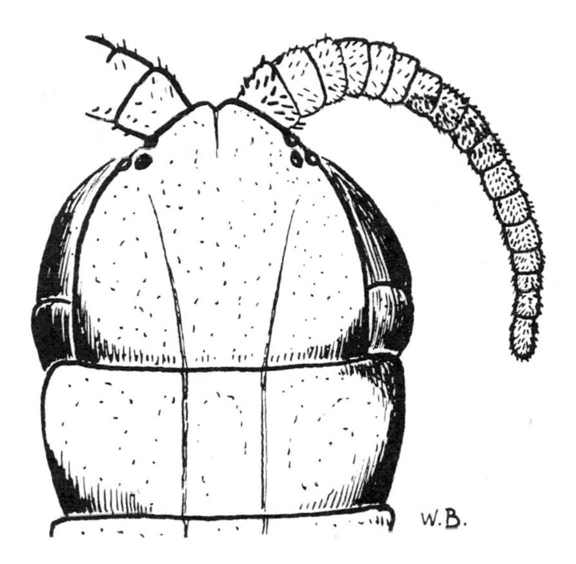

FIG. 2. *Rhysida brasiliensis*. Dorsal view of the head with eyes, antennae, and cephalic sulci.

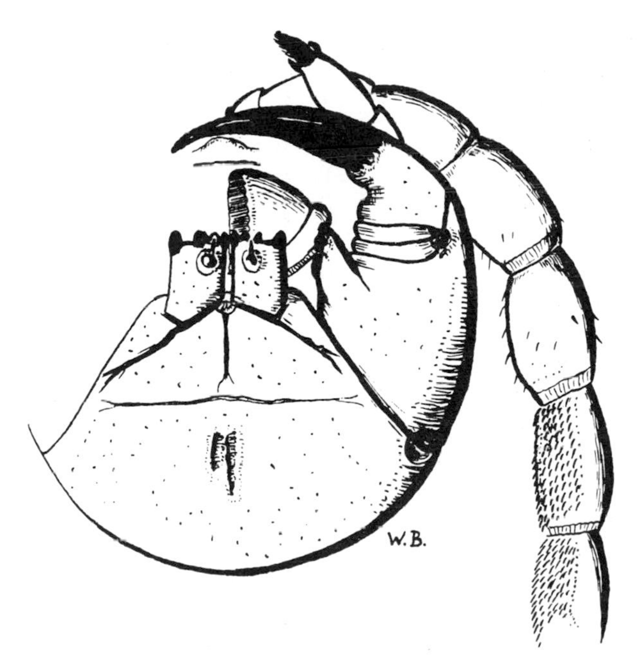

FIG. 3. *Scolopendra viridicornis*. Ventral view of the postcephalic segment, showing the coxosternum, the dental plates, the telopodite with the venom jaw and antennae.

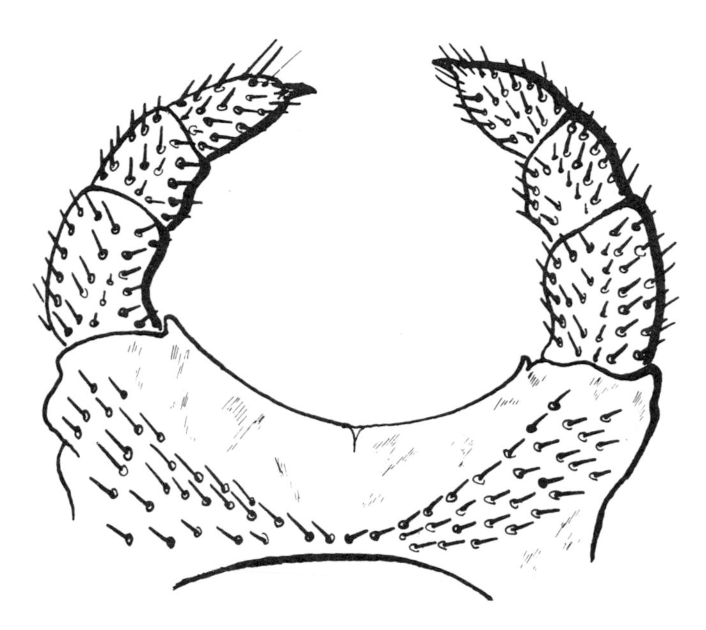

FIG. 4. The second pair of maxillae of a *Cryptops*.

the cephalic plate, and in others they extend to the second or sixth tergite or even more. The eyes are of the simple ocelli type. Commonly there are eight eyes, four on each lateral side, behind the antennae. All the representants of the family Cryptopidae are blind, without eyes. The anterior ventral portion of the head is named the clypeus. Affixed to it at the middle line and separated from it by a transparent suture is the labrum, generally strangulated in the middle. On each side of the labrum and united to the clypeus with transparent sutures there is a small sclerite, the coclypeus. The mandibles show a true articulation, inserted laterally in an excavation near the coclypeus. The mandibles bear a dental block with five teeth at the right and four at the left side. In the young the teeth are separated, but in adults they are fused, with serated ends. Between the articulation and the teeth is a thin velvety brush of very fine hairs; behind the teeth are several series of longer curved hairs and on the ventral side a series of long stout setae. The first pair of maxillae consists of a basal portion and of a telopodite. The basal part (coxae) shows two processes, united only partially, which form the interior part of the mouth. The telopodite consists of two articles, the first short and disclike, the distal one a little longer, rounded, and covered with dense hairs and a fine brush of setae. The sternite is absent in adults. The second pair of maxillae (Fig. 4) shows the two coxae soldered in the

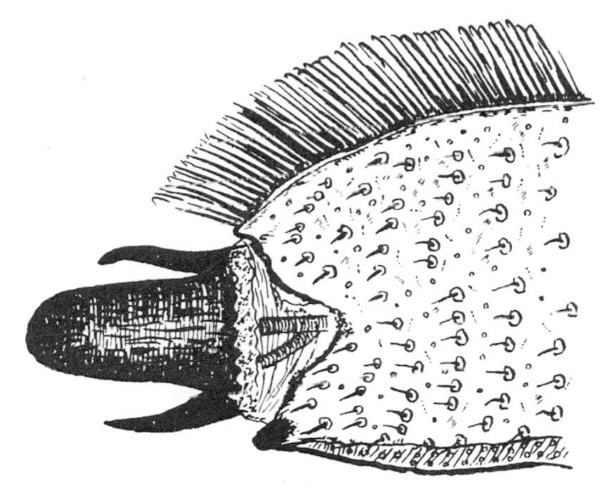

FIG. 5. *Scolopendra subspinipes*. The second maxillae showing the claw and the "cleaning" brush.

middle by a small bridge. The telopodite bears three articles, the first broad, the second smaller and the last showing distally a robust claw with two lateral smaller spines and an efficient "cleaning brush," consisting of long serated setae (Fig. 5).

The trunk (Fig. 1) consists of the postcephalic segment and of 21 or 23 segments with legs. In the last segment are the genital and the anal region, commonly not visible. The postcephalic segment (Fig. 3) shows a very small dorsal sclerite, ordinarily completely covered by the cephalic plate and the first tergite of the trunk. The two coxae are completely fused in a prosternum or coxosternum, which extends cephalad in two processes, separated commonly from the coxosternum by divergent sutures. These so called teeth plates may or may not bear teeth. When teeth are present, they are robust, three to five in number, the inner commonly partly fused and the outer more isolated. On each tooth plate there usually exists a fine isolated seta. The telopodite shows four articles; the basal one is very powerful and robust and generally bears a prominent distal process on the inner side, often provided with two or three rounded teeth; the second and third articles are very small, broad, and ringlike; the last article forms the important poison jaw with a curved pointed end, densely chitinized. Near the pit of each poison claw the venom duct opens; the venom gland is situated in the cylindrical internal lumen of the basal article, surrounded by several powerful muscle bundles.

On the segments with legs we can distinguish a dorsal tergite (Fig. 6), a ventral sternite, and at each side the pleurites with the insertions of the legs. Covered by the preceding segments there generally exist pretergites and

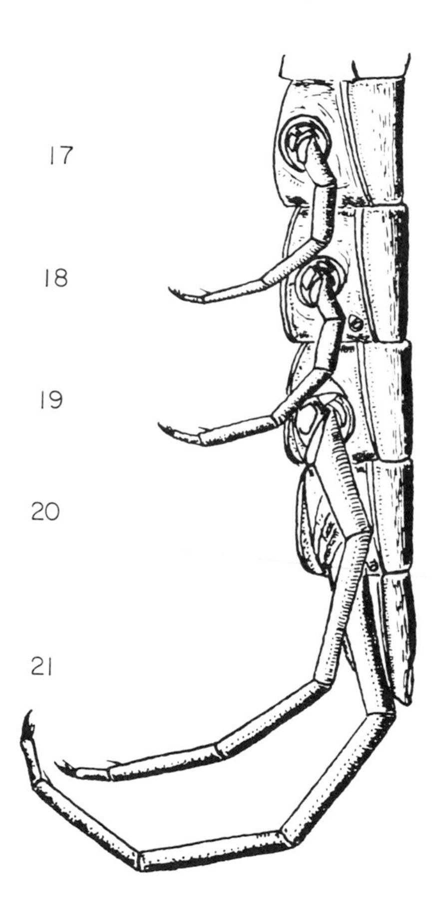

FIG. 6. *Otostigmus tibialis.* The last five body segments with legs.

presternites, totally or partially invisible. The tergites generally bear two fine sutures, bifurcated or not, called paramedian sulci (Fig. 2) or longitudinal sulci. They may be complete or shortened anteriorly or posteriorly. The lateral margins of tergites may be rebordered (Fig. 7). On the tergites are fine pores, with or without small setae or a median elevated keel; the whole area is smooth and shining or has several longitudinal keels with or without small spines. The sternites (Fig. 8) also show paramedian furrows, complete or shortened, or only one median furrow, or even a furrow-cross; they may also lack sulci (Fig. 8), showing several rounded depressions. The legs consist of the following parts: coxa, trochanter, prefemur, femur, tibia, metatarsus and tarsus with a principal and two lateral claws. The last body

FIG. 7. *Scolopendra subspinipes.*

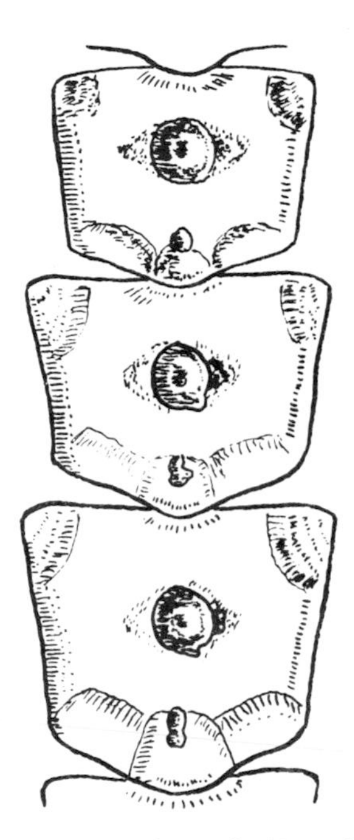

FIG. 8. *Otostigmus scabricauda.* Ventral sternites.

segment with the last pair of legs is generally very different from the preceding ones. The coxa of anal legs is absent or fused with the pleurae, forming thus the so-called "pseudopleura" or "coxopleura"; the trochanter is also absent or rudimentary; the pseudopleurae are strongly developed and are more or less produced caudad at the mesal side into the so-called "coxopleural process." This process often bears a number of spines near the pit which may also exist on the caudal margin ectad. The coxopleurae are commonly densely porous. The tarsus is always biarticulated, even in those species in which the other legs show only an unarticulated tarsus. The representatives of the genera *Newportia* and *Tidops* show the second last tarsus subarticulated in numerous more or less distinguishable articles. In the tropical large *Scolopendra* species (Fig. 1) the last legs also have a prehensile function (Fig. 9) which is even more evident in *Cryptops*, in which the tibia and the metatarsus bear a row of curved pointed black spines, a very efficient prehensile apparatus.

The respiratory openings of the tracheae, the so-called stigmata, are present in all Scolopendromorpha. They may be slitlike, triangular, or oval in aspect, with or without a proper closing apparatus; they are situated at

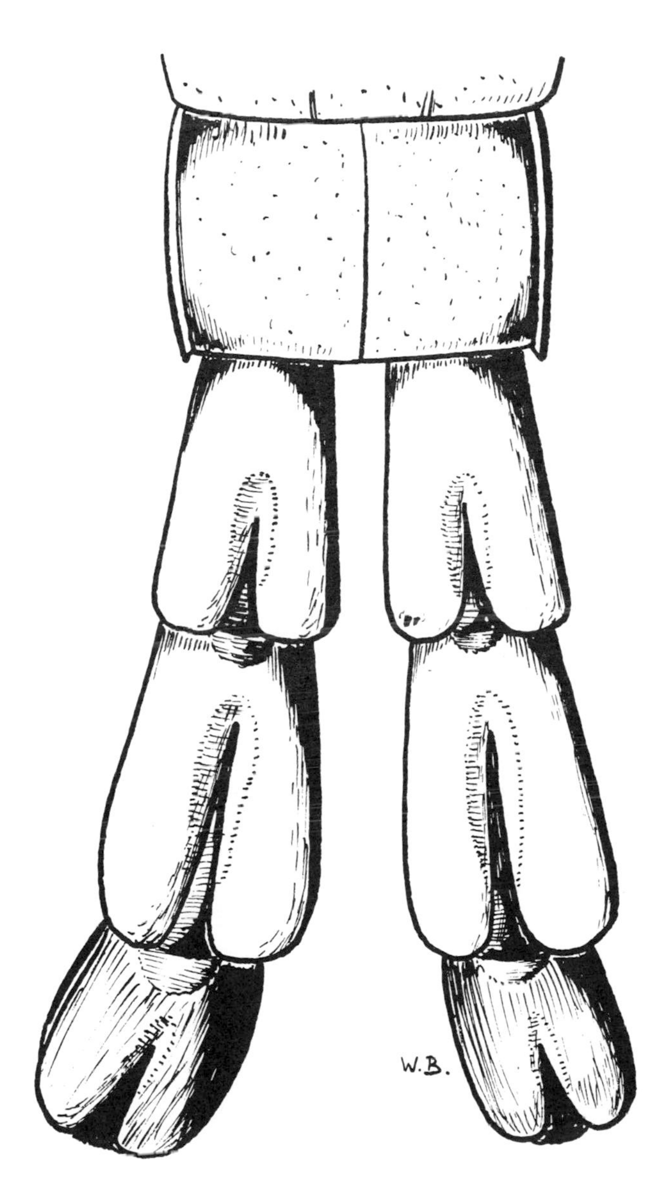

FIG. 9. *Scolopendra morsitans*. Last body segment with prefemur, femur and tibia of anal legs.

each side on the pleural region of the trunk, near the margin of tergites in segments 3, 5, (7), 8, 10, 12, 14, 16, 18, 20 (22).

B. Classification and Distribution

Key for families:

With four ocelli on each side of the cephalic plate. Sternites without sulci or with two parallel sulci, rarely with only one median sulcus, never with transverse sulcus .... 1. Family Scolopendridae
Without ocelli; blind. Sternites generally with one median sulcus, rarely with two paramedian sulci, often with a transverse sulcus........... 2. Family Cryptopidae

1. Family Scolopendridae

With four ocelli on each side. Antennae from 17 to more than 30 articles, two or more basal articles smooth and shining; coxosternum of the post-cephalic segment with two dental plates, absent only in the genus *Arrhabdotus*; all tarsi biarticulated; sternites with two paramedian sulci or without sulcus, only in *Arrhabdotus* with one median sulcus, never with transversal sulcus.

2 subfamilies, 4 tribes, 16 genera, more or less 250 species and about 45 uncertain species.

Key for subfamilies:

Spiracles of stigmata angular, triangular or narrowly slitlike, parallel to the long axis of the body ....1. Subfamily Scolopendrinae
Spiracles oval or circular, oblique to the long axis of the body....... 2. Subfamily Otostigminae

Subfamily Scolopendrinae

Respiratory stigmata on segments 3, 5, 8, 10, 12, 14, 16, 18, 20 (22); 21 pairs of legs, only in *Scolopendropsis* 23 pairs; tarsal spines absent or only one in number (rarely the first pair with two tarsal spines).

2 tribes, 9 genera, and 135 certain and 38 uncertain species.

Key for tribes:

Coxopleura of the last legs with pores. Antennae extending to the third tergite or longer .... 1. Tribe Scolopendrini
Coxopleurae without pores. Antennae short, not extending over the first tergite ....2. Tribe Asanadini

Tribe Scolopendrini

Key to most important genera:

1 { 23 pairs of legs—*Scolopendropsis* Brandt 1841—Brazil: Bahia—1 species
21 pairs of legs 2

2 The first tarsus of anal legs shorter than the second. Pseudopleurae of anal legs without posterior processus—*Rhoda* Meinert 1886- 2 species—Brazil: Santarém, Recife, Bahia

First tarsus of the legs distinctly longer than the second, rarely as long as 3

3 All legs without tarsal spines on the ventral side; lips of spiracles not divided 4

The most legs with distinct tarsal spines 5

4 Stigmata not covered by pleurites—*Cormocephalus* Newport 1844/5—more than 60 species—tropical, subtropical cosmopolites; Mediterranean region; South Africa

Stigmata more or less covered by pleurites—*Campylostigmus* Ribaut 1923—5 species—New Caledonia

5 Head overlapping the first tergite; labrum without setae; last legs with a claw and two distinct basal spines—*Scolopendra* Linn. 1758.

First tergite slightly overlapping the head; labrum with setae 6

6 Lips of spiracles undivided; last legs without basal spines at the claw—*Arthrorrhabdus* Pocock 1891—4 species—South Africa, Mexico, N.W. Australia and Brazil: Pará

Lips of spiracle slitlike or triangular; last legs with or without basal spines at the claws—*Trachycormocephalus* Kraepelin 1903—8 species—East Africa, South Africa, Congo, Somalia, Syria, Israel, Mesopotamia, India.

Genus *Scolopendra* Linnaeus 1758

21 pairs of legs. Head with or without sulci, without basal plates, overlapping the first tergite. Antennae from 17 to 31 articles, long, cylindrical. Clypeus without setae. Claws of the second maxillae with two robust basal spines. Dental plates of prosternum with three to numerous teeth; on the area generally one single seta; basal sulci of dental plates distinct, often branched. Tergites 2, 4, 6, 9, 11, 13, 15, 17 and 19 shorter than the others; the first tergite with or without a deep, transverse, slightly curved impression; the paramedian furrows of tergites generally well visible, rarely indistinct or absent; they may begin over the first or over one of the subsequent plates. The lateral margins of tergites bordered from the third tergite or only in the last one or from the sixth or seventh; the last tergite with or without a middle elevation or depression; sternites with or without two longitudinal furrows. Pseudopleura with a short or a longer caudal process with spines. The claws of last legs with two basal spines. Tarsal spines always present, one or two on the tarsus of first legs, one on the other legs, generally absent on the 21st pair. The first tarsus longer than the second. Stigmata slitlike or triangular.

About 40 species. The most important are the following:

Scolopendra morsitans Linnaeus 1758 (Fig. 9):

From about 80 to 120 mm in length. Color extremely variable, from lemon yellow to reddish yellow, with or without greenish caudal borders to the tergites, or olive green to dark green, with the head and the last segment brownish. Head without sulci. Antennae from 17 to 23 articles, generally from 19 to 20, six or seven smooth. Tergites with paramedian sulci from the second or the third; lateral borders variable, from the third or the 19th tergite. Prosternal plates with four or five teeth. Sternites 2 to 20 with two furrows. Legs 1 to 19 or 20 with one tarsal spine. Pseudopleural process with three, four or five spines and one spine near the posterior border. Prefemur of anal legs ventrally with three rows of three spines, mesally without spines, dorsally with four to six spines in two rows; apical process bearing 3–8, generally four spines.

One of the commonest centipedes. Cosmopolitan in tropical, subtropical, and temperate regions.

Scolopendra subspinipes Leach 1815 (Fig. 7):

Length of about 200 to 230 mm; one of the largest scolopendrids. Uniformly brown on the head and the first tergite lighter than the darkish green body. Head porous, without sulci. Antennae from about 17 to 20, generally 18 to 19 articles, 6 basal articles smooth. Paramedian sulci beginning from the second to the ninth tergite, delicate or often indistinct. The bordering of the tergites beginning from the fifth to the sixteenth. Last tergite without furrow. Prosternum without transverse sulcus; prosternal plates with four to nine teeth each. Sternites with two longitudinal furrows. One tarsal spine of the legs. All leg claws with two basal spines. Pseudopleurae with a long processus with one to three spines. The porous area extending to the tergite. Anal legs long, the prefemur rounded dorsally, armed beneath with from one to three spines, mesally with zero to two and dorsally with zero to three spines; apical process with one or three spines.

Range: Cosmopolitan in tropical and subtropical regions of the earth.

Scolopendra gigantea Linnaeus 1758:

The largest scolopendrid species we know, from about 250 to 265 mm in length. Lighter or darker olive-brown to deep reddish-brown; legs olive-green with yellowish articulations; second tarsi yellow. Head finely punctuated with two delicate sulci. Antennae with 17 articles, 9 to 12 basal articles smooth. Prosternal plates with four teeth each, the two mediales fused. First tergite with a deep transverse rounded impression; second with two paramedian sulci, shortened in advance; tergites from third to twentieth with two complete sulci, from the fourth to the eighteenth also with a median sulcus; from the fourth to the fifth with lateral borders. Last tergite abruptly elevated on the anterior half. Sternites with two paramedian furrows, distinct in the middle of each and indistinct at the anterior and posterior borders. The first legs with one or two, from the second to the nineteenth or the twentieth pair with only one tarsal spine; prefemur of the legs, dorsally distad, with two, four, or five small spines; femur of all legs or only of a few posterior pairs with one small spine dorsally. Pseudopleural process with four to nine spines distad and laterally often also with a few spines, with zero to one small spine at the posterior border; prefemur of anal legs ventrally with two to three rows of two to four spines, mesally and dorsomesally with 12 to 15 spines; apical process robust, bearing six to eight spines; femur dorsally distad with or without spines.

Range: Jamaica, Trinidad, Venezuela, Colombia, North Chile, Equatorial Brazil.

Scolopendra angulata Newport 1844:

Attaining a maximum length of about 170 mm. From olive-brown to deep green or reddish-yellow with deep green borders of tergites. Antennae with 17 articles, 4 basal ones smooth. Prosternal plates with four teeth each, the two or three mesales fused. First tergite with anular depression. Paramedian sulci from the third tergite; lateral borders from the sixth or the seventh; last tergite without elevation or furrow. Furrows on the sternites present only at the anterior half. First pair of legs with two, from the second to the twentieth pair with one tarsal spine; prefemur 19 dorsally, distad, with one or two, femur with one spine; prefemur 20 above, distad, with two, femur with one or two, and mesally with zero or one spine. Pseudopleural process with three to four spines and one at the posterior border. Prefemur of anal legs ventrally with seven, mesally with four, above mesally also with four spines; apical process with three to four spines; femur at the middle, mesally, with two to four spines, distad, above, with zero to one and mesally often with a few spines.

Range: Little Antilles, Venezuela, Colombia, Ecuador, Bolivia, Tropical Central Brazil.

Scolopendra valida Lucas 1840:

160 mm. Olive green, olive-brown or chestnut brown, often with deep green posterior border of tergites or head; antennae, first tergites and anal legs deep green and the other parts of the body olive-brown. Head with pores and two sulci. Antennae from 17 to 27 articles. Four to six basal articles smooth. Prosternum with median sulcus; prosternal plates with four teeth each, the mesales more or less fused, in the area a round depression with a small seta. First tergite with deep annular furrow and two paramedian sulci; lateral borders visible from 5th or 8th tergite. Sternites from 2nd to 20th with two deep paramedian furrows. Pseudopleural process with two, three, or four spines; at the posterior border one spine. The first pair of legs with two, the 2nd to 20th with one tarsal spine; all legs with two lateral

spines near the claw, the mesal one often absent on the posterior legs; prefemur of the 20th pair of legs, distad, dorsally, with two to four, of the 19th pair with one to three, of the 18th pair with one spine; prefemur of anal legs with 14–20 spines, ventrally; apical process with three to five spines.

Range: Mediterranean region, Atlantic Islands of African west coast, North and East Africa, West Asia.

Scolopendra viridicornis Newport 1844 (Fig. 1):

Length of about 150 mm, attaining a maximum length of 220 mm. From yellow-red to brown, light green to deep green, often with grayish borders of tergites. Head with pores and two sulci, often finely ramificated behind. Antennae with 17 articles, four basal are smooth. Prosternal plates with four to five teeth each, the two mesales partly fused; in the area a round depression with a light bristle. First tergite with annular furrow and two sulci, ramificated in front, forming a "W"-like figure; tergites 3 to 20 with two paramedian sulci; last tergite with a median elevation in front and often a depression behind. Sternites with two paramedian sulci. Legs with one tarsal spine; prefemur of legs dorsally, distad, with one, two, or three spines; 20th often with one small spine mesally; all legs with two basal spines near the claws. Pseudopleural process with one to three or more spines distad. Prefemur of anal legs beneath with four to seven, mesally with one, dorsally mesally with four to six spines; apical process with three to five spines.

Range: From Antilles to Paraguay and Argentina. One of the commonest giant scolopendrids of Brazil.

Scolopendra heros Girard 1853:

With a maximum length of about 210 mm. Color variable, from dark green to olive or chestnut brown. Head and first tergite finely punctuated; head with two delicate sulci. Antennae with 24 to 26 articles, $4\frac{1}{2}$ to 6 basal articles smooth. Prosternal teeth four to five on each dental plate, the three inner ones more or less united at base; without depression in the area. First tergite with a deep semiannular furrow and generally with two paramedian sulci. The following tergites with two sulci and with lateral borders form the sixth or the twelfth; last tergite without, rarely with, a median sulcus. All sternites with two paramedian furrows. First pair of legs with two, the following with one tarsal spine; all legs with two basal spines near the claw; pseudopleural process bearing 5 to 11 spines, and zero or two on the caudal margin near the posterior border; prefemur of anal legs mostly with 11 spines ventrally and mesally, five spines on the inner surface and above; apical process bearing 5 to 11 spines; disposed around a central spine.

Range: Mexico and southern states of U.S.A.: Georgia, Alabama, New Orleans, Kansas, Texas, Arizona, etc.

Other species, attaining a length of more than 100 mm, are the following

S. hirsutipes Bollman 1893—West Indies; *S. spinosissima* Krpln. 1903—Philippines; *S. hardwickei* Newport 1844—Sumatra, Java, Sunda Islands; *S. alternans* Leach 1815—Cuba, Haiti, Puerto Rico, St. Thomas, St. Croix, Guadelupe, Antigua, Florida, Venezuela, tropical Brazil; *S. armata* Krpln. 1903—Venezuela, Brazil: Amazonas; *S. crudelis* C. L. Koch 1847—West Indies: Bartholemy Island; *S. galapagoensis* Bollmann 1890—Galapagos Islands; *S. robusta* Krpln. 1903—from Mexico to Colombia; *S. sumichrasti* Saussure 1860—from Mexico to Honduras; *S. viridis* Say 1821—U.S.A.: southern states, Mexico, Central America, tropical Brazil.

Minor species from about 40 to 80 mm in length are *S. laeta* Haase 1887—Australia, *S. metuenda* Poc. 1895—Solomon Islands, *S. gracillima* Attems 1898—Java, *S. pinguis* Poc. 1891—Burma, Java, *S. cingulata* Latr. 1829—Central Europe and Mediterranean region, *S. calcarata* Porat 1876—China, *S. madagascariensis* Att. 1910—Madagascar, *S. gordulana* Att. 1909—Southern Ethiopia, *S. clavipes* C. L. Koch 1847—Eastern Mediterranean region, *S. dalmatica* C. L. Koch 1847—Dalmatia, *S. canidens* Newport 1844—Persia, *S. explorans* Chamberlin 1914—Central Brazil.

Tribe Asanadini

Includes only a few small species, from about 25 to 45 mm in body length, belonging to the two genera *Asanada* Meinert, 1886– India, New Guinea, Philippines, Palestine, South Africa, and *Pseudocryptops* Pocock 1891—Africa and India, both genera without medical importance.

Subfamily Otostigminae

Spiracles large, rounded; cephalic plate generally overlapped by the first tergite; without basal plates. 21 pairs of legs. 2 tribes, 7 genera, several subgenera, 120 species, more or less 40 subspecies, 8 uncertain species.

Tribe Arrhabdotini

Sternites with one median furrow; without dental plates on the prosternum; anal legs short and broad; prefemur without spines; legs without tarsal spine; five basal articles of the antennae smooth; last tergite triangular behind, long, covering the front part of the prefemur; with 10 pairs of stigmata on segments 3, 5, 7, 8, 10, 12, 14, 16, 18 and 20.

Only one genus with one species: *Arrhabdotus octosulcatus* (Tömösváry) 1882—Borneo.

Tribe Otostigmini

Sternites without furrows or with two parallel sulci, never with only one sulcus. Coxosternum always with dental plates with several teeth each. Anal legs long, slender, prefemur generally with spines, on the males often with sexual processes. Tarsal spines generally present, absent only in a few species of neotropical *Otostigmus*. Two to four basal articles of antennae smooth, the others finely haired. From nine to ten pairs of spiracles.

6 genera: *Digitipes* Attems 1930, with only one small species—Congo; 9 pairs of spiracles; *Alipes* Immhoff 1854, with 5 small species found in South and Central Africa; *Ethmostigmus* Pocock 1891, with 10 pairs of spiracles; about 15 small species of the Indo-Australian and Ethiopian region; *Rhysida* H. C. Wood 1862, also with 10 pairs of spiracles; about 20 species are known, from 60 to 120 mm of body length, found in the Indo-Australian, Ethiopian, and Neotropical regions: Brazil; *Alluropus* Attems 1930, *nomen novum* for the older *Edentistoma* and *Anodontostoma*, with only one species of Borneo. The most important genus is *Otostigmus* Porat 1876, with 21 pairs of legs, 9 pairs of stigmata, including about 75 species, found in all tropical and subtropical regions. *Otostigmus scabricauda* (Humbert and Saussure) 1870 reaches a length of from 60 to 110 mm. Olive green, legs yellow with greenish to blue tarsi, anal legs, head, and body often blue.

Range: Brazil, Colombia, Venezuela.

2. Family Cryptopidae

Blind, without ocelli; antennae with 11 or 13, generally 17 articles, the basal articles with setae; prosternum often without dental plates; tarsi uniarticulated; sternites with one median furrow, absent only in *Mimops*, often also with a transverse furrow, forming a "furrow-cross"; stigmata small, rounded. Without basal plates on the head. Pseudopleurae with pores, with or without process behind. Pretergites often visible.

Key to the three subfamilies:

1 { 23 pairs of legs. Pseudopleurae with a long conical process, with only one spine behind—Subfamily Dinocryptopinae
  { 21 pairs of legs 2

2 { Anal segment normal. Dental plates on the prosternum wanting or, if present, without teeth—Subfamily Cryptopinae
  { Anal segment enlarged. Prosternum with two long toothed dental plates. —Subfamily Theatopsinae

Subfamily Cryptopinae

Prosternum rounded cephalad. 9 pairs of spiracles. Anal legs long. Endosternites generally visible. Legs with numerous spiniform setae. Sternites generally with cruciform impressions.

Four genera are known. *Paracryptops* Pocock 1891, with 4 species—India, New Guinea and British Guiana; *Anethops* Chamberlin 1902—only one species—California; *Mimops* Krpln. 1903 with 2 species, *M. orientalis* Krpln. 1903—China, and *M. occidentalis* Chamb. 1914—Brazil: Rio de Janeiro; *Cryptops* Leach 1815 with about 60 species. Tropical, subtropical, and temperate cosmopolites. One of the commonest South American species is *C. iheringi* Brölemann 1902.

Subfamily Theatopsinae

Includes the two genera, *Theatops* Newport 1844, with three small species—*T. spinicauda*, *T. postica*, and *T. erythrocephala*—from southern U.S.A., the Mediterranean region, and Mexico, and *Plutonium* Cavanna 1881 with the species *P. zweierleini* of about 140 mm of body length found in Sicily.

Subfamily Dinocryptopinae

23 pairs of legs. Without ocelli. Antennae generally with 17, rarely with 11 or 13 articles. Without dental plates, or, if present, without teeth or with two elongated processes.

2 tribes, 6 genera.

Key for the tribes:

With 11 pairs of spiracles—Dinocryptopini Crabill.
With 10 pairs of spiracles Scolopocryptopini

Tribe Dinocryptopini

Key to genera:

1 { Tarsus of anal legs biarticulated—*Dinocryptops* Crabill 1953
2 species China, South, Central, and North America
Tarsus of anal legs with numerous articles. 2

2 { Dental plates cylindrical, with teeth—*Tidops*—British Guiana
Dental plates absent or with two very small, not toothed plates—*Newportia* Gervais 1847—about 30 species—South and Central America. Figs. 10, 11, and 12.)

Tribe Scolopocryptopini

Antennae with 11 articles—*Kartops* Archey 1923—one species *K. guiannae*—British Guiana.

Antennae with 17 articles; prefemur, femur, and tibia of anal legs with numerous small spines; cephalic plate with two sulci—*Kethops* Chamberlin 1912—one species, *K. utahensis* Chamb. 1909—U.S.A.: Utah and New Mexico.

Antennae with 17 articles. Prefemur of anal legs with one or two strong spines, femur and tibia without spines. Cephalic plate without sulci—*Scolopocryptops* Haase 1881. This is by far the most important genus of this whole subfamily, including larger and venomous species, found in the warmer regions of North America, Central and chiefly South America, China, Japan, Korea, Philippines, West Africa. Five species have been well described, *S. ferrugineus* (l.) 1767 being the most frequent from West Africa, southern states of U.S.A., Mexico, Central America, Antilles, South America. Tropical representatives with a body length of 120 mm are common.

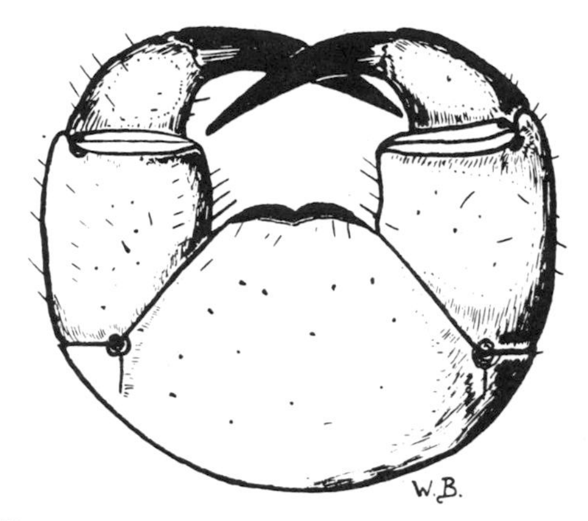

FIG. 10. *Newportia ernsti*. Postcephalic segment with venom jaws.

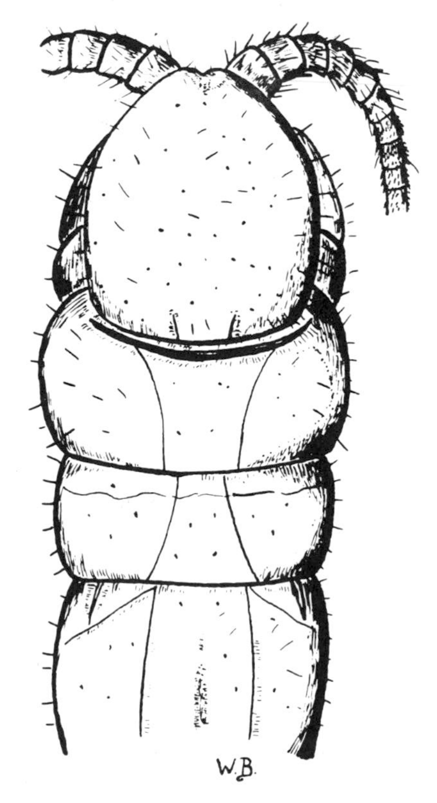

FIG. 11. *Newportia ernsti*. Dorsal view of the head and the first three body segments.

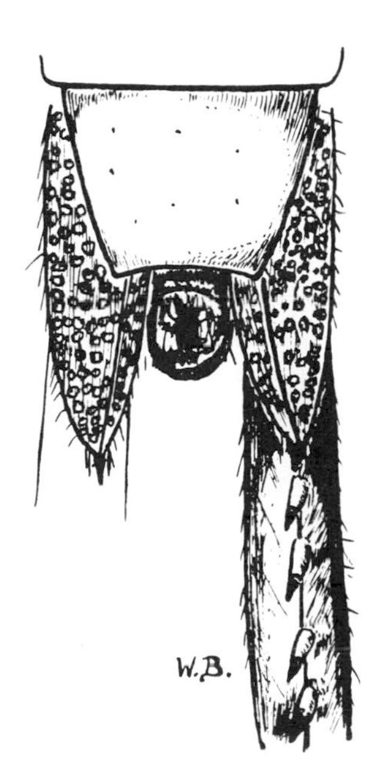

FIG. 12. *Newportia ernsti.* The last sternite, coxopleurae, and ventral view of the prefemur of anal legs.

C. Biology

In scolopendromorphs the sexes are always separated. In a few species of *Scolopendra* it is very difficult, even in adults, to distinguish between males and females. Secondary sex characters are present only in some species of *Otostigmus*, *Scolopendra*, and other genera.

The males produce spermatophores, which, during the mating act, are introduced in the spermathecae of the females. The female then fertilizes her oocytes herself when the eggs are deposited in a dark, protected place. The eggs, studied by us, from 15 to 30 in *Otostigmus tibialis*, *Otostigmus scabricauda*, or *Scolopocr. ferrugineus*, are more or less transparent, yellowish, spherical, not agglutinated, from about 1 to 2 mm in diameter. The mother always guards these until the young ones come out. The embryonal and postembryonal development obeys the so-called "epimorph" evolution: an early embryonal phase, late embryonal phase, and a fetus stage may be distinguished. We have observed that *S.v.viridicornis* is ovoviviparous.

All young and adult scolopendrids have a solitary independent life. They always rest during daylight and capture their prey during night. The specimens of *Scolopendra* and *Otostigmus*, *Cryptops* and others attack their prey with the last prehensorial anal legs; then the head is rapidly curved behind and the venom claws deeply and firmly buried in the body of the prey. The prey is immobilized by several other legs, until it dies from the rapid action of the venom. All scolopendres are voracious creatures, eating worms, spiders, and insects. The larger ones capture also newborn birds or mice. All can go without food for several weeks or months. They take water every day.

Scolopendrids periodically shed the cuticle of the body. The young molt several times a year, but the adults only once a year. After the old cuticle is loosened, it splits first along the sides of the cephalic plate and then the whole animal works out through this opening, the old exuvia being abandoned on the ground. The exuvia of an adult *Scolopendra viridicornis* with 200 mm of body length is only 4 cm long, the head and anal legs very visible and the segmental legs juxtaposed.

I observed that the large *Scolopendra* species became adults only in the third or even fourth year of age. They can live more than 10 years, changing skin only once a year. The small scolopendromorphs, like the representatives of *Cryptops, Newportia, Otostigmus,* are very sensitive to moisture. Nearly all groups burrow into the ground during the dry season or wintertime or even during their entire life, coming out only occasionally. All members of the group are essentially nocturnal, during the daytime lying concealed in holes in the ground, under stones, bark, logs, fallen leaves, in canals.

III. VENOM APPARATUS AND TOXICITY OF SCOLOPENDROMORPH VENOMS

A. Venom Apparatus

Figures 3 and 10 show, respectively, the venom apparatus of a *Scolopendra* and of a *Newportia* species situated on the ventral side of the head. The venom apparatus consists of a venom gland, a venom duct, and the venom-injecting curved, pointed jaw. The venom gland is situated in the interior of the basal articles of the telopodite; the venom duct traverses the terminal needlelike jaw and leads near the tip. Large and powerfully developed muscle bundles, the abductors and adductors of the jaw and of the basal article, move the apparatus horizontally. By contraction of the adductors the venom is pressed out from the central lumen of the gland, forced through the canal, and may be rapidly ejected and inoculated in the body of a victim. All the muscle fibers are striated.

We have studied the venom apparatus in several large neotropical *Scolopendra* species, as well as in *Otostigmus caudatus* and in *S.f. ferrugineus.*

In adult *Scolopendra viridicornis* the venom gland is from about 7 to 7½ mm long and about 1 mm wide in the largest part; the duct is about 7 mm long. In *Otostigmus scabricauda, Cryptops iheringi,* and *S.f. ferrugineus,* these respective measurements are about 4 mm, 3.4 mm, and 3 mm.

After staining the venom glands *in toto* in borax carmine or sectioning and staining them with hematoxylin eosine or with van Guison, we found four main layers on the gland: the central canal (Fig. 13), the secreting cells, the basement membrane, and the muscular layer (Figs. 13, 14). The muscularis (Fig. 13) seems to be covered externally by a very delicate homogeneous sheet of connective tissue. The muscularis itself consists of delicate circular and longitudinal bundles, the fibers extending from one end of the gland to the other and showing cross striation. The basement membrane of

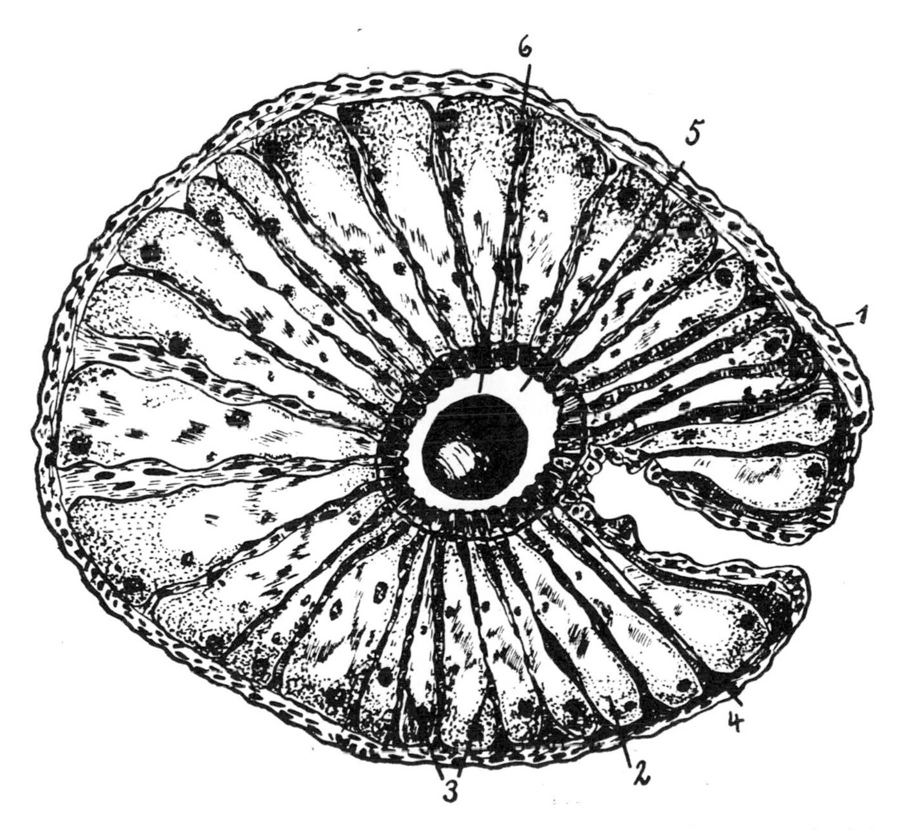

FIG. 13. *Scolopendra viridicornis*. Schematic cross section through a venom gland. 1, muscularis; 2, secreting cell; 3, basement membrane septa; 4, central canal; 5, central lumen; 6, venom.

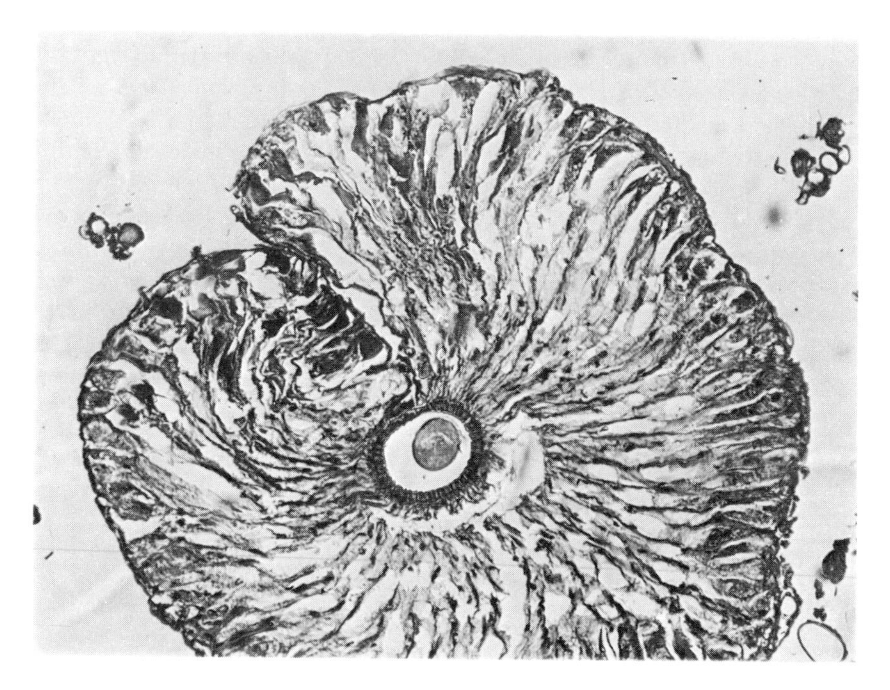

FIG. 14. *Scolopendra viridicornis*. Venom gland, cross section.

connective tissue (Fig. 13) is a transparent layer from which several concentric thin septa (3) extend between the glandular cells, ending on the central canal. The basement membrane seems to be a noncellular sheet; the nuclei, visible in Fig. 13, may be interpreted as nucleal elements of the glandular cells. The glandular or venom-secreting layer consists of numerous polygonal prismatic or columnar cells (Fig. 13-2), each with a subbasal nucleus, secreting granules, and a few larger masses in their supranucleal portion. After using Mallory stain, one may observe all stages of formation of venom granules and their condensation in the apical portions of the cells. The glandular cells show a broad basal portion near the basement membrane; their lateral walls are "S" curved and distally, near the central canal, they are much smaller. The central canal (Fig. 13-4), ringlike in cross sections of the gland, and of yellowish color after staining with Mallory, consists really of the true chitinous cuticular elements of the skin surface, extending from the pit of the jaw to the central portion of the venom gland. On the deeper portion of the gland this canal is enlarged to store the venom droplets extruded from the secreting cells. The wall of the ring is not homogeneous, but subdivided and interrupted by numerous small pores; through these the venom granules migrate from the apical portion of the cells into the central lumen (Fig. 13-5), where they are accumulated (Fig. 13-6). Undoubtedly the venom glands of Scolopendromorphs are of the merocrine type, the cells remaining undestroyed after secretion.

B. Toxicity of Scolopendromorph Venoms

There is very little literature on venomous scolopendromorphs and the effect of their venoms on mammals and man. Some documents are often contradictory. Plateau, D'Herkulais, Briot, MacLeod, Dubosq and others think that these kinds of venom may produce only local effects, burning, swelling, a small necrosis, without serious consequences. Briot wrote in 1904, the *Scolopendra* "fait des morsures très doloureuses chez l'homme avèc oedéme de la partie atteinte". Dobosq (1898) stated that man is extremely sensitive to the venom. During the cold wintertime the effects are local only and transient, but after a bite in summertime the inflammation may progress during 3 days. After a bite on a finger, the hand and the half of the arm may swell. Faust, Calmette, Martini, Eysell and Heymons are of the same opinion. B. de Castro (1921) described the effects of a bite from *Scolopendra heros*; Sebastiany (1870) cited two accidents with *Scolopendra morsitans*. An 8-year-old child and a 49-year-old man have been bitten. Intense local pain, vomiting, headache, swelling of a large area around the bite with a blackish center and lymphangitis were the symptoms. Wood, in 1866 and Chalmers (1919) described several human deaths from *Scolopendra* bites, the poison causing local and general symptoms; at first there is itching, but this is quickly followed by intense pain, which extends all over the limb. A red spot appears at the side of the bite, which enlarges and becomes black in the center, and sometimes there is lymphangitis and lymphadenitis. The general symptoms are great mental anxiety, vomiting, irregular pulse, dizziness, and headache. Hase (1926, 1928), Owano (1917) Cornwall (1915) and Faust (1906) made tentative assays of some chemical components of scolopendromorph venoms, but without success, because nobody has been successful in obtaining sufficient quantities of venom. Stanton Faust, in 1928, related several fatal human accidents which occurred in India, probably from bites from *Scolopendra subspinipes*. Pawlowsky in 1927 described initially severe cases of poisoning; after experiences in 1935 with venoms from *Sc. cingulata*, he concluded that these venoms may produce only local effects on the human body, such as pain, inflammation, edema, and superficial necrosis, all symptoms disappearing after a few days. Venzmer (1932) refers to one fatal case, a child of 7 years, bitten on the head by a Philippine *Scolopendra*. Barthmeyer and Schmalfuss (1935) wrote about several severe accidents with *Scolopendra* bites. Porter (1941) thinks that a bite from

Scolopendra gigantea certainly may be deadly for humans, whereas Machado (1944) and others attribute to the *Scolopendra subspinipes* venom only local pain, edema, and necrosis.

To elucidate the action of the venom on laboratory animals we have collected (Bücherl, 1946) twenty venom glands of the larger and most common Brazilian scolopendromorphs: *Scolopendra viridicornis, Sc. subspinipes, Otostigmus scabricauda, Cryptops iheringi,* and *S.f. ferrugineus,* representants of the two families of this order. Each portion of 20 glands was crushed, suspended in physiological water, filtered and standardized so that 10 ml of physiological water contained exactly the venom of 20 glands. The results of injections of these venoms were the following:

Scolopendra viridicornis:

0.500 gland, injected intravenously, kills all mice of 20 gm body weight in 20 to 32 seconds;

0.250 gland, intravenously, kills all mice in 30 to 49 seconds;

0.061 gland, intravenously, kills all mice in 1 to 2 minutes;

0.050 gland, intravenously, kills all mice in 8 to 13 minutes;

0.040 gland, intravenously, kills all mice in 3–7 hours;

0.030 gland, intravenously, is the median lethal dose, killing half the number of mice.

After injecting the same quantities of venom in mice of 20 gm body weight, but intramuscularly, we found that 0.250 gland is the median lethal dose.

There was no significant local action in these animals, the *S. viridicornis* venom acting very powerfully on the nervous system, with activation of all glands with smooth musculature, respiratory acceleration, abundant sweating, chiefly on the neck, loss of stability, vomiting, respiratory failures, progressive paralysis of the respiratory centers, convulsions and death, followed by immediate tetanus. Undoubtedly this is a neurotoxic venom, acting chiefly on the bulbar centers.

Thirty-six white mice, with more or less 20 gm of body weight, bitten by different specimens of *S. viridicornis* or by the same specimen, but a few days later, were killed in 5 to 8 minutes. One specimen, on my arm, has bitten me and injected the venom: there were pains for about 8 hours, without irradiation, no general effects on nervous system or blood. Thirty-six hours afterward, a small superficial necrosis developed, healing after 12 days.

Scolopendra subspinipes:

0.047 gland is the median lethal dose, intravenously, established in 30 white mice of 20 gm, more or less, of body weight, and *1.2 gland* intramuscularly. *2.5 gland* kill an adult pigeon and also an adult guinea pig within 8 to 16 hours, The symptoms are the same as those of *S. viridicornis.*

Otostigmus scabricauda:

0.012 gland constitutes the median lethal dose, intravenously, established in 25 mice of 20 gm of body weight and 0.070 gland the median lethal dose, intramuscularly.

Cryptops iheringi:

0.150 gland and 0.340 gland are respectively the median lethal doses of adult mice of 20 gm weight, intravenously and intramuscularly.

Scolopocryptops f. ferrugineus:

0.160 gland and 0.390 gland are respectively the 50% mortal doses of adult mice of 20 gm of weight, intravenously and intramuscularly.

C. Conclusions

Human mortality cases after Scolopendra bites have been reported, but it seems that they may not stand up against exact analysis. On the other hand, all scolopendromorphs are venomous; but as they are essentially nocturnal, their first instinct, when emerging into daylight, is to escape to the dark. Under these circumstances even the largest tropical and subtropical scolopendrids seem to be relatively harmless to man, bites occurring very rarely and without serious consequences in man. No serum exists against scolopendromorph bites; no chemical, biochemical, or toxicological newer studies have been made of the constitution of their venom.

REFERENCES

Archey, G. (1923). *Records Canterbury Museum (New Zealand)* **2**, 113–116.

Attems, C. G. (1930). *Tierreich* **54**, No. 2, 1–308.

Auerbach, St. I. (1951). *Ecol. Monogr.* **21**, 97–124.

Barthmeyer, A., and Schmalfuss, H. (1935). *Bronn's Klassen* **5**, No. 2, 317–321.

Bollmann, C. (1893). *Bull. U.S. Nat. Museum*, p. 205.

Brandt, J. F. (1840). *Bull. Sci. Acad. St. Petersbourg* **8**, 12.

Brölemann, H. W. (1902). *Rev. Mus. Paulista* **5**, 35–237.

Bücherl, W. (1946). *Mem. Inst. Butantan (Sao Paulo)* **19**, 181–198.

Bücherl, W. (1969). *Beitr. zur. neotrop. Fauna* **1(3)**, 229–242.

Chamberlin, R. (1901). *Proc. U.S. Natl. Museum* **24**, 21–25.

Chamberlin, R. (1914). *Bull. Museum Comp. Zool. Harvard* **57**, No. 2, 39–104.

Chamberlin, R. V. (1962). *Bull. Univ. Utah* **12(4)**, 1–29.

Faust, E. S. (1928). *In* "Lehrbuch der Toxikologie," pp. 128-129. Fleury, Zanger, Berlin.

Crabill, R. E. (1955). *Bull. zool. Nomencl.* **11**(4), 134–136.

Girard, A. (1880). *Bull. Sci. Dept. Nord.*

Haase, E. (1881). *Z. Entomol.* **8,** 12–17.

Heymons, R. (1898). *Sitzber. Deut. Akad. Wis. Berlin.* pp. 244–251.

Humbert, S. (1894). *Mem. Soc. Geneve* **32**, 1–92.

Immhoff, L. (1853, 1954). *Verhandl. Naturforsch. Ges. Basel* **1**, 35.

Koch, C. L. (1847). *In* "Hans Schaffer's Krit. Revis. d. Insektenf.," Deutschlands, Vol. 3. Regensburg.

Kohlrausch, E. (1881). *Arch. Naturgesch.* **47**, 11.
Kraepelin, K. (1903). *Mitt. Naturhist. Museum Hamburg* **20**, 275.
Kraus, O. (1966). *Abh. senckenb. naturf. Ges.* **512**, 1–143.
Latreille, P. A. (1802–1805). "Histoire Naturelle des Crust. et d. Insects." Paris.
Lawrence, R. F. (1966). *Zoologica Afric.* **2**(**2**), 225–262.
Lewis, J. G. E. (1968). *J. Linn. Soc.* (*Zool.*) **48**, 49–57.
Linnaeus, K. (1758). "Systema Naturae," 10th ed. Holmiae.
Lucas, H. (1840). *In* "Blanchard's Hist. Nat. des anim. articul.," Vol. I. Paris.
Machado, O. (1944). *Bol. Inst. Vital Brazil*, **27**, 5–7.
MacLeod, J. (1878). *Bull. Acad. Belg. Cl. Sci* [2] 45.
Newport, G. (1844). *Ann. Mag. Nat. Hist.* 13.
Pawlowsky, E. N. (1927). "Gifttiere und ihre Giftigkeit." Jena.
Pawlowsky, E. N., and Stein, A. K. (1933). *Med. Parasit. Moskowa* **4**, 1–2, 88–90.
Pocock, R. J. (1902). *Quart. J. Microscop. Sci.* **45**, 3, 417–448.
Porter, C. (1941). *Imprensa. Univ. Santiaga, Chile.*
Sundara Rajulu, G. (1967). *Curr. Sci.* **36**(**9**), 242–243.
Venzmer, G. (1932). "Giftige Tiere und tierische Gifte," Berlin.
Verhoeff, K. W. (1902–1925). *Bronn's Klassen* **5**, No. 2.

Chapter 51

Spiders

WOLFGANG BÜCHERL

INSTITUTO BUTANTAN, SÃO PAULO, FELLOW OF THE CONSELHO NACIONAL DE PESQUISAS, RIO DE JANEIRO, BRAZIL

I. Introduction 197
II. The Morphology of Spiders 199
III. The Classification of Venomous Spiders 209
IV. Description, Distribution, and Biology of Dangerous Species 223
  A. Description 223
  B. Distribution 238
  C. Biology 238
V. Defense against Dangerous Spiders 246
VI. The Venom Apparatus of Spiders 247
  A. The Musculature of the Venom Apparatus 252
  B. The Venom-Inoculating Apparatus 254
  C. The Venom Glands 257
VII. Extraction and Toxicity of Spider Venoms 262
  A. Extraction of Spider Venoms 263
  B. The Toxicity of Spider Venoms 265
  C. Treatment of Spider Bites 273
  References 273

I. INTRODUCTION

Most spiders are actively venomous, with one pair of venom glands, efferent ducts, and the proper apparatus to inject venom into the body of their prey. Their venom quantity and the activity of the venoms introduced into the human body are usually so small, feeble, or low that their bite is of little or no medical importance. The large majority of spiders are very useful animals for man, killing Acarina, insects, and other spiders. Even the so-called "venomous" spiders are commonly shy animals, running away whenever they can. However, most people the world over are deeply convinced that all spiders, without exception, are very poisonous, and that their bite can be deadly for man and higher animals.

On the other hand, many famous arachnologists seriously don't believe that deadly spiders exist. In fact, of more than 25,000 different spider species, only a few do exist with very powerful venom.

There are nearly a thousand publications about the medical importance of dangerous spiders, their venoms, the clinical treatments, the pharmacology, chemistry, and biochemistry of their toxic constituents. Human death can occur after a bite from *Trechona, Atrax, Harpactirella, Loxosceles, Latrodectus, Phoneutria* and perhaps *Mastophora, Chiracanthium,* and *Lithyphantes*; superficial wounds may be common after venom injection from certain species of *Lycosa, Araneus, Argiope, Nephila, Tegenaria,* and *Dendryphantes*; deeper and larger wounds may result from the *Loxosceles* bite and from some Orstognatha, like *Acanthoscurria, Megaphobema, Xenesthis, Theraphosa,* some *Avicularia, Phormictopus, Pamphobeteus*; intravascular hemolysis with hepatic and renal complications result from the bite of *Loxosceles.*

The *Phoneutria* species of South America, the *Atrax* of Australia and New Zealand, the *Harpactirella* of the Cape Town province, and several species of tropical tarantulas are very irritable with defensive-offensive attitudes. None of these attack without provocation, but the simple movements of a sleeping child in its bed are often "interpreted" by the spiders as "aggressive," and they bite. Two little children, sleeping in the same bed in a rural house near São Sebastião on the Atlantic Coast of the state of São Paulo, Brazil, were killed a few years ago on the same night by only one spider, a *Phoneutria* sp., brought to me for identification.

It is curious enough that in the same spider family, even within the same genus, one species may be harmful to men, while another is not. The *Lycosa erythrognatha* from southern Brazil is handsome and harmless during warm weather, but irritable and aggressive when there is no sun for several days. The loxoscelids are not at all aggressive, and will bite only when circumstances force them to defend themselves, as when squeezed against a human body, when they are under clothing. The *Latrodectus* species can crawl delicately over a human hand, but they can bite and will suddenly do so when a human hair is too large to crawl over.

All female spiders become more aggressive when they take care of their egg sacs or young spiders. No male of any spider species is capable of injecting a deadly dose of venom into the human body.

Under these circumstances it is very difficult to distinguish a priori between harmless, dubious, and notoriously dangerous spiders. It is necessary to obey popular belief and to investigate a posteriori the spiders indicated as dangerous. People can identify latrodectism, loxoscelism, and phoneutriism accurately, and can point out with several spectacular names the spiders responsible.

At the "International Symposium on Animal Venoms," held in August

1966 at Butantan Institute, more than twenty original research works on "dangerous spiders" were discussed by Brazilian and other visiting scientists. In several countries spiders constitute a far more serious public health problem than the venomous snakes.

II. THE MORPHOLOGY OF SPIDERS

In the animal kingdom the spiders are classified within the subkingdom Metazoa, division Coelomata or Bilateralia, phylum Protostomia, branch Arthropoda, class Arachnida or Arachnomorpha, order Araneida or Araneae. These latter differ from other orders of the class in having the body subdivided into cephalothorax (Fig. 2a), pedicel (Fig. 2b), and a commonly unsegmented abdomen (Fig. 2c), with six, four, or only two spinnerets, in having one pair of clawlike chelicerae, one pair of leglike pedipalps, four or two book lungs and one, two or four slits for the tubular tracheae.

Their abdomen is commonly saclike and almost without traces of segmentation, except in the very small caudal area. On the ventral side of the abdomen exists the so-called epigastric furrow, separating the basal portion from the other region. In this furrow there are the openings of the lungs, the lung slits, and in the middle line the opening of the reproductive organs, covered in most adult females of the "true" spiders by an external, median, chitinous

FIG. 1. *Avicularia avicularia.* (All photographs of spiders show actual size.)

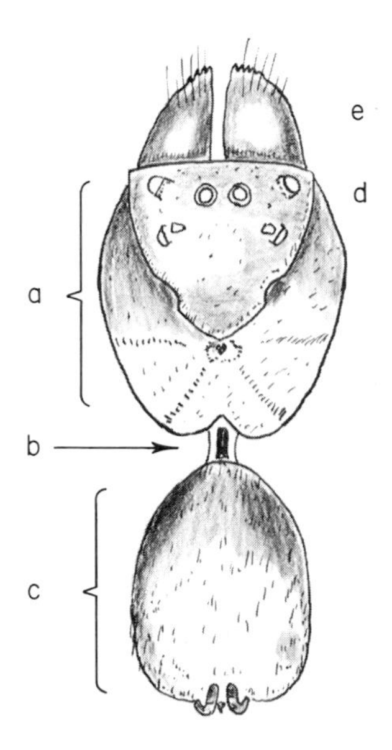

FIG. 2. *Actinopus crassipes*: (*a*) cephalothorax; (*b*) pedicel; (*c*) abdomen; (*d*) eyes; (*e*) chelicerae with rake.

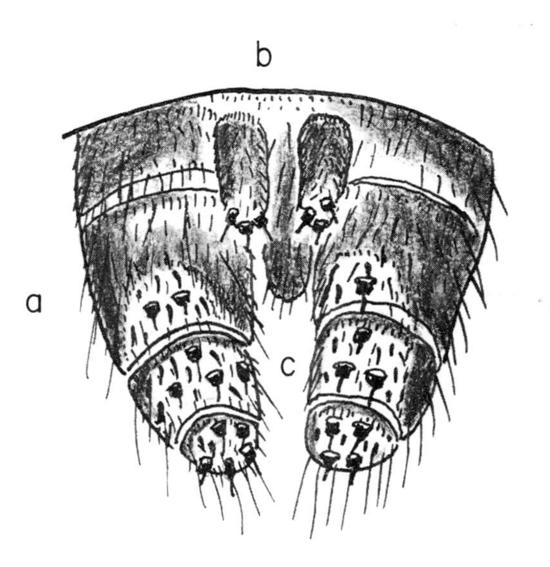

FIG. 3. *Harpactirella lightfooti*: the spinnerets.

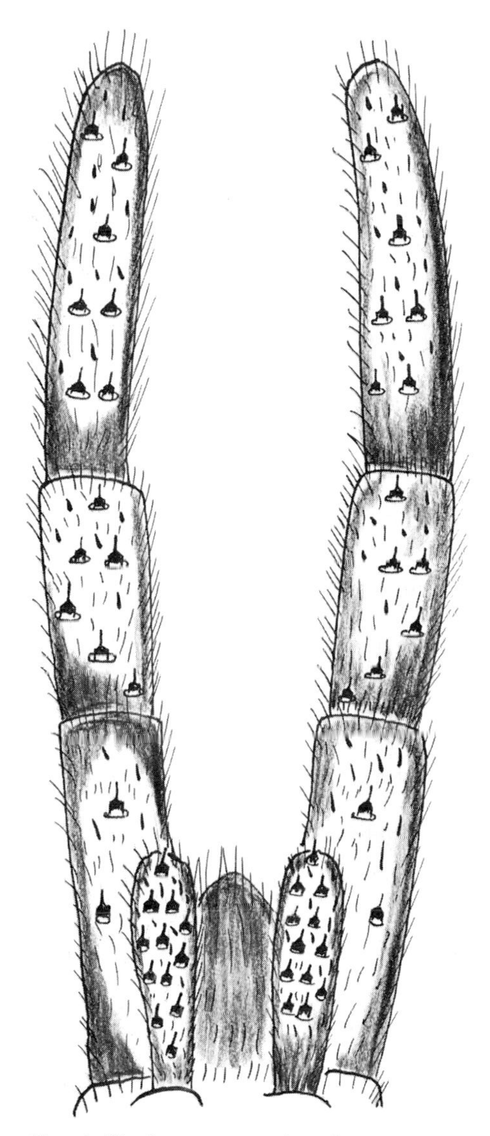

FIG. 4. *Trechona venosa*: the spinnerets.

sclerite, the epigynum. The exact morphological study of the epigynum of adult females is undoubtedly the best morphological characteristic to rapidly determine the spider genus. The epigynum is often absent, even in true spiders, in all mygalomorph spiders, the young, and the males. In all Mygalomorpha or Orthognatha spiders and in some families of Labidognatha or "true" spiders there is a second pair of lung slits, behind the first. On the middle line of the abdomen, a short distance in front of the spinnerets, usually is situated a very small, single spiracle, the opening of the tracheae; or the spiracle is missing or situated near the spinnerets. There are six, or four, or rarely only two spinnerets; the first and the third pair is commonly three-segmented (Figs. 3–5); the second pair is small and not segmented. The length of the posterior pair and the length of the last article of the posterior pair are very important for distinction of the venomous representatives of Mygalomorpha. *Harpactirella* (Fig. 3) bears a very short terminal article, whereas the whole posterior pair of spinnerets in *Trechona* (Fig. 4) and *Atrax* (Fig. 5) is nearly as long as the abdomen. The spinnerets contain the spinning fields and the fusulae or spinning tubes, which are usually two-segmented (Figs. 3 and 4). In the venomous loxoscelids, in the middle line in front of the spinnerets, is a little pointed appendage, the colulus, fingerlike in form and a homologous organ to the cribellum of the orb-weaving spiders. Just behind the spinnerets is situated the posterior opening of the alimentary canal. Different kinds of short or long hairs, brushes, and setae protect the abdomen.

The abdomen is joined to the cephalothorax by a slender stalk, the pedicel or petiolus (Figs. 2b and 6) with either a divided or undivided dorsal sclerite, the lorum and, in some groups, another ventral sclerite, the plagula.

The cephalothorax (Figs. 2a and 6) is unsegmented, although frequently the head or pars cephalica (Fig. 2d) and the thorax or pars thoracica are more or less slightly separated by a median V-shaped groove, the stria thoracica (Fig. 6c), and by radiating depressions, the striae radiantes (Fig. 6c). The cephalothorax is covered dorsally by the chitinous carapace, ordinarily convex in form and more elevated in front. The mouth parts consist of one pair of chelicerae, the rostrum, the epipharynx, the labium, and one pair of pedipalps. Only the rostrum and the epipharynx have no systematic importance. The chelicerae (Fig. 7) are situated in front of and above the mouth and are of vital importance for all spiders in capturing, killing, or holding

FIG. 5. *Atrax robustus*: abdomen and spinnerets.

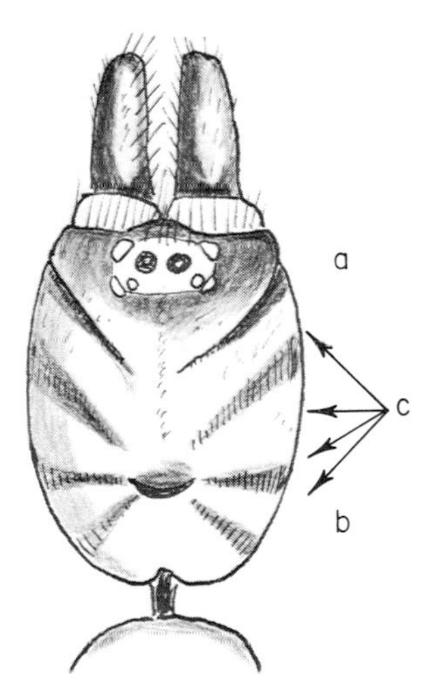

FIG. 6. *Trechona venosa*: cephalothorax. (*a*) eye tubercle: (*b*) thoracic groove; (*c*) striae radiants.

their prey. Each chelicera consists of two segments, a generally thick and powerful basal segment and a curved claw, ulcus, or unguis, folded back into a groove on the ventral side of the basal segment (sulcus subunguealis). In the true spiders the two venom glands are situated in the cephalothorax (loxoscelids, black widows, lycosids, ctenids); in the Orthognatha or "tarantulas," as in the *Trechona, Atrax* (Fig. 5), *Harpactirella, Actinopus* (Fig. 2), and all Theraphosidae (Fig. 1) the venom glands are just on the basal segment of the chelicerae. The poison duct traverses each claw and opens near the tip with an oval slit. In many families of true spiders there is an articulation at the base of each chelicera, on the external face, named the condylus, which is most evident in the genera *Lycosa* and *Phoneutria*. The chelicerae can be moved simultaneously or independently. Only in a few genera of true spiders, such as in *Loxosceles*, are they joined at their base; in Orthognatha and in the genera *Latrodectus, Lycosa,* and *Phoneutria* their independence is remarkable. In certain Orthognatha which burrow in the ground, like the trapdoor spiders (Fig. 7,1) the extremity of the basal segment is armed with several teeth, or stout and rigid spines or setea, which are used for excavation; they are named the rake or rastellum. Along the longitudinal furrow for the reception of the claw, generally a series of teeth exists, which in the Orthognatha form an important instrument for maintaining and squeezing its prey. The venom-injecting claws are very hard, curved, and pointed, with microscopic teeth on the extreme internal side (Fig. 7,3a). The position of the chelicerae in the body of the spiders is of fundamental importance for the distinction of the two suborders of Araneida. Chelicerae

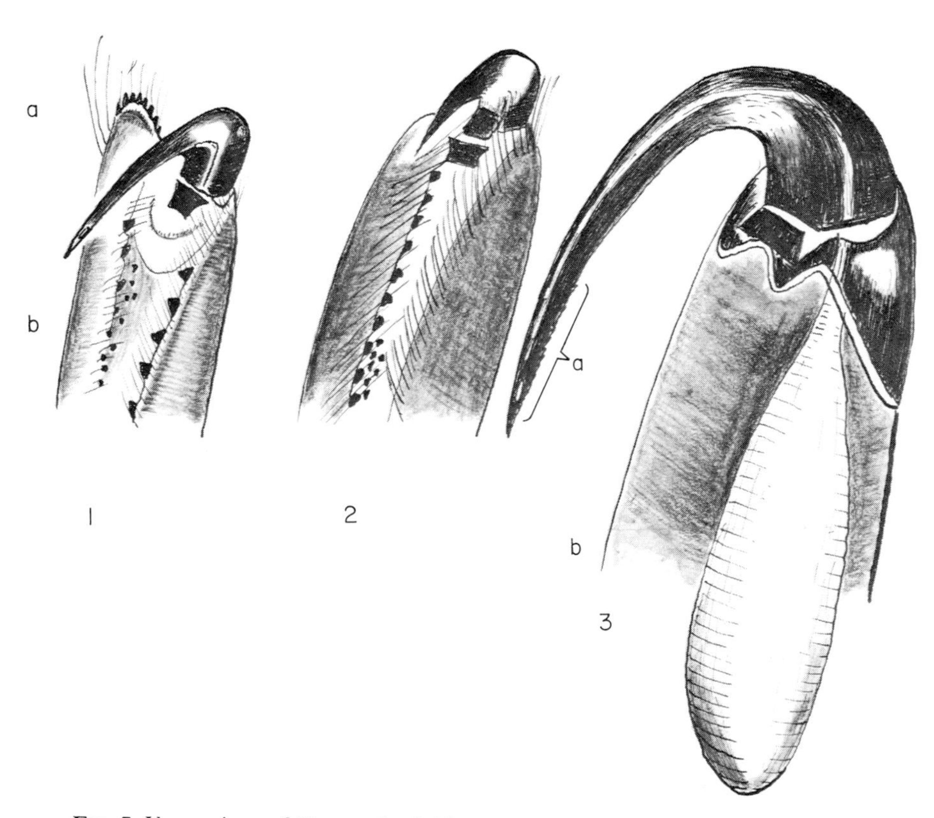

FIG. 7. Venom jaws of "tarantulas." (1) *Actinopus*; (2) *Trechona*; (3) *Lasiodora* (with venom gland).

paraxial to the body length and claws moving vertically define the suborder Orthognatha or "tarantulas"; chelicerae diaxial to the body length and claws moving horizontally define the Labidognatha or "true" spiders.

The labium (Figs. 8 and 9) forms the ventral wall of the mouth. More or less movable and separated from the sternum by a suture (Fig. 8) or immovable and forming the anterior part of the sternum (Fig. 9), the labium is a chitinous, rounded, or subquadrate sclerite. On its front part are brushes of hairs or setae and other differentiated structures like the "cuspulae" (Fig. 8). The pedipalps (Fig. 10) are situated one on each side of the mouth between the labium and the chelicerae. They are commonly leglike in aspect, chiefly in the females and in young spiders of both sexes. In females each pedipalp consists of six segments, and in adult males there are seven, named the coxa, trochanter, femur, patella, tibia, tarsus, and, in the males, the bulbus. On the inner face of the coxae, near the labium, the endite or maxillary plate, can

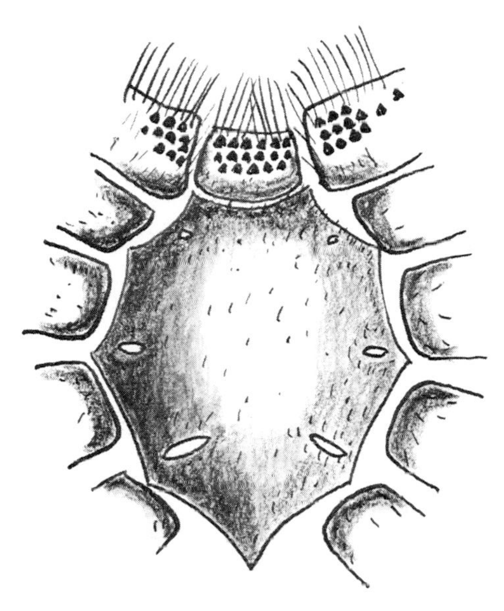

FIG. 8. "Tarantula"; sternum and lip.

be seen, usually with finely toothed keels, the serrula, so very important for lacerating the prey and thereby injecting its juices. In several spider groups the endites may be poorly developed or even absent. On the anterior face of the coxae in some spiders a stridulating lyra (Fig. 11) may exist, with a series of vibrating parallel chords. The form and aspect of the copulating organ or bulbus (Fig. 12) in the males—as well as the epigynum of the females—is of fundamental importance for the exact classification of the spider genera. Even for the distinction of the species they may be very important and useful. Complete revision of spider collections throughout the world, using the epigyna of females and the pedipalps and bulbs of the males as a taxonomic guide, certainly would improve many systematic difficulties.

The six or eight eyes (Figs. 2 and 6) are situated on the anterior median part of the carapace, behind the front or clypeus. All spider eyes are of simple structure, resembling occlli of insects. In some genera they are situated upon a tubercle, the comorus (Fig. 6), in others they occupy either half or the whole width of the head (Fig. 2). Number, arrangements, and distances of the eyes are important for classification of spiders. The so-called nocturnal eyes —as in loxoscelids— are commonly pearly white, the diurnal eyes are yellow, blackish, green, or variously colored. The same spider can have eyes of both types or only those that are diurnal or nocturnal. The anterior median eyes are postbacillary and the others commonly prebacillary in their internal

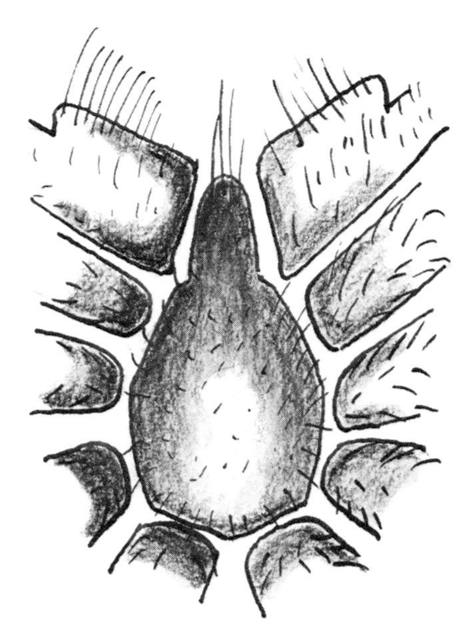

FIG. 9. *Actinopus*: sternum and lip.

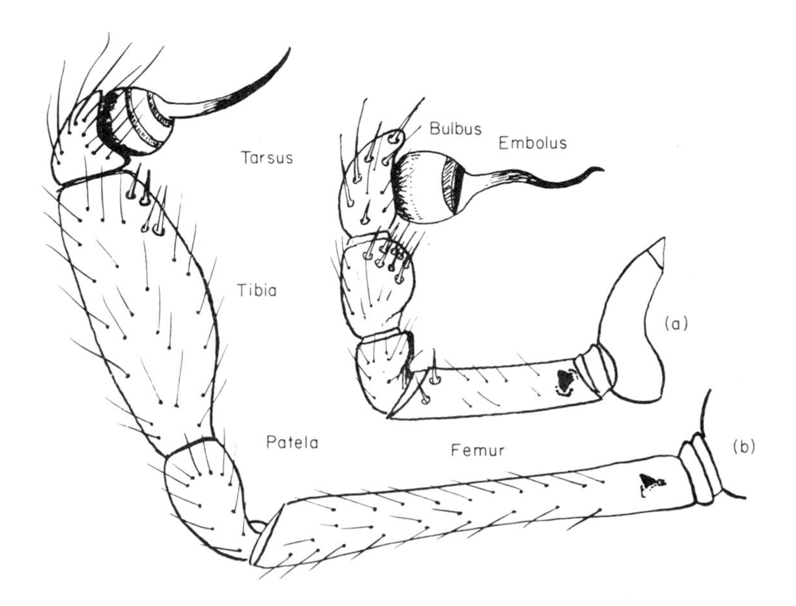

FIG. 10. Male palpus of *Loxosceles*: (*a*) *L. similis*; (*b*) *L. rufipes*.

anatomical structure. All eyes are generally arranged in two or three transverse rows, or as in the six-eyed spiders, in three diads (*Loxosceles*). In eight-eyed spiders the following names and abbreviations are used in systematic works: anterior median (A.M.); anterior laterals (A.L.); posterior median (P.M.); and posterior laterals, (P.L.). The two rows of eyes can be procurved, recurved, or nearly straight.

The thorax bears the four pairs of walking legs. Each leg consists of seven segments named after those of the pedipalps, except the penultimate article called the metatarsus. The tibiae and metarsi are generally armed with long, stout, erectile spines (Fig. 13, 1). The number and the position of these over the articles is often very important for the correct determination of a specimen. Hairs of different kinds and colors, longer or shorter setae or spinules may cover the leg segments. The lower surfaces of tarsi, and often also the metatarsi are generally covered by soft, short, velvety hairs named the scopulae or tenent brushes. (Fig. 13,2 and 3). The tarsal scopulae are undivided, or are

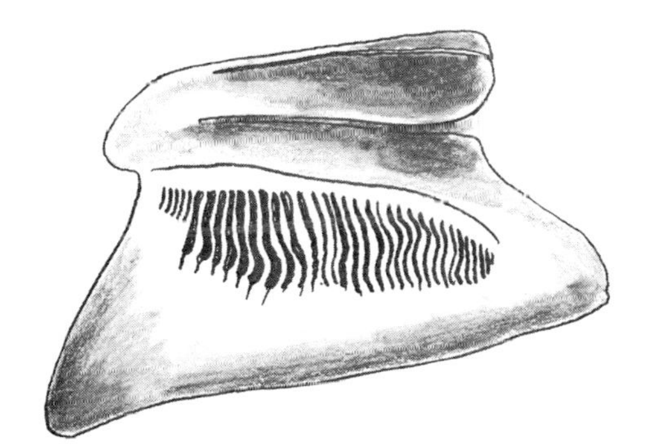

FIG. 11. Stridulating apparatus (lyra of *Trechona venosa*).

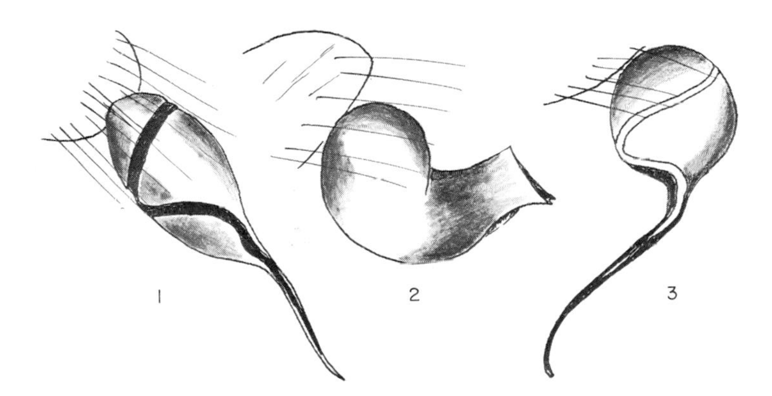

FIG. 12. Mating bulbs of tarantulas. (1) *Grammostola*; (2) *Pamphobeteus*; (3) *Avicularia*.

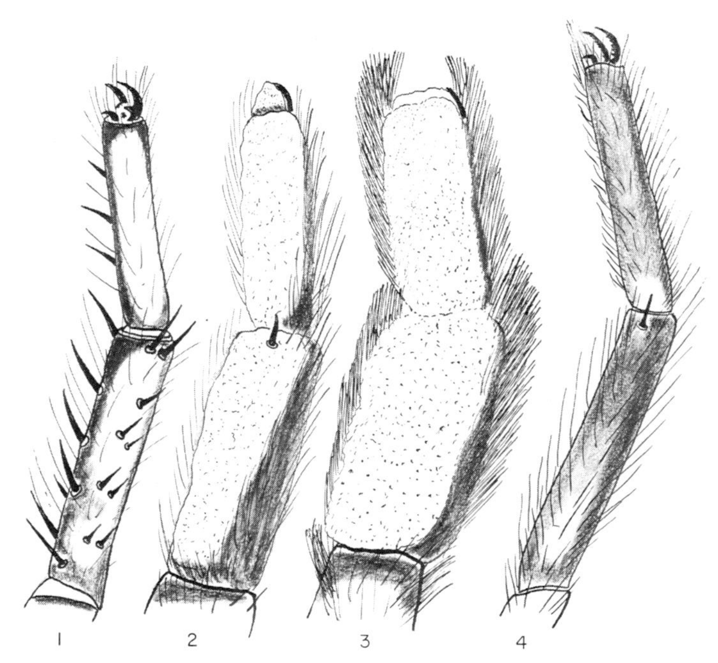

FIG. 13. Metatarsus and tarsus of four kinds of tarantulas. (1) *Actinopus*; (2) *Lasiodora*; (3) *Avicularia*; (4) *Trechona*.

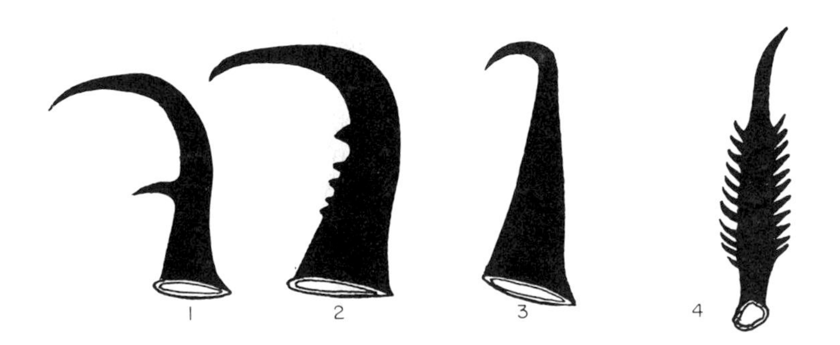

FIG. 14. Leg claws of tarantulas. (1) *Actinopus*; (2) *Lasiodora*; (3) *Avicularia*; (4) *Trechona*.

bisected in two areas by a ventral median, longitudinal strip of longer setae (Fig. 13, 4). The end of the tarsi bears two or three claws (Fig. 13,1 and 4) situated over a little movable terminal article, named the onychium. The two superior claws are the largest and may bear one or two series of fine teeth or a single tooth only (Fig. 14). Spurious claws often may be present, as in black widows. The claw tufts or fasciculi subungueales generally are two-bundled and are present in many groups of Theraphosidae completely covering the claws (Fig. 13,2 and 3). Claw tufts and tarsal scopulae serve to aid the spider in clinging to very smooth surfaces, even to glasses in vertical position.

The sternum (Figs. 8 and 9) is the plate forming the ventral wall of the thorax. Each lateral margin bears four notches for the coxae of the legs. In many Orthognatha the sternum shows one, two, or three pairs of lateral impressions, the sigillae, which are often considered in the classification of genera, and species.

III. THE CLASSIFICATION OF VENOMOUS SPIDERS

Notoriously dangerous spiders known the world over may be classified as follows:

Order Araneida or Araneae

Key for families

1

Chelicerae paraxial; venom glands in the basal segment of the chelicerae; four book lungs near the epigastric furrow; suborder: Orthognatha, Mygalomorphae, or "Tarantulas". 2

Chelicerae diaxial; venom glands in the cephalothorax; generally only one, rarely two pairs of book lungs; suborder: Labidognatha, Araneae Verae, or "True Spiders" . . 4

2

Claw tufts wanting; all tarsi generally with three claws; the tarsi of the third and fourth pair of legs generally without scopulae or only with short apical scopulae; chelicerae without a rastellum; lip free; posterior spinnerets long—longer or as long or only slightly shorter than the abdomen. (1) Family: Dipluridae.

Claw tufts present; all tarsi generally with two claws; posterior spinnerets much shorter than the abdomen 3

3

Terminal joint of the posterior spinnerets as long as or distinctly longer than the preceding segment. (2) Family: Theraphosidae.

Terminal joint of the posterior spinnerets distinctly shorter than the preceding article. (3) Family: Barychelidae.

4

Chelicerae soldered at base; six nocturnal, pearly white eyes in three diads; cephalothorax more or less depressed; brown, scanty-haired, strictly nocturnal spiders from about 1 cm body and leg length. (4) Family: Sicariidae or Scytodidae.
Chelicerae free; eight eyes, homogeneous or of different color 5

5

Three claws; claw tufts usually wanting 6
Two claws; claw tufts usually present 8

6

Fourth tarsus with a comb of curved, ventrally serrated, blackish bristles; abdomen generally globose, blackish, with red, yellow, or grayish markings; scanty-haired spiders with a body length from about 0.8 to 1.3 cm and a leg length from about 1.3 to 1.6 cm. (5) Family: Theridiidae.
Fourth tarsus without a comb, but with common hairs and setae or with spurious claws; abdomen oblong, striped or diversely colored 7

7

Tarsi with at least one pair of spurious claws in addition to the three true claws; eyes usually homogeneous; small orb-weaving spiders. (6) Family Araneidae or Argiopidae.
Spurious claws lacking, chelicerae with a distinct boss or condylus; third claw smooth or with a single tooth; eight eyes in three rows: 4-2-2. (7) Family: Lycosidae.

8

Eye formula: 4-2-2, the A.M. by far the largest and directed forward, eyes of the second row by far the smallest, often indistinct; small, vagrant, jumping, house and garden spiders. (9) Family: Salticidae or Attidae.
Eye formula: 2-4-2, the median of the second row larger than that of the first row and the eyes of the third row by far the largest. (10) Family: Ctenidae.
Two claws; eyes in two rows with the formula: 4-4; maxillary scopula delimited, not extending over the total external surface; small spiders of about 15 to 20 mm of body length. (8) Family: Clubionidae.

1. *Diagnoses, Keys for Classification, and the Names of the Most Important Species.*

Suborder Orthognatha or "tarantulas": This suborder, called also Mygalomorphae or Avicularioidea by some authors, includes equally large and small "tarantulas", distinguished from the true spiders by vertical movements of their chelicerae and by the fact that the two venom glands are situated in the basal segment of the chelicerae. All the venomous representants of this suborder have two pairs of book lungs, always easily visible and situated in front and behind the lateral sides of the epigastric furrow. Each lung has a single posterior slit. In most of them there are two pairs of spinnerets, the first being unsegmented and of small size, situated ventrally, and the superior pair longer, three-segmented. The adult females do not have external genitalia, but after careful inspection of the genital groove one may distinguish the simple genital opening and two very small chitinous seminal receptacles. The copulating organs of the adult males at the ends of the tarsi of the pedipalps are of simple structure, with only a dilatable calix, an enlarged bulbus and a smaller, long, often serpentinform embolus or stylus. The form of the stylus is very important in distinguishing the different genera (Fig. 12).

Notorious venomous "tarantulas" exist in the families Dipluridae, Theraphosidae, and Barychelidae, found in several species in Australia and New Zealand, in South and Central America and in South Africa, respectively.

(1) Family Dipluridae: "Funnel-web spiders" or "Mygales fileuses"

Key for subfamilies with venomous representants:

Upper claws pectinated in a double row; a stridulating apparatus on the basal segment of chelicerae and the coxae of the pedipalps (Fig. 11). Subfamily: Diplurinae, tribe Trechonini.

Upper claws pectinated in a single row. Subfamily: Macrothelinae — Tribe Atraxini.

The most important genus of the tribe Trechonini is the South American *Trechona* C. L. Koch 1850, with the type-species *T. venosa* (Latreille) 1830, described by Latreille (1830, 1832) as *Mygale venosa*; by Walckenaer (1835–1837) as *Mygale zebra*; by Koch (1842) as *Trechona zebra*; by Pocock (1895) as *T. zebrata*; by Karsch (1879), Simon (1892), de Mello-Leitão (1923), Vellard (1924), and others as *Tr. venosa.*

In the subfamily Macrothelinae, tribe Atraxini, is situated the famous genus *Atrax* O. P. Cambridge 1877 with the following species: *Atrax robusta* O. P. Cambridge 1877, *A. formidabilis* Rainbow 1914, *A. versuta* Rainbow 1914, *A. tibialis* Rainbow, 1914, *A. modesta* Simon 1891, *A. pulvinator* Hickmann 1927, *A. venenatus* Hickman 1927, and *A. valida* Rainbow and Pulleine 1918.

(2) Family Theraphosidae or Aviculariidae, according to some authors, or "tarantulas," are Bird spiders, *Vogelspinnen, Mygales errantes, Mygales chasseuses, Araignées-crabes, Araña pollito, a. peluda, a. de caballo, a. matacaballo, a. revienta caballos, Aranhas caranguejeiras; Nandú-guassú, nandú-cavallú, ana-rymbá.*

The representants of Theraphosidae have no rastellum over the end of the basal segment of chelicerae; their maxillary lobes of the coxae of pedipalps are more or less rudimentary; the lip is always free and more or less movable; the tarsi of legs are armed with only two claws, slightly or abundantly pectinated in only a single series with claw tufts well-developed and always present as a single or a double bundle; four spinnerets; the last joint of the superior pair being usually distinctly longer than the preceding joint or only a little longer or, in a few cases, as long as the median segment, but never distinctly shorter, except in the young. The thoracic groove is transverse, pro- or recurved, or straight. The absence of a third claw and of the rake and the presence of claw tufts and tarsal scopulae, the short spinnerets, mark the difference between the "tarantulas" and the specimens of the preceding family.

To this family belong the largest spiders known, those which are most feared because of their large size, their strange behaviour, and hairy appearance. In tropical and subtropical regions from Amazonia, the West Indies to the highest Andes in Chile are several species with a body length from 8–12 cm whose legs expand to more than 10–20 cm. The venom fangs extend from 6–9 mm and can be deeply introduced into the body of their victims.

Key to subfamilies with venomous representants:

1

Legs, chiefly the tibia and the metatarsus of the fourth pair, with numerous, strong spines; none of the tarsal scopulae in adults bisected by setae 2

Legs without spines or only with very small spines at the distal end of the metatarsi. (1) Subfamily: Aviculariinae.

2

Fourth femur with a thick, velvety pad of short hair on the retrolateral surface. (2) Subfamily: Theraphosinae.

Fourth femur without velvety pad of hairs, but with longer semierected setae. (3) Subfamily: Grammostolinae.

The Aviculariinae include the genera *Avicularia* Lamarck 1818, *Psalmopoeus* Pocock 1895, *Pachystopelma* Pocock 1901, and *Tapinauchenius* Ausserer 1871. The most important genus is *Avicularia* with 32 different species. The commonest and best known are *Avicularia avicularia* (Linnaeus) 1758. *A. caesia* (C. L. Koch) 1842, *A. detrita* (C. L. Koch) 1842, *A. laeta* (C. L. Koch) 1842, *A. metallica* Ausserer 1875, *A. rutilans* Ausserer 1875, *A. velutina* Simon 1889, and *A. versicolor* (Walckenaer) 1775.

The subfamily Grammostolinae includes the genera *Aphonopelma*, Pocock 1901, *Brachypelma* Simon 1891, *Chaunopelma* Chamberlin 1940, *Clavopelma* Chamberlin 1940, *Citharacanthus* Pocock 1901, *Delopelma* Petrunkevitch 1939, *Dugesiella* Pocock 1901, *Grammostola* Simon 1892, *Pterinopelma*

Pocock 1901, *Eurypelma* C. L. Koch 1850 and others. The genus *Grammostola* is the most important, with very large representants, such as the species *G. actaeon* (Pocock) 1903, *G. gossei* (Pocock) 1900, and *G. mollicoma* (Ausserer), 1875.

The subfamily Theraphosinae includes several genera of large, venomous, and more or less aggressive species. The genera may be separated as follows:

1

Without stridulating apparatus 2
With stridulating apparatus 6

2

Tibiae and metatarsi of the fourth pair of legs are thicker than those of the first pair; tibia is as large as or even larger than the femur; fourth tibia and metatarsus with numerous, long, erect hairs, setae or spines: *Eupalaestrus.*
Tibiae and matatarsi of the fourth pair of legs are more slender than those of the first pair, tibia is thinner than femur, metatarsus is thinner and longer than tibia 3

3

The metatarsi of the fourth pair of legs with dense, velvety scopulae from the apex to the base: *Xenesthis*
Metatarsi IV with apical scopulae only 4

4

Femur of the third pair of legs very thickened; the fourth pair of legs distinctly longer than the first pair; the male with two apical apophyses on the apex of the tibia I, the apophysis inferior longer and curved inside: *Megaphobema.*
Femur III normal, not thickened; the fourth pair of legs as long as, or only insignificantly longer or even shorter than the first pair; the males without apophyses or with two apophyses, but the inferior curved outside 5

5

The tibiae of the first pair of legs from the male without apophysis; patella and tibia of the fourth pair of legs as long as those of the first pair: *Sericopelma.*
Tibia I of the male with two apophyses, the inferior one longer and curved to the outer side; patella and tibia IV somewhat longer than that of the first pair–*Pamphobeteus.*

6

A distinct stridulating apparatus is present on the anterior side of the coxae of the first pair of legs, above the suture, consisting in claviform, erectile setae; males with two apophyses on tibia I, the inferior larger but more or less straight: *Lasiodora.*
Stridulating apparatus on the coxae and the trochanters or only on the trochanters . . . 7

7

Stridulating apparatus on the coxae and the trochanters of the first pair of legs; the last pair of sigillae far from the margins of the sternum 8
Stridulating apparatus situated on the trochanters of the first pair of legs and the pedipalps 9

8

Eye tubercle depressed, the eyes minute and distant; scopulae on metatarsus II apical only, on metatarsus IV absent; apex of tibia I of the male without apophysis: *Theraphosa.*

Eye-tubercle more wider than long, eyes larger and narrower; scopulae present on metatarsi III and IV, but not complete; with two apophyses on the apex of tibia I of the males: *Phormictopus.*

9

Thoracic groove large, transverse; last pair of sigillae near the margin of the sternum; tibia I of the males only with one lateral, rakelike apophysis: *Acanthoscurria.*
Thoracic groove very small, procurved; last pair of sigillae far from the margins: *Trasiphoberus.*

Eupalaestrus Pocock 1901 includes the following species: *E. campestratus* (Simon) 1891, *E. pugilator* Pocock 1901, E. *spinosissimus* de Mello-Leitão 1923, E. *tarsicrassus*, and *E. tenuitarsus* Bücherl 1947.

Xenesthis Simon 1891 includes *X. immanis* (Ausserer) 1875. *X. intermedius* Vellard, Gerschmann and Schiapelli, 1945, and *X. monstruosus* Pocock 1903.

Megaphobema Pocock 1901 includes the species, *M. robusta* (Ausserer) 1875.

Sericopelma Ausserer 1875 includes the species *S. communis* F. Cambridge 1897, *S. fallax* de Mello-Leitão 1923, and *S. rubronitens* Ausserer 1875.

Pamphobeteus Pocock 1901 includes more than 20 species. The most common are *P. antinous* Pocock 1903, *P. augusti* (Simon) 1887, *P. benedenii* (Bertkau) 1880, *P. ferox* (Ausserer) 1875, *P. fortis* (Ausserer) 1875, *P. insignis* Pocock 1903. *P. isabellinus* (Ausserer) 1871, *P. vespertinus* (Simon) 1887, *P. roseus*, and *P. tetracanthus* de Mello-Leitão 1923.

Lasiodora C. L. Koch 1850 encompasses also more than 20 species. The most important are *L. klugi* (C. L. Koch) 1842, *L. saeva* (Walckenaer) 1837, *L. spinipes* Ausserer 1871, and *L. striatipes* (Ausserer) 1871.

The genus *Theraphosa* Thorell 1870 includes only the species *Th. blondi* (Latreille)1804.

Phormictopus Pocock 1901 includes 13 species of which the most important are *Ph. cancerides* (Latreille) 1806, *Ph. cubensis* Chamberlain 1917, *Ph. hirsutus* Strand 1906, and Ph. *meloderma* Chamberlain 1917.

Acanthoscurria Ausserer 1871 encompasses 38 species. The most important are *A. antillensis* Pocock 1903, *A. atrox* Vellard 1924, *A. brocklehursti* F. Cambridge 1896, *A. convexa* (C. L. Koch) 1842, *A. ferina* Simon 1892, *A. geniculata* (C. L. Koch) 1842, *A. gigantea* Tullgreen 1902, *A. insubtilis* Simon 1892, *A. minor* Ausserer 1871, *A. musculosa* Simon 1892, *A. sternalis* Pocock 1903, *A. suina* Pocock 1903, *A. tarda* Pocock 1903, and *A. theraphosoides* (Ausserer) 1871.

Trasiphoberus Simon 1903 includes only the species *T. parvitarsis* Simon 1903.

(3) Family Barychelidae

Lip free; maxillary lobes rudimentary; chelicerae generally with a rastellum; two claws pectinate in a single row; claw tufts present; eight eyes; terminal joint of posterior pair of spinnerets very short. This family is subdivided into the subfamilies Diplothelinae, with only two spinnerets, Sasoninae, with four spinnerets, eyes dispersed over the head, chelicerae without rastellum, Barychelinae, with four spinnerets, eyes in a compact group, chelicerae usually with a rastellum at least in the female, eyes not on a tubercle, lateral eyes separated at least by their diameter, and Leptopelmatinae, with four spinnerets, eye group on a distinct tubercle; lateral eyes separated by less than their diameter. To this subfamily belongs the famous genus *Harpactirella* Purcell 1902, with 11 venomous species: *H. karrooica*, *H. lightfooti*, *H. longipes*, *H. treleaveni* Purcell 1902, *H. domicola*, *H. helenae*, *H. magna*, Purcell 1903, *H. schwarzi* Purcell 1904, *H. lapidaria*, *H. spinosa* Purcell 1908, *H. flavipilosa* Lawrence 1936.

Suborder—Labidognatha or Aranae Verae or True Spiders

To this suborder belong all the notoriously venomous spiders from strictly diversified structures, life habits, and of relatively small size. Of many families, several hundreds of genera and many thousands of species distributed throughout the world, only the families Sicariidae, Theridiidae, Lycosidae, and Ctenidae include very venomous species and the families Argiopidae, Salticidae, and Clubionidae suspect species.

(4) Family Sicariidae or Scytodidae

With one pair of book lungs; chelicerae fused together at their base; without cribellum and calamistrum; colulus present, fingerlike in aspect; six spinnerets; usually six eyes only, pearly white, nocturnal, arranged in three diads; if eight eyes, they are disposed in two rows and the lateral ones of each side are on a tubercle; promargin of chelicerae, on the inner side, with a transparent lamina; chelicerae on the outer side without condylus, often with microscopical stridulating ridges; lip longer than broad, immovable or free; maxillary lobes of pedipalps converging around the lip; adult males with a simple copulating bulb and a simple embolus; adult females of haplogyn type, without external sexual organs or epigynum, the receptacula seminalia being covered by the chitinous integument; with a single barely visible tracheal spiracle near the spinnerets; small spiders.

Only in the subfamily Loxoscelinae are spiders of medical importance found. This subfamily may be easily distinguished from the other six subfamilies by the following characters: retroclaws pectinated in a single row; lip immobile; six eyes in three diads; tarsi with an onychium; fourth coxae approximated together; third tarsal claw generally absent.

The only genus of Loxoscelinae is *Loxosceles* Heinecken and Lowe 1833, 35, which includes very venomous species of medical importance, commonly called "brown spiders," *araña de los rincones, araña homicida, La temible or huayruro*, etc. About 50 different species have been described, 18 from North and Central America and the West Indies, 17 from South America, and about 14 or 15 from southern Europe, North and Central Africa, and Asia. The systematic classification of loxoscelids continues in a confused state. The former authors have described many species in a very limited manner, without distinction of the sexes, many type-species have been lost, the ubiquitous nature of some species have contributed to confusion in the classification of these "monotone spiders." Many *Loxosceles* have been introduced on products shipped from continent to continent. Gertsch, in 1958, was the first to base the classification of the species exclusively on the microscopical structures of the female receptacle and the male pedipalps and copulating organs. Certainly this is an important step. The following species are the most important and are of medical interest: *Loxosceles rufescens* (Dufour) 1820, *L. rufipes* (Lucas) 1834, *L. spadicea* Simon 1907, *L. reclusa* Gertsch and Malaik 1940, *L. decemdentata* Franganillo 1926, *L. distincta* (Lucas) 1846, and *L. unicolor* von Keyserling 1887. *L. laeta* (Nicolet) 1849 was synonymized by Bücherl in 1961 with *L. rufipes*. From all other species, described by different authors, we know only one male, one female, or one young. Most of the species described by Gertsch in 1958 have possibly only subspecific or populational range.

(5) Family Theridiidae

One pair of book lungs; chelicerae free; cribellum and calamistrum absent; has a single tracheal spiracle near the spinnerets; with three tarsal claws and the fourth tarsus of legs with spurious claws or with a ventral series of large, stout, blackish, ventrally serrated bristles; cololus present; six spinnerets, the anterior pair approximated, shorter, and stouter than the posterior pair; eight heterogeneous or even homogeneous eyes, in two rows, the A.M. pair alone diurnal, or only six or even four nocturnal eyes; sometimes eyes wanting; lip free; chelicerae without condylus; maxillary lobes slightly converging, with a fringe of long hairs; legs with few or without spines; upper claws smooth or commonly pectinate in a single row.

The Theridiidae include more or less 1500 species, separated into subfamilies. The suspect or proved venomous species belong to the two subfamilies *Asageninae* and *Lactrodectinae*.

The representants of Asageninae show a soft abdomen; claws with well-developed teeth of almost equal length; abdomen more or less oval, with a stridulating organ in the adult males. The suspect venomous species belong to the genus *Lithyphantes* Thorell 1870. The commonest are *L. anchoratus* (Holmberg) 1876 and *L. andinus* von Keyserling 1884, the famous "Cirari" of Bolivia, Paraguay, and Chile.

The subfamily Latrodectinae includes representants with a soft globular abdomen; on their superior claws exist well-developed teeth, disposed in a single row and of equal length; without the stridulating organ in adult males and females, at least in the *Latrodectus*.

The subfamily Latrodectinae includes extremely important poisonous spiders, perhaps the most famous and dangerous arachnids we know. They can easily be distinguished by the following morphological characters: abdomen is soft, rounded, globose; body and legs are without visible hairs; the upper claws bear well-developed teeth, arranged in a single row and of the same length.

Latrodectus Walckenaer 1805

"Black widow – *Schwarze Witwe – viuda negra-viuva negra*"—universal names
"Karakurt" – *"Tchim"* – southern Russia
"Tendaraman" – Marroccos
"Mur Kion"– by Avicenna
"Malmignathe" – Corsica
"Katipo" – "Red back spider" – Australia, New Zealand
"Knoppie" – or "Button-spider" or *Knoppiespinnekop* – South Africa
"Araña capulina" – *"Chintatlahue"* or *"Po-ko-moo"* – native people of Mexico
"Coul-rouge" or *"24-horas"* – in the Antilles
"Hour-glass spider" – people in the United States
"Araña naranja" – Venezuela
"Lucacha" – in Peru
"Huyruro" – in Bolivia

"Araña brava" – *"Pallu"* – *"araña del rastrojo"* – *"rastrojera"* – *araña de lino"* – *"del trigo"* – Chile, Argentina, Uruguay, Paraguay

The genus *Latrodectus* includes, according to F. P. Cambridge "those very interesting spiders which, under various local names, have been notorious in all ages and in all regions of the world where they occur on account of the reputed deadly nature of their bite."

The type-species is *Latrodectus 13-guttatus* Rossi 1790 (= *L. mactans*) by subsequent designation of Latreille 1810. A revision of the black widow genus was made by O. P. Cambridge in 1902. Dahl in 1902 and Badcock in 1932 created many new names for older species. Chamberlain and Ivie, in 1935, studied the widow spiders from Northern Mexico with descriptions of several subspecies. These names were subsequently used by nonarachnologists, by physiologists and toxicologists and great confusion between the species was created in the United States, in South America, and in other continents. As a result, the toxicological and medical literature on black widows, chiefly on the American continent, confuses the spider species. Roewer's *Katalog der Araneae*, in 1954, lists 21 species of *Latrodectus* all over the world; Bonnet's *Bibliographia Araneorum*, in 1957, also lists the same 21 species. Since that time *L. rivivensis* has been described by Shulow, in 1948, in Palestine. The definitive solution of taxonomic problems essentially depends on further collections of specimens. Immature spiders cannot be identified with certainty. Adults are best classified by comparing their genitalia (females) or bulbi (males) under a dissecting microscope, and by also comparing the thickness of legs, the length of the first and fourth legs, the shape of the clypeus, the measures of the carapace.

Levi, in 1958 described only three species of *Latrodectus* on the whole American continent: *L. geometricus, L. mactans,* and *L. curacaviensis;* in 1959, after a carefully, comparative study, chiefly after very accurate comparison of the genitalia of the adults, and after a complete literature review he will admit to only six species in the world:

1. *L. geometricus* C. L. Koch 1841 – cosmotropical. Synonyms: *L. zickzack* Karsch 1878 (Zanzibar), *L. obscurior* Dahl (Madagascar), *L. concinnus* O. P. Cambridge 1904 (Cape Town), *L. g. subalbicans* Caporiacco 1949 (Kenya). *L. g. modestus* Pa. 1949 (Kenya), and *L. g. obscuratus* Cap. 1949 (Kenya)

2. *L. mactans* (Fabricius) 1775 – cosmotropical and temperate zones, but most widespread in America. There are five subspecies (2a – 2e).

2a. *L. mactans mactans* (Fabricius) 1775 – From United States to Argentina. Synonyms: *L. formidabilis* Walckenaer 1837 (Georgia), *L. perfidus* Walck. 1837 (Georgia), L. *variolus* Walck. 1837 (Georgia), *L. intersector* Walck. 1837 (Georgia), *Theridion verecundum* Hentz 1850 (U.S.A.),

Th. lineatum Hentz 1850 (U.S.A.), *L. insularis insularis* Dahl 1902 (Haiti), *L. i. lunulifer* Dahl 1902 (Porto Alegre, Rio Grande do Sul: Brazil), *L. sagittifer* Dahl 1902 (Bismarck Archipel), *L. ancorifer* Dahl 1902 (Luzon), *L. hahli* Dahl 1902 (East Africa), *L. luzonicus* (Luzon), *L. stuhlmanni* (East Africa), *L. renivulvatus* Dahl 1902 (South West Africa), *L. albomaculatus* Franganillo 1930 (Cuba), *L. m. texanus* Chamberlin and Ivie 1935 (Texas), *L. m. hesperus* Chamb./Ivie 1935 (Utah), *L. m. mexicanus* Gonzalez 1945 (Mexico)

2b. *L. mactans tredecimguttatus* (Rossi) 1790 – Mediterranean region, Central Asia, Abyssinia, Arabia. Synonyms: *Theridion lugubre* Dufour 1820 (Egypt), *Aranea tredecimguttata* Rossi 1790 (Toscana: Italy), *L. erebus* Audouin 1827 (Egypt), *L. argus* Aud. 1825 (Egypt), *L. venator* Aud. 1825 (Egypt), *Meta hispida* C. L. Koch 1936 (Greece), *M. schuchii* C. L. Koch 1836 (Greece), *L. malmignathus* Walck. 1837 (Italy), *L. martius* Walck. 1837 (Italy), *L. occulatus* Walck. 1837 (Egypt). *L. intersector* Walck. 1837 (Egypt), *L. quinqueguttatus* Krynicki 1837 (southern Russia), *L. conglobatus* C. L. Koch 1838 (Greece), *L. revivensis* Shulow 1948 (Palestine)

2c. *L. mactans cinctus* Blackwell 1865 – Abyssinia, East and South Africa. Synonyms: *L. concinnus* and *indistinctus* O. P. Cambridge 1904 (Capland), *L. incertus* Lawrence 1927 (Cape Town), *L. indistinctus karooensis* Smithers 1944 (South Africa)

2d. *L. mactans menavodi* Vinson 1863 – Madagascar. Synonym: *L. obscurior* Dahl 1902 (Madagascar)

2e. *L. mactans hasselti* – From India to Australia and New Zealand. Synonyms: *L. h.* Thorell 1870 (Australia), *L. scelio* Th. 1870 (Australia), *L. katipo* Powell 1870 (New Zealand), *Theridion melanoxantha, zebrina, katipo, atritus* Urquhart 1889 (New Zealand), *L. scelio indica* Simon 1897 (Arabia), *L. hasselti aruensis* Strand 1911 (Indies), *L. h. elegans* (Thorell) 1898 (Burma, Tonking)

3. *L. pallidus* Cambridge 1872 – Russia, Syria, Palestine (type place), Iran, Libya, Tripolitania. Synonyms: *L. p. immaculatus* Caporiacco 1933 (Lybia), *L. p. pavlovskii* Charitonov 1954 (Russia)

4. *L. curacaviensis* (Müller) 1776 – From southern Canada to Patagonia. According to Levi apparently absent in Mexico, Central America. Synonyms: *L. variegatus* Nicolet 1849 (Chile), *L. thoracicus* Nic. 1849 (Chile), *Theridion mirabile* Holmberg 1876 (Patagonia), *L. malmignathus tropica* von Hasselt 1860 (Curacau), *L. carolinum* Buttler 1877 (Galapagos), *L. apicalis* B. 1877 (Galapagos), *L. geographicus* von Hasselt 1888 (Dutch Guiana), *Chacoana antherata, flavodorsata, distincta, carteri, cretaceus* Badcock 1932 (Chaco paraguayan), *L. mactans bishopi* Kaston 1938 (*Florida*), *L. foliatus* de Mello-Leitão 1940 (Buenos Aires)

5. *L. hystrix* Simon 1890 – Aden and Yemen. The male is unknown

6. *L. dahli* Levi 1959 – Iran, Sokotra. The male is unknown.

(6) Family Araneidae (Argiopidae)-Orb-weaving spiders

One pair of book lungs; chelicerae not fused together at the base; cribellum and calamistrum absent; has a single tracheal spiracle near the spinnerets, or spiracle wanting; three tarsal claws, fourth tarsus with spurious claws; lip free; eight homogeneous eyes.

This family includes several subfamilies, with thousands of different species. Only in the subfamily Araneinae do there exist a few species that are poisonous to men. They were incorporated into the genus *Mastophora,* described by Holmberg in 1876. The species of this genus can be easily recognized by the shape of its cephalothorax, which bears prominent chitinous outgrowths on each side and numerous little tubercles; two conical humps exist on the abdomen. *Mastophora cornifera* and *M. extraordinaria* have been described by Holmberg, in 1876, in Buenos Aires, Brazil, and in Uruguay. The most famous is *M. gasteracanthoides*, described by Nicolet as *Epeira g.* in 1849 and redescribed by Simon in 1895 as *Glyptocranium g. M. bisaccata*, described by Emerton as *Cryrtarachne b.*, in 1884, and *M. cornigera* described by Hentz in 1850, as *Epeira c.* are the two most frequent species in the United States. Several authors use the name *Glyptocranium* for *Mastophora.*

(7) Family Lycosidae

Only one pair of book lungs; chelicerae not fused together at their base; cribellum and calamistrum absent; has a single tracheal spiracle close to spinnerets; three tarsal claws, without spurious claws; chelicerae with a distinct lateral condyle at the outer side of the basal segment; all trochanters deeply notched; the superior claws with a few teeth in a single row and the third claw with one or no teeth; colulus wanting; six spinnerets; eight homogeneous, diurnal eyes arranged in three rows: 4-2-2, the anterior eyes usually much smaller than the posterior; lip free; margins of chelicerae toothed; maxillae of the pedipalps more or less parallel with scopula at the anterior margin; legs prograde, with long black erectile spines, long, blackish, erectile setae and short hairs.

More than two thousand species have been described and arranged into five subfamilies. The classification of many genera and species is incorrect.

The poisonous lycosids are included by some authors in the subfamily Lycosinae or Tarantulinae. All the Lycosinae have the four anterior eyes situated directly over the front, not on tubercles; the retromargin of the chelicerae has two or three teeth; the anterior row of eyes is nearly straight or only gently pro- or recurved and the anterior median eyes are not much larger than the anterior laterals; the sides of the face are slanting; the lip is longer than broad; the fourth metatarsi are shorter than tibia with patella;

the terminal joint of the posterior pair of spinnerets is short and rounded and the posterior pair is not much longer than the anterior. Only the genus *Lycosa* includes poisonous and dangerous species:

Lycosa Latreille 1804 – *Tarentula* according to some authors. The following species are the best known:

L. narbonensis Latreille 1806 – Syn. *Tarentula melanogaster* Thorell 1870:
L. tarentula (Rossi) 1790 – the type-species – Syn. *Tarentula apuliae* Koch 1850.
L. capensis Simon 1898 (Capeland)
L. albopilata Urquhart 1893 (Tasmania)
L. hilaris L. Koch 1877 – New Zealand, Tasmania
L. auroguttata (von Keyserling) 1891 – Syn. *Tarentula a.*
L. erythrognatha Lucas 1836 – Syn. *L. raptoria* Walck. 1837
L. nychthemera (Bertkau) 1880
L. ornata Perty 1833
L. poliostoma (C. Koch) 1848 – Syn. *Tarentula p.*
L. thorelli (von Keyserling) 1876 – Syn. *Tarentula th.*

(8) Family Clubionidae

Only one pair of book lungs; chelicerae not fused together at their base; cribellum and calamistrum absent; has a single tracheal spiracle near the spinnerets; only two tarsal claws; eight homogeneous eyes, dark in color, arranged in two more or less parallel rows; tarsal claws toothed; legs prograde, the first pair directed forward; colulus absent; six spinnerets, the anterior pair approximated; chelicerae with a distinct boss or condylus on the outer side of the basal segment, the subunguinal margins transverse and toothed; lip free; maxillae parallel, often widened at their ends, with well-delimited scopulae; sternum wide, pointed behind; claw tufts well developed.

The numerous species occurring all over the world in tropical, subtropical, and temperate zones are arranged into four subfamilies. Only in the subfamily Clubioninae are there dangerous spiders. They can be easily distinguished by the distinctly cone-shaped terminal joint of posterior spinnerets; their maxillae are constricted in the middle and their lip is long.

Cheiracanthium C. L. Koch 1838 or *Chiracanthium: Ch. punctorium* (Villers) 1789—Europe, Turkestan. *Ch. brevicalcaratum* and *diversum* L. Koch 1873—Australia, Fiji Islands, Hawaii. *Ch. inclusum* (Hentz) 1874—from Canada to Mexico and the West Indies.

(9) Family Salticidae (Attidae) = Jumping Spiders

One pair of book lungs; chelicerae not fused together at their base; cribellum and calamistrum absent; colulus wanting; six spinnerets, the anterior pair close more strongly than the others; a single tracheal spiracle close to spinnerets; two tarsal claws, pectinate in a single row; claw tufts well developed; eight homogeneous, diurnal eyes, arranged in three rows: 4–2–2, the anterior median pair the largest, third pair usually very small; chelicerae without boss, with scopula and toothed margins; lip free; maxillae more or less parallel, with scopula; legs prograde, adapted for jumping, with spines.

The Salticidae form a very large group found the world over and arranged in more than twenty subfamilies. All of these are harmless to men, except a few species of the genus *Dendryphantes* from the subfamily Dendryphantinae.

Dendryphantes Simon 1901

Eyes in three rows, the smallest of the second row situated midway between the anterior lateral and the posterior; tibia and patella of the third pair of legs shorter than those of the fourth pair; sternum not much narrowed in front, the anterior coxae being separated by a distance greater than the width of the labium; quadrangle of eyes occupying less than one half of the length of the cephalothorax and as wide behind as in front or wider; abdomen with transverse white bands; cephalothorax much longer than wide; with only one tooth on the inner margin of the chelicerae. *D. noxiosus* Simon 1886.

(10) Family Ctenidae or Wandering spiders, Banana spiders

With one pair of book lungs; chelicerae not fused together at their base; Cribellum and calamistrum absent; with one tracheal spiracle near the spinnerets; tarsi of the legs with only two claws, pectinate in a single row; in *Cupiennius* a third, obsolete claw is present; claw tufts present; colulus wanting; six spinnerets; eight eyes, homogeneous and diurnal. In *Phoneutria* the two lateral eyes of the second row often pearly white in color, nocturnal; eyes in three rows: 2–4–2 (rarely 4–2–2); chelicerae with a distinct condylus; both margins toothed and with scopula; lip free; maxillae parallel, with a dense scopula; thoracic groove longitudinal.

The Ctenides are subdivided into three subfamilies, with representants from 3 to 5 cm in length, armed with powerful chelicerae. The most important species belong to the subfamily Phoneutriinae. In the Phoneutriines the lip is longer than wide with lateral excavations and reaches the middle of maxillae; there are from three to five, rarely six, pairs of stout, robust, moderately long, blackish, erectile spines on the ventral side of the first tibiae; the eye formula is 2–4–2. The most important genus of this subfamily is *Phoneutria*, described by Perty in 1833. Walckenaer, in 1837 synonymized this genus with the older genus *Ctenus* Walckenaer 1805 and was followed by von Keyserling (1891), F. Cambridge (1897, 1902), Simon (1897),

Petrunkevitch (1911), Bonnet (1956), Roewer (1954), and by several other authors. C. Koch reviewed the genus *Phoneutria* in 1848, as well as did de Mello-Leitão, in 1936, and Bücherl in 1956 and 1969 (new subfamily: *Phoneutriinae*).

Phoneutria can easily be separated from *Ctenus* by the genital organs in females and males that are completely different in the two genera. In *Phoneutria* the inner face of the segments of pedipalps shows scopulae, formed by short, dense, velvety hairs, non-existent in *Ctenus*. The size and the life habits of *Phoneutria* are very different from those of *Ctenus*. The most frequent species of *Phoneutria* are as follows:

An Amazonian Group—*Phoneutria fera* Perty 1833; type: a female, from Rio Negro, Amazonas, Brazil; described again by E. Strand at the Museum in Berlin. *Ph. rufibarbis* Perty 1833; another female; type lost; may probably be a synonym for *fera*, because Spix and Martius collected both at the same place along the Rio Negro. *Ph. reidyi* (F. Cambridge) 1897—Santarém, Pará. Syn. *Ph. andrewsi* (F. Cambridge) 1897. *Ph. sus* (Strand)—Surinam;

A South Brazilian Group (Uruguay, Argentina)—*Phoneutria nigriventer* (Keyserling) 1891—locality: Rio Grande do Sul; range: Argentina, Uruguay, Brazil from Rio Grande do Sul to Rio de Janeiro;

A Bolivian Andean Group—*Ph. boliviensis* (Cambridge) 1897; locality: Madre di Dios, Bolivia; *Ph. nigriventroides* (Strand) 1907; Sorata, Bolivia; *Ph. colombiana* Schmidt 1954: Colombia.

IV. DESCRIPTION, DISTRIBUTION, AND BIOLOGY OF DANGEROUS SPECIES

A. Description

All the representatives of the genus *Trechona* (Orthognatha) have a nearly straight, low cephalothorax, The four eyes of the first row are equal or subequal in size and are equidistant. The eye-tubercle is two times as broad as it is long. The atridulating apparatus (Fig. 11) is situated partly on the chelicerae and partly on the coxae of the pedipalps, the catching spines of black color are on the ventral side of the basal segment of chelicerae and the "chords" of the lyra on the anterior side of the coxae of pedipalps. The superior spinnerets are only somewhat longer than half the length of the abdomen, their segments equal in length. The legs are long and robust, tarsi and metatarsi of the two anterior pair of legs have dense scopulae; the third and fourth metatarsi are scopulated only at the apex; in the males the fourth metatarsus is often without scopula. Body length is from about 30 to 45 mm; leg length from about 60 to 70 mm. They are grayish-brown or blackish; the

FIG. 15. *Grammostola mollicoma.*

abdomen has several yellowish, parallel, and divergent stripes. These spiders are sedentary, found in holes, or paths in tropical forests, and on vegetation near the Atlantic Coast. When disturbed, they assume characteristic positions of attack, distending their venom chelicerae enormously, and they bite very quickly (Fig. 4).

In all species of the genus *Atrax* the cephalic part is long and convex, the thoracic groove is very remote, deep, and procurved; the eye-tubercle is depressed, slightly more broad than long, sternum longer than broad, lip distinctly broader than long, and is very convex bearing numerous cuspids; posterior spinnerets are generally shorter than the abdomen with the last segment longer than the preceding one (Fig. 5). The small funnel-web tarantulas are sedentary but notable for their aggressiveness and the strong action of their venom on human beings. *A. robustus* and *formidabilis* have caused human deaths. They are brown, with clearer or darker stripes on the sides of their abdomen. During most of their life they are settled in their funnel-web, but when mating time arrives they may begin a vagrant life, occasionally entering human dwellings. When disturbed by man, they assume an offensive-defensive position, forelegs held high with their body supported on the last legs, their chelicerae widely opened with the venom claws widely distended. They may jump as far as 10 cm and bite very actively, injecting their venom into the victim. According to Wallace and Sticker (1956), *A. robustus* inhabits a very restricted area which includes most of suburban Sydney. A large number infect the densely populated districts of the city. The other species, already cited, are much more rare.

The representatives of the genus *Avicularia* (Theraphosidae, Aviculariinae) are large "tarantulas" with eight eyes on a tubercle, the legs, chiefly the tarsi and metatarsi are very densely scopulated (Fig. 1). The mating bulb of the males is spherical with a thin, long, serpentiniform embolus (Figs. 12 and 13); on the tibiae and metatarsi of its legs only hairs and setae are present, not blackish spines. Terrestrial and arboreal species are about 50 to 90 mm body length and 60 to 100 mm leg length. Their color varies from reddish-brown to grayish-red, with red hairs on the abdomen and the chelicerae. Generally not at all aggressive, they are yet much feared by the Amazonian people. They occasionally eat small birds, surprised in their nests.

The male of *Grammostola mollicoma*, with its enormously enlarged forelegs (Theraphosidae, Grammostolinae) is the largest "tarantula" I know; it is about 21 to 27 cm from the apex of the first pair of legs to the apex of the fourth, and has a body of 7 to 10 cm (Fig. 15). Regardless of their large size, all *Grammostola* may be considered harmless. Many species are peaceful and can be held in the hands.

In the *Eupalaestrus* species (*Theraphosidae*, *Theraphosinae*), the cephalothorax is more long than wide; the thoracic grove is transverse or slightly procurved; the eye-tubercle much wider than long, first row of eyes procurved, its lip is as long as it is wide; sternal sigillae is near the margin; the legs are thickened on the femura, tibiae, and the basal parts of the metatarsi, with long, stout setae on the last pair. Body length is from about 4 to 6 cm and legs from 8 to 10 cm. It is dusky brown, the ventral side of the abdomen greenish-gray, the legs blackish-brown with yellowish-red transversal bands on the femura and patellae. Not aggressive at all, it is perhaps the most harmless of the "tarantulas."

The representatives of the genus *Xenesthis* have little eyes, the first row procurved; its legs are very long and robust, patella and tibia of the fourth pair in females of the same length as those of the first pair, in the males they are longer; the scopulae on the fourth metatarsi extend nearly to the base in females. Very robust species of large size, about 8 to 10 cm in body length, they are much feared.

Megaphobema robusta shows a procurved thoracic grove; striae radiantes very distinct; sternum as long as wide; first tibiae of the male with two apical

FIG. 16. *Lasiodora klugi.*

apophyses, the inferior twice as long as the lateral; third femura thickened; cephalothorax, pedipalps, abdomen, and femura with dense short hairs intermixed with longer, semierected, red setae on the abdomen. Very big and robust and feared by many people because it is aggressive and very venomous with from about 8 to 9 cm of body length and 10 to 12 cm of leg length.

The representatives of *Sericopelma* show dense scopulae to the base on metatarsi I and II, at the middle on metatarsus III, and apical scopulae only on the last metatarsi or even scopulae absent; cephalothorax somewhat longer than wide, depressed; thoracic groove deep, transverse; eye-tubercle elevated, convex, more wide than long; anterior eyes subsequal in size, forming a strongly procurved line; A.M. nearer than to the A.L.; lip as wide as long; densely cuspulated; sternum as wide as long; posterior sigillae minute, submarginal; first tibia of the males without apical spur. They are blackish, with short black hairs and longer reddish setae over the chelicerae, abdomen, and legs. On the anterior face of the coxae of the first pair of legs and over the suture is a bundle of plumose hairs intermixed with very small spines. Body length is from about 5 to 6 cm and leg length from 7 to 8 cm.

Pamphobeteus shows a cephalothorax somewhat longer than wide, thoracic groove straight or curved, eye-tubercle as long as it is high and wide, lip more wide than long, armed with numerous cuspulae, sternum somewhat longer than wide, posterior sigillae submarginal; without stridulating apparatus; all tarsi with velvety scopulae, metatarsi also totally or partially with scopulae, chiefly by females; first tibiae of males with two apical spurs. Brown, blackish-brown, they have brown hairs and clearer setae; fang grooves have long red hairs. *P. tetracanthus* often is very aggressive and bites often. The females of the other species may be aggressive when they are taking care of their young. Males are harmless. They are about 5 to 8 cm in body length and 6 to 9 cm in leg length.

All the representatives of *Lasiodora* show a stridulating apparatus on the front face of coxae I, over the suture, and on the posterior face of the palpal coxa; first tibiae of males have two apical spurs. They are big, robust, and agile "Tarantulas". Several species become aggressive when disturbed, and they bite whenever they can do so. They are about 5 to 8 cm in body length and from 7 to 9 cm in leg length and are blackish-brown, entirely covered with grayish short hairs and with numerous, long, erected, reddish setae on the chelicerae, the abdomen, and the legs (Fig. 16).

Theraphosa blondi (Theraphosidae, Theraphosinae) has a cephalothorax as long as it is wide; in the females it is somewhat longer than wide, the lateral margins are rounded and much smaller in front; thoracic groove is deep and transverse; the eye-tubercle is minute, anterior eyes minute and of

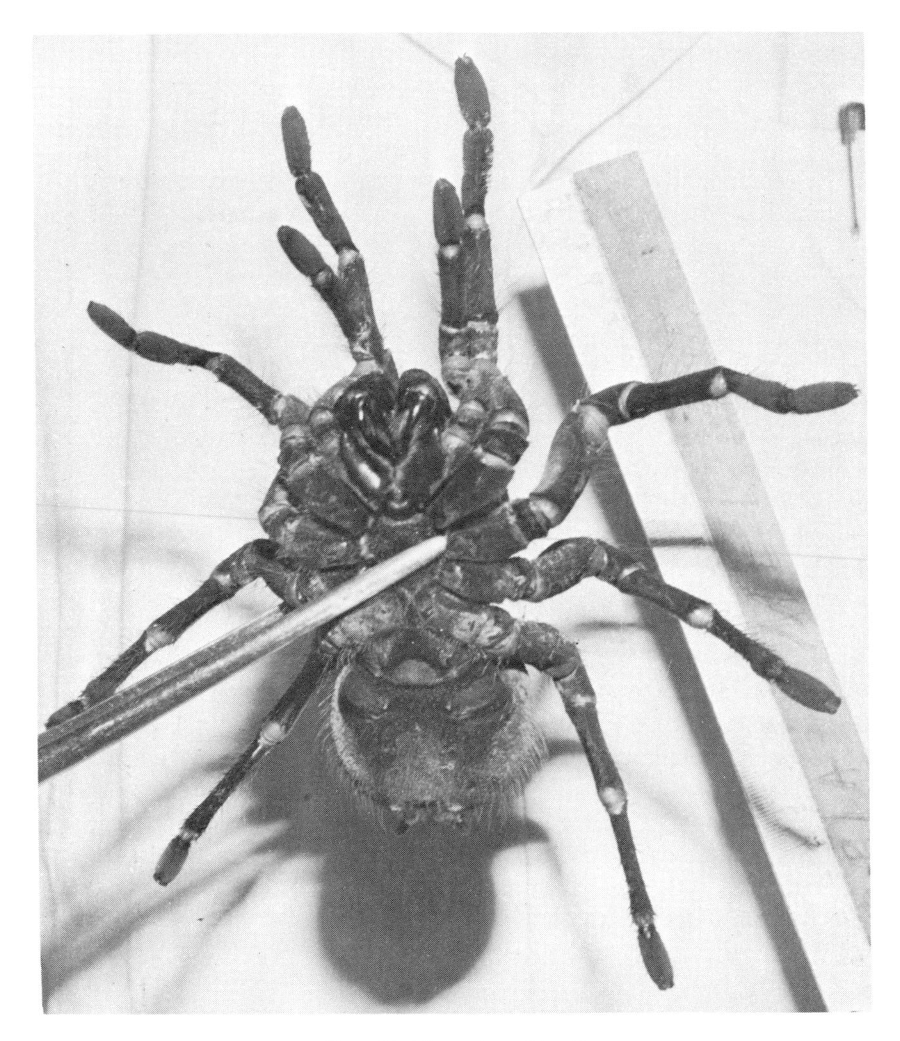

FIG. 17. *Theraphosa blondi.*

the same size and in a very procurved row; the lip has only a few large cuspulae; legs are long, strong, and armed with black, robust, erectile spines; metatarsi of the two first pairs of legs have scopulae extending to the base, on the third apical scopulae only; males have no spur at the apex of tibia I. This species has the largest stridulating apparatus of all "tarantulas," I know, extending between the coxae, trochanters, and the femura of palps and the first pair of legs. Their stridulating "music" is perfectly audible to humans

FIG. 18. *Loxosceles similis*.

from 2 to 3 meters distance. In our laboratories there is one female with 9 cm in body length and a leg length from about 11 cm, one of the largest "tarantulas." When she is nervous she stays in an aggressive position and may bite (Fig. 17).

The *Phormictopus* spiders show a "musical" apparatus on the coxae and trochanters of the front side of the first pair of legs and on the posterior side of coxae and trochanters of palps; tibiae I of males has two spurs. They are about 5 to 6 cm in body length; more or less aggressive, and feared by many people.

The *Acanthoscurria* spiders show "musical setae" on the trochanters of the palps and the first pair of legs. The adult males may be easily distinguished by the single, lateral, rakelike spur on the apex of tibia I *A. juruenicola* is very large and greatly feared by the Indians of Mato Grosso because of its deadly bite. *A. sternalis* is much smaller, from about 5 to 6 cm in body length, but is also very aggressive and bites whenever she can do so. *A. violacea* and *A. atrox* are also aggressive.

Figure 24 shows a *Trasiphoberus parvitaris* spermatic bridge. *Trasiphoberus*

FIG. 19. *Latrodectus curacaviensis.*

has the musical apparatus on the same articulations as those of *Acanthoscurria,* but the "chords" are on the posterior face of the trochanter and the "catching pins" on the front face of the trochanter of the first pair of legs.

The venomous, smaller "tarantulas" of the genus *Harpactirella* (Barychelidae, Leptopelmatinae) are feared by most people in Cape Town, because they are very venomous and aggressive and frequently enter homes. The existence of venomous spiders in South Africa and the oceanic islands of the west coast of the Cape Province has long been known, but it was not until 1939 that Finnlayson and Smithers described *Harpactirella lightfooti* as the cause of "tarantula" bites producing severe illness in men. The fangs of this species are from about 1/8 inch long and turned upward, with claw tufts on the apices of the legs. These spiders are able to climb over smooth surfaces. Brown, with a yellowish border around the cephalothorax, they have clear stripes on the dorsolateral sides of the abdomen, and are about 20 to 30 mm in body length. They actively hunt and capture and kill their prey with their venom fangs. Their nests consist of silk-lined tunnels under logs, stones, or other debris, or under loose boulders. When surprised by men or

FIG. 20. *Latrodectus curacaviensis* with an egg sac.

animals, they immediately respond by assuming the attacking posture, forwardly erect with their venom jaws widely opened.

The six-eyed true spiders (suborder Labidognatha) of the genus *Loxosceles* (Fig. 18) are exclusively nocturnal. They may be found under stones, stored bricks and tiles, under the bark of trees, dry stumps of bamboos, and other ground objects either far from or near human dwellings. They are often unknowingly carried into houses or stables on these objects and during the night may crawl into garments. While dressing in the morning, a person may be bitten in those parts of the body where clothes hide the spider, under the arms or on the lower parts of the abdomen. All the known species are usually tawny, light brown, or grayish on the cephalothorax and olive-greenish on their abdomen. Their abdomens and legs are sparsely covered with short grayish hairs, and there are fine setae on the tarsi. In the dark they are sedentary on their white, flocky, and adhesive webs. When the web is destroyed, they may run rapidly, but rest when they again find a dark place. They are never aggressive and flee whenever able. Accidental bites occur only when the spiders are squeezed against the human body, when one is dressing or in bed. Adult females vary from about 7 to 12 mm, rarely reaching 20 mm in total length, with a leg length from about 16 to 24 mm. The carapace is smooth or covered with short thin hairs; grayish small setae are dispersed over the head in longitudinal rows; at each side of the carapace there are from three to four dusky patches, often indistinct.

The "comb-footed" true spiders of the genus *Latrodectus* (Figs. 19 and 20) are sedentary types that stay or hang below leaves, or the dry threads of irregular webs, and retreat to denser web or holes in the soil. All are small spiders, from about 8 to 15 mm in body length, with the legs the same length. The adult males are from four to five times smaller, but their leg length is only somewhat shorter than that of the females. Their web is often so fine that it may be invisible to human sight. Frequently the *Latrodectus* find fissures in the soil, or in debris, or even empty tin cans, boxes, or car tyres for their nests. Their abdomen is globose in shape, black with red stripes, spots, or markings; the legs are blackish or grayish, with a row of blackish, curved, and toothed setae on the ventral side of the fourth tarsi, forming a "comb"; they have three claws and spurious claws; the ventral side of the abdomen has an hourglasslike red or yellowish spot; the eight eyes are in two rows. They are active in evening, morning, and during the night. In addition to their central web with a retreat, *L. curacaviensis* weaves two or more long guy lines 4–10 cm above the ground. When a prey is caught in the silk, the spider approaches cautiously, presents its long back legs to the victim, weaves a sticky silk, and with its combs involves the legs of the prey until a strong band is formed; then it bites into the body of the victim and injects the venom. The dead prey is often several times bigger than the *Latrodectus* spider. Prey, small stones, parts of leaves are ingeniously lifted up to the retreat by several lines (Fig. 20).

All the species of this genus show a depressed cephalothorax, a transverse thoracic groove, the carapace is narrowed in front, posterior row of eyes are somewhat recurved, the front eyes slightly procurved; the L.A. and L.P. separated; clypeus no smaller then the ocular area; sternum longer than broad, narrowed on the back side; maxillary lobes parallel; lip free, broader than long; chelicerae are without teeth; colulus large; epigynum of the females is wider than long; the microscopic structures of the receptacula seminalia in females and of the bulbus and embolus in males are the only decisive factors for correct species determination. *Latrodectus geometricus*: carapace yellow-brownish; sternum with a longitudinal yellowish stripe; palps and legs yellowish to grayish with darker transverse spots in all segments; abdomen yellowish to grayish with three triangular spots and an oval spot on each side, and the three oval markings have brown margins followed by four smaller markings with the second two bigger. Color variations may be found.

L. mactans mactans is black with reddish or red-yellowish spots on the abdomen. The spots vary in number and form on each specimen, the most constant marking being the hourglass like one on the ventral side of the abdomen.

L. mactans tredecimguttatus also has a blackish abdomen with several red

FIG. 21. *Lycosa erythrognatha.*

spots in front, four red spots in the middle around a larger triangular spot, and several reddish spots of smaller size; two red spots are on the ventral side.

L. mactans cincus, m. menavodi, m. hasselti are geographical subspecies. *L. curacaviensis* cannot be distinguished from *L. mactans* by the color pattern (Fig. 19). We have collected more than a thousand on the beaches of Rio de Janeiro. They are easily collected, for they will feign death when disturbed and fall into our collecting glasses.

The representants of the genus *Mastophora* have repudiated the typical orb web of Araneinae. They are known as "*podadoras*" or "*bolas-spiders*" and are much feared. Their abdomen, from about 20 to 39 mm of average length in the adult females is adorned with two sharp "horns," also present on the cephalothorax. They live over bushes or in smaller trees, from 1 to 3 m above the ground, between the leaves. The females build several egg sacs, brown in color and spherical, about 8 to 11 mm in diameter. When handled during the daylight, they remain immovable; in evening or during the night they weave a hanging line and a globule. When a nocturnal insect approaches, the spider swings the "bola" like a lasso and the insect is held with the sticky ball.

All "wolf spiders" of the genus *Lycosa*, family Lycosidae of the suborder of true spiders, are very agile hunters. Their eyesight seems to be useful to them only in an area of from 15 to 20 cm. Generally they notice prey only

FIG. 22. *Phoneutria nigriventer* in an aggressive position.

when it is very near. Then they suddenly stop and jump directly and vigorously on it, holding it with the claws of the first pair of legs and the palpi. Their chelicerae are largely distended and firmly dug into the body.

Among the representants of the genus *Lycosa* the labium is longer than wide, with basal excavations on each side; the face is much wider below than above, with the sides strongly convex. The anterior tibiae are armed with three pairs of long blackish spines. In *Lycosa erythrognatha, L. pampeana, L. carolinensis. L, punctulata, L. thorelli, L. nordenskiöldi*, etc., the body length of an adult female varies from about 30 to 37 mm and the leg length from about 40 to 50 mm. They are found in gardens, running over the ground, under stones, in fields, at the edges of woods, or near water. In their subadult stages and when they are not taking care of young, they often actively wander during the night, occasionally entering human dwellings. As a rule, all wolf-spiders are shy and rapidly run away when disturbed. But since they are frequent in the higher regions of the tropics, chiefly in South America, they

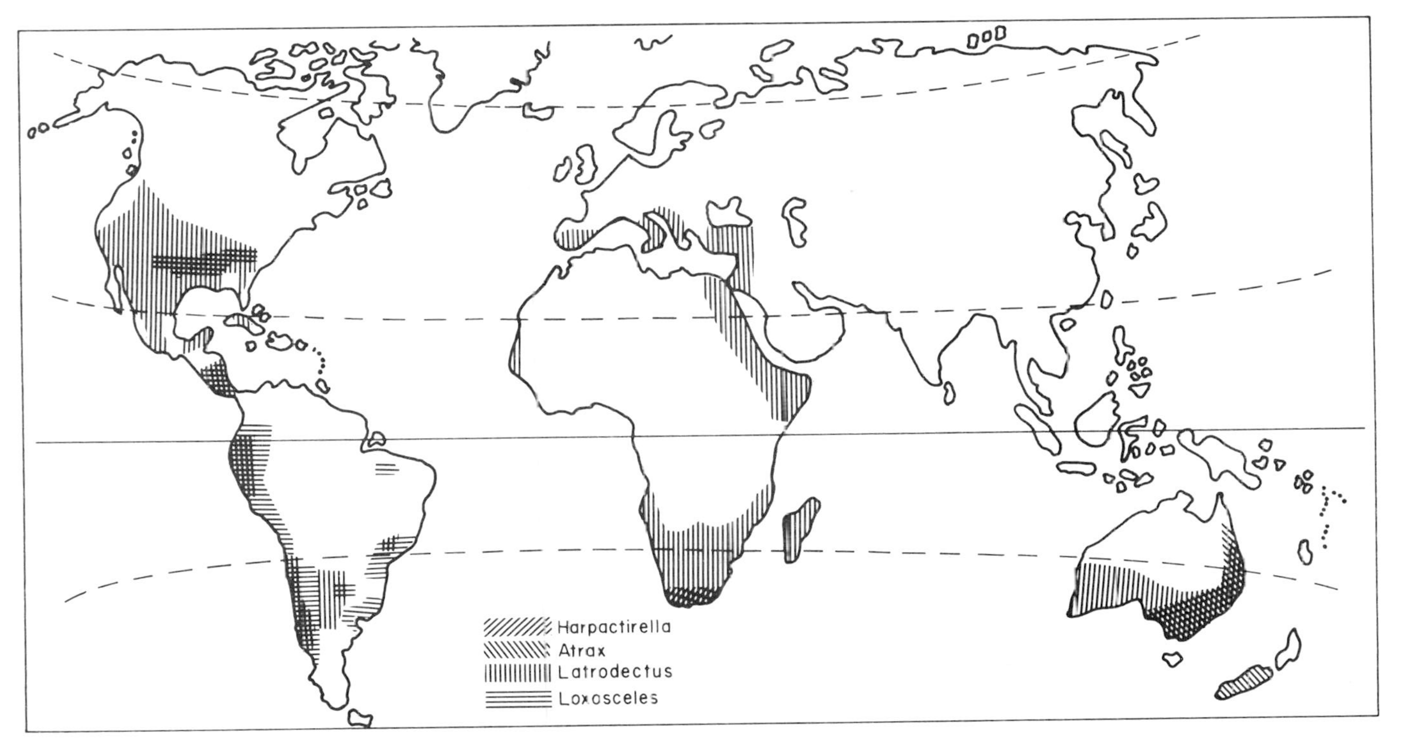

FIG. 23. Map of distribution.

TABLE I

DISTRIBUTION OF PROVED DANGEROUS SPIDERS

Genus or species	Country
Atrax	Australia, New Zealand
Harpactirella	South Africa,
Loxosceles	South, Central and North America
Latrodectus m. mactans	From United States to Argentina, Hawaii
Latrodectus m. tredecimguttatus	Mediterranean region, Arabia, Abyssinia, South Russia
Latrodectus m. cinctus	South and East Africa, Abyssinia
Latrodectus m. menavodi	Madagascar
Latrodectus m. hasselti	From India to Australia, New Zealand
Latrodectus pallidus	Russia, Syria, Palestine, Iran, Libya, Tripolitania
Latrodectus curacaviensis	From South Canada to Patagonia
Lycosa	South and Central America
Phoneutria	Rio Grande do Sul

TABLE II

DISTRIBUTION OF NONPROVED DANGEROUS SPIDERS

Genus	Country
Orthognatha	
Trechona	South America
Xenesthis	Colombia, Venezuela, Panama
Megaphobema	Colombia
Pamphobeteus	South America
Lasiodora	Brazil
Theraphosa	Venezuela, Guyana, Amapa
Phormictopus	Central America
Acanthoscurria	South America
Labidognatha	
Lithyphantes	Chile, Bolivia, Argentina, Brazil
Latrodectus geometricus	Cosmopolitan
Mastophora or *Glyptocranium*	Peru, Chile, Bolivia
Chiracanthium	Hawaii, Peru
Dendryphantes	Bolivia, Chile, Brazil

encounter people frequently. Upon direct contact with the human body they bite immediately and flee (Fig. 21).

The cephalothorax of the representants of the genus *Chiracanthium*, family

TABLE III

DISTRIBUTION OF DUBIOUSLY DANGEROUS SPIDERS

Genus	Country
Heteropoda venatoria	Seaports of South and Central America
Sericopelma	Central America, Panama
Eurypelma	Southern United States, Central America
Segestria	Argentina, Brazil, Paraguay
Filistata	Argentina, Brazil

Clubionidae, is convex, without a visible median furrow. The inferior margin of chelicerae is armed with two or three teeth; the first pair of legs is distinctly longer than the fourth. These small spiders are found in tubes of white silk; they are not aggressive, occasional bites are not very severe except in Hawaii where they are greatly feared.

Several, small, jumping spiders of the genus *Dendryphantes*, family Salticidae, are thought to be harmful to man in Bolivia.

Phoneutria nigriventer the most common representant of *Phoneutria*, family Ctenidae, is the most robust, aggressive and largest true spider of South America, known as "aranha armadeira" or "wandering spider" or "banana-spider." It weaves no nests, but wanders in the shade in evening, morning, and night. With light eyes in three rows –2–4–2, the last two eyes the largest with the two laterals of the second row the smallest. Adult females vary from about 25 to 50 mm and average about 35 mm in total body length. Their leg length varies from about 45 to 60 mm. The male's body is smaller and the legs proportionally longer. The thoracic groove is a deep, longitudinally linear, depression. The abdomen is more long than broad and highest in front; the six spinnerets are surrounded by a distinct wall; chelicerae has a distinct boss: the inferior margin of the fang groove has five black teeth, the upper margin with only three teeth; labium free, longer than broad, with lateral excavations; pedipalps with a brush of velvety short erected hairs on the ventral and the inner surfaces of the femur, patella, tibia, and the tarsus; legs and pedipalps with long, robust, black, erectile spines, on the posterior legs, inserted in a round grayish basis; with two claws, with one series of four to five teeth. Grayish to brownish-gray, chelicerae with a red brush of long hairs; abdomen dorsally has whitish marks, forming a longitudinal band, with or without several divergent side bands. They are very active hunters, but during the daylight stay in dark places, and at evening or during night run several hundred meters to capture their prey. They frequently enter human dwellings, and when morning comes often hide in clothes or shoes. As they are very aggressive, no one approaches without being bitten. Frequently the same spider will bite furiously several times (Fig. 22).

B. Distribution

Table I may elucidate the distribution of genera with proved dangerous representatives. For further distribution see Fig. 23.

C. Biology

The members of the different families of spiders with venomous representatives vary in some ways in their modes of life; even in the single families there frequently exist generic and specific variations and peculiarities of life habits.

1. The Pairing of Spiders

All venomous spiders are separated into two sexes. Until they reach maturity the sexes cannot be distinguished externally, young males and females behaving in the same manner. *Araneus, Trechona, Loxosceles, Latrodectus, Mastophora,* and *Chiracanthium* make webs; *Atrax,* some *Theraphosinae,* and *Lycosa* are semivagrant; *Harpactirella, Dendryphantes,* and *Phoneutria* are vagrant; *Trechona, Atrax,* some *Theraphosidae, Harpactirella,* and *Lycosa* burrow in the soil. When the mating season arrives, generally in autumn, although exceptions may exist, the two sexes show a

FIG. 24. *Trasiphoberus parvitarsis* with spermatic web.

profound differentiation in their biological habits. All males begin a vagrant life. They construct a small, "spermatid" web in dark retreat (Fig. 24), in a more or less horizontal position. This "spermatic bridge" is often of wonderful regularity in *Lycosa, Phoneutria*, and some Theraphosinae, consisting of a closely woven whitish sheet extending in a single plane. The male emits its spermatic liquid over this web, and immediately fills its two pairing bulbs situated at the ends of its pedipalps with the liquid. Each male searches for a female of the same species. In the sedentary weaver females, such as *Mastophora*, and in *Loxosceles* and *Latrodectus*, the males are found over the webs of the females, at first generally a certain distance from her; in the funnel web tarantulas, *Trechona* and *Atrax*, the males stay on the outer borders of the female funnel; in the vagrant Theraphosinae, the *Lycosa, Dendryphantes* and *Phoneutria* the males wander actively at evening or during the night to find a female. The male generally approaches with very great care. The *Dendryphantes* male dances in front of the female, the *Phoneutria*, and *Lycosa* males stop a distance four to five times longer than their body length, then jump abruptly over the females; the males of *Pamphobeteus, Lasiodora, Acanthoscurria*, and other genera of Theraphosidae stay in front of the females, head against head, the males moving their forelegs and pedipalps very rapidly and catching with them the female's head. For the transmission of the spermatic liquid the males introduce the emboli of the bulbs into the two canals of the female receptacles. For this purpose the male lifts the front of the body of the female to a nearly vertical position. The *Phoneutria* and *Lycosa* males remain directly over the bodies of the females, the females being completely passive, and transmit the sperm within a few minutes by alternating the bulbs into the receptacle. In *Latrodectus curacaviensis*, the male, about six times smaller than the female, stays directly on the ventral side of the female abdomen near her genital openings to introduce his bulbi into these.

The separation of the two sexes after mating is not always peaceful: the large "tarantula" females commonly bite and kill the males immediately after pairing. I have seen only one exception to this rule: the male of *Grammostola mollicoma*. As he has enormously long forelegs, he may escape. In other cases of Theraphosidae both fight after pairing, killing each other, but commonly the smaller males are killed by the larger females. This murdering of the spouse is difficult to explain. I never have seen it in the venomous true spiders. The males of *Loxosceles, Lycosa*, and *Phoneutria*, we have observed, freely leave after mating, thereby constructing a new spermatic bridge a few days later and again fill their bulbs to seek out another female. Even in *Latrodectus curacaviensis* I have found that females tolerate the males on their webs, give them food, and even do so when taking care of their egg sacs. A few weeks after pairing the males die naturally and their bodies hang on the female's webs. I have never seen females kill the males.

After pairing, the females may keep the spermatic liquid in their spermathecae for several days, weeks, months, or even years depending on longevity and climatic factors.

2. The Motherhood of Venomous Spiders

All venomous spiders, without exception, show a very highly developed position of the females in taking care of their descendants. The first step is the care with which the fertilized females build a chamber for their egg laying. Funnel-web spiders *Atrax* and *Trechona* enlarge the lower part of the funnel, sometimes into the soil; the large Theraphosidae begin to live underground, enlarging natural holes and covering them carefully with protecting web layers; the arboreal Aviculariinae weave a wonderful nest with several tree leaves; the *Lycosa* females burrow a large chamber in the soil for their motherhood; the *Phoneutria* locates a proper, dark, protected place and begins a sedentary life; the *Loxosceles, Latrodectus, Mastophora,* and *Chiracanthium* construct proper retreats on their webs or near to them. As a rule, all vagrant or semivagrant females reside at the same place during the period of motherhood or always stay near it at this time. The second step is the very careful construction of the egg sac where it is dark or during the night.

In the dark the *Loxosceles* weave one of the simplest cocoons, which is a mesh of threads of such delicate structure that the eggs may be seen through it; the egg sacs of *Latrodectus mactans* and *L. curacaviensis* are strictly spherical, from 7 to 10 mm in diameter and of white color; *L. geometricus* also constructs a spherical cocoon, but with numerous small conelike protuberances on the outer face; the Lycosae have a resistant spherical cocoon from 10 to 15 mm in diameter; the saucerlike white egg sac of *Phoneutria* has a diameter of from about 30 to 40 mm long and from 7 to 9 mm high. The subspherical cocoons of Theraphosidae are from 25 to 45 mm in diameter, and are also carefully woven. First the inner layer of cocoon is constructed, with elevated surrounding borders; then the eggs are deposited and an outer, more dense layer added. After this process the mother turns the eyeball under her spinnerets and weaves a second and a third layer around the ball until it is spherical. Many spiders, such as Theraphosidae, protect the cocoon with an external layer, variously colored, and frequently intermixed with hairs and foreign substances. The egg sacs of strictly nocturnal spiders are white (Fig. 25).

The third step of motherhood is the proper preservation of the egg sacs: *Loxosceles* and *Phoneutria* affix the cocoon under stones, bark, or in suitable dark places, the mother sitting over them or supporting them between her pedipalps and chelicerae (*Phoneutria*); *Latrodectus* affixes the egg sacs to the underside of leaves or on a retreat in her web; *Lycosa* affixes the sac to her spinnerets; the Theraphosidae females hold the large cocoon between their first pair of legs and the chelicerae.

FIG. 25. *Phoneutria nigriventer* female with egg sac.

The number of eggs varies greatly from genus to genus: 700 to 800 among *Pamphobeteus* and *Acanthoscurria,* 400 to 500 in *Eupalaestrus, Grammostola,* and *Lasiodora.* I have also observed certain "tarantulas" with more than 1,000 eggs included in one egg sac; the *Loxosceles* seem to bear only 40 to 70 eggs per cocoon; *Latrodectus* from 100 to 150 per cocoon. The most fertile spiders I know are the South American *Lycosa* and *Phoneutria* females. All *Latrodectus* species weave from 3 to 4 or even 5 egg sacs, more or less simultaneously, *Lycosa* and *Phoneutria* one by one, with nearly 1000 eggs in the first, 700 in the second, 400 in the third, 40 in the last, all intermixed with unfertile oocytes. In *Lactrodectus* the whole number of eggs in the 4 or 5 cocoons may be from 400 to 600.

All species of spiders have eggs that are spherical, white, grayish or yellowish, and not agglutinated. The oocytes are fertilized by the female just at the moment when they pass from the ovary to the genital opening into the so-called "uterus." Here open the spermatic ducts from the receptacula seminalia. Through several observations we have established that more than 95% of the oocytes are fertilized.

The "tarantulas" construct only one egg sac for a period of 1 year. In the next year they mate again with a young adult male. However the *Loxosceles*, the *Latrodectus, Lycosa,* and *Phoneutria* females show only one mating during their entire life. Their first cocoon is made with great care, and is wonderfully built; the second one is also perfect, however the third, fourth, or even the fifth are not so perfectly finished, and are much smaller. One can establish the progressive exhaustion of the female, apparently by the smaller size of cocoons, their less perfect construction, fewer number of fertile eggs, and mixed in the last oviposition with unfertile oocytes. This is evident chiefly in *Lycosa* and *Phoneutria* which spend 3–4 whole months with one or two successive egg sacs, yet this also clearly appears in *Latrodectus.*

All female spiders are extremely conscientious mothers. They defend the egg sac, carry the cocoons into the sun in the morning, protect them against rain and cold weather.

When the young are born, they climb over the mother's back and use her body without any ceremony for their first exercises into nature. The *Lycosa* mothers, become so feeble and weak after caring successively for several cocoons that they ,frequently die, just as spiders of the last birth climb on their back.

3. The Development of Spiders

Five periods in spider development may be observed: the embryonal development in the eggs, "larval" development in the egg sac, "postlarval" development from egg sac to an independent life, adolescent phase until the "sexual" molt, and the phase of maturity.

The eggs are richly supplied with yolk and covered externally by a thin chorion. The embryo develops superficially over the yolk surface, being enclosed in a embryonic cuticule. In the last embryonal stage, 15 to 24 days after fertilization of the egg, it occupies the total region of the egg around the central yolk. Finally, the chorion is broken off around the upper pole by extreme pressure of the embryonal cephalothorax, the spiderling comes out, changing the embryonic cuticule, which remains with the chorion. The newborn is transparent, not at all hairy, without pigment except the two large bluish eyes; it's chelicerae are provided with a pointed tooth, the so-called "egg-tooth" used to open the chorion; cephalothorax and abdomen are spherical full of yolk, the tarsi of legs are uniarticulate, without claws.

The newborn stay for several days in the dark and firmly closed egg sac, which must be interpreted as a second chorion with more space and freedom for larval movements. In 3–5 days the larvae complete their morphology and their adaptability to daylight; they then change their skin a second time. Their body now shows short, discolored hairs, the legs become divided in metatarsus and tarsus with rudimentary claws, the chelicerae form a claw still without a venom canal, and the spinnerets are relatively well developed.

The young on the inside and the mother on the outside work to open the egg sac at some small point. The spiderlings emerge, one, ten, several hundred, with their body already like the adult's but with the abdomen swollen and full of yolk, with darker hairs, four eyes or with the definitive number of eyes, their chelicerae with a superficial venom canal, but without venom glands. The young *Lycosae*, *Phoneutria* and Theraphosidae climb over their mother's back, and the *Loxosceles* and *Latrodectus* stay on the outer side of the egg sac. Soon all begin to weave single silk lines from the egg sac to the mother, from the mother's back onto stones, leaves, or small trees situated a few centimeters away. When disturbed they fly rapidly to the mother's back or into the egg sac; when cold or rainy weather occurs they return into the egg sac for days, weeks, and sometimes months in cold climates. Their generic habits vary greatly, depending chiefly on the climates of both hemispheres. This "postlarval" period may last from 7 to 12 days according to our studies of venomous spiders of tropical and subtropical climates. In several species, like *Phoneutria* and *Latrodectus*, a negative geotropism forces the young to weave a dense sheet, layer by layer, higher and higher until they are about 10 to 20 cm above the soil. Finally all the hundreds of brothers and sisters, still closely packed together, become immovable and during the night they change their skin, not simultaneously, but the more vigorous leaders change earlier, the smaller spiders 1 or 2 days later. In this period of life active cannibalism begins: those young ones, who have earlier changed their skin and whose body aspect is now completely like that of pre-adults, without any ceremony, eat their brothers and sisters who are immobilized expecting their own molting. In *Lycosa*, *Phoneutria,* and Theraphosidae that we studied, the cannibalism must be considered as a "normal" vital habit; without it no spider can become an adult.

After the third skin molt the survivors leave as far as possible from each other, and actively disperse themselves, wandering over silk lines built in dark places or, as in *Latrodectus*, in an aeronautic way. The young climb high into the tips of stems and shrubs, impelled by a strange urge of negative geotropism. They then expel several threads from their spinnerets and await a favorable wind; by this method they can elevate themselves several hundred, or even more than a 1,000 meters high, flying farther than 10, 100, or even several hundred kilometers. It is for this reason, and other observers feel, that the *Lactrodectus curacaviensis* from the Atlantic beaches of Caravelas, in southern Bahia, Brazil, is found from 400 to 600 kilometers from the coast, around the city of Teófilo Otoni, and that the representants of the same species from the beaches of Campos, Cabo Frio, Itaipuassú, Itacoatiara, Itaipú, and Piratininga, are found several months later up to 50 or 100 or more kilometers to the west, on the islands of Guanabara Bay in Rio de Janeiro, Jacarepaguá, etc. Aeronautic acrobats, really experts in ballooning, they seem to be very common in the true spider families, but we have never found them in Orthognatha.

The "adolescent phase," from the third molting, or the beginning of independent life, until "sexual" molting, may extend from 6 to 8 months in *Loxosceles* and *Latrodectus* in tropical climates; with winter from 12 to 14 months in the *Lycosa* species and some *Latrodectus* species (the so-called "winter-time" variations); from $1\frac{1}{2}$ to 2 years in *Phoneutria*, and 3 to 4 years in the Orthognatha species, *Grammostola actaeon, Gr. iheringi, G. mollicoma, Lasiodora klugi, Acanthoscurria atrox, A. sternalis, Pamphobeteus tetracanthus, P. socacabae, P. roseus,* and other "tarantulas." During this period mortality is exceedingly high; rain, sun, cold, other spiders, predator insects (such as ants, solitary wasps, and scorpions), the natural risks of molting, etc., may extinguish several hundreds of these solitary, predaceous creatures, so that only two or three survivors, males and females, even reach maturity.

The "maturity phase" separates the two sexes. After mating the males die within a few weeks or some months later, or are killed by the females (the Orthognatha species), with only a few exceptions; the females of *Loxoscles, Latrodectus, Lycosa,* and *Phoneutria* spend from 2, 4, or 5 weeks taking care of their descendants and die after this period.

The following scheme outlines the development of *Grammostola* of South America:

Mating

November: the males are commonly killed by the females; construction of the egg sac and oviposition.

December: eclosion from the eggs and first ecdysis—22 days after oviposition.

January: opening of the egg sac, second ecdysis, the young over the open cocoon.

January–February: third ecdysis, dispersion of young, ecdysis of the mother itself.

From four to six moltings in the first year of age; from three to four moltings in the second year of age; two moltings in the third year of age in March and October. (The molting in October may be called the "presexual molting," because the tarsi of males are swollen and enlarged.) "Sexual-molting" in March of the fourth year of age; mating in November, etc.

The adult females of South American tarantulas change their skin once a year, in March generally, mate in October, have young in February, and may naturally die from 2 to 3 years later, at 5 to 6 years of age.

4. The Molting Processes

To increase their size, since the chitinous carapace is not elastic, the spiders must completely change their skin, the molting process exactly duplicating the moment of birth, the breaking off of the cephalothorax corresponding to the egg chorion broken by the embryo. There also exists a strange coincidence between the date of birth and that of molting in the adults in February or March for the "tarantulas." The spiderlings change their skin even when sick, without food, or vigor, even when this process will cause them to die or emerge with fatal lesions, or even when it will in result a reduction in their size. The more vigorous leaders take food and change their skin in shorter intervals during adolescent phase, and grow up more rapidly to become mature a few weeks or months earlier, yet their complete life cycle is also shorter and they die earlier. The study of several hundreds of *Lycosa erythrognatha*, *Phoneutria*, and some species of Orthognatha which were kept in our laboratories from birth to death under the same temperature, humidity, and food conditions, and which were descendants of one egg sac only in each case, led us to the following conclusions: the leaders complete the period from birth to maturity in about 8 to 9 months in *Lycosa* and the slower spiderlings spend from 11 to 12 months in the adolescent phase with the same number of moltings. *Phoneutria* and "tarantulas" show a difference of from 2 to 3 months. The mature phase is not significantly changed, except that unfertilized females live much longer, more than 2 or 3 months. The black widow, *Latrodectus mactans*, becomes adult after five to seven molts, *Phoneutria* show from about twelve to thirteen molts in 2 years.

The molting process takes place in the dark or during night. A few days before it occurs the spiders refuse food. The large Theraphosidae stay immovable, with their backs on the ground, their legs symmetrically erected; the cephalothorax breaks off around the front part and at the sides; the "new" cephalothorax comes out, as do the chelicerae, the palps and the legs; finally the abdomen opens laterally. The eyes, hairs, setae, spines, the claws, even the venomous claw, the epipharynx, the book lungs, and spinnerets are all changed; the "new" spider is freshly colored. The exuviae form a true cast of the former spider.

5. Regeneration

The very interesting ability to regenerate a whole leg or a tarsus lost at the articulation or a leg segment artificially amputated at the middle was studied on several species of South American Theraphosidae: legs or apical segments lost just on the articulation are regenerated "explosively" by the next molt, when the molting period occurs 4 to 5 months later. When a leg is broken off only a few days or weeks before molt time, the molt will be retarded. The new leg often is smaller, and is regenerated to whole size in the following molt. Regeneration is only possible as long as the spiders have not stopped molting. When the middle of an articulation is wounded so that blood runs out and necrosis occurs, the spider will not molt until the piece of leg is lost just at the base of the former articulation. If the molting occurs with wounding in the middle of articulation, the spider cannot change the exuvia at this part.

6. Longevity

Latrodectus and *Loxosceles* seem to live from about 13 to 17 months; under laboratory conditions, to 24 months. *Lycosa* species live 16 to 20 months, in the laboratory 24–29 months. *Phoneutria* live from $3\frac{1}{2}$ to $4\frac{1}{2}$ years. The large Theraphosidae of tropical and subtropical climates live from 4 to 5 and 6 years and in laboratories up to 8 years. In general, all females live longer than do males. All spiders of venomous nature, not orb-weavers, live much better in laboratories than in their natural conditions.

7. The Food of Venomous Spiders

All venomous spiders are for the most part predatory creatures, however only *Atrax*, *Loxosceles*, and *Latrodectus* spin webs to get their prey. *Latrodectus* captures small locusts, ants, flys, and mosquitoes and kills them with their venom. Even siblings are attacked when they invade the domicile, and are killed and eaten. The *Loxosceles* are peaceful, feeding on all small insects that live under stones, barks, and leaves. They are extremely egoistic (like black widows) and do not tolerate other individuals—except males during mating time—using their webs. The lycosas are skillful hunters, running or jumping over their prey—all kinds of live insects with soft bodies—and killing them with their venom. The *Phoneutria* seem to be the most predatory and most aggressive "monsters" we know, actively hunting during darkness, as the *Lycosa* do, killing insects, small amphibians, young birds, and small rats in their nests. All Orthognatha spiders are very active hunters, killing insects, small amphibians, other spiders, even small rats and birds. *Theraphosa blondi* eats a newborn rat weekly. *Lycosa* can go without food for about 4 months, *Phoneutria* 6 months, and the large Theraphosidae about 7 to 10 months. Water is essential for the life, chiefly in warm, dry weather. A newborn rat of 3 gm of body weight can be completely eaten by a large Theraphosid within 14 to 20 hours, and is entirely digested in about 1 week.

V. DEFENSE AGAINST DANGEROUS SPIDERS

The thousands of spider species that exist in the world are very important as control agents in eliminating destructive insects. Therefore, it makes no sense to indiscriminately kill all spiders. It would be far more important to exactly know the very small number of spiders which are dangerous to man.

Contact insecticides are only effective against the smaller spiders, such as *Loxosceles*, black widows, and the young of other species. Its effect is conditional with direct contact to spiders, and is not always guaranteed. Rain and sunlight may neutralize the contact insecticides in a very short time; many spiders even develop resistance against BHC and DDT, as proved in Rio de Janeiro with *Latrodectus curacaviensis*.

Toads, frogs, solitary wasps, house hens, and the large "tarantulas" destroy many venomous spiders. In regions where venomous spiders are frequent and where there are human accidents, it is recommended that windows and doors of homes be kept tightly shut; the surroundings of the house should be clear; weekend houses must be carefully examined cleaned, and no boards, planks, tables, bricks, and tiles should be kept near the house or in the gardens.

VI. THE VENOM APPARATUS OF SPIDERS

The venom apparatus of spiders consists of one pair of chelicerae and one pair of venom glands. The paraxial position of chelicerae and vertical movements of fangs are characteristic for the venomous "tarantula" of the suborder Orthognatha, including the genera *Atrax* of Australia and Tasmania, *Harpactirella* of South Africa, the "trap-door spiders," *Actinopus* and the funnel-web spiders *Trechona* of Central and South America, and all other genera of Theraphosidae with large representants.

The diaxial position of chelicerae to the body length and horizontal movements of fangs are proper for all representants of the so-called true spiders of the suborder Labidognatha, to which belong several species of "brown spiders" (*Loxosceles*) of both hemispheres, the "black widows" (*Latrodectus*) abundant in tropical and subtropical zones all over the earth to temperate climates in Canada, several tropical and subtropical species of "wolf-spiders" (*Lycosa*), and the South American representants of "*armadeiras*" or "banana-spiders" (*Phoneutria*).

The morphology of the venom apparatus and the histology of poison glands in spiders have interested only a few specialists in this century. McLeod, in 1880, pointed out that the histological structures of venom gland vary with different kinds of spiders. A transparent and smooth basement membrane may exist as a continuous layer or with several incomplete sections, the glandular epithelium being firmly attached to this membrane, the epithelium is formed by caliciform cells; there are also other types of cells without definite structure and which vary in different species. Bordas, as cited by Vital Brazil, has studied the glands of *Latrodectus tredecim-guttatus*, the black widow of the Mediterranean region, and found a basement membrane with venom-secreting cells of the prismatic type; near the base of these cells a granular protoplasm may exist; at the apex the same cells show a whitish hyaline zone and the venom secretion may be realized by cell diffusion. Brazil and Vellard, with assistance of Dr. Camargo, in 1925 studied the glands of *Phoneutria* by staining with hematoxylin eosine or hematoxylin van Gieson and found two external muscle layers, one circular, the second longitudinal; the thick basement membrane bears several "processus" penetrating into the central lumen, the secreting cells being attached to the basement membrane and also to the processus; the form of the epithelial cells varies from low cuboidal to simple columnar and high columnar epithelium, with all transitions. In the active secreting phase the nuclei of the cells stay near the basement membrane with abundant chromatin as in prophase of mitosis, the cytoplasm being granular; in another phase the cells show a greater volume, the nuclei being compressed against the peripheral membrane and the cell body becomes full of amorphous granules; in other cells, in addition to these granules, there are bigger granules, variable in form and aspect, like colloidal substances; with the continuation of the secreting process the cells are broken off and their fluid runs into the central lumen. The whole venom gland is saclike in form. The amorphous granular substance in the lumen appears reddish after eosine staining and yellowish with van Gieson. A second substance of colloidal type, also present in the central lumen in addition to the remains of broken cells, is of intensive red color with eosin or intensive yellow with van Gieson stain. The authors conclude that the venom of the Brazilian "*armadeira*" may be of complex structure and that the venom gland is of the holocrine type, the entire cells with their secretory material being extruded and the cells dying. Attached to the basement membrane the authors have seen low cuboidal epithelial cells.

Millot, in 1931, published the study of venom glands of several spider species, without new progress. In the same year, there appeared the investigation of the venom glands of the "capulina" spider, *Latrodectus mactans*, of Mexico. In a cross section of the venom gland three tissues may be distinguished, the muscular, the conjunctive, and the glandular. The muscle fibers surround the glandular body in a serpentinelike form and are attached on a double chitinous ring. The conjunctive basement membrane emits several processus into the central lumen of the gland. The glandular tissue consists of spherical or cuboidal cells, the smaller ones with small nucleus, the larger cells without nucleus or with nuclei compressed against the wall. The venom is elaborated by the cell plasm. In other groups of secreting cells the nuclei seem to excrete the toxic substance. The author concludes that two or more substances elaborated by different secreting cells may constitute the so-called venom.

Sampayo (1942) has studied the morphology, anatomy, and histology of the glands of *Latrodectus mactans* (or *L. curacaviensis*) and *L. geometricus* of the province Santiago del Estero, Argentina: the glands of both species are saclike, with a smaller portion on the middle, with a length and breadth of about 1.4 and 0.4 mm, respectively; the glands are of whitish color. The fibers of the surrounding muscularis are striated, begin on the neck of the gland, attached on a chitinous ringlike formation, run over the whole gland, surround the distal portion, attaching again on the chitinous ring on the neck portion of the gland, but on the other side. In cross sections the external muscle layer shows transversal striation with several peripherial nuclei. The muscle fibers are interpenetrated by reticular fibrils of the second layer, the basement membrane, which forms perimuscular sheaths. The epithelial cells, attached on the inner surface of the basement membrane, show a serpentiniform arrangement around the central lumen. The cells are of the simple polygonal prismatic or columnar type. In each secreting cell two portions may be distinguished: the basal, infranuclear portion and the apical, supranuclear portion. After staining with phosphotungstic hematoxylin one may observe in different cells all stages of formation of venom granules. At first, numerous infranuclear granules are born which increase in size while they migrate to the apical portion; on the middle of cells one may separate distinctly the cytoplasm and the granules; in the apical portion the granules become swollen, enlarging the cell volume, and dropletlike. They are of blue color after Mallory staining. The mass of droplets is extruded into the central lumen, the apical portion of cells becoming clear and transparent, but the cell itself begins again the process of active formation of new granules. The author describes the secreting cells of both *Latrodectus* species as of merocrine type.

Reese, in 1944, studied the venom glands of *Latrodectus mactans* of West Virginia, after staining them *in toto* with borax carmine or after staining the sections in Lyons blue: the external muscular fibers extend from one end of the gland to the other and show little or no cross striation, the distal surface of the fibers being covered by a sheath of tissue, from which thin septa extend between the fibers to join with the much thicker connective tissue of the basement membrane. This seems to be of noncellular structure. Inside the basement membrane is a thin layer with very fine fibers, mixed with granules and with small, darkly stained cells. At intervals the fibrogranular layer projects into the central lumen. To these fibrogranular elevations and to the basement membrane are attached the oval or pear-shaped secreting cells. They vary greatly in size and in shape. At the base of the larger cells, and at intervals in the fibrogranular layer, are seen smaller cells of generally round or oval outlines. Whether these cells are the ones from which the larger swollen secreting cells arise could not be determined, but there seems to be no other source of origin for the large cells, which apparently break down to form the secretion. It seems strange that in all of the hundreds of sections examined, the exact origin of the swollen cells could not be definitely determined.

Vellard, in 1936, studied the venom glands of 12 species of Ortognatha, and 18 species of Labidognatha species of South America. He found the same histological structure of venom glands as that described by Brazil and Vellard in 1925: the venom droplets are acidophiles, the secreting cells are broken off and the cell nuclei are extruded into the central lumen with the venom and the apical cell portions; the secreting cells die from the centrum to the peripheral basal portions; after the secreting stage is finished, the gland shows only the external muscularis and the basement membrane, with internal processes, but without a proper glandular tissue. Along the basement membrane small, low cuboidal cells may exist. The mass, stored in the central lumen, consists really of three portions: the rests of broken epithelial cells; one venom portion, finely granular, without vestiges of cell nuclei, of red color with eosine or brown color with van Gieson, and a second venom portion of colloidal nature, of intensive red color with eosin. Amyloidal rests may be mixed with the venom. After a period of rest, the same epithelial cells may begin a new phase of venom production. In other species a new epithelial tissue may be formed by the low cuboidal cells around the basement membrane. The cell function is of the holocrine type.

Barth, in 1962, described the histology of the venom glands of *Latrodectus curacaviensis*, captured in the Guanabara Bay of Rio de Janeiro, Brazil: the gland is saclike, about 1.7 to 2 mm in length and about 0.3 to 0.35 mm in width. The venom-excreting canal is about 50 to 60 μ in diameter and shows a simple epithelium. The external muscle layer is formed of from four to forty-five muscle bundles, inserted on tonofibrils at the distal end of the gland and contrainserted near the neck of the gland. All muscles are striated and surrounded by a thin sacrolemma. The chelicerae-moving muscles are inserted externally on the sarcolemma with a tendon, the muscle-tendon junctions being of chitinous nature. The basement membrane, which is about 3 to 5 μ in diameter, consists of two very thin sheaths, an external and an internal one, the external forming the sacrolemma of muscular fibers, the internal forming the basement membrane, called by Barth the "sarcoperitoneal" membrane. The secreting epithelium is divided into two distinct portions: the so-called "lipocrine cells," situated on the neck portion of the gland, behind the excreting canal, and the "ragiocrine gland-cells" in the saclike body of the gland. The ragiocrine cells are of two types: prismatic or columnar cells, in active or in rest stages of secretion, and low, cuboidal auxiliary cells, both attached at the membrane propria, the cuboidal at the base of the columnar cell. The venom-secreting phase is activated by the auxiliary cells, their granules being acidophiles and passing with diffusion processes through the basal wall of the larger prismatic cells. After this acidophile stimulus, the principal cells beginning their proper function: fine granules are formed around the nucleus, which increase in size while they migrate to the central and apical portions of the cell; in the apical portion the secreted masses are stored, the cell is swollen, and finally the cell wall splits off and the mass is extruded into the central lumen, while a second and a third mass of venom may be formed in the basal portion of the same columnar cell, always with the help of cuboidal auxiliary cell. A total number of five to eight venom masses may be produced by both cells before they die definitely.

The lipocrine cells are also of the simple epithelium type with columnar cells attached to the basement membrane, with oval nuclei, situated on the basal zone; but the secreted granules of these cells are very fine and do not agglutinate to a mass, like those of ragiocrine cells, and pass through cell walls with diffusion.

Barth thinks that the lipocrine cells of *Latrodectus* are of the merocrine type, and both the low cuboidal and high columnar cells of the apocrine type, the first functioning permanently during the whole life of the spider and the second being destroyed after a relatively short period of venom production. Consequently three different types of venoms may be produced and mixed in the central lumen of the venom gland: two mixtures of both cell types of principal epithelium of the sac and one of the lipocrine cells of the neck portion, the mixture of the two first forming colloidal venom masses, stored in the central lumen, and of albuminoid character, while the secretion of the neck cells is of lipoid nature. Both types of venoms may be mixed only at the anterior portion of the gland, the lipoid product dissolving only a small portion of albuminoid masses. This may be the liquid mixture injected into the body of a victim when the spider bites. The author concludes that the albuminoid masses stored in the central lumen of the gland may be sufficient for the whole life of *Latrodectus*. In cross sections of glands of older spiders he has seen that nearly all excreting cells of the saclike portion of the glands are dead, the lumen being full of old venom masses.

We have studied in the last 2 years the venom apparatus of several South American spider species, such as *Actinopus crassipes* and *Trechona venosa*, this funnel-web spider being of the same family as the very feared Australian representatives of the genus *Atrax*, *Diplura bicolor* from the forests of Rio de Janeiro and the Serra dos Orgaos, *Neostothis gigas*, which is of the same family of *Barychelidae* as the very venomous *Harpactirella* of extreme South Africa, *Magulla symmetrica*, *Grammostola mollicoma*, *Gr. actaeon*, *Gr. iheringi* and *Gr. pulchripes*, *Pamphobeteus sorocabae*, *P. roseus*, *P. tetracanthus*, *Lasiodora klugi*, *Acanthoscurria geniculata*, *A. atrox*, *A. juruenicola*, *A. sternalis*, and *A. violacea*, and *Avicularia avicularia* of the state of Amazonas, Brazil. In the suborder of *Labidognatha* we have studied the venom apparatus of two Loxosceles species, the most frequent Brazilian representatives, which may cause deadly cases of poisoning in humans, of *Latrodectus curacaviensis*, the same species of black widows as studied by Barth, *Latrodectus geometricus*, of *Lycosa erythrognatha*, one of the most common wolf-spiders around the São Paulo city and of "armadeira" the most dangerous and aggressive species of the state of São Paulo, with several serious or deadly cases of poisoning.

In the Orthognatha species the chelicerae were carefully dissected, the chitinous layer was opened, the muscles and venom glands exposed; in the Labidognatha it was necessary to open the front part of the cephalothorax as well as the basal segment of chelicerae. The muscle system of all was studied *in situ* and from the venom glands several hundred longitudinal, tangential, and cross sections, stained with hematoxylin eosine, van Gieson, or Mallory stains have been made.

One may distinguish three parts of the venom apparatus in spiders: the musculature of the venom apparatus, the venom-inoculating apparatus, and the venom glands.

A. The Musculature of the Venom Apparatus

The musculature of the venom apparatus is situated in the basal segment of chelicerae ("tarantulas") and partly in the cephalothorax (true spiders). All the muscle bundles are of striated nature, their contractions or distensions operating only at the volition of the spider. The muscle bundles may be divided in abductors and adductors of chelicerae of the venom canal and the venom gland.

The claw or fang segment has no muscle bundles, except on its posterior border. On this border, outside in the true spiders, on the upper, convex side in the "tarantulas," two vigorous abductor muscle bundles are inserted. The contraction of these opens the fangs; when the fangs are closed or in a position of rest they are distended. Each or both bundles insert on a whitish tendon which ends on the membrana basalis of the subcutaneous epidermis. From these tendons extend the thin white-yellowish muscle fibers, which soon expand fanlike, penetrating into the basal segment of chelicerae and running along the dorsal convex side, immediately under the epidermis. In *Actinopus* and *Trechona* there exist only two abductor bundles; in several other "tarantulas" as well as in all the studied true spiders both bundles seem to be subdivided into two secondary bundles, each one with its proper insertion tendon. The contrainsertions of these abductors are not performed by tendons, but each muscle fiber ends, one by one, directly and independently on the membrana basalis of the epidermis of the second segment, at the convex side. In *Actinopus, Trechona, Neostothis*, and all the other genera of Orthognatha the venom gland is situated also dorsally in the second segment of chelicerae, immediately along these abductors, partly under the epidermis and partly covered by abductors. The form of the abductors is spindlelike, with the smaller ends in front and the larger portions behind.

The fang-adductor bundles are more numerous and even more robust than the abductors. By their abrupt contraction the fangs may be rapidly closed in the moment of the bite; by slow contraction the fangs are folded back into the fang groove, in a rest position. The insertions of the fang-adductor bundles are curious: at the inner side in true spiders, on the under side in the "tarantulas" where is a nearly quadrangular, chitinous sclerite, called the "interarticular sclerite." A careful study shows that this sclerite bears two very elevated inner borders, firmly attached to the inverse border of the posterior portion of the fang. On these borders are inserted the tendons —four or six—of the adductor bundles. A little traction on these borders is sufficient for the mechanical movement of the fang article. In *Actinopus* and *Trechona* as well as in the representatives of Theraphosidae we have seen four adductor muscles with four tendons; in several genera it seems that two bundles are subdivided into four bundles, resulting in six adductors. The same seems to be the rule in true spiders, as in *Latrodectus, Loxosceles, Lycosa,* and *Phoneutria*, where six insertion tendons exist. On these adductors a ventral pair runs along the inner surface of the basal segment of chelicerae, beginning at the interarticular sclerite with two tendons, expanding fanlike and contra-inserting, fiber by fiber independently, at the membrana basalis of the epidermis, just at the opposite face of the abductors. The four or two inner adductor bundles, borne also at the borders of the interarticular sclerites, run along the first adductor pair, but nearer the central lumen and contrainsert also on the membrana basalis of the epidermis of the basal segment, at the inner side in true spiders, at the under side in the "tarantulas." While in the dorsal abductors and the ventral adductors the insertion tendons run along the outer side of these bundles and the fibers expand fanlike in only one direction, the tendons of the inner adductor muscles are central in position and the fibers are borne on secondary shorter tendons and run in both directions. As the violent contraction of the adductors may increase their thickness, of course they press in tarantulas against the body of the venom gland and contribute very efficiently to the venom-expelling process. In the "tarantulas" this is proved also by the form of the abductors and adductors as well as of the venom gland: all chelicerae muscles are spindlelike in aspect, the small part being in front and the large part behind, while the venom glands are carrotlike in aspect with the large portion just in front and the spindlelike small portion behind. The venom is ejected at the volition of the spider by violent contractions of the muscularis of the venom gland and of the adductors.

In the suborder of true spiders, such as the representatives of *Loxosceles, Latrodectus, Lycosa,* and *Phoneutria*, the poison glands, first observed by A. van Leeuwenhoeck, two and a half centuries ago, are situated in the front part of the cephalothorax of the spider; the venom canal runs through the fang and the second segment of chelicerae; consequently the musculature of the apparatus is different in some way from that of "tarantulas." The insertions of abductors and adductors are nearly on the same places at the borders of the fang article; this is also the case for tendons and the contrainsertions on the membrana basalis of the epidermis of the basal segment. In this group we found a powerful and elastic tendon, inserted into the fang article and running just along the venom channel in the basal segment of chelicerae and penetrating into the cephalothorax of the spider. From both sides of the tendon are borne muscle fibers, one part of them contrainserting at the membrana basalis of the epidermis and the other part contrainserting at the outer wall of the venom canal—into the chelicerae—and at the sacrolemma of the anterior portion of the venom gland into the front part of the spider cephalothorax. This muscle seems to have nothing to do with adduction or abduction of chelicerae, but exerts a function as a "venom-canal enlarging muscle" and also as a "traction muscle" of the venom gland.

The basal segment of chelicerae of both suborders Orthognatha and Labidognatha shows also abductor and adductor muscle bundles. They are not as strong as those of the fangs. They are also inserted with tendons just on the posterior border of this segment, on the inner face of the dorsal and the ventral side, respectively. At the outer and inner face of the same border exist also an abductor and an adductor, which may operate the lateral mobility of the basal segment. The contrainsertions of all these bundles are situated in both suborders in the cephalothorax, the abductors contrainserting at the membrana basalis of the cephalothoracic epidermis, the adductors contrainserting partly on the musculature of the stomach. All contrainsertions are fiber by fiber.

The posterior chitinous border of the basal segment of the chelicerae is in true spiders very much constricted, so that the resulting opening is so small that the saclike body of the venom gland cannot pass through it. This rounded posterior border is covered also with a strong circular muscle, which contracts the opening even more. This constriction appears in *Lycosa* and *Phoneutria*, and may exist also in *Loxosceles* and *Latrodectus*. When the long traction muscle is contracted, the venom gland itself will be compressed against this muscular ring, with rapid ejection of the venom.

B. The Venom-Inoculating Apparatus

As we have seen before, the venom-inoculating apparatus consists in both suborders of spiders of two segments: the claws or fangs and the basal cylindrical segment. Both are hollow; in both on the convex side of the fangs, at the end of their tip, there are small openings for venom ejection.

The approximate measurements of this wounding apparatus may be seen in Table IV.

It may be of interest to know that spiders such as *Loxosceles* and *Lactrodectus*, with fangs only about 0.30 mm in length, can perfectly puncture the human skin and inject, under certain circumstances, deadly doses of venom. Fangs and the basal article of *Loxosceles* (Fig. 26) may be distended by the spider at the moment of bite at a maximum of 0.8 mm, perforating human skin at a maximum of 0.4 mm only (Fig. 27). *Latrodectus m. mactans* of Santiago del Estero, Argentina, of Filadelfia, Paraguay, examined by us,

TABLE IV

DIMENSIONS OF THE WOUNDING APPARATUS

Species	Length of fangs (mm)	Length of basal segment (mm)
Orthognatha		
Actinopus crassipes	3.00– 4.00	4.00– 5.00
Trechona venosa	3.00– 4.00	4.00– 5.00
Diplura bicolor	2.50– 3.80	4.50– 5.50
Neostothis gigas	3.50– 4.00	4.10– 6.20
Magula symmetrica	3.50– 4.80	5.90– 6.00
Grammostola mollicoma	6.20– 7.50	6.50– 7.30
Grammostola actaeon	5.90– 6.00	6.60– 7.30
Grammostola iheringi	5.80– 6.20	6.50– 7.40
Grammostola pulchripes	5.70– 6.80	7.50– 8.00
Pamphobeteus sorocabae	4.70– 5.20	5.30– 6.30
Pamphobeteus roseus	4.60– 5.30	6.20– 7.40
Pamphobeteus tetracanthus	4.70– 5.70	5.20– 7.00
Lasiodora klugi	7.90– 8.30	8.30– 8.20
Acanthoscurria atrox	7.20– 8.40	7.40– 8.60
Acanthoscurria geniculata	7.50– 8.70	7.70– 8.20
Acanthoscurria sternalis	5.80– 6.60	7.20– 7.30
Acanthoscurria violacea	5.30– 6.40	5.80– 7.10
Acanthoscurria juruenicola	6.30– 8.20	7.50– 8.60
Avicularia avicularia	7.40– 8.40	7.30– 8.20
Theraphosa leblondi	9.80–12.60	11.40–13.60
Labidognatha		
Loxosceles similis	0.32–0.40	1.40–1.70
Loxosceles rufipes	0.32–0.40	1.40–1.70
Latrodectus m. mactans	0.38–0.45	0.90–1.10
Latrodectus geometricus	0.37–0.43	0.85–1.10
Latrodectus curacaviensis	0.39–0.45	0.90–1.15
Lycosa erythrognatha	2.50–3.20	4.50–5.80
Phoneutria nigriventer	4.20–4.80	6.20–7.80

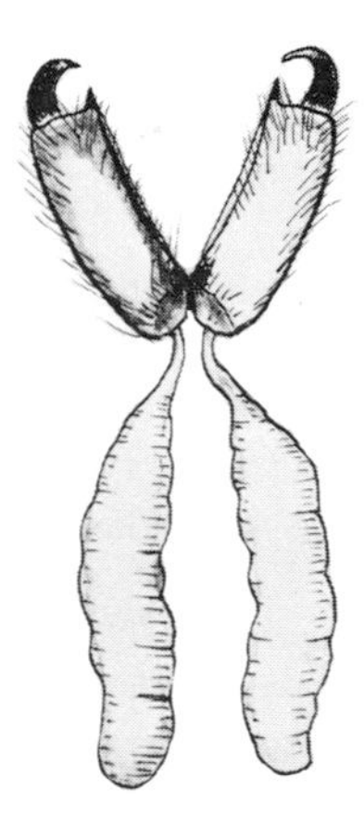

FIG. 26. *Loxosceles similis*: chelicerae and venom glands.

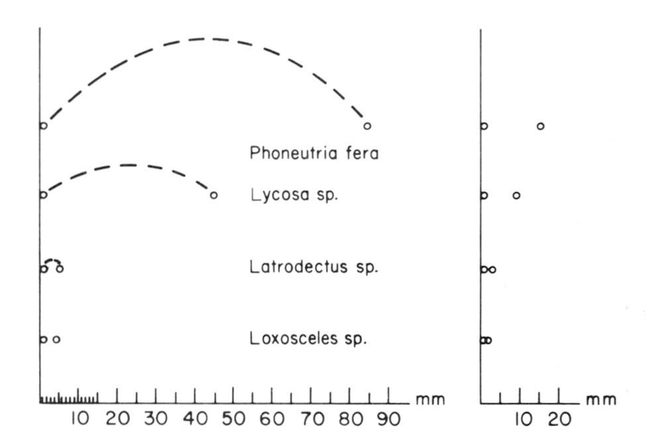

FIG. 27. Distances between the pits of the fangs, in biting spiders. At left magnified 5 times; at right, natural size.

and *L. curacaviensis* of Guanabara Bay and the beaches of Rio de Janeiro, Brazil, show a maximum distension of chelicerae and fangs from about 1 to 1.2 mm with a possible penetration into the skin of about 0.4 mm (Figs. 27 and 28). The bigger tropical and subtropical species of *Lycosa* (*L. erythrognatha, L. thorelli, L. nychthemera, L. pampeana*, etc.) may impress their fangs in human skin up to 2.5 mm, with a distance between the tips of both fangs of about 9 mm at maximum (Figs. 27 and 29). The aggressive *Phoneutria sp.* can impress its fangs up to 4 mm with a distance of the two pits of fangs of about 17 mm (Fig. 27).

The maximum distance of the penetrating teeth may be of interest for

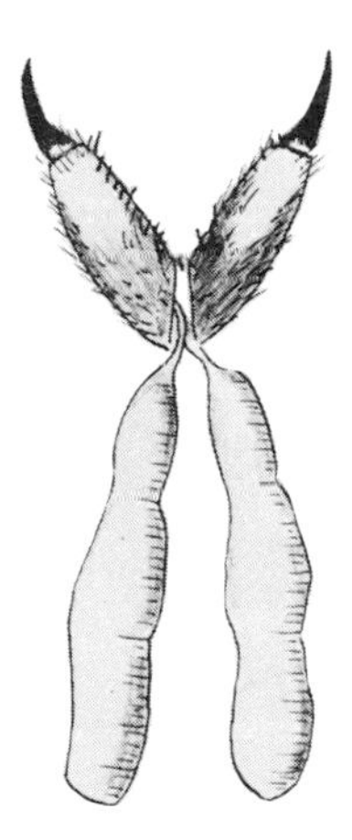

FIG. 28. *Latrodectus curacaviensis*: chelicerae and venom glands.

differential diagnosis of spider bites, chiefly in regions where representatives of the four venomous genera exist.

The giant representatives of "tarantulas," with a length of their fangs from about 3 to 12 mm and a vertical movement of chelicerae, may inject their venoms very deeply in the human skin, often as deep as an adult venomous snake. Fortunately their venoms, except those of *Atrax, Trechona,* and *Harpactirella,* seem to have little effect on humans.

C. The Venom Glands

According to earlier authors, such as Herms, Bailey and Mc Ivor, "There is good evidence, that the poison sacs are not glandular in nature but function as absorptive organs which take up the poisonous constituents from the body of the spider."

Of course this is a tremendous mistake. In all representatives of the suborder of tarantulas the venom glands are situated in the basal article of chelicerae, between the abductor and adductor muscles; the glands are carrotlike in form, with the largest portion just in front and ending behind like a spindle. From the end of the gland a strong elastic muscle fiber extends and inserts on the membrana basalis of the epidermis at the posterior border on the upper side of the basal segment; this muscle fiber comes from the muscularis of the gland; it holds the gland firmly in its position.

In all representatives of the suborder of true spiders the venom glands are situated in the cephalothorax. The abductor and adductor and the traction muscles hold the gland in its position. The gland is saclike or cylindrical in form.

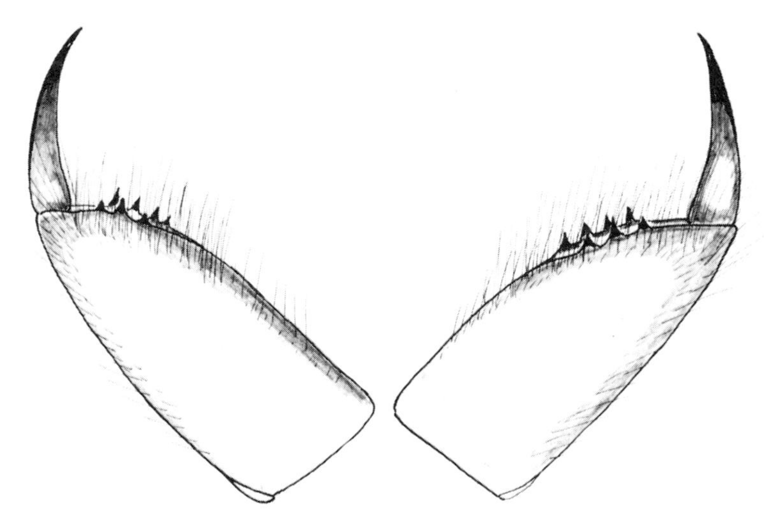

FIG. 29. *Lycosa erythrognatha*: chelicerae opened.

On the venom glands of all spiders the following portions may be distinguished: the excreting canal, the carrotlike or saclike body of the gland.

The excreting canal is always a very small, but long, whitish elastic tube. Its length corresponds in "tarantulas" to the length of fangs and in the true spiders—*Loxosceles, Latrodectus, Lycosa, Phoneutria,* etc.—to the length of both articles of chelicerae. In some true spiders *Lycosa* and *Phoneutria* we have seen in several specimens that the canal bears a spherical ampule just at the level of the apical portion of the basal segment of chelicerae, which may be called the "interarticular ampule." In the lumen of this ampule venom is stored, perhaps ready for use. The canal does not show a proper muscularis. It seems unnecessary in "tarantulas" such as *Atrax, Harpactirella*; where the channel is short and the venom is ejected very efficiently by a squeeze of the adductor muscles; in true spiders with a much longer channel, there is a proper "venom canal enlarging muscle."

The canal is of exodermic origin, with an endocuticular, transparent, and elastic sheath in its center. On it are attached the simple cuboidal epidermis cells, followed by the basement membrane. The central lumen is closed and may be opened only by the pressure of the venom just extruded in "tarantulas" or by the venom canal enlarging muscle in "true spiders."

On the body of the venom gland of spiders three fundamental portions must be distinguished: the outer muscularis, the basement membrane, the epithelial venom-excreting cells.

Length and width of the body of the venom glands in the different species studied are shown in Table V.

The venom glands are always whitish in color. Cross and longitudina sections of the glands, stained with hematoxylin-eosine, with Mallory or van Gieson, show very good comparative figures, the Mallory method being preferable for the cross striation of the muscularis, the van Gieson method for nuclear elements and their chromatin formations as well as for the excreted venoms, and hematoxylin-eosine offering very good differentiation in the glandular cells, their walls, nuclei, and different phases of venom formation.

The muscularis of the gland surrounds the whole gland beginning on the neck of the gland and covering the whole body. The muscularis itself is surrounded at its outer side by the sacrolemma, a very fine, transparent, homogeneous sheath of intensive blue color with Mallory, yellow with van Gieson and more or less violet with hematoxylin-eosine. It covers the muscularis as a whole and also bundle by bundle. From space to space a small dense nucleus may be seen. On the inner face of the muscle bundles the sacrolemma seems to be attached to the basement membrane, forming what Barth has called a double basement membrane. The muscularis consists of several muscle bundles, all of striated nature. By careful examination of several cross, longitudinal and tangential sections one may see that the muscle bundles do not obey a rigorous orientation around the gland. They may cover the gland in a more or less serpentiniform way, being often nearly circular, at the ends of the gland longitudinal, etc. At several places of the same gland the muscle bundles may form a simple, or a double or even a triple sheath, varying from gland to gland in the same species. All the muscle fibers insert and contrainsert at the basement membrane, along the whole body of the gland.

As a rule the muscularis is more powerful on the dorsal side of the gland and much smaller on the ventral side. *Lycosa erythrognatha* shows about 40 bundles in cross section and 120 bundles in longitudinal section; *Phoneutria* has about 160 bundles of serpentiniform aspect; *Loxosceles similis* shows about 38 bundles in cross sections and about 40 bundles in longitudinal sections; in the Theraphosidae the muscularis is very powerful, forming a sheath three times bigger than that of the secreting zone. The fibrils are attached at the basement membrane by tonofibrils.

The basement membrane forms a continuous layer inside the muscularis. In the glands of younger specimens it emits several concentric processes into the central lumen to form a larger surface for the excreting cells. As the excreting cells undergo degeneration, the basement membrane processes become shorter and shorter, and the free central lumen increases in size; in older venom glands the basement membrane consists of a nearly straight homogenous layer. After using Mallory stain one can see exactly the external zones of the membrane, at which the muscle fibers are attached by tonofibrillar structures. Small nuclei may be seen, intensively colored.

The venom-excreting cells form a simple epithelium. Two kinds of epithelial

TABLE V

LENGTH AND WIDTH OF THE VENOM GLANDS

Species	Length (mm)	Width (mm) Width in front
Orthognatha		
Actinopus crassipes	3.0–5.0	0.8–1.0
Trechona venosa	3.0–4.0	0.7–0.9
Diplura bicolor	4.0–5.0	0.7–0.9
Neostothis gigas	4.0–6.0	0.6–1.1
Magula symmetrica	4.0–5.0	0.7–1.0
Grammostola mollicoma	6.0–6.9	0.9–1.3
Grammostola actaeon	6.0–6.0	0.8–1.2
Grammostola iheringi	6.0–7.0	0.8–1.2
Grammostola pulchripes	5.5–6.7	0.7–0.9
Pamphobeteus roseus	6.0–7.0	0.9–1.2
Pamphobeteus sorocabae	6.0–7.0	1.0–1.2
Pamphobeteus tetracanthus	6.0–7.8	0.9–1.3
Lasiodora klugii	6.5–8.0	0.8–1.4
Acanthoscurria atrox	6.4–7.5	0.9–1.3
Acanthoscurria geniculata	6.3–7.0	1.1–1.4
Acanthoscurria sternalis	4.0–5.0	0.7–0.9
Acanthoscurria violacea	4.0–5.0	0.7–0.9
Acanthoscurria juruenicola	5.0–6.8	0.8–1.1
Acivularia avicularia	4.6–6.3	0.7–1.0
Theraphosa leblondi	8.5–12.0	1.2–1.6
Labidognatha		Width behind
Loxosceles rufescens	1.7–2.0	0.3–0.35
Loxosceles rufipes	1.7–2.0	0.3–0.35
Latrodectus m. mactans	1.6–2.0	0.3–0.32
Latrodectus curacaviensis	1.6–2.1	0.3–0.33
Latrodectus geometricus	1.5–1.7	0.2–0.25
Lycosa erythrogenatha	4.5–6.3	1.0–1.20
Phoneutria nigriventer	7.6–10.4	2.4–2.70

cells are present in "tarantulas," three kinds in the true spiders, both the first and the second being of the same morphological structure and of the same physiological properties: the first layer is formed by very small and low, subcuboidal cells, attached to the basement membrane, and with an oval nucleus, which seems to be in a resting stage. The layer may be continuous, but as the subcuboidal cells often are covered by the second layer, they can be seen only from space to space, just at the base of the high columnar cells. We have seen these cells in *Trechona, Avicularia, Grammostola, Pamphobeteus, Acanthoscurria,* as well as in *Loxosceles, Latrodectus, Lycosa,* and *Phoneutria.* These cells will substitute the excreting columnar cells after their degeneration, so that the spider always has venom-elaborating cells, at least during its principal life cycle. Only in the last months of its life do all excreting cells seem to be degenerated.

The second layer of epithelial cells is formed by very high columnar cells, attached with their basal portion at the basement membrane and, in younger individuals, also at the basement processes. In these cells one may distinguish the nucleus, the cytoplasm, the cell wall and the venom in different aspects. The columnar cells and also the subcuboidal cells are disposed in a serpentiniform manner, so that on the same section they may appear cross, tangential or longitudinally bisected. It seems to be the reason for the differences between different authors, like Brazil, Vellard, Millot, Ancona, Sampayo, Reese, Barth, and others. In younger cells the nuclei are disposed in the basal portion of cells; during the venom production phase the nuclei increase in size and migrate to the middle of the cell; in older cells, with the apical portion of the cell wall broken off, the nuclei remain more apical, and finally in a later phase of venom production they may be extruded into the central lumen with the last portion of venom, and the cell lacking its nucleus will degenerate. The cytoplasm is nearly transparent and covers the whole cell body; in younger cells it is more colored in the basal portion and nearly transparent in the apical part; in older cells the middle and the apex of the cell are more intensively colored; after the apical cell wall is broken off, portions of cytoplasm may also be extruded into the central lumen with the venom and one may see apical cell portions without cytoplasm. The cell walls appear very clearly in different sections; when the venom is ejected into the central lumen, pieces of the apical cell wall which are broken out travel also into the central lumen. They do not dissolve and may be ejected with the venomous mixture and may also penetrate into the body of a victim, as well as the extruded nuclei and the degenerated cytoplasm. The venom itself, produced by the columnar cells, presents four aspects—fine granules at the basal portion of the cell, around the nucleus; bigger granules at the middle of the cell; droplets on the apical portion of the cell, and bigger droplets, often condensed to a mass, when they break off the cell wall and enter into the central lumen. These phases of venom production and venom ejection, of course, form a continuous process.

The same columnar cell will produce venom masses and extrude these into the lumen, several times, perhaps four to seven times, until its nucleus will be also extruded. Then it degenerates, being replaced by a subcuboidal substitution cell.

This type of epithelial cell is of the apocrine type and occurs in all spiders examined by us. We have not succeeded in confirming the interpretation of Barth of the *Latrodectus curacaviensis* of a ragiocrine type of both columnar and subcuboidal cell layers.

In *Loxosceles similis*, and *L. rufipes*, *Latrodectus curacaviensis* and *Lycosa erythrognatha*, we have seen the third type of epithelial cells, situated just around the neck of the gland. The zone of these cells can be easily determined, because the muscularis ends at this portion of the gland. In several sections we have seen a homogeneous, thin membrane, separating these cells from the other epithelial cells. They are of prismatic aspect, attached with their basal portions to the basement membrane. On this region we have not seen low, subcuboidal cells. The nuclei of these columnar cells are not so regularly distributed along the basal portion of the cells, but are situated indiscriminately in the cell body, the base, middle, or apex. The cell walls are continuous. We have never found apically broken cells and the studied cell bodies always have been full of cytoplasm. The substances produced by these "neck-cells" are much more homogeneous than that of principal cells and show in the whole cell body the same, fine granular structure. Under these circumstances one may conclude, as Barth has found in *Latrodectus curacaviensis*, improperly called by him *L. mactans*, that the toxic substances of these cells may be extruded through the cell wall by "transfusion" or "transmigration." On the other hand, as we have not found subcuboidal substituting cells and the cell nuclei never are extruded, the cell walls never are broken off, these glands must be considered as merocrine and of permanent function during the whole life of the spider.

It is interesting to know that the venom injected by these kinds of spiders into the body of a victim is really a mixture of at least two types of substances, elaborated by two different epithelia, those of the body and those of the neck of the gland. Several families of true spiders show two types of epithelial cells: apocrine and merocrine, while the "tarantulas" seem to have only apocrine epithelial cells.

VII. EXTRACTION AND TOXICITY OF SPIDER VENOMS

The division of spiders into groups proved dangerous, nonproved but dubious, and suspect, cannot be considered as very strict. Exceptions may exist. But, in a general view, one can relate the problem of venomous and dangerous spiders to *Atrax, Harpactirella, Loxosceles, Latrodectus, Lycosa, Phoneutria*, and some tropical "tarantulas." This problem increases in importance according to whether the spiders are more or less frequent in a geographical region. Table VI shows the number of some spiders captured and sent to Instituto Butantan.

All these, except "tarantulas" and *Latrodectus*, have been captured around São Paulo. *Latrodectus curacaviensis* was captured by myself and my staff on the beaches of the state of Rio de Janeiro, from Cabo Frio to the

TABLE VI

SPIDERS SENT TO INSTITUTO BUTANTAN

Year	Spider names				
	"Tarantulas"	*Loxosceles*	*Latrodectus*	*Lycosa erythr.*	*Phoneutria nigriventer*
1958	438			9,621	571
1959	515			499	352
1960	422	3,057	3,045	3,179	591
1961	538	9,548	969	4,653	944
1962	169	11,547	118	7,942	285
1963	237	4,357	727	11,673	704
1964	786	2,680	12	843	427
1965	580	1,300	2,255	602	743
1966	150	315	2,603	689	441
1967	658	1,907	13	1,298	1.117
	4,493	34,711	9,742	41,009	6,175

Piratininga beach. The "tarantulas" came from Rio Grande do Sul, Parana, Santa Catarina, São Paulo, East Mato Grosso, the south of Minas Gerais and Goiás, and ports of Rio de Janeiro.

A. Extraction of Spider Venoms

In the large "tarantulas," *Actinopus, Atrax, Trechona,* the Grammostolinae, Theraphosinae (*Xenesthis, Megaphobema, Theraphosa, Pamphobeteus, Lasiodora, Acanthoscurria, Eupalaestrus,* etc.), and *Harpactirella,* the pure, natural, liquid venom is relatively easy to obtain; one seizes the "tarantula," fastened to the table, by the cephalothorax, with a long forceps. This way, it is not hurt, for its cuirass in the cephalothoracic region is very strong. Lift the spider, holding it with the thumb and forefinger of the left hand. Then the forceps is taken away. As the spider does not appreciate all this, it tries to revenge itself, stretches and separates the poison fangs, and often two small, round drops of venom come out of the pits of the fangs, which can be sucked with a capillary pipette. With a little practice, one can collect from 10 or 20 "tarantulas" of the same species a surprisingly large quantity of venom; after the first drop is collected, one needs to press only a little on the sternum; as a consequence, the spider becomes irritated, tries to bite, and a fresh flow of venom comes out and can be collected again.

The venoms of all "tarantulas" consist of a transparent liquid, forming round drops easily soluble in water.

In order to methodize the extraction of larger quantities of venoms, several hundred "tarantulas" of different genera are kept in Butantan, each one in its proper case. In 1950 we built an electric apparatus: six flashlight batteries were fastened together and led upwards by means of the two last batteries, provided with two poles, a positive and a negative one. In front of it we placed a small glass dish. Both fangs of the tarantula were placed so that they would touch both poles; immediately there was a muscle contraction and a vertical movement of the fangs, and the venom drops flowed into the glass. Afterward, we improved the apparatus, introducing an interruption lever; then it was possible to stimulate the same spider twice or three times and gather more venom. This process does not harm them at all, the spiders continue eating and can be "milked" every 2 or 3 weeks. With about 500 "tarantulas," 30 or 50 from each species, which are milked every 3 weeks, we are able to collect quantities of pure, fresh venom, that will aid chemists, physiologists, and pharmacologists in further studies.

Pure spider venom is very important because it alone makes possible a high qualitative raw material for serum production and for testing the activity of the venoms in experimental animals, avoiding the kind of discrepancies that occur when the experimenter uses the direct spider bite into experimental animals. With this simple method it is also possible to establish the danger or harmlessness of a determined species with relatively few specimens. With this method the greatest quantity of venom that a single specimen can give and also inject when it bites can be established; if this quantity is sufficient to harm or to kill an adult or a child, a domestic animal, or any other precious life, the amount can be exactly determined with injections of standardized solutions of the venoms into experimental animals.

The representatives of "true" spiders, chiefly those of larger size, such as adult lycosids and *Phoneutria*, are also handled with pincers and "milked" by electric shock every 3 weeks. Smaller species, such as *Loxosceles*, black widows, and representatives of *Cheiracanthium* and *Glyptocranium*, may also be extracted electrically. The problem is how to handle them without harming them. The venom, running out of the fangs after electric stimulus, is not deposited "from the spider" to the glass, but is collected from the fang pits with a capillary pipette.

The liquid venoms are immediately dried in vacuum, forming whitish or whitish-yellowish or whitish-gray slices of lamellae. The venoms of true spiders dry very well and may be stored and conserved *in vacuo* with the exclusion of daylight for several years. The venoms of "tarantulas" are much more hygroscopic; they dry only after a few days, repeating the vacuum or with a higher vacuum, and they can be stored only for a few months *in vacuo*. In contact with the air, they will liquefy in a short time.

During 1963, "armadeiras" 115 have been milked 21 times and gave 2.830 gm of dried venom; 115 *Lycosa erythrognatha* milked 8 times gave

0.605 gm, 300 *Loxosceles similis* gave 0.025 gm, and 140 *Latrodectus curacaviensis* gave 0.020 gm. Table VII shows the venom quantities in dryform obtained from *Ph. nigriventer*, the most dangerous Brazilian spider.

TABLE VII

VENOM QUANTITIES OBTAINED FROM *Phoneutria nigriventer*

Year	Quantity of spiders	Number of extractions	Dry venoms (gm)	Mean quantity venom (mg/spider)
1967	112	14	2.703	1.086
1966	92	12	1.736	1.572
1965	140	14	3.122	1.600
1964	110	12	2.350	1.780
1963	115	21	2.830	2.050
1962	95	20	2.247	1.183
1961	155	10	2.600	1.678
1960	75	22	2.555	1.547
1959	53	21	1.050	0.943
1958	103	15	2.420	1.570
1957	70	21	1.121	0.762
1956	85	21	1.612	0.963
1955	71	20	1.206	0.847
1954	80	20	0.950	0.593
1953	40	26	0.1587	0.564
	1,396		29,089	1,245

The quantities of dried venom obtained from a single spider vary widely and cannot be expressed with absolute accuracy. After 10 years of experience with the electric method of venom extraction of several hundred specimens of each species, we have collected the approximative data about the minimum, mean, and maximum dried venom quantities of spiders which are shown in Table VIII.

The minimum, average, and maximum quantities of venoms obtained from spiders and shown in Table IX are very important. If somebody is bitten by a spider, the serum therapy should be sufficient to neutralize the maximum quantities, to prevent all possibilities of intoxication. According to Kaire, in 1963, the minimum, mean, and maximum venom quantities of *Atrax robustus* are 0.08, 0.31, 1.6 mg, respectively.

B. The Toxicity of Spider Venoms

For testing the activity of spider venoms, in general, one needs to establish their effects on man, on higher domestic mammals, on the so-called "experimental" animals. such as white rats, mice, hamsters, rabbits, guinea pigs, and

TABLE VIII

MINIMUM, MEAN, AND MAXIMUM DRIED VENOM QUANTITIES

Species	Dry venom quantities (electric methods)		
	Minimum (mg)	Mean (mg)	Maximum (mg)
Orthognatha			
Actinopus crassipes	0.04	0.09	1.20
Trechona venosa	0.06	1.00	1.70
Avicularia avicularia	0.90	1.30	6.50
Grammostola actaeon	0.90	3.70	5.20
Grammostola iheringi	1.00	3.80	5.20
Grammostola mollicoma	1.30	4.00	6.20
Grammostola pulchripes	0.70	2.90	4.50
Acanthoscurria atrox	0.20	2.40	8.90
Acanthoscurria musculosa	0.40	2.30	4.20
Acanthoscurria rhodothele	0.30	1.80	3.90
Acanthoscurria sternalis	0.30	1.00	3.10
Acanthoscurria violacea	0.20	0.60	1.50
Eupalaestrus tenuitarsus	0.40	1.30	1.80
Eurypelma rubropilosum	0.60	2.00	6.00
Lasiodora klugi	0.20	2.40	3.60
Pamphobeteus roseus	0.90	1.60	3.00
Pamphobeteus sorocabae	0.50	0.80	2.80
Pamphobeteus tetracanthus	0.80	2.20	2.70
Pamphobeteus platyomma	0.50	2.00	3.40
Pterinopelma vellutinum	0.50	1.70	3.00
Sericopelma fallax	0.90	1.10	2.10
Labidognatha			
Loxosceles similis	0.10	0.70	1.5
Loxosceles rufipes	0.10	0.70	1.5
Latrodectus m. mactans	0.10	0.60	1.3
Latrodectus curacaviensis	0.10	0.60	1.3
Latrodectus geometricus	0.10	0.30	0.50
Polybetes pithagoricus	0.30	0.60	1.00
Nephila brasiliensis	0.20	0.70	3.00
Nephila clavipes	0.30	0.60	1.80
Lycosa erythrognalha	0.30	1.00	2.05
Phoneutria nigriventer	0.30	1.25	8.00

pigeons, for the LD_{50} determinations. Very important also is the activity of venoms on animals with cold blood and on other spiders of the same or of different species.

For the correct situation of the health problem in relation to the notoriously dangerous spiders, a more exact terminology should be used. The medical literature is often disappointing in this aspect, as words such as "arachnoidism, araneismo, tarantulism, and arachnidism," are used indiscriminately by older and modern authors.

The convenient word to indicate all dangerous spiders, the "tarantulas" as well as the "true" spiders, who may cause human accidents, is "Araneism," from the word Araneida, which includes the whole order. Words like "arachnoidism" or "arachnidism" includes the whole class Arachnida, with scorpions, acarina, etc. Consequently the correct names for the symptomatology of accidents with the notoriously dangerous spiders in the world are, respectively, atraxism, harpactirellism, loxoscelism, latrodectism, lycosism, and phoneutriism, etc., always derived directly from the scientific name of the genus of the spider. These words may be explained with the more important symptoms of intoxication in human beings, as follows: Loxoscelism with cutaneous ictero-hemolytic action; latrodectism with neuromyopathic action; phoneutriism with sharp local pains and neurotoxic action, and so on.

Table IX shows the venom quantities of spiders which kill a 20-gm white mouse when the venom is injected either intravenously or subcutaneously (LD_{50}) in one experiment only.

We have used only fresh venoms, recently dried. In each experiment 20 white mice, weighing about 20 gm each, have been used, and the results have been calculated according to the technique described by Reed and Muench. We are convinced that the data have to be interpreted as approximate only; that pigeons, rats, rabbits, guinea pigs show different sensitivity to the same venom.

It seems that the most active spider venom we know is that of *Phoneutria spp.* After intravenous injection all experimental white mice, which received more than the LD_{50}, died very rapidly, in $\frac{1}{2}$ to 1 hour. After subcutaneous injection the first symptoms of intoxication were seen in 10 to 20 minutes, increased in 30 to 40 minutes, and usually terminated in death in 2 to 5 hours. In the survivors, elimination of the venom or neutralization began a few hours after injection and was complete with recovery of the animal in 24 hours. *Phoneutria* venom is essentially neurotoxic in action on the central and peripheral nervous system; it is also painful; the experimental animals die by paralysis of respiration, followed by spasm.

The *Lycosa* venoms are cytotoxic, with local necrotic action only, without general symptoms. The local pain is never exaggerated.

In relation to the frequency of spiders, the accidents with representatives of *Loxosceles* and *Latrodectus* are the most important. The *Loxosceles* venom is cytotoxic and hemolytic; congestive hemorrhagic lesions, particularly in the liver and kidneys, may be present as well as extensive zones of hemorrhagic infiltrations to the point of local necrosis and epithelial destruction around the bite and even in the stomach and intestines in dogs, as stated by Vellard.

TABLE IX

VENOM QUANTITIES WHICH KILL A 20-gm MOUSE

Spider species	Intravenous dose (mg)	Subcutaneous dose (mg)
Trechona venosa	0.030	0.070
Grammostola mollicoma	0.500	1.000
Grammostola iheringi	0.450	1.000
Grammostola actaeon	0.490	1.150
Grammostola pulchripes	0.480	1.200
Eurypelma rubropilosum	0.350	0.850
Eupalaestrus tenuitarsus	0.950	2.100
Pamphobeteus roseus	0.850	1.700
Pamphobeteus sorocabae	0.700	1.500
Pamphobeteus tetracanthus	0.600	1.400
Pamphobeteus platyomma	0.800	1.500
Acanthoscurria sternalis	0.300	0.620
Acanthoscurria violacea	0.280	0.610
Acanthoscurria atrox	0.300	0.850
Acanthoscurria musculosa	0.210	0.450
Acanthoscurria rhodothele	0.280	0.650
Lasiodora klugi	0.640	1.200
Loxosceles rufipes	0.200	0.300
Loxosceles similis	0.130	0.250
Latrodectus m. mactans	0.110	0.200
Latrodectus curacaviensis	0.170	0.240
Latrodectus geometricus	0.230	0.450
Lycosa erythrognatha	0.080	1.250
Phoneutria nigriventer	0.006	0.0134

The medical literature on loxoscelism in human beings is abundant, published chiefly in Chile, Uruguay, Argentina, Peru, and the United States, and concludes that human beings are extremely sensitive to this venom.

The *Latrodectus* venom is neurotoxic in action, affecting chiefly the spinal cord; in addition, it may cause generalized injuries in the liver, kidneys, spleen, lymph nodes, thymus, and adrenals, as demonstrated by Marzan, in 1955. The principal lesions in experimental animals consisted of a parenchymatous necrosis affecting blood vessels and epithelial cells, as well as nervous tissue and lymphoid cells. Experimental liver necroses in rats have been described by D'Amous, Becker, and Van Ripes in 1936; Keegan, Hedden, and Whittemore, in 1960, reported striking pathological changes in the liver, with fatty infiltration, hemorrhages and necrosis in cats, guinea pigs, and mice, injected with venom of *L. geometricus*. According to Vellard, in 1936, latrodectism leads to severe pain, muscle spasm, and profuse perspiration, together with nervousness and anxiety to the point that several patients felt as if they were going insane. The South African black widow causes, according to Villiers, violent pains and fantastic perspiring; several deaths have been reported. Meek, Kalawey, McKay, Lethbridge, Rodway, and others wrote that the black widow of Australia, the "night stinger" or "katipo," is always very feared and often common in certain regions. *L. m. menavodi* of Madagascar causes also severe intoxication in human beings. A bite of *Latrodectus m. mactans* kills guinea pigs from 250 to 500 gm in intervals of from $1\frac{3}{4}$ to 4 hours, according to Herms, Railey, and McIvor (1933/1934). The venom solution, obtained from 8 cephalothoraxes of *L. m. tredecimguttatus*, injected intravenously in cat of 2450 gm, killed it within 28 minutes, wrote Kobert, in 1901; Castelli, in 1914, stated that a cat was dead after 38 minutes as well as rabbits and guinea pigs injected with *Latrodectus m. tredecimguttatus* extract; according to Kellog (1915), the venom of *L. mactans* injected subcutaneously kills cats within 10 minutes, after convulsions of the clonic type, followed by tonic spasm. In white rats bitten by *L. mactans*, severe intoxication takes place, but with recovery. Bogen (1932) injected extracts of *Lactrodectus mactans* venom into white mice (compare our Table IX), rats, guinea pigs, rabbits, and chickens, without convincing results. Blair, in 1934, found that mice die, while rabbits, rats, cats, dogs, and sheep seemed little affected. Gray, in 1935, stated that the venom of two glands of an adult female of *Latrodectus mactans* kills guinea pigs after extreme excitement and pain, paralysis of the hind quarters, which soon involved all the muscles, including those of respiration. The second animal registered only severe illness. According to D'Amour, Becker, and Van Riper (1936), the LD_{50} of the American black widow is from about a half gland in rats; in 1938 they stated that 0.064 mg of *L. mactans* venom kills about 90% of injected rats; on a dry weight basis this venom is 15 times more potent than the venom of the prairie rattlesnake.

The literature on "black widows" and latrodectism is worldwide and relates hundreds of severe cases. According to Thorp and Woodson (1945) from 1726 to 1943 there have been 207 reports with 1726 accidents, with 55 deaths in 48 states of the United States. The references of Australia, Madagascar, South Africa, Europe, Chile, Peru, Argentina, Uruguay may be more than 300, with more than several hundred deadly cases and thousands of accidents. These spiders may be very frequent in certain places and under certain circumstances. Several hundreds of specimens may be found in an area of only a few hundreds of quadratmeters; they may inhabit old, unused buildings as well as fields with corn; they may be found under stones, pieces of wood on the ground, in holes in the ground, etc. When disturbed they often will feign death; they may crawl into the victims underclothing during the night. "Ninety% of cases of latrodectism," wrote Horen, in 1963, "occurred in Texas in men who were bitten while using outdoor privies." *L. mactans tredecimguttatus* caused the virulent epidemics in Spain during 1833 and again in 1841, and in Sardinia in 1833 and 1839, when men and domesticated animals fell victims to the bite of this spider. The same spider called by the people of southern Russia "*Karakurt*" or "*Schim*," caused in 1838 the death of many cattle. *L. m. mactans* is also very common in Hawaii.

Vellard has pointed out the potential toxicity of *Trechona* spider as follows: white rats bitten by *Tr. vensosa* die immediately; 2 mg of venom injected intramuscularly may cause contractures, paralysis and death; a small quantity of venom, 0.4 mg, intramuscularly, may also cause death with tetanism; 0.03 mg, intravenously and 0.07 mg, intramuscularly, may constitute the minimum lethal dose for white rats. From about 0.01 mg to 0.007 mg, intravenously may be the minimum lethal dose, for pigeons. As the spider is sedentary and not frequent, no human accident is known.

I know nothing about the toxic qualities of the representatives of *Xenesthis*, *Megaphobema*, and *Theraphosa*. Regarding *Acanthoscurria*, Vellard pointed out that 3 mg of venom intravenously will cause death in pigeons, 6 mg subcutaneously cause death in guinea pigs, and 5 mg the death of white rats.

In white rats 5 mg of venom intravenously constitutes the minimum lethal dose for *Phormictopus*, *Pamphobeteus*, and *Grammostola*, while the same dose of venom of *Lasiodora* may cause severe intoxication but not death in the same experimental animal; subcutaneously or intramuscularly the deadly doses are from about two and a half to three times higher. As Vellard has tested venoms of various South American "tarantulas" on pigeons, guinea pigs, rabbits, mice, white rats, amphibians, snakes, and lizards without sufficient quantities of venoms and, consequently, without sufficient laboratory animals for each experiment, his conclusions certainly are of value, but not conclusive.

The representatives of *Sericopelma* and *Eurypelma* of the suborder Orthognatha may produce on man, according to Thorp and Woodson, moderate swelling only and rather severe pain; Berg insists that a transient pain is the only effect of bites of *Eurypelma* of Central America and the southern states of the United States; Ewing writes that there is no swelling, but some smarting and soreness of the arm, disappearing after a few hours; P. W. Fatting relates that the bite of these kinds of "tarantulas" is without visible sign of swelling, the pain being like that of two or three honeybee stings, disappearing about 2 hours after.

Death is reported to have occurred after the bite of a Texas trap-door spider, probably of the genus *Bothriocyrtum*, according to Thorp and Woodson.

According to Kaire (1963), the LD_{50} of *Atrax robustus* venom in mice is from 0.2 to 0.25 mg intravenously and 0.4 mg subcutaneously.

Igrams, Musgrave, Beasly, McKeown, and others have reported fatal cases in New South Wales, in children and adults, dead from within $1\frac{1}{2}$ or $1\frac{1}{4}$ hours to 11 and 13 hours after the bite. Cases of accidents with recovery are more frequent.

The very poisonous representatives of "tarantulas" in the Union of South Africa belong to the genus *Harpactirella*. The most important species is *H. lightfooti*, related by Finnlayson, Hollow, Smithers, Mac Pherson, and others as the cause of severe spider bite.

According to Vellard, the venoms of *Olios rapidus*, the representatives of *Segestria* and *Dysdera* of Chile and Argentina, and of *Nephila* must be considered as more or less ineffective in human beings. These results are in contradiction with older reports about *Segestria ruficeps* and *S. perfida*; these are common six-eyed tube weavers, from about 17 to 22 mm in body length and with a cylindrical body, the first three pairs of legs directed forward, chelicerae metallic green in females, brown in males. Holmberg, in 1876, observed that a boy bitten by this spider on his hand had edema, fever, and a slight necrosis. This observation may be possible, but that the same kind of spider has killed a boy, as related by Weyenberg, in 1877, with hepatorenal and neurotoxic symptoms, cannot be accepted; Mazza and Argerich, in 1910 refer to a case of a man of Buenos Aires, bitten by the same kind of *Segestria*, with ictero-hemolytic consequences. In the same year del Pino reported that in a white rat, bitten by *Segestria ruficeps*, nothing occurred except local equimotic symptoms and a small and transient edema. Confusion with *Loxosceles* is evident. The *Filistata hibernalis* is a very common house spider, found in southern American states as well as in Central and South America to Patagonia. Its webs are often on the darker outside walls of human houses and are constructed during the night. The web is often more than 20 cm in diameter and is composed of more or less regular, radiating lines of dry silk and many lines of thicker bands. The female is about 2 cm long with 6 cm of leg length. The spiders are from dark brown to light brown or even velvety black. The male is much smaller, his body slender, and legs relatively much longer. Dorsally the trochanters present a light pearly color. Argerich, in 1908, Mazza, in 1910, Solari, in 1911, and others have described several accidents with *Filistata*, with local necrosis, hepatic disturbances, hemoglobinuria, and icterus. The spider has never been determined; *Loxosceles* is frequent in these regions.

Nothing has been done to confirm the toxicity of *Lithiphantes*. Vellard wrote about the representatives of *Mastophora* or *Glyptocranium* that deadly cases in human beings may be very rare; the "Podadora" generally cause local affections, skin lesions with necrosis, with more or less intensive systemic intoxication: general pains, edema, perspiration, fever, albuminuria, hematuria, anuria, and death with convulsions. According to Horen, in 1963 severe cases existed in Hawaii with *Chiracanthium.* Vellard was not able to obtain any toxic effect in animals with the venoms of the "mico," *Dendryphantes.*

Heteropoda venatoria, a relatively frequent spider in all seaports of tropical and subtropical maritime climates, has bitten accidentally, as Thorp and Woodson relate, P. W. Fatting of Emory University, Atlanta, Georgia; his finger was swollen to twice its natural size within 2 minutes; sharp pains were felt to his shoulder for about 2 hours. According to Vellard, white mice, bitten, show perspiration, dyspnea, paralysis of legs, and die within a few hours; in other animals the venom shows a very feeble local action, without necrosis; the minimum lethal dose, intramuscularly, in guinea pigs of 300 gm of weight is about 4 mg of venom, for white rats 2 mg intramuscularly and 1 mg intravenously.

Vellard tested also the venom of *Polybetes maculatus* and found that the toxicity is similar to that of *H. venatoria*, but less active; 1.5 mg of venom constitutes the minimum lethal dose, intravenously, in white rats, and from 4 to 5 mg, intramuscularly.

In 1968 Lucy Arvy found that in *Dysdera crocata*, *Pholcus phalangioides*, *Teutana grossa*, *Araneus diadematus*, *Clubiona terrestris*, *Tegenaria parietina*, and *Tegenaria derhami* two histochemically different substances may be distinguished in the venom glands: one is granular, rich in aromatic amino acids and indolic compounds (it gives the alloxan-Schiff reaction, the Danieli tetrazo reaction, and the rosindole reaction of Glemmer); the other, which is not granular, contains sulfhydryl groups.

Summarizing, we may conclude that the venom of *Phoneutria*, *Trechona*, *Atrax*, and *Harpactirella* chiefly act over the peripheral and central nervous system; those of the Theraphosinae over the central nervous system and, in some genera, may be cytolytic; those of *Lycosa* are cytolytic only; those of *Latrodectus* are neurotoxic and spasmodic, and those of *Loxosceles* are cytolytic and hemolytic.

The interested reader may find profound discrepancies between the experimental works of different investigators; these discrepancies are often so subtle that the same investigator may not be capable of repeating his own results. Differences in experimental animals used, different amounts of venom expelled, fresh or old venom samples used, error of interpretation of results, and chiefly the variations of the habitats of *Latrodectus mactans* with its several subspecies (Mediterranean, Australian, South African, Madagascarian, North, Central, and South American, etc.) may explain it. It is a fact that the *Latrodectus mactans,* recognized by the taxonomists as only one species, found in the United States as well as in Mexico, in several countries of Central America, in Peru, Chile, and Argentina, and in several places of Brazil, shows different venom activity on the same experimental animals.

It is a fact also that the "black widows" all over the world, the *Loxosceles* species in South America and in certain states of the United States, the *Lycosa* and *Phoneutria* representatives of Brazil have been studied for more than a half century by several biologists, chemists, pharmacologists, and immunologists, and that the final results have not been cleared completely up to the present. The studies on all the other dubious or suspect spiders, except perhaps the Australian *Atrax* and the South African *Harpactirella*, are far from the final stage. It may be clear also that the giant "tarantulas" of the tropics, so much feared by the native people, may be considered as harmless to men.

C. Treatment of Spider Bites

All members of the genera *Latrodextus, Loxosceles,* and *Phoneutria* and their subgenera may produce in humans severe, even fatal accidents. The areas of distribution of these spiders have been discussed above. Several hundreds of fatalities owing to the bites of these three genera occur even at the present time; several thousands of injured people are saved every year by specific or polyvalent serum therapy.

Antisera against *Lactrodectus* venom are manufactured in the United States, Mexico, Argentina, South Africa, Australia, Istria, Israel, and Russia.

The Butantan Institute carries the following sera:

Anti-aracnídico against *Phoneutria, Loxosceles,* and scorpions
Anti-ctenídico-licósico against *Phoneutria* and *Lycosa*
Anti-ctenidico against the *Phoneutria* species
Anti-loxoscélico against all American species of *Loxosceles*
Anti-licósico against the wolf-spiders of the genus *Lycosa*

The serum dose in the case of bites by *Latrodectus* or *Loxosceles* should be sufficient for neutralization of 2 mg dry venom. One-half of the serum should be given intravenously, the other half subcutaneously or intramuscularly.

In case of *Phoneutria* stings, the serum dose should neutralize approximately 6 mg venom although that means that the contents of 5 to 10 ampules of serum must be given as a single dose. This massive single dose serum treatment is vitally important, particularly in small children.

The antisera against the venoms of *Atrax* and *Harpactirella* in Australia or South Africa are still in the experimental stage.

REFERENCES

Abalos, J. W., and Báer, E. C. (1967). *In* "Animal Toxins," pp. 59–74. Pergamon, Oxford.
Aguilar, P. G. (1969). *An. Cient.* **6**, 46–51.
Ancona, L. (1931). *Anales Inst. Biol.* (*Univ. Nacl. Mex.*) **2**, 77–84.

Arvy, L. (1968). *Mem. Inst. Butantan* (*Sao Paulo*) **33**, 711–724.
Audouin, V. (1827). *Exptl. Planch. Arachn. Savigny Descr. Egypt,* p. 353.
Ausserer, A. (1871). *Verhandl. Zool. Botan. Ges. Wien.* **21**, 1939.
Ausserer, A. (1875). *Verhandl. Zool. Botan. Ges. Wien* **25**, 142.
Badcock, H. D. (1932). *J. Linnean Soc. London* (*Zool.*) **38**, 12.
Banks, N. (1905). *Proc. Entomol. Soc. Wash.* **7**, 95.
Barth, R. (1963). *Mem. Inst. Oswaldo Cruz* **60**, No. 2, 275–292.
Barrio, A. (1968). *Mem. Inst. Butantan* (*Sao Paulo*) **33**, 865–868.
Barrio, A., and Ibarra-Grasso, A. (1968). *Mem. Inst. Butantan* (*Sao Paulo*) **33**, 809–820.
Becker, L. (1897). *Compt. Rend. Soc. Entomol. Belg.* p. 114.
Bertkau, P. (1880). *Verz. Bras. Arachn.* pp. 19–37.
Biasi, P. (1970). *Rev. Brasil. Biol.* **30**, 233–244.
Bogen, E. (1926). *A.M.A. Arch. Internal Med.* **38**.
Bogen, E. (1932). *Ann. Internal Med.* **6**.
Bogen, E., and Loomis, R. N. (1934). *Calif. Western Med.* **45**.
Bogen, E. (1956). *In* "Venoms," Publ. No. 44, pp. 101–105. Am. Assoc. Advance. Sci., Washington, D. C.
Bonnet, P. (1957). "*Bibliographia Araneorum,*" **3**, 2364.
Brazil, V., and Vellard, J. (1925). *Mem. Inst. Butantan* **2**, 24–25.
Brethes, J. (1909). *Annaes Museu, Buenos, Aries* **12**, 48.
Brignoli, P. M., (1969). *Fragment. Entomol.* **6**, 121–166.
Bücherl, W. (1947). *Mem. Inst. Butantan* (*Sao Paulo*) **20**, 297.
Bücherl, W. (1951). *Monograf. Inst. Butantan* **1**, 1–204.
Bücherl, W. (1952). *Mem. Inst. Butantan* (*Sao Paulo*) **24**, No. 2, 127–148.
Bücherl, W. (1956). *In* "Venoms," Publ. No. 44, pp. 95–97. Am Assoc. Advance. Sci., Washington, D.C.
Bücherl, W. (1960). *Bol. Chileno Parasitol.* **15**, No. 4, 73–77.
Bücherl, W. (1961). *Bol. Chileno Parasitol.* **15**, 1–4.
Bücherl, W. (1963). *Cienc. Cult.* (*Sao Paulo*) **15**, No. 3, 243.
Bücherl, W. (1964). *Mem. Inst. Butantan* (*Sao Paulo*) **31**, 77–84.
Bücherl, W. (1968). *Rev. Brasil. Pes. Med. Biol.* **1**, 83–88.
Bücherl, W. (1969). *Mem. Inst. Butantan* (*Sao Paulo*) **34**, 25–32.
Cambridge, F. (1896). *Proc. Zool. Soc. London,* p. 728.
Cambridge, F. (1897). *Ann. Mag. Nat. Hist.* [6], 19, 76.
Cambridge, O. P. (1877). *Ann. Mag. Nat. Hist.* [4], 19,26.
Cambridge, O. P. (1892). *Biol. Centr. Am.* **1**, 93.
Cambridge, O. P. (1904). *Ann. S. African. Museum* **3**, No. 5, 152.
Cambridge, O. P. (1902). *Proc. Soc. Zool. London* **1**, 247.
Caporiacco, C. (1933). *Ann. Museo Civico Storia Nat. Giacomo Doria* (*Genoa*) **56**, 323.
Caporiacco, C. (1949). *Comm. Port. Acad. Sci.* **13**, 376.
Chamberlin, R. (1916). *Bull. Museum Comp. Zool. Harvard* **60**, 202.
Chamberlin, R. (1917). *Bull. Museum. Comp. Zool. Harvard,* pp. 29–60.
Chamberlin, R. (1925). *Proc. Col. Acad. Sci.* **14**, 107.
Chamberlin, R. (1920). *Brooklyn Museum Sci. Bull.* No. 3, **2**, 39–40.
Chamberlin, R. (1940). *Bull. Univ. Utah* No. 30, **13**, 1–40.
Clerck, C. (1757). *Aran. Svec.* p. 76.
Comstock, J. H. (1948). "The Spider Book," Cornell Univ. Press (Comstock), Ithaca, New York.
Costa, A. T. (1960). *Bol. Museu. Nacl.* (*Rio de Janeiro*) **216**, 1–11.
Crome, W. (1956). In "Die Neue Brehm Tierbuch," pp. 36–51. Wittenberg.

Dahl, F. (1902). *Sitzber. Naturforsch. Freunde, Berlin,* **36**, 42.
de Mello-Leitão, C. (1917). *Broteria,* **15**, 74.
de Mello-Leitão, C. (1921). *Ann. Mag. Nat. Hist.* **9**, 8, 339.
de Mello-Leitão, C. (1923). *Rev. Museu Paulista,* **13**, 1–438.
Delgado, A. (1968). *Mem. Inst. Butantan (Sao Paulo)* **33**, 683–687.
Dessimoni von Eickstedt, V. (1969). *Mem. Inst. Butantan (Sao Paulo)* **34**, 33–36.
Dessimoni von Eickstedt, V., Lucas, S., and Bücherl, W. (1969). *Mem. Inst. Butantan (Sao Paulo)* **34**, 67–74.
Diaz, M. O., and Saer, F. A. (1968). *Mem. Inst. Butantan (Sao Paulo)* **33**, 153–154.
Diniz, C. R. (1962). *Acta Physiol. Lat. Amer.* **12**, 211.
Dufour, L. (1820). *Ann. Soc. Gen. Sci. Phys.* **5**, 203.
Espiñoza, N. C. (1968). *Mem. Inst. Butantan (Sao Paulo)* **33**, 799–808.
Fabricius, J. C. (1775). *System Entomol.* p. 432.
Fergus, J. O'Rourke (1935). *Can. Dept. Agr. Proc. Publ.* p. 127.
Fergus, J. O'Rourke (1956). *In* "Venoms," Publ. No. 44, pp. 89–90. Am. Assoc. Advance. Sci., Washington, D.C.
Finlayson, M. H. (1936). *S. African Med. J.* **10**, 43–45 and 735–736.
Finlayson, M. H. (1937). *S. African. Med. J.* **11**, 163–167.
Finlayson, M. H. (1956). In "Venoms," Publ. **44**, No. 85–87. Am Assoc. Advance. Sci., Washington, D.C.
Finlayson, M. H., and Hollow, K. (1945). *S. African Med. J.* **19**, 431–433.
Finlayson, M. H., and Smithers, R. (1939). *S. African Med. J.* **3**, 808–810.
Fischer, F. G., and Bohn, H. (1957). *Ann. Chem.* **603**, 232.
Franganillo, B. P. (1926). *Biol. Soc. Entomol. Espan.* **9**, 71.
Furlanetto, R. S. (1961). Thesis, Sao Paulo, Brazil.
Gajardo-Tobar, R. (1966). *Mem. Inst. Butantan (Sao Paulo)* **33**, 45–54.
Gajardo-Tobar, R. (1968). *Mem. Inst. Butantan (Sao Paulo)* **33**, 689–98.
Gertsch, W. J. (1949). "American Spiders," Van Nostrand, Princeton, New Jersey.
Gertsch, W. J. (1958). *Am. Museum Novitales,* 1907.
Gertsch, W. J., and Mulaik, C. (1940). *Bull. Am. Museum Nat. Hist.* **77**, 314–320.
Girard, G. (1854). *Marcy's Rept. Red. River Louis* p. 262.
Gonzalez, C. (1954). *Adates Inst. Biol. (Univ. Nacl. Mex.)* **24**, 455.
Guérin-Menéville, E. F. (1838). *Arachnol. Voyage Favorite, Cl.* **8**, 3.
Heinecken, H., and Lowe, R. T. (1835). *Zool. J.* **5**, No. 19, 320–336.
Hentz, L. (1842). *Boston Soc. Nat. Hist.* **4**, 227.
Hickmann, V. V. (1927). *Papers Proc. Roy. Soc. Tasmania* **70**.
Holmberg, E. L. (1876). *Anales Agricult. Rept. Arg.* **4**, 25.
Holmberg, E. L. (1881). *Ann. Soc. Cient. Arg.* **11**, 171.
Horen, W. P. (1963). *J. Am. Med. Assoc.* **185**, No. 11, 839–843.
Ingram, W. W., and Musgrave, A. (1933). *Med. J. Australia* **2**, 10.
Kaiser, E. (1967). *Mem. Inst. Butantan (Sao Paulo)* **33**, 461–466.
Kaire, G. H. (1961). *Med. J. Australia* **2**, 450.
Kaire, G. H. (1963). *Med. J. Australia,* **50**, 307–311.
Karsch, E. (1879). *Z. Ges. Naturfiss.* **3**, No. 4, 545.
Karsch, E. (1886). *Berlin Entomol. Z.* **30**, 93.
Kaston, B. J. (1938). *Florida Entomologist* **21**, No. 4, 60.
Koch, C. L. (1842). *Arachniden* **9**, 102.
Koch, C. L. (1850). *Ubersicht Arachn. Syst.* **5**, 75.
Krynicki, J. (1837). *Bull. Soc. Nat. Moscou* **5**, 75.

Lamarck, J. B. (1804). *Hist. Nat. Animaux. Sans. Vert., Paris* **5**, 88–108.
Latreille, P. A. (1804). *Hist. Nat. Crust.* **7**, 159 and 249.
Latreille, P. A. (1806). *Gen. Crust. Ins.* **1**. 83.
Latreille, P. A. (1830). *Anal. Trav. Acad. Sci.* p. 80.
Levi, H. W. (1958). *Science* **127**, 1055.
Levi, H. H. (1959). *Trans. Am. Microscopol. Soc.* **78**, No. 1, 7–43.
Linnaeus, C. (1758). "*Systema Naturae*," **1**, p. 622.
Lucas, H. (1834). *Ann. Soc. Entomol. France* [3], 1, 381.
Lucas, H. (1845). *Ann. Soc. Entomol. France* [3], 71.
Lucas, H. (1843). *Rev. Zool.* **3**, 57.
Maretic, F. (1963). *Toxicon* **1**, 127–130.
McCrone, J. D., and Netzlof, M. L. (1965). *Toxicon* **3**, 107–110.
McCrone, and Porter, R. J. (1965). *Quart. J. Acad. Sci.*, **27**, 307–310.
Millot, J. (1931). *Ann. Sci. Nat. Zool.* [10], 14, 113–147.
Müller, P. S. (1776). *Reg. Bd.* (*Aran.*) p. 342.
Musgrave, A. (1927). *Records Australian Museum* **16**, 33.
Nicolet, A. C. (1849). *In* "Gay's Hist. fis. y pol. de Chili," **3**, pp. 322–541.
Perty, M. (1833). "Delectus Animalium." p. 197.
Petrunkevitch, A. (1911). *Bull. Am. Museum Nat. Hist.* **29**, 473.
Petrunkevitch, A. (1939). *Trans. Conn. Acad. Arts Sci.* **33**, 133–338.
Petagna, V. (1792). *Inst. Entomol.* p. 436.
Piza, S. T. (1938). *Folia. Clin. Biol.* (*Sao Paulo*) **1**, 21.
Piza, S. T. (1939). *Rev. Agr. Piracicaba* **14**, Nos. 7 and 8, 1–15.
Pocock, R. I. (1895). *Ann. Mag. Nat. Hist.* [6], 16, 195–199; [7] 8, 547.
Rainbow, W. J. (1914). *Records Australian Museum* **10**, No. 8, 256–260.
Rainbow, W. J., and Pulleine, R. H. (1918). *Records Australian Museum* **12**, No. 7, 165–170.
Reed, C. F. (1968). *Mem. Inst. Butantan* (*Sao Paulo*) **33**, 645–650.
Reese, A. M. (1944). *Trans. Am. Microscop. Soc.* **63**, 170–174.
Rocha e Silva, M. (1967). *Mem. Inst. Butantan* (*Sao Paulo*) **33**, 457–460.
Roewer, K. F. (1954). "Katalog der Araneae," **2**, Sect. A., 650.
Rosenberg, P. (1967). *Mem. Inst. Butantan* (*Sao Paulo*) **33**, 477–508.
Rossi, B. (1790). *Fauna Etrusc.* **2**, 132.
Rothschild, A. M. (1967). *Mem. Inst. Butantan* (*Sao Paulo*) **33**, 467–476.
Sampayo, R. L. (1942). Thesis, No. 5864. Universidad Nacional, Buenos Aires. Facultad de Ciencias Médicas.
Schenone, H. (1966). *Mem. Inst. Butantan* (*Sao Paulo*) **33**, 207–212.
Schenberg, S., and Pereira Lima, F. (1967). *Mem. Inst. Butantan* (*Sao Paulo*) **33**, 627–638.
Shulow, A. (1948). *Ecology* 29, 209.
Shulow, A. (1966). *Mem. Inst. Butantan* (*Sao Paulo*) **33**, 93–100.
Simon, E. (1888). *Acta Soc. Linnean Bordeaux,* **41**, 407; **42**, 400.
Simon, E. (1889). *Ann. Soc. Entomol. France* **6**, No. 9, 175; **6**, No. 10, 99; and 60, 303.
Simon, E. (1892–1898). *Hist. Nat. Araign.* Vols. I and II. Paris.
Simon, E. (1898). *Ann. Soc. Entomol. Belg.* **60**, 311.
Soares, B. (1944). *Papers Avuls. Dept. Zool.* (*São Paulo*) **4**, 151.
Strand, E. (1907). *Jahresber. Ver. Vertl. Naturk. Wiesbaden,* p. 63.
Suzuki, T. (1967). *Mem. Inst. Butantan* (*Sao Paulo*) **33**, 389–410.
Thorell, T. (1870). *European Spiders Oefers. vet. ak. Forkandl.* **27**, No. 4, 369.
Thorell, T. (1890). *Ann. Museo Storia Natl. Genova* **2**, No. 8, 339.
Thorell, T. (1894). *Bihang Svenska Vetenskapsakad. Handl.* **20**, No. 4, 31.

Thorp, R. W., and Woodson, W. D. (1945). "Black Widows," Univ. of North Carolina Press, Chapel Hill, North Carolina.
Tinkham, E. R. (1956). *In* "Venoms," Publ. No. 44, pp. 99–100. Am. Assoc. Advance. Sci., Washington, D.C.
Vellard, J. (1924). *Arch. Inst. Vital. Brazil, Niteroi,* **2**, 158.
Vellard, J. (1936). "Le venin des Araignées." Masson, Paris.
Vellard, J. (1947). "Los Animal venenoessos." Univ. Nacl. Major de San Marcos, Lima, Peru.
Vellard, J., Gerschmann, P., and Schiapelli, R. (1945). *Acta Zool. Lilloana* **3**, 165.
Vinson, A. (1863). *Aran. Rinn. Maurit. Madagascar,* p. 306.
von Hasselt, W. M. (1888). *Tijdschr. Entomol.* **31**, 166.
von Keyserling, E. (1882). *Verhandl. Zool. Botan. Ges. Wien.* **32**, 221.
von Keyserling, E. (1891). *Spinn. Am., Brasil. Spinnen.*
Walckenaer, C. A. (1805). *Tabl. Aran.* p. 81.
Walckenaer, C. A. (1837). *Ins. Apteres* **2**, 211–272.
Walckenaer, C. A. (1847). *Ins. Apteres* **4**, 379.
Wallace, A. L., and Sticka, R. (1956). *In* "Venoms," Publ. No, 44, pp. 107–109. Am. Assoc. Advance. Sci., Washington, D.C.
Wiener, S. (1957). *Med. J. Australia* **2**, 377.
Wiener, S. (1959). *Med. J. Australia* **2**, 679; Thesis, Univ. of Melbourne.
Wiener, S. (1960). *Med. J. Australia* **1**, 449.
Wiener, S. (1961). *Med. J. Australia* **2**, 693.
Witt, P. N. (1967). *Mem. Inst. Butantan (Sao Paulo)* **33**, 639–644.

Chapter 52

*Phoneutria nigriventer** Venom—Pharmacology and Biochemistry of Its Components†

S. SCHENBERG AND F. A. PEREIRA LIMA

SERVIÇO DE FISIOLOGIA DO INSTITUTO BUTANTAN, SÃO PAULO, BRAZIL

I. Crude Venom Pharmacology 280
A. Effects on Dogs 280
B. Effects on Mice 284
C. Effects on Other Rodents 287
D. Accidents in Man 287
II. Biochemistry of the *Phoneutria nigriventer* Venom 287
A. Characteristics of Venom Components 287
B. Venom Fractionation 291
References. 297

The spider *Phoneutria nigriventer* (Keyserling, 1891) is of the Araneida order, Labidognatha suborder, Ctenidae family, and Phoneutriinae subfamily. The adult spider is relatively large in size (3 cm body length and 10–12 cm between the legs); it is nocturnal, vagrant, very aggressive, and constructs no web. The female does not kill the male after mating as she does in many other species.

The *Phoneutria nigriventer* venom used in these investigations was extracted by manual or electrical stimulation (Bücherl, 1953a) of specimens taken predominantly in the City of São Paulo and environs, where these spiders are relatively common. The venom was vacuum dried over $CaCl_2$ at room temperature; after drying it turns a grayish-white color. According to Bücherl (1953b, 1956) the amount of dried venom obtainable in winter

*After recent reconsideration (Bucherl *et al.*, 1969), *Phoneutria nigriventer* is the Southern Brazil species; *P. fera* is that of Amazonas State, whose venom remains unstudied.

†This work was supported in part by grants from the Fundacão de Amparo à Pesquisa do Estado de São Paulo, Fundo de Pesquisas do Instituto Butantan, and CNP_q.

from single specimens may reach 1.8 mg, and in summer 2.5 mg. Though toxicity is comparable to that of *Crotalus* and *Bothrops* ophidian venoms, the small amount of venom available in individual spiders is generally only lethal in children.

I. CRUDE VENOM PHARMACOLOGY

A. Effects on Dogs

The venom of *Phoneutria nigriventer* injected subcutaneously in male dogs in low lethal doses (180–200 μg/kg body weight) successively produces: intense local pain, violent sneezing, lacrimation, excessive salivation, adynamia, ataxia, prostration, drowsiness, dyspnea, vomiting, priapism, ejaculation, sanguinolent feces, and, in some cases, death. In dogs, sublethal subcutaneous doses generally are insufficient to induce priapism and doses which are effective subcutaneously, when injected intravenously are very toxic and produce a sharp drop in blood pressure. Except for priapism (Schenberg and Pereira Lima, 1962), all the aforementioned effects have been previously reported by Vital Brazil and Vellard (1925).

1. Local Pain

The venom is excruciatingly painful subcutaneously injected; it makes dogs yelp for nearly an hour and forces them to keep the injected hind leg contracted for periods longer than an hour. The specific antivenin neutralizes the pain-producing component, which excludes the possibility of this effect being caused by histamine or serotonin contained in the venom. Since the pain factor is dialyzable, its molecule must be relatively small; for this reason it is unlikely that the factor is a bradykinin-releasing enzyme.

2. Sneezing

In dogs, sneezing constitutes one of the first signs of envenomation. The sneezes occur intermittently, being generally observed for more than 24 hours. Violent attacks occur within the first 2 hours, during which period the animal throws its head violently and uncontrollably towards the floor, and, very often, hurts its nose and lips.

3. Lacrimation and Mydriasis

During envenomation, lacrimation is constant, even in anesthetized animals. Mydriasis occurs early after venom injection and persists for many hours. The dogs present visual disturbances as a consequence of mydriasis, as exteriorized by characteristic movements of the paws over the eyes, and the rubbing of the face against the floor as if to remove foreign bodies from the eyes.

FIG. 1. Salivation induced by *P. nigriventer* venom. It appears before any other sign of intoxication; the dog is already adynamic, prostrated, and drowsy.

4. Salivation

The venom-provoked salivation is abundant and resembles that induced by pilocarpine (Fig. 1). Unswallowed saliva drips continuously from the animal's mouth for hours. This is blocked by atropine and doses of eserine and hexamethonium, which, though they increase the signs of poisoning slightly, do not seem to interfere with this effect (Schenberg and Pereira Lima, 1962). Subcutaneous doses which produce salivation fail to produce it when injected into dogs under chloralose, chloroform, and barbituric anesthesia; however, intravenously, the same doses are effective.

5. Priapism

The venom-induced priapism is also of an intermittent and long-standing character. It occurs repeatedly for hours (very often over 24 hours), in which cases edema generally occurs at the penis distal extremity correspondent to the glans. The priapism develops slowly, taking more than 15 minutes from its beginning stage (Fig. 2) to its final stage (Fig. 3). It generally occurs at a more advanced envenomation phase, when the dog has already presented signs of adynamia, ataxia, drowsiness, mydriasis, visual disturbances,

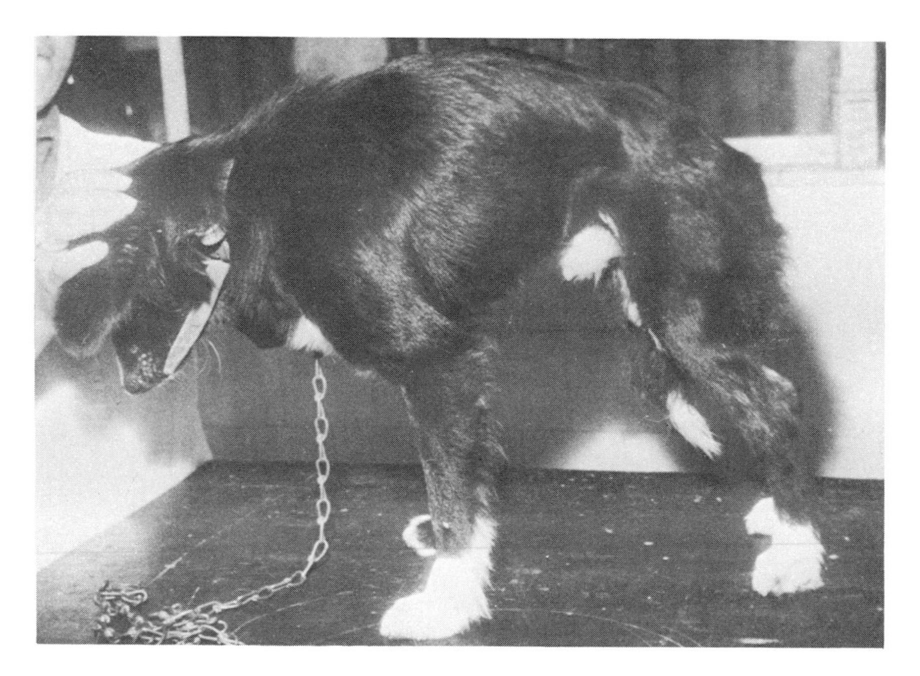

FIG. 2. Beginning of priapism induction by *P. nigriventer* venom. The dog displays intoxication signs, drowsiness, adynamia and ataxia.

dyspnea, and vomiting. Priapism could not be induced in anesthetized dogs even when the venom was applied intravenously in larger doses than the dose found to be effective subcutaneously.

Priapism induction does not seem to depend on the excitation of higher nervous system centers, since it could be provoked in dogs in which the medulla had been cut at DXII (Schenberg and Pereira Lima, 1963). Unlike cantharidin priapism, the priapism seen in arachnidian envenomation does not seem to result from reflexes caused by irritation of the urinary tract. In dogs, the venom induces priapism before micturition occurs, and priapism is seen in animals in which urine flow through the urethra has been prevented by implanting both ureters in the skin; it is not induced by perfusing venom solutions through the urethra into the bladder.

6. Ejaculation

In dogs, the *P. nigriventer* venom induces ejaculation during or after the onset of priapism. This effect seems to result from excitation of organs other than the seminal vesicles since these organs are not present in the dog (Schenberg and Pereira Lima, 1963).

FIG. 3. Fully developed priapism, 15 minutes after its first appearance.

7. Toxicity

Phoneutria nigriventer venom is approximately 4 times more toxic in dogs than in mice. When 200 $\mu g/kg$ of venom is injected subcutaneously, some dogs survive for several hours before death, with sneezing, salivation, and priapism occurring during the severe intoxication phase.

8. Hypotensive Response

Small doses of venom injected intravenously provoke a sharp fall in blood pressure. The histamine content of venom at these doses is too low to affect the blood pressure.

9. Tachyphylaxis

Sampayo (1942) has reported that black widow spider (*Latrodectus mactans*) venom provokes tachyphylaxis in dogs. Tachyphylaxis phenomena were not observed with *P. nigriventer* venom, neither of those phenomena of short duration nor for long standing and intermittent effects. The latter effects would not appear for hours, if tachyphylaxis was involved. Also, dogs, which had recently recovered from the effects of a venom injection, showed the same response to venom when another dose was injected. These facts seem to show that tachyphylaxis is not an universal characteristic of spider venom constituents; they also point up the different molecular structures of the active principles involved.

10. Guinea Pig Ileum Contraction

Diniz (1963) separated 2 polypeptides, which contract the guinea pig ileum from *P. nigriventer* venom; according to his findings, the fraction containing one of these polypeptides was also responsible for the toxicity of the venom.

B. Effects on Mice

Diniz was able to reproduce, in albino mice, the *P. nigriventer* venom priapism first observed in dogs (1963). Except for lacrimation, mydriasis, sneezing, ejaculation, and vomiting, which are difficult to follow in mice, all the other *P. nigriventer* venom effects can easily be seen in these small rodents (Schenberg and Pereira Lima, 1963). Mice were shown to be suitable assay animals for this venom, permitting qualitative and quantitative study of envenomation with low venom consumption. Another advantage is that the same animal may be used for several assays over a period of 30 days, before immunization has a quantitative effect on the determinations. The reproduction of venom effects induced in dogs, on mice, permitted us to pursue our investigations on a biochemical level.

Strain and weight control is essential for reproductive quantitative assays. The best responses, for mice of the Instituto Butantan strain [Strain C, Statens Seruminstitut, Copenhagen (1956)], are obtained by using animals in a 22–25 gm weight range. In mice weighing less than 20 gm, toxic effects occur more often than priapism. The method of Reed and Muench (1938) has been shown to afford a statistically valid estimate of venom-produced priapism in mice.

1. Local Pain

Though pain in mice is not as easily followed as in dogs, it is exteriorized by contraction of the injected hind leg. The animals also bite the injection area. These manifestations are not seen after the onset of distensive paralysis.

2. Salivation

As Fig. 4 shows, salivation can be easily followed in mice. First, small bubbles of saliva accumulate at the mouth. Later on, depending on the venom dosage, the animals may have a large part or nearly all of its fur wetted by saliva. Small doses of venom provoke salivation without any sign of toxicity. With larger doses, salivation appears before toxic manifestations and continues on their appearance. The venom ED_{50} for salivation is 0.43 mg/kg body weight.

3. Priapism

Figure 5 clearly shows priapism in mice. The priapism dose-effect relationship in mice is maintained until a maximum of response in a group of animals is attained; thereafter, increasing doses of venom produce a decrease in the

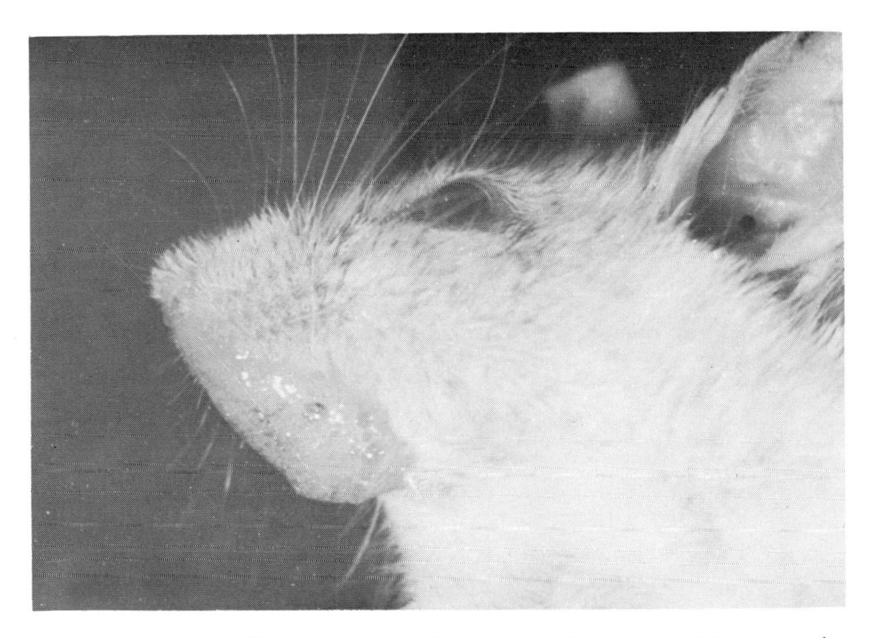

FIG. 4. The first manifestation of *P. nigriventer* venom is salivation. The mouse shows no signs of intoxication.

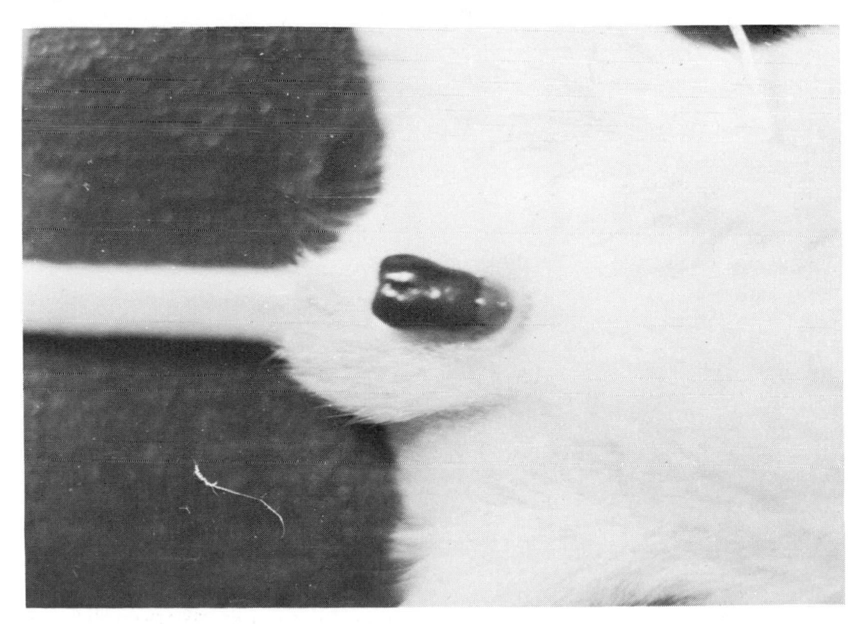

FIG. 5. Priapism produced in the mouse by 10 μg of *P. nigriventer* venom. This dose level produces priapism and salivation but not intoxication.

number of animals which present priapism, and death occurs in some of them before this effect appears. Toxicity would thus exert a sort of functional antagonism to priapism. Similarily, with salivation, small doses of venom can induce priapism in mice without any toxic manifestations; in this point mice react differently from dogs. The priapism ED_{50} for mice is 0.25 mg/kg body weight.

4. Toxicity

The toxic responses to venom in mice are dyspnea, prostration, distensive paralysis, and death; the LD_{50} is 0.76 mg/kg body weight.

5. Distensive Paralysis

As can be seen in Fig. 6, crude venom in mice provokes a distensive paralysis of the hind legs and tail; and the tail, during a certain period, remains bent upon the animal's back. Hind leg paralysis takes place after salivation and priapism appear and continues until death, when the animals already present a premature sort of rigor mortis.

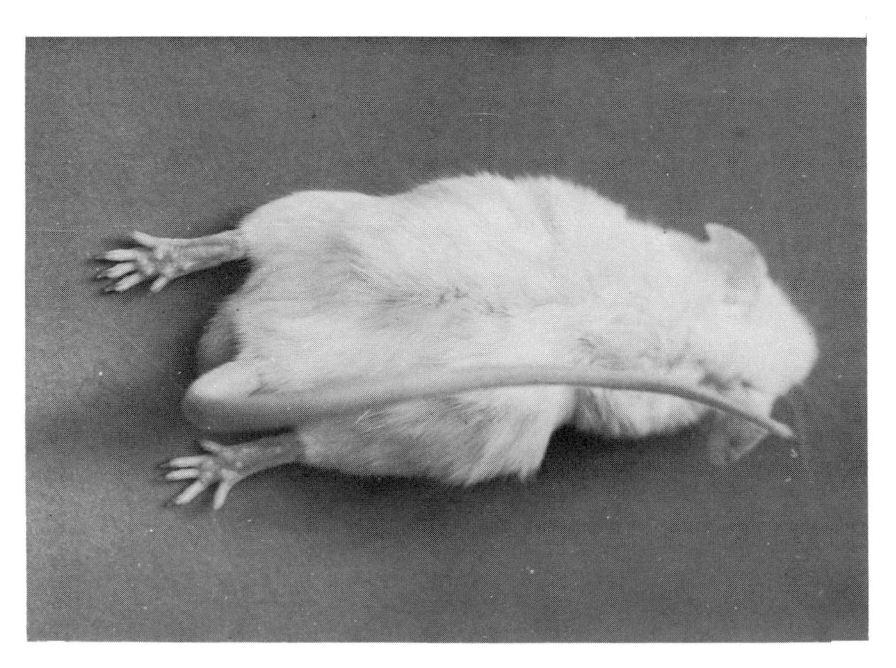

FIG. 6. Distensive paralysis produced by *P. nigriventer* venom. Note the characteristic position of the hind legs and tail; the mouse dies in a premature rigor mortis.

C. Effects on Other Rodents

Guinea pigs display nearly all the responses to envenomation seen in dogs and mice, but they have been investigated less than have mice due to their larger venom consumption; nevertheless, the gelatinous consistency of their semen makes them very appropriate for the study of the semen ejaculation effect (Schenberg and Pereira Lima, 1962, 1963). Rats and rabbits are very resistant to this venom; a 500 μg dose injected in 150-gm rats provokes secretion of the glands of Harder, and a 1000-μg dose injected in 2000-gm rabbits induces a moderate salivation and slight intoxication (Schenberg and Pereira Lima, 1962).

D. Accidents in Man

A pattern which resembles that of dog envenomation is also noticed in humans bitten by *Phoneutria nigriventer*: local unbearable pain, salivation, visual disturbances, sweating, prostration, priapism, and death. Priapism is observed predominantly in boys under 10 years old, who present signs of severe intoxication. Generally patients do not complain of priapism, and it is usually not perceived by the clinician. In the last few years, having become aware of this venom effect, clinicians have paid more attention to this symptom of *P. nigriventer* bites, particularly in cases brought to the Instituto Butantan Hospital.

II. BIOCHEMISTRY OF THE PHONEUTRIA NIGRIVENTER VENOM

Barrio has reported on the neuromuscular and muscular effects of the electrophoretic fractions of the venom (1955). Diniz demonstrated by electrophoresis and chromatography that it contains histamine, serotonin and 2 guinea pig ileum-contracting polypeptides (1963). Fischer and Bohn (1957) separated its histamine by electrophoresis, demonstrating that the venom also contains free glutamic acid (23.6%), aspartic acid (1.0%), and lysine (0.2%). Welsh and Batty (1963) studied its serotonin content, while a hyaluronidase and a proteolytic enzyme of this venom were reported by Kaiser (1956). The venom histamine content varies, according to different reports, from 0.06 to 1.0%; the serotonin content varies from 0.03 to 0.25%.

A. Characteristics of Venom Components

1. Antigenicity

Except for histamine and serotonin, all the *P. nigriventer* venom active constituents are neutralized by the specific antivenin; this could be considered proof that these constituents are large molecules, probably protein in nature (Schenberg and Pereira Lima, 1962). The composition of venom antigens determined by double-diffusion agar immunoprecipitation (Ouchterlony method), gave 15 immunoprecipitation lines, of which 13 were differentiated by agar immunoelectrophoresis (Pereira Lima and Schenberg, 1964). Thus, the venom might have several distinct components, each one associated with one or more of its effects. It is also possible that the toxic factors might be separable from the nontoxic factors and that other principles could be isolated and their pharmacological properties investigated.

2. Flaccid Paralysis Component

A venom fraction, separated by electrophoresis on agar plates at pH 5.0, induced flaccid paralysis in mice when injected subcutaneously (Fig. 7). The same component was also separated on cellulose acetate strips at pH 5.0 (Pereira Lima and Schenberg, 1963). The paralysis produced by this fraction differs from the distensive paralysis already discussed: the hind legs are paralyzed in a flaccid, not distensive, form, the tail is flaccid, and the animal is motionless; when movement is attempted the forelegs are used to pull the body. This fraction is very toxic, and when death occurs the mice present a flaccid body as compared to death produced by the crude venom in which

FIG. 7. Flaccid paralysis induced by Sephadex G-50 fraction of *P. nigriventer* venom heated at 100°C for 6 minutes in 0.05 *M* phosphate buffer at pH 8.0. The hind legs are paralyzed and one of the mice is trying to move by using the forelegs; both presented flaccid bodies after death.

the animals die in rigor mortis. Fractions containing the principle responsible for this effect have been separated by other methods and will be described later. The action of this fraction is probably masked by the dominant distensive paralysis seen when the crude venom is assayed.

3. Heat Inactivation

The pharmacologically active components of the venom are resistant to heating. Venom solutions in physiological saline can be heated at 100°C for 20 minutes without loss of activity. However, inactivation occurs when these solutions are heated at the same temperature for 6 minutes at pH 8.0 (Table I) (Schenberg and Pereira Lima, 1966). Under these conditions the guinea pig ileum-contracting polypeptides show themselves to be the most heat resistant of all the components; these polypeptides were 100% active after this treatment, while activity after heating of some of the other components were decreased as follows: priapism 39.2%; salivation, 34.2%; and toxicity, 19.1% (Table I). These data and other data to be reported seem to show that the venom has at least 2 toxic components, which vary in their resistance to heat (Schenberg and Pereira Lima, 1966, 1968).

4. Chemical and Enzymic Inactivation

Treatment of *P. nigriventer* venom with concentrated acetic acid for 1 hour greatly diminishes its capacity to induce priapism, salivation, and death (Table I). The fractions responsible for these effects and for the guinea pig ileum-contracting effect are more resistant to acids (0.1 *N* HCl) than to alkalis (0.1 *N* NaOH); also they are not completely inactivated by heating at 100°C for 6 minutes in 0.1 *N* HCl. Stronger alkaline solutions (1.0 *N* NaOH), however, inactivate these 4 components. The ileum-contracting polypeptides are the most resistant, being more resistant than the other 3 to 0.1 *N* NaOH, 0.4% formaldehyde, ethanol, and to heating at 100°C for 6 minutes in 0.1 *N* HCl (Table I). Sulfuric ether, acetone, chloroform, and butanol do not inactivate any of these components (Table I), though the first 3 solvents extract, from the crude venom, a significant amount of an inactive material together with some histamine and serotonin (Pereira Lima and Schenberg, 1964). Except for histamine and serotonin, all the other components of the venom are inactivated by trypsin, chymotrypsin, and pepsin, thus demonstrating their protein nature (Pereira Lima and Schenberg, 1968; Schenberg and Pereira Lima, 1966). The proteolytic enzyme of *P. nigriventer* venom inactivates all its active polypeptides; guinea pig ileum-contracting polypeptides are the most rapidly inactivated of all. It is of interest that this enzyme also degrades casein much more rapidly than albumin (Schenberg and Pereira Lima, 1967, 1968).

TABLE I

PROPERTIES OF ILEUM CONTRACTION, PRIAPISM, SALIVATION, TOXICITY, AND FLACCID PARALYSIS FRACTIONS

Treatment	Activity				
	Ileum contraction	Priapism	Salivation	Toxicity	Flaccid paralysis
Antivenin	−	0	0	0	0
Heating (100°C for 6 minutes; 500 μg/ml at pH 8.0)	++++	++	++	+	+++
Heating (100°C for 30 minutes)	++++	++	++	+	−
Concentrated acetic acid (1 hour)	−	+	+	+	−
0.1 *N* HCl (2 hours)	++++	++++	++++	++++	−
0.1 *N* HCl (5 minutes at 100°C)	+++	++	++	++	−
0.1 *N* NaOH (2 hours)	+++	+	+	+	−
1.0 *N* NaOH (2 hours)	+	0	0	0	−
Trypsin, chymotrypsin, pepsin	0	0	0	0	0
Sulfuric ether, acetone, chloroform, butanol	++++	++++	++++	++++	−
Ethanol	+++	++	++	++	−
0.4% Formaldehyde (2 hours)	+++	++	++	++	−
Dialysis—Visking tubes No. 24/32 (48 hours)	−	0	0	0	0
Dialysis—Visking tubes No. 18/32 (48 hours)	−	+++	+	+	−
Dialysis—Visking tubes No. 8/32 (48 hours)	−	++++	++++	+++	−
Electrophoresis migration (pH 8.6)	−	Cathode	Cathode	Cathode	Cathode
Electrophoresis migration (pH 5.0)	−	Cathode	Cathode	Cathode	(separates)
$(NH_4)_2SO_4$ precipitation (%)	60–75	45–65	45–65	45–75	70–75
Sephadex G-25	—	Excluded	Excluded	Excluded	Excluded
Sephadex G-50 (heated at pH 8.0 100°C/6 minutes)	Diffuses	Diffuses	Diffuses	Diffuses	Diffuses
(Peaks No.)	(2)	(2)	(2)	(2)	(1) (separates)
DEAE-cellulose (Peaks No.)	(3)	(4)	(4)	(2)	(1)
CM-Sephadex G-50	(4)	(3)	(4)	(2)	—

5. Dialysis

The Visking tube 24/32 (impermeable to insulin) is impermeable to all the large active factors; tube 18/32 (impermeable to insulin) is more permeable to components which induce priapism and salivation than to those that induce intoxication (Table I). All the active constituents of the venom dialyse rapidly through tube 8/32, while insulin dialyses slowly (Pereira Lima and Schenberg, 1964; Schenberg and Pereira Lima, 1962, 1966). Since these components dialyse through tube 18/32 (impermeable to insulin), their molecular weight must be less than that of insulin (5733). They can be considered polypeptides since (1) they are also degraded by proteolytic enzymes and (2) they are immunogenic.

B. Venom Fractionation

1. Ammonium Sulfate and Electrophoresis

The flaccid paralysis component and the ileum-contracting polypeptides were separated out by 65–75% ammonium sulfate saturation (Table I). The flaccid paralysis component was precipitated either from venom solutions in physiological saline or from phosphate buffer solutions at pH 7.0, but not from buffer solutions at pH 5.0. Electrophoresis of the *P. nigriventer* venom was performed with different supporting media: agar plates (immunoelectrophoresis), starch blocks, filter paper, and cellulose acetate strips; the latter give better resolution. All active components are positively charged and migrate to the cathode. The flaccid paralysis component was the only one separated by electrophoresis (agar plates or cellulose acetate strips) in acetic acid ammonium acetate buffer at pH 5.0 (Pereira Lima and Schenberg, 1963; Schenberg and Pereira Lima, 1966).

2. Gel Filtration Columns

All the large molecule components are excluded in Sephadex G-25, except for the ileum-contracting polypeptides which have not yet been assayed. These data indicate that the molecular weight of these polypeptides are higher than 5000, but less than 10,000 since they diffuse in Sephadex G-50; this confirms other results previously obtained by dialysis (Pereira Lima and Schenberg, 1968; Schenberg and Pereira Lima, 1962, 1966). Both sets of data seem to indicate that the larger pharmacologically active molecules of the *P. nigriventer* venom have a molecular weight between 5000 and 5733. These components are also immunogenic and subject to hydrolysis by proteolytic enzymes and are considered to be polypeptides.

Figure 8 shows a chromatogram of a 20-mg *P. nigriventer* venom run in a Sephadex G-50 column (760 × 22 mm), eluted with phosphate buffer at pH 7.5 (0.05 and 0.1 *M* NaCl) (Pereira Lima and Schenberg, 1964). Except for the flaccid paralysis component, all the other components which are

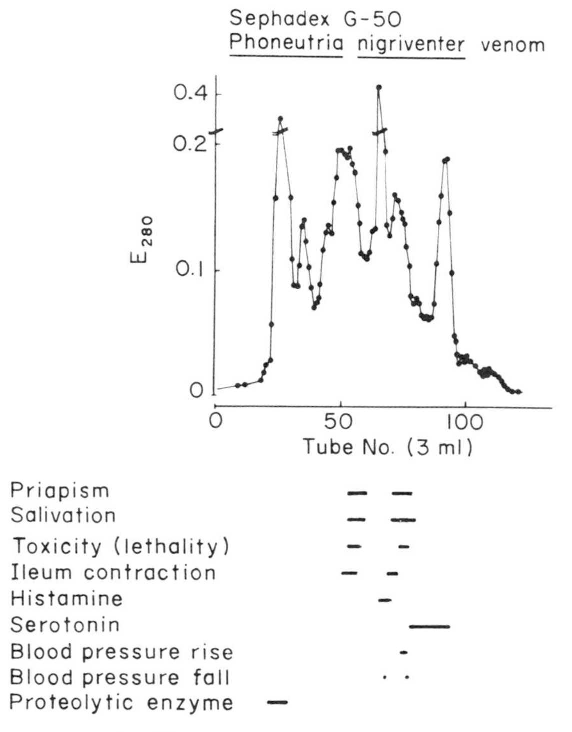

FIG. 8. Chromatogram of a 20-mg *P. nigriventer* venom run on a Sephadex G-50 (fine) column (760 × 22 mm): 13 pharmacologically active components, distributed in 2 zones are represented; the proteolytic enzyme appears in the first tubes; flaccid paralysis effect was not produced by any of the effluent fractions.

already identified, are found in the effluent fractions. The activities are distributed in 2 zones; they are not associated with the 280 mμ absorption peaks, which seems to indicate that the venom active polypeptides contain little or no tyrosine, tryptophan, and phenylalanine. Some of the same effects are produced by components in the 2 activity zones, where they can be differentiated by their molecular size. These data indicate that venom contains components of different molecular weight which have, however, similar pharmacological activity.

The proteolytic enzyme can be distinguished from the pharmacologically active polypeptides by its larger molecular weight. The ileum-contracting polypeptides, histamine and serotonin, were obtained free from the other active components; this is considered evidence that toxicity and ileum contraction are produced by different venom fractions. The polypeptides responsible for priapism, salivation, and toxicity are not separated by Sephadex G-50. The toxicity components were not detected in all the tubes from the second zone, which contained the salivation and priapism polypeptides. Probably, because of their low concentration in the tubes corresponding to the descending parts of the activity peaks, and also, because very little solution was left after the other assays, the fraction might not have been present in large enough quantities for a toxicity assay which requires a higher dose.

Flaccid paralysis could not be produced by these column fractions, since the polypeptide responsible for its induction probably is associated with that responsible for the dominant distensive paralysis. These fractions always induced a characteristic distensive paralysis in mice, with death occurring in rigor mortis.

Two other venom components were separated on this column. One, administered intravenously in dogs, induced a rise in blood pressure, and the other a fall in blood pressure (Schenberg and Pereira Lima, 1966, 1967, 1968). Neither has been well studied, though it has been found that some of these fractions contain neither histamine nor serotonin. In crude venom the hypotensive effects dominate and mask the hypertensive ones (Schenberg and Pereira Lima, 1966, 1967, 1968).

Figure 9 shows a chromatogram of a column on which 20 mg of venom, which had been heated at 100°C for 6 minutes in 0.005 *M* phosphate buffer at pH 8.0, was run. The first 280 mμ absorption peak of Fig. 8, represents venom components, which are excluded in Sephadex G-50, including the venom proteolytic enzyme, were destroyed in heating and are absent in Fig. 9. The other 280 mμ peaks components, which are probably polypeptides, are heat stable and unchanged. As mentioned before, about 80% of the toxicity is lost by this heat treatment, and this seems to be represented (Fig. 9) by the absence of the distensive paralysis polypeptides from a large number of tubes. Its absence revealed the flaccid paralysis component which is also toxic, since the 2 components ordinarily overlap (Schenberg and Pereira Lima, 1968). The venom seems to contain a third toxic component which does not produce paralysis.

3. Ionic Columns

Figure 10 shows a chromatogram of the fractionation of 20 mg of venom in a DEAE-cellulose column (300 × 20 mm), equilibrated with 0.005 *M* glycine buffer at pH 8.35. The same buffer was used as a solvent for the venom and also as the first eluent solution; for stepwise elution, its molarity was increased.

Similar to Sephadex G-50 columns, the pharmacologically active components are distributed in 4 zones on the DEAE-cellulose column. In the fourth zone the 280 mμ absorption peak seems to overlap the peaks associated with priapism and salivation fractions. These fractions are free of toxic components in the third and fourth zones, which would indicate

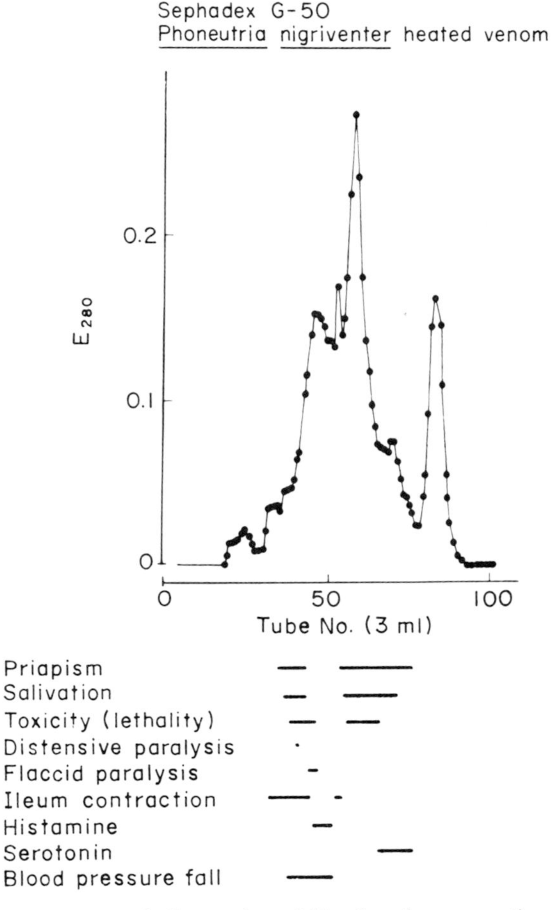

FIG. 9. A chromatogram, similar to that of Fig. 8, only representing a run of venom heated at 100°C for 6 minutes in 0.05 *M* phosphate buffer at pH 8.0. The proteolytic enzyme was denatured; the flaccid paralysis fraction is separated from that of distensive paralysis, in one tube only.

that toxicity, priapism, and salivation are not produced by a single component. However, these findings cannot be considered definitive, since priapism and salivation are both induced by lower minimal effective doses than is toxicity.

The flaccid paralysis component is separated from the distensive component in fractions from DEAE-cellulose columns. Distensive paralysis, not represented in Fig. 10, was induced by part of the first toxicity fraction and by all of the second fraction. According to these data, the venom would contain 14 pharmacologically active components, including histamine and serotonin, not assayed in the effluent fractions. Two of the 3 fractions

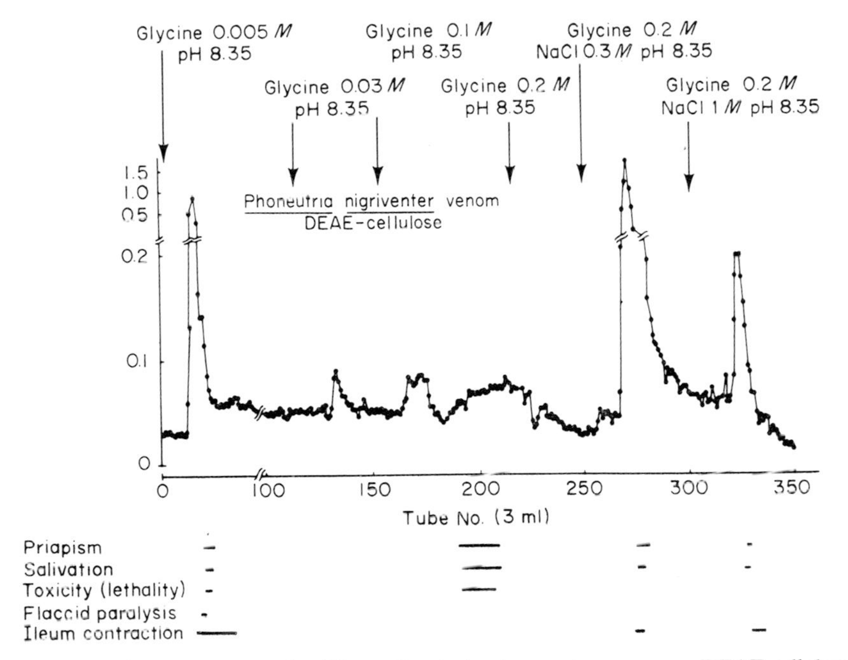

FIG. 10. Chromatogram of a 20 mg *P. nigriventer* venom run on a DEAE-cellulose column (300 × 20 mm). Some of the same effects are distributed in 4 zones; flaccid paralysis is separated from distensive paralysis.

attributed to the ileum-contracting effects in Fig. 10 probably correspond to these amines.

Figure 11 shows the fractionation chromatogram of 20 mg of venom in a CM-Sephadex G-50 column (270 × 10 mm), in gradient and stepwise elutions. Overlapping of the effluent activities with the 280 mμ peaks does not occur, and, again, 4 activity zones occur in this venom fractionation. Histamine and serotonin were not assayed in these fractions, and 2 of the 4 fractions which contract the ileum are thought to correspond to them. The flaccid paralysis fraction was not separated from that of distensive paralysis on this column, and flaccid paralysis was not demonstrated. Toxicity is not represented in the graph for priapism and salivation activities for the first and third activity zones; however, as discussed above, large volumes are necessary to induce these effects so that toxicity components might be present in these fractions and still not be seen by assay.

The absence of priapism in the fourth zone is very significant, since the minimal effective dose is much smaller than that of toxicity, which was induced without any sign of priapism; thus, it is possible that priapism and toxicity might be produced by 2 distinct polypeptides.

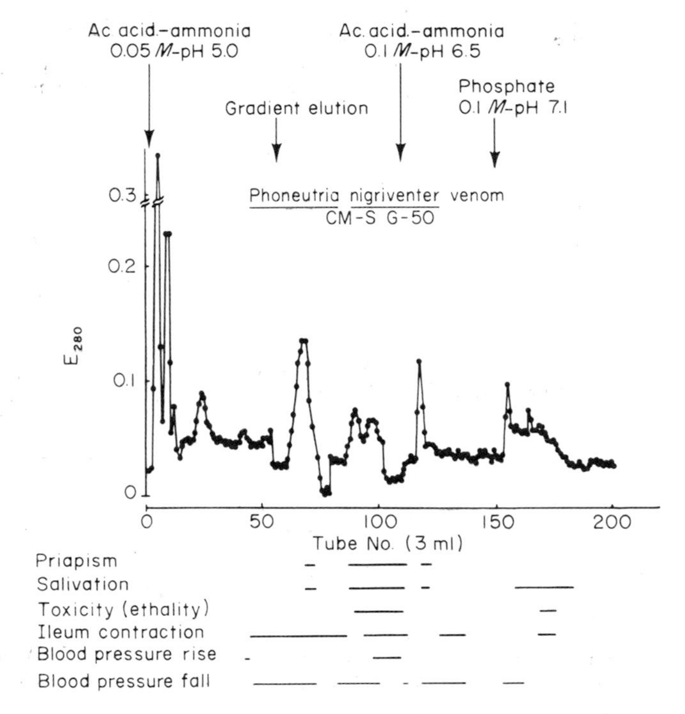

FIG. 11. Chromatogram of a 20 mg *P. nigriventer* venom run on a CM-Sephadex G-50 column (270 × 10 mm); gradient and stepwise elutions. Some of the same effects are distributed in 4 zones. The priapism fraction is absent in the fourth zone, indicating the possibility that its molecules might differ from those responsibles for salivation and toxicity.

These experimental findings furnish evidence which seems to demonstrate that the molecules of each activity zone differ from those of other zones in size and electrical charge. The molecules of each zone also differ one from the other in their pharmacological properties.

The problem increases in complexity when it is observed that 4 different polypeptide molecules can produce the same effects. It would appear unlikely that the veneniferous gland would secrete a number of components which would duplicate effects of others (Schenberg and Pereira Lima, 1967, 1968).

Further study of the venom proteolytic enzyme might explain this paradox. As mentioned earlier, all *P. nigriventer* venom active polypeptides are inactivated by trypsin, chymotrypsin, pepsin, and by the venom's proteolytic enzyme. The former 3 enzymes differ one from the other by their substrate linkage specificity. In spite of this, they are unable to distinguish between the pharmacologically active components of the venom polypeptides, since all these components are inactivated by these enzymes. The venom proteolytic enzyme, however, can not only distinguish albumin from casein, but also the ileum-contracting polypeptides from the other active venom protein components. Thus, it seems probable that the veneniferous gland would secrete the proteolytic enzyme together with a few larger molecules, with one or more of these particular linkages. Depending on the cascade degradation of a molecule by the venom proteolytic enzyme, the venom accumulated in the veneniferous gland might contain molecules having the same pharmacologically active components bound to larger or smaller chains, which would affect their size and electrical charge. The degradation of these large molecules would have to be interrupted at a certain equilibrium point, otherwise all the venom would be broken down to amino acids.

The free amino acids in the venom seem to represent degradation process residues; this also supports the above hypothesis. The concentration of the 3 amino acids would provide some indication of the structure of large molecules secreted by the veneniferous gland. The high glutamic acid content (23.6%) of the venom might show: (1) that the proteolytic enzyme preferentially attacks the glutamic acid linkage and (2) that glutamic acid is the terminal amino acid of one or several of these large molecules.

REFERENCES

Barrio, A. (1955). *Acta Physiol. Latinoam.* **5**, 132.

Brazil, O. V., and Vellard, J. (1925). *Mem. Inst. Butantan* **2**, 1-70.

Bücherl, W. (1953a). *Mem. Inst. Butantan* **25**, No. 1, 133.

Bücherl, W. (1953b). *Mem. Inst. Butantan* **25**, No. 2, 1.

Bücherl, W. (1956). *In* "Venoms," Publ. No. 44, pp. 95–97. Am. Assoc. Advanc. Sci., Washington, D.C.

Bücherl, W., Lucas, S., and Dessimoni von Eickstedt, V. (1969). *Mem. Inst. Butantan* **34** 47–66.

Diniz, C. R. (1963). *Anais Acad. Brasil. Cienc.* **35**, 283.

Diniz, C. R. (1963). Personal communication.

Fischer, F. G., and Bohn, H. (1957). *Z. Physiol. Chem.* **306**, 265.

Kaiser, E. (1956). *In* "Venoms," Publ. No. 44, pp. 91–93. Am. Assoc. Advanc. Sci., Washington, D.C.

Keyserling, E. (1891). *In* "Die Spinnen Amerikas—Brasilianische Spinnen" G. Marsc, ed.), Vol. 3, p. 144. Bauer and Raspe, Nürnberg.

Pereira Lima, F. A., and Schenberg, S. (1963). *Ciencia Cult.* (*São Paulo*) **15**, 268.

Pereira Lima, F. A., and Schenberg, S. (1964). *Ciencia Cult.* (*São Paulo*) **16**, 187.

Reed, L. J., and Muench, H. (1938). *Am. J. Hyg.* **27**, 493.

Sampayo, R. L. (1942). "*Latrodectus mactans* y latrodectismo," pp. 108–111. "El Ateneo," Buenos Aires.

Schenberg, S., and Pereira Lima, F. A. (1962). *Ciencia Cult.* (*São Paulo*) **14**, 237.

Schenberg, S., and Pereira Lima, F. A. (1963). *Ciencia Cult* (*São Paulo*) **15**, 267.

Schenberg, S., and Pereira Lima, F. A. (1966). *Abstra. 3rd Intern. Pharmacol. Congr., São Paulo*, 1966, p. 189.

Schenberg, S., and Pereira Lima, F. A. (1957). *Ciencia Cult.* (*São Paulo*) **19**, 456.

Schenberg, S., and Pereira Lima, F. A. (1966). *Mem. Inst. Butantan* **33**, 627.

Welsh, J. H., and Batty, C. S. (1963). *Toxicon* **1**, 165.

Chapter 53

Latrodectism in Mediterranean Countries, Including South Russia, Israel, and North Africa

ZVONIMIR MARETIĆ

THE MEDICAL CENTER, PULA, CROATIA, YUGOSLAVIA

I. Introduction and Symptomatology 299
II. Kinds of *Latrodectus* in Mediterranean Areas 301
III. History 302
IV. Epidemiology of Latrodectism 304
V. Spread of Latrodectism in Single Mediterranean Countries 306
VI. Folklore Treatment of Latrodectism in Mediterranean Areas 308
References 309

I. INTRODUCTION AND SYMPTOMATOLOGY

Latrodectism means envenomation by the bite of the spider from the genus *Latrodectus* Walckenaer, a term which is specific and more adequate than the general term arachnidism which is used for this poisoning in most of North America. The *Latrodectus* derives from the Greek words "λᾶθρα" = secretly and "δήκτης" = biter, thus "secret biter," a name which was in ancient Greece given to mad dogs.

Latrodectism is a syndrome, a separate medical entity with its own characteristics.

The bite itself is very slight, is often unperceived, and local symptoms are often hardly visible, i.e., two punctures on the site of the bite, a slight redness, edema, urtical plaque, hypesthesia or anesthesia dolorosa. The time of latency is from 10 to 60 minutes. First there appears an early pain in the regional lymphatic nodes, most often axillary or inguinal, which are often a little swollen. The pain soon becomes more intense and grips the lumbar region, belly, and thighs, increasing till paroxysm, so that this unsupportable

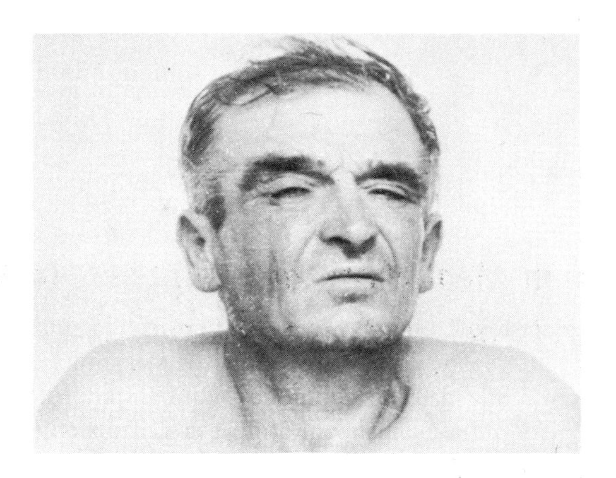

FIG. 1. "Facies latrodectismica," with redness, edema of eyelids, blepharoconjunctivitis, and painful grimace.

pain is the most prominent symptom of latrodectism. The patient shows a characteristic motor restlessness, cramps, has an oppression in the chest and feels *pavor mortis*. His general condition is very poor, sometimes with manifest signs of shock. Characteristic is a profuse sweating, lacrimation, disturbances in secretion of saliva (hyper or hyposecretion) and especially the "Facies latrodectismica" (Dameski and Masin, 1960; Maretić, 1955), a reddened sweat-covered face with edema of eyelids, a strong blepharoconjunctivitis, painful grimace, eventually trismus of masseters. (Fig. 1). There is an increase of tendon reflexes and pressure of the cerebrospinal fluid, a hypertension of convergent type, mostly a tachycardia in the first phase of intoxication and later, bradycardia, and changes in ECG (high P_2 and P_3 waves, devalvation of ST segment, low T waves and prolongation of QT interval) (Maretić, 1963). The patient is tachypneic, feels nauseated, vomits, and suffers from anorexia, obstipation, and sleeplessness. In the acute phase there is an anuria or oliguria, often with albuminuria and sometimes with erythrocytes and granulated casts in the sediment. Priapism and ejaculation were described. The rigidity of the musculature is typical, especially of the abdominal walls, which together with leukocytosis, neutrophilia, and eosinolymphopenia can be confused diagnostically with abdomen acutum. Other characteristic laboratory findings are increasing values of blood sugar, urea, and hematocrit, and changes in electrolyte levels (hyperpotassemia, hyposodemia).

Further acute psychoses in patients may occur, especially in severe cases. After some days a scarlatiniform, morbilliform, papulous, or vesiculous exanthem may appear, generalized or localized at the site of the bite. (Fig. 2).

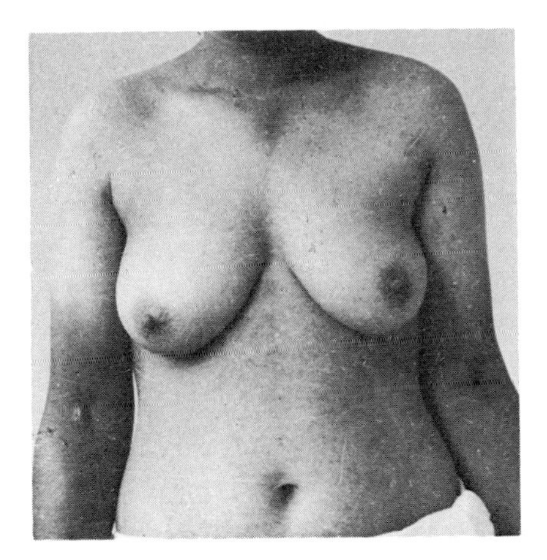

FIG. 2. Generalized scarlatiniform exanthem, a few days after the bite.

Histopathologically in experimental animals, exhaustion of hyphophysis was found, enlargement of adrenal cortex with changes typical for alarm reaction, caryorhexis in thymus, decay of lymphocytes in spleen and lymphatic nodes, hyperemia and necrosis in the liver and ulcerations in the mucosa of the stomach and intestines (Maržan, 1955).

In untreated patients, the next day the pains in the lumbar region and belly decrease, but increase in the lower extremities, especially the soles of the feet where often a violent burning is felt. The duration of illness in these people is up to 8 days and the convalescence is long, characterized by a psychasthenic syndrome. The prognosis is in general favorable, but fatalities do occur.

II. KINDS OF *LATRODECTUS* IN MEDITERRANEAN AREAS

The species *Latrodectus tredecimguttatus* Rossi, 1790, or according to new terminology by Levi, the subspecies *Latrodectus mactans tredecimguttatus* is found in all Mediterranean countries. It is velvet black and has about 13 red spots (tredecimguttatus = lat. "with 13 spots") on its round abdomen. The number of spots can vary from 17 to 0 in the black variety. Its abdominal spination is strong. The embolus of the palpus has three coils and the female has four or five loops dorsally of her internal genitalia (Levi, 1959) (Fig. 3).

The black variety, formerly *Latrodectus erebus* Walck. 1837 (syn. *L. lugubris* Mochulsky 1849, *L. erebus* Audouin 1825, *Theridion lugubre* Dufour 1820) is found in Brittany, southern France, Corsica, the Balkans, southern Russia, and northern Africa (Levi, 1959; Kobert, 1901).

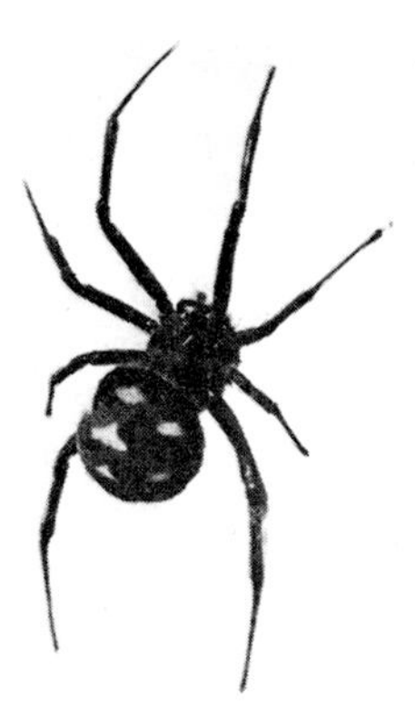

FIG. 3. *Latrodectus m. tredecimguttatus* Rossi 1790, female. Body length about 1.5 cm.

Latrodectus tredecimguttatus, like other spiders of the same genus, has many synonyms. These were described: *L. malmignatus* Walck. 1837—Italy, *Meta hispida* C. L. Koch 1836, *Meta schuchi* C. L. Koch 1836 and *L. conglobatus* C. L. Koch 1936—Greece, *L. argus* Audouin 1825, *L. oculatus* Walck. 1837, *L. venator* Audouin 1825—Egypt, and *L. quinqueguttatus* Krynicky 1837—southern Russia. *Latrodectus revivensis* Shulov 1948 is a distinct species found only in the Negev desert in Israel (Levi, 1966).

Another species, *Latrodectus pallidus* O. P. Cambridge 1872, lives in southern parts of Asian Russia, Syria, Israel, Egypt, Libya, and Tripolitania. The surface of its abdomen is smooth without hairs, looks leathery, and is of pallid yellowish color with dark red spots. The epigynum and internal sex organs are like those in *L. m. tredecimguttatus.* The venom of *L. pallidus* is of less toxicity than that of *L. m. tredecimguttatus* (Bettini, 1964).

III. HISTORY

Latrodectism in the Mediterranean area was known in the most ancient times (Fig. 4). Xenophon in the fourth century B.C. wrote that Socrates discussed "phalangia," spiders which were no bigger than half an obolus but were able to cause pains so severe that a human would lose his mind. Aristotle in 4 B.C. wrote about the venomous bite of a small mottled spider. He described the courtship and alleged that it also attacked animals bigger than itself, such as lizards (Kobert, 1901). Nikander of Klaros around 136 B.C. wrote on nine kinds of spiders and phalangia, some of which, like the "Rhox" or perhaps also another, "Asterion," were said to be dangerous to life; these are considered by Kobert (1901) to be identical with *Latrodectus.* He described erection and ejaculation as a consequence of the spider bite (see Kobert, 1901).

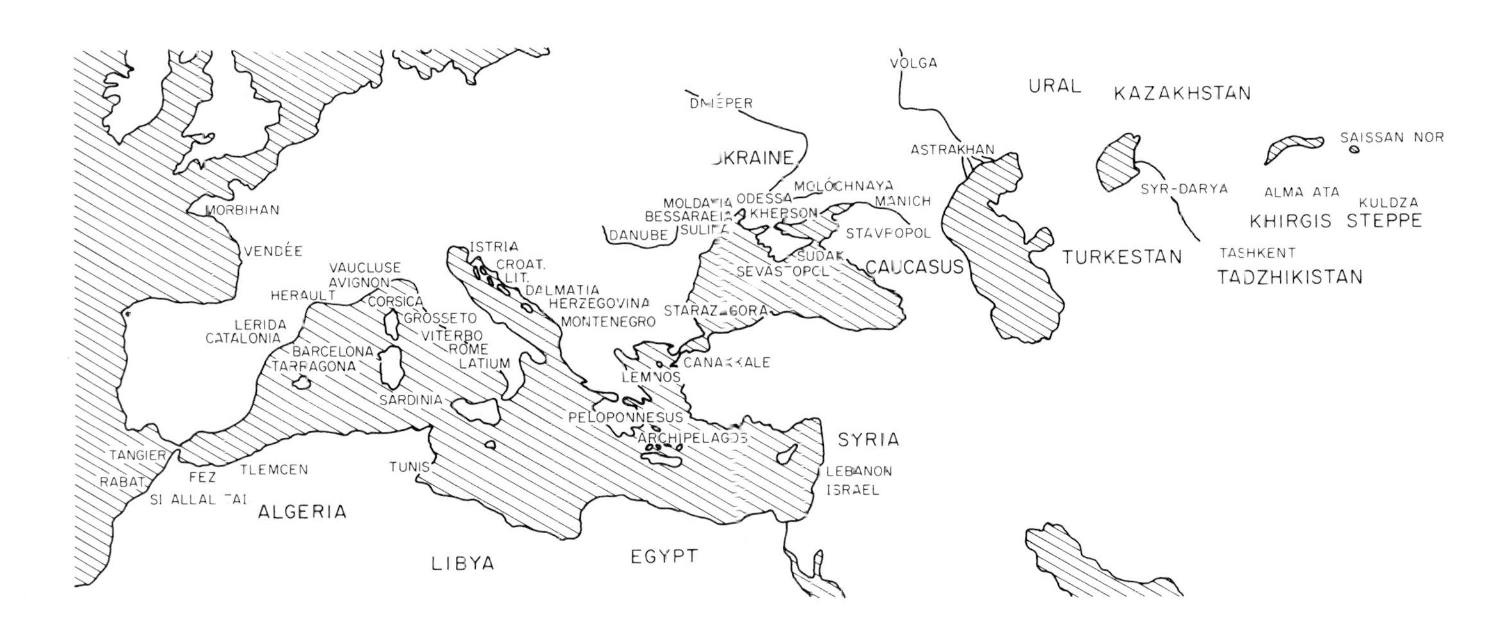

FIG. 4. Map of the Mediterranean and southern parts of Russia with localities where *Latrodectus* and latrodectism have been described.

In the second century B.C. Pedanius Dioscorides wrote on venomous spiders and described in detail the characteristic symptoms of latrodectism (Kobert, 1901). Celsius in the first century A.D. also wrote on venomous spiders, among which was surely included *Latrodectus* (Kobert, 1901). In the same century, Strabo wrote on arachnidism in general and on latrodectism in particular (Kobert, 1901). The same subject was treated by Caius Plinius Secundus in the second century A.D., Caius Iulius Solinus, about 250 A.D., and Aelianus (200–222 A.D.), asserted that in Crete lives a little spider which can kill a man with its bite (Kobert, 1901). The troops of the German emperor Ludwig in Calabria were reported by the monks Alberich, 866 A.D., and Regino, 867 A.D. (Kobert, 1901), to have been decimated by spiders. These reports, if at all reliable, may have arisen from the bite of *Latrodectus*, rather than the more generally suspected "tarantula," a spider of the genus *Lycosa* or *Hogna*. Rases, Avicenna, and Maimonides in the 10th, 11th, and 12th centuries also described *Latrodectus* and latrodectism (Kobert, 1901). Authors in later centuries, like Perotti (1412–1480) (Grevinus, 1571) and many others confused the symptoms of *Latrodectus* bite with hysterical "tarentism." Boccone wrote on latrodectism in Corsica in 1697 and Bourienne wrote about it 70 years later (Kobert, 1901). Marmocchi treated about 30 patients for latrodectism in Volterra in Italy (1786) (see Marmocchi, 1800), and the Italian author Toti performed the first experiments on dogs, hens, and pigeons (1786–1789) (see Vellard, 1936). Rossi described the spider in his *Fauna Etrusca* and gave it the name *Latrodectus tredecimguttatus* (1790). Graëlls wrote on great epidemics of latrodectism in Catalonia in 1830 and near Barcelona in 1834 (Kobert, 1901). Cauro in Corsica (1833) and Dugés from Montpellier (1836) gave data on *Latrodectus* and latrodectism (see Vellard, 1936). Dax (1881) and Guibert (1895) described latrodectism in southern France (see Vellard, 1936). Mochulsky and Becker (1838), Shchensnowich (1870), Rossikow (1898), and Shcherbina (1903) described epidemics of latrodectism in man and cattle in the steppes of southern Russia caused by the bite of the *L. erebus*.

IV. EPIDEMIOLOGY OF LATRODECTISM

In Mediterranean areas, except parts of Israel, *Latrodectus* is not urbanized as in some other parts of the world; it is found only outdoors, and so latrodectism has a rural character. *L. m. mactans*, for instance, in North America is more or less urbanized; it has been found in cellars, garages, even in flats on high floors. The most characteristic in the United States is a bite in the genitals of the victims by *L. mactans*, which often made its web under the boards of privies (Bogen, 1926). The same circumstance is described in

FIG. 5. A typical biotope for *Latrodectus m. tredecimguttatus:* a Mediterranean landscape with cornfield, stone wall, olive trees, and brambles.

Portorico, Cuba, Mexico (Vellard, 1936) with *L. m. curacaviensis* and in Australia with *L. m. hasselti* (Kellaway, 1930).

L. m. tredecimguttatus spins its web preferably in wheat, along borders of trenches, below and beside big stones, in bushes, in excavations of hollow trees and similar places. The typical ecological situation for this spider is a sunny field with various sized masses of rocks which stick out from the dry red earth overgrown with sharp grass, low scrubs, and brambles. Another typical biotope is the cornfield in which the spider makes its web among the stalks. (Fig. 5). *Latrodectus* is not aggressive at all; it does not have to leave its web, so that usually there are not many occasions for a meeting between man and spider. The best occasion for it is, for instance, during harvest when man violently enters *Latrodectus's* biotope, destroys its cobwebs, and the frightened spiders spread around and bite in self-defense when they are carelessly pressed by the hands and arms of harvesters or by their clothes and footwear when the spiders climb into their trousers, shoes, or sleeves.

Of 177 patients who were treated in the period of 1948–1965 in the Medical Center in Pula, all except two were bitten outdoors. These two, however, were also bitten by spiders which were obviously carried into the house with firewood a short time before.

Latrodectism in Yugoslavia, as in other Mediterranean countries, has an agricultural and professional character. Most of our patients were bitten during field work, especially harvesting and threshing. An interesting observation was made in connection with this. Of 137 patients from whom data are available, 31 or 22.6% have been bitten on the same side, i.e., in the lower part of the left forearm. This arm is more exposed to the danger of being bitten while collecting corn into sheaves; for instance in Istria (Croatia, Yugoslavia) farmers collect the wheat in such a manner that the ears are cut by the sickle which is held in the right hand and are caught by the left arm and hand. In binding sheaves the left forearm is pressed against the sheaves while the right hand is binding. The statement of the old Italian author Chellini (1728) that the harvesters are often bitten in their arms is in accord with this. In Istria, cases of latrodectism occur only during summer months and the spider does not hibernate, which is opposite to the warmer countries of the Mediterranean.

In Israel, contrary to other Mediterranean areas, Leffkowitz *et al.* say that besides seeking hollow tree trunks, and rocks, "the spider [*L. tredecimguttatus*] lives also in dark corners, especially in privies where most bites occur" (Leffkowitz *et al.*, 1962). According to this, it would seem that in Israel signs of urbanization of *Latrodectus* are present.

The epidemiology of latrodectism is characterized by great oscillations in number of spiders and their bites. During some years *Latrodectus* overflows the fields; accordingly, the number of cases of latrodectism increases greatly. This lasts several years and then the number of spiders and their bites decline, even practically disappear for many years, so that often, as happened in Istria, the population forgets them; then when latrodectism appears again it is sometimes considered to be a "new disease (1948)." The reasons for these cyclical variations is not clear, but probably they are caused by disturbances in the balance between the spiders, their food, and their natural enemies.

V. SPREAD OF LATRODECTISM IN SINGLE MEDITERRANEAN COUNTRIES

Latrodectus exists in Portugal (Colaço, 1954).

In Spain, among other places, *Latrodectus* and latrodectism were recorded around Taragona and El Plor in Catalonia, near Barcelona and in the province of Lerida (Sales Vásquez, and Biosca Florensa, 1944).

In France it exists in the south around Avignon, Vaucluse, Hérault, in Corsica as well as in Brittany (Morbihan) and in Vendée (Kobert 1901; Vellard, 1936).

In Italy spiders and their bites are known from ancient times, especially in Volterra so that *Latrodectus* was called "*Ragno rosso volterrano*" or "Red spider of Volterra." The spider was well known also in the southern parts of the country. In the fifth decade of this century its appearance in great number was recorded around Rome, in the middle parts of the western Italian coast in the provinces of Grosseto, Viterbo, Rome, and Latina, and in southern parts of Sardinia. In the period of the epidemic of 1946 to 1950 there were 492 cases recorded in Italy (Bettini, 1963, 1964).

In Yugoslavia *Latrodectus* and latrodectism were described in Istria,

Croatian littoral, in Dalmatia with its islands, Herzegowina, Montenegro, and Macedonia (Maretić and Stanić, 1954; Ramzin, 1947; Grujić, 1959; Vanovski, 1962). Although it was reported in several regions of Yugoslavia, latrodectism was mostly described in the northwestern peninsula of Istria. Previous to 1948, data for latrodectism in this area were not available. The last epidemic, in Istria, or better, in some areas of Istria which altogether has hardly about 30,000 inhabitants, began in 1948, as in nearby Italy, and reached its climax in 1952 with 42 cases during the summer months of this year and gradually declined. Out of 177 patients treated in the Hospital of Pula in the period of 1948 to 1965, only a few were bitten during the last years.

In Greece latrodectism was said to occur in the islands of Archipelagos, in Peloponnesus (Hadjissarantos, 1940) and the Greek part of Macedonia. Savoura described 43 cases which occurred on the island of Lemnos during the period 1930 to 1961 (Savoura, 1962).

In Bulgaria in 1925 to 1926 in the area of Stara Zagora, 87 cases of bite in humans and many in cattle were recorded. Latrodectism was described also in Rumelia (Kobert, 1901; Vellard, 1936).

In Romania there was a case of latrodectism near Sulina in the Danube delta (Vintilã *et al.*, 1963).

In Russia (European part) *Latrodectus* and latrodectism were described in the Ukraine (Kobert, 1901; Pawlowsky, 1927), Moldavia (Maksianowich, 1939), in the Crimean peninsula, especially around Sudak and Sevastopol, in Bessarabia about Odessa, Kherson, and Stavropol, and in general in the whole coastal area of the Black and Azovian Sea, in Tauria, in the districts of Dnijeper around Molochnaya, the Don and Manich, in Podolia, Powolzhie, Caucasus and Ural (Pawlowsky, 1927; Kobert, 1901; Lepolova, 1955; Yarovoy and Shewchenko, 1957). In the Asian part of the U.S.S.R. latrodectism was described around Astrakhan, the Kirghis steppe, in Chokand, around Kouldza, Saissan Nor, in Kazahstan around Syr Darya, Alma Ata, Transcaucasia, Turkestan, Uzbekistan, particularly about Tashkent, and in southern Tadzhikistan (Kobert, 1901; Pawlowsky, 1927; Lepolova, 1955; Arustamyan; 1955; Blagodarny, 1957).

In Turkey, according to personal communications, the spider was found about the Dardanelles and Çanakkale (Oytun, 1953; Bettini, 1964). Its bite occurs further in Syria and Lebanon.

In Israel, where three species of Latrodectus, i.e., *L. m. tredecimguttatus, L. revivensis, and L. pallidus* are found, *Latrodectus* and latrodectism were known a long time ago; but only after 1942 was more attention given to this problem, thanks to the works of Shulov, Seidman, Kriegel, and others (Leffkowitz *et al.*, 1962).

Latrodectus and latrodectism extend over the whole of North Africa, in Egypt, Libya, Tunisia, Algeria, and Morocco (Levi, 1959). In Algeria latrodectism was reported near Tlemcen (Bouisset and Larroy, 1962), and in

Morocco about Gharb, Si Allal Tazi, Rabat, Fez, and Tangier (Gaud and Delesalle, 1949).

VI. FOLKLORE TREATMENT OF LATRODECTISM IN MEDITERRANEAN AREAS

It has been published in other places that the best therapy for latrodectism is specific antivenin and calcium, respectively; the treatment of choice is the combination of both drugs which encourages an early recovery. (Maretić and Stanić, 1954).

The attempts to cure latrodectism are very ancient and are preserved in folklore medicine. If one discards the superstition and fantasy in ancient methods of treating latrodectism, as is often true for other fields of folk medicine, one can see faintly principles which have a medical justification. In such treatment of latrodectism there are two principal constituents; (a) exposure to physical exertion and (b) provocation of vasodilatation by means of heat, and in other ways.

With regard to physical exertion there is no doubt that the ancient "tarantism" was a manifestation of mass hysteria, but it is true that it originated from the early knowledge that dancing to exhaustion, i.e., physical strain, helps in curing the bite of the spider of the genus *Latrodectus*. That physical exertion, which represents a stress resulting in an alarm reaction with a strong countershock, has a positive influence upon latrodectism could be proved also experimentally in recent times (Maretić and Stanić, 1954). The origin of the layman's confusion of the tarantula and *Latrodectus* is unknown. *Latrodectus* and *Tarantula* are also nowadays mixed up in the mind of the people in southern Italy and in some parts of Istria, so that both are called the same name: "Tarantola" and "Tarantism." But the motor restlessness and movements of the patient who searched for relief from the intense pains of latrodectism had a suggestive influence upon the psychically labile environment, and so was born the hysterical tarantism which raged in the Middle Ages and later for many centuries in southern and middle Europe. Tarantism accompanied by melodies of the vivacious "tarantella" exists also nowadays in some remote parts of southern Italy, Sardinia, and Spain. Another method of inducing physical strain was swinging the patient on a stretched rope or in a net while conjuring and singing; this was still performed in the Croatian coastal region 40 years ago. In Greece vasodilatation is provoked by putting the patient in a heated bread oven (Mercier, 1952). According to Bourienne and Cauro, a similar practice existed in Corsica in the last century (Kobert, 1901), but it is not impossible that it might be preserved somewhere even today. In Morocco, Gaud and Delesalle reported (1949) that the bitten person was put into a deep hole dug in the earth which was previously heated with burning wood. Wine and other alcoholic drinks having a known vasodilatating action are also used as popular remedies for latrodectism.

REFERENCES

Arustamyan, A. T. (1955). *Med. Parazitol. i Parazitarn. Bolezni* **24**, 355.
Bettini, S. (1963). *Riv. Parassitol.* **24**, 106.
Bettini, S. (1964). *Toxicon* **2**, 93.
Bettini, S. (1964). Personal communication.
Blagodarny, Y. A. (1957). *Klin. Med.* **35**, 76.
Bogen, E. (1926). *A.M.A. Arch. Intern. Med.* **38**, 623.
Bouisset, L. and Larroy, G. (1962). *Presse Med.* **70**, 1019.
Colaco, A. F. (1954). Personal communication.
Dameski, D., and Masin, G. (1960). *Maked. Med. Pregl.* **15**, 389.
Gaud, J., and Delesalle, D. (1949). *Bull. Inst. Hyg. Maroc.* **9**, 233.
Grevinus, J. (1571). "De venenis libri duo." Antwerpen.
Grujić, I. (1959). *Ziv. i Zdr.* **13**, 12.
Hadjissarantos, H. (1940). "The Spiders of Attica." Athens.
Kellaway, C. H. (1930). *Med. J. Australia* **1**, 41.
Kobert, R. (1901). "Beitr. zur Kenntn. d. Giftspinnen." Enke, Stuttgart.
Leffkowitz, M., Kadish, U., and Stern, J. (1962). *Dapim Refuiim* **21**, 2.
Lepolova, A. S. (1955). *Klin. Med.* **33**, 80.
Levi, H. W. (1959). *Trans. Am. Microscop. Soc.* **78**, No. 1, 7.
Levi, H. W. (1966). *J. Zool. (London)* **150**, 427.
Maksianowich, M. I. (1939). *Med. Parazitol. Parazitarn. Bolezni* **8**, 51.
Maretić, Z. (1955). *Med. Glas.* **9**, 159.
Maretić, Z. (1963). *Toxicon* **1**, 127.
Maretić, Z., and Stanić, M. (1954). *Bull. World Health Organ.* **11**, 1007.
Marmocchi, F. (1800). *Atti Accad. Sci. Sienna* **8**, 218.
Maržan, B. (1955). *A.M.A. Arch. Pathol.* **59**, 727.
Mercier, P. (1952). Personal communication.
Nikander. Theriaka, (quoted by Kobert, 1901).
Oytun, H. S. (1953). Personal communication.
Pawlowsky, E. N. (1927). "Gifttiere und ihre Giftigkeit," pp. 150–173. Fischer, Jena.
Perotti, N. (1536). "Cornucopiae latinae linguae," Basel.
Plinius Caius Secundus. (Second Century A.D.). "Naturalis Historia," 6th ed. Mayhoff, Leipzig, 1897.
Ramzin, S. (1947). *Vojnosanit. Pregl.* **4**, 267.
Regino. (1521). "Chronicorum sive annalium Regionis Prussiensis coenobii Abbatis a Christi nativitate usque ad Ottonem secundum finis." Schoeffer, Mainz.
Rossikow, K. N. (1898). "The Venomous Spider Karakurt" (in Russian).
Sales Vásquez, M. and Biosca Florensa, M. (1944). *Med. Clin. N. Am.* **12**, 244.
Savoura, A. (1962). Thesis, University of Athens.
Shchensnowich, M. (1870). *St. Petersburgh Med. Wechschr.* p. 54 (quoted by Kobert, 1901).
Shcherbina, A. (1903). *Arb. Entomol. Bur. St. Petersburg* **4** (quoted by Pawlowsky, 1927).
Thorp, R. W., and Woodson, W. D. (1945). "Black Widow, America's Most Poisonous Spider." Univ. of North Carolina Press, Chapel Hill, North Carolina.
Vanovski, B. (1962). *Liječnički vjesnik.* **84**, 131.
Vellard, J. (1936). "Le venin dès Araignées." Masson, Paris.
Vintilă, I., Fuhn, I. E., Vintilă, P., and Popescu, V. (1963). *Microbiol., Parazitol. Epidemiol. (Bucharest)* **8**, 231.
Xenophon. (Fourth Century B.C.). *In* "Memorabilia Socratis," Lib. I, Chap. 3 (transl. into German by J. Irmscher, Phil. Studientexte, Berlin, 1955).
Yarovoy, L. V., and Shewchenko, M. S. (1957). *Klin. Med.* **35**, 143.

Chapter 54

Chemical and Pharmacological Properties of Tityus Venoms

CARLOS R. DINIZ

DEPARTAMENTO DE BIOQUÍMICA, FACULDADE DE MEDICINA,
UNIVERSIDADE FEDERAL DE MINAS GERAIS, BELO HORIZONTE, BRAZIL

I. Introduction 311
II. Toxicity and Symptoms of Intoxication 312
III. Chemistry of the Venoms 312
IV. Pharmacological Properties 313
V. Conclusion 314
References 314

I. INTRODUCTION

Studies of the properties of scorpion venoms from the South American genus *Tityus* C. L. Koch 1836 have been done mainly with the venoms obtained from the most common species found in Brazil—*Tityus bahiensis* (Perty) 1834 and *T. serrulatus* Lutz e Mello 1922. Pioneer work by Brazil (1918), Houssay (1919), Magalhães (1925), and Carvalho (1945) demonstrated that the scorpion venom produces neurotoxic, circulatory, and muscular effects when injected in man or experimental animals.

Human accidents from scorpion sting had relatively high incidence in some urban areas of Brazil, chiefly in Belo Horizonte in the State of Minas Gerais and in Ribeirão Preto and Aparecida in the State of São Paulo. At Belo Horizonte, an average of 843 accidents per year, with an average of 16 deaths per year in the same period was reported by Magalhães (1946) between 1938 and 1945.

The treatment of human accidents by serum therapy was introduced by Vital Brazil, and the Instituto Butantan, São Paulo, still produces an average of several thousand serum ampules per year.

II. TOXICITY AND SYMPTOMS OF INTOXICATION

(*a*) *Toxicity*. Toxicity of the scorpion venom assayed in terms of LD_{50} in mice confirmed the clinical findings that the stings of *T. serrulatus* are commonly of greater severity than those inflicted by *T. bahiensis*. More recent work done by Bücherl (1953) demonstrated that the LD_{50} of these venoms in mice differs from batch to batch and according to the procedure of collecting the venoms. The average values when the venom was injected subcutaneously ranged from 6 to 12 μg for the *T. serrulatus* and 25 to 27 μg for the *T. bahiensis* venoms per 20 gm mouse. Based on the average possible amount of venom obtained when *T. serrulatus* is electrically stimulated, this author estimates that a single sting is able to kill 7–10 kg of mice. He considers human beings more sensitive to the venom than mice.

(*b*) *Symptoms of the Intoxication*. The inflicted sting of these scorpions in man is followed by severe pain which persists for several hours and in some cases is accompanied by edema of the affected area. Constant features of the poisoning are sialorrhea, rhinorhea, lachrymation and pronounced sudoresis, pallor, muscular twitchings, convulsion, opisthotonus, and even abortion. Tachycardia, hypertension, glycosuria, and hyperglycemia have been observed in experimental animals as well as from accidents in human beings (Magalhães, 1946).

III. CHEMISTRY OF THE VENOMS

The main components of scorpion venom from *T. serrulatus* and *T. bahiensis* obtained by electrical stimulation are proteins. Seven different protein components were separated by zone electrophoresis, all of basic character at pH 7.8 (Fig. 1). Toxic, smooth-muscle contractile, capillary permeability-increasing, and hyaluronidase activities were also separated by this method. Other substances such as histamine, serotonin, acetylcholine, adrenaline or cholinesterase inhibitors were not found in this venom (Diniz and Gonçalves, 1960).

More recently in this laboratory, in collaboration with Gomez (Diniz and Gomez, 1968), we were able to fractionate the venom from *T. serrulatus* in several components by a combination of extraction and chromatographic techniques using gel filtration in dextran (Sephadex G-25) and ion exchange in carboxymethyl cellulose. The separation of two components with toxic activity was obtained and toxic protein homogeneous by electrophoretic criteria in acrylamide gel and by chromatographic procedure was obtained. This substance, called *Tityus-toxin I,* when injected in mice reproduces the intoxication symptoms of the whole venom and its toxicity is 16–20 times

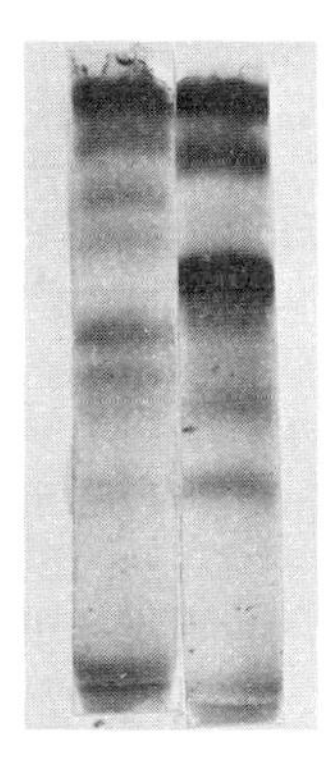

FIG. 1. Disc electrophoresis of the same amount of venom from two different specimens of *T. bahiensis* collected in different regions of Brazil. Note qualitative and quantitative differences of these venoms.

higher than the whole venom. The LD_{50} of this substance ranges from 0.05 to 0.12 μg/gm/mice. NH_2 terminal amino acid residue of the *Tityus*-toxin I determined by Sanger's method showed lysine as the N-terminal residue. Total amino acid analysis of the toxin hydrolyzate showed 18 amino acids.

IV. PHARMACOLOGICAL PROPERTIES

The study of the mechanism of action of scorpion and other animal venoms has been handicapped by the use of total venomous secretions, which may contain several active substances. The study of the effects of the *Tityus* venom suggests the participation of the autonomic nerves in the genesis of the symptoms of intoxication. It is interesting that symptoms of either cholinergic and adrenergic origin are described when effects of the venom are studied upon whole animals. Tachycardia, hypertension, hyperglycemia, bradycardia, hypotension, and muscular twitchings are described (Magalhães, 1925; Ramos and Corrado, 1954; Freire-Maia *et al.*, 1959). The authors in general attempt to interpret these phenomena in terms of separate mechanisms. They may be interpreted of course as manifestations of central action affecting the autonomous center. The venom acts also on isolated organs (Carvalho, 1945; Diniz and Valeri, 1959). An analysis of the effects of *Tityus* venom in the guinea pig ileum suggests that the contraction of the smooth muscle produced by the venom is an indirect effect through the release of an acetylcholinelike substance (Diniz and Torres, 1964). On the other hand, analysis of the effects in the perfused and isolated guinea pig heart indicated the participation of both adrenergic and cholinergic mechanisms (Fig. 2). Catecholamines and acetylcholine are probably

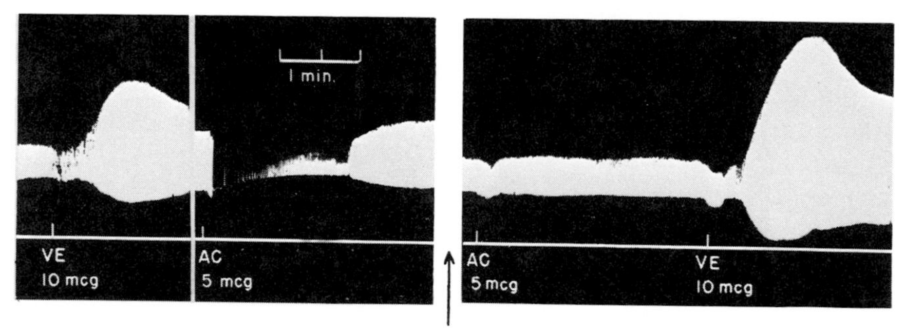

FIG. 2. Normal guinea pig heart showing effects of 10 μg of scorpion venom (VE, final concentration 2×10^{-6} gm/ml) and acetylcholine (AC, final concentration 1×10^{-6} gm/ml) before and after the simultaneous administration of prostygmine (4×10^{-6} gm/ml) and atropine (2×10^{-5} gm/ml). Notice the increase of the cardiostimulant effect of the venom after treatment of the atropinized heart by prostygmine. (From Antonio *et al.*, 1968.) Prost, prostygmine; Atrop, atropine; mcg, micrograms.

released by the venom, since the adrenergic effects are abolished in reserpinized animals and by inderal on the one hand, and cholinergic effects, such as bradycardia and the decrease of strength of the contraction, are potentiated by eserine, inhibited by atropine, or hemicholinium on the other (Antonio *et al.*, 1968).

V. CONCLUSION

The purification of the toxic components of *Tityus* venom and the analysis of the effects of pure toxic substances on biological structures present an interesting field of interdisciplinary studies. These studies can appropriately be organized in the broad discipline of molecular biology.

A common denominator can be accepted to explain the mechanism of action of the venom, if we accept the Burn and Rand hypothesis (1962) that postulates a previous release of acetylcholine as the initial chain of reactions responsible for the release of adrenaline. Another possibility is that the venom contains different substances with both cholinergic and adrenergic actions. This hypothesis, however, probably can be discarded because the recent work in our laboratory with the purified "*Tityus*-toxin I" indicates that this substance reproduces all the effects of the whole venom.

REFERENCES

Antonio, A., Corrado, A. P., and Diniz, C. R. (1968). *Mem. Inst. Butantan* **33**(3), 957–960.

Brazil, V. (1918). *Mem. Inst. Butantan* (*São Paulo*) **1**, No. 1, 47–50.

Bücherl, W. (1953). *Mem. Inst. Butantan* (*São Paulo*) **25**, 83.
Burn, J. H., and Rand, M. J. (1962). *Nature (London)* **184**, 163.
Carvalho, P. (1945). *Arquiv. Inst. Benj. Bapt.* **4**, 21.
Diniz, C. R., and Gomez, M. V. (1968). *Mem. Inst. Butantan* (*São Paulo*) **33**(3), 899–902.
Diniz, C. R., and Goncalves, J. M. (1960). *Biochim. Biophys. Acta* **41**, 470.
Diniz, C. R., and Torres, J. M. (1964). *Ciencia Cult.* (*São Paulo*) **16**, 197.
Diniz, C. R., and Valeri, V. (1959). *Arch. Intern. Pharmacodyn.* **121**, 1.
Freire-Maia, L., Ferreira, M. C., and Silva, S. G. (1959). *Mem. Inst. Oswaldo Cruz* **57**, 105.
Houssay, B. A. (1919). *J. Physiol. Pathol. Gen.* **18**, 305.
Magalhães, O. (1925). *Compt. Rend. Soc. Biol.* **113**, 35.
Magalhães, O. (1946). *Monographias Inst. Oswaldo Cruz* **3**, 220.
Ramos, H., and Corrado, A. P. (1954). *Anais. Fac. Med. Univ. São Paulo* **28**, 81.

Chapter 55

Classification, Biology, and Venom Extraction of Scorpions

WOLFGANG BÜCHERL

INSTITUTO BUTANTAN, SÃO PAULO, FELLOW OF THE CONSELHO NACIONAL DE PESQUISAS, RIO DE JANEIRO, BRAZIL

I. Introduction 317
II. Morphology, Classification, and Distribution 318
A. Morphology 318
B. Classification of Scorpions 322
C. Geographical Distribution 331
III. Biology of Scorpions 333
A. The Life of Scorpions 333
B. The Venom Apparatus of Scorpions 339
IV. Extraction and Toxicity of Scorpion Venoms 341
A. Extraction of Venoms 341
B. Toxicity of Some Scorpions 342
C. Treatment of Scorpion Poisoning 345
References 346

I. INTRODUCTION

The most frequent scorpions are confined to tropical and subtropical climates with dispersions into the temperate zones of both hemispheres. In South America they are found in the highest Andes in Chile, Bolivia, Peru, and Colombia, in Europe in the Alps of Switzerland, while in North America they have been collected in Washington, Oregon, and Montana as well as in southern Canada.

Scorpions belong to a relatively small order of the subclass Arachnoidea, class Arachnomorpha, with only about 650 different species. The smallest adults may range from approximately 2 to 3 cm and the largest between 15 to 25 cm.

The scorpion shows eight legs, a pair of large pinchers or pedipalps at the front of the cephalothorax and another pair of small mouth pinchers, the chelicerae or mandibulae, and on its underside, just behind the base of the last pair of legs, the pectines, which form a pair of comblike structures; its main body is divided into a large front segment, called cephalothorax, followed by the large preabdomen with seven segments, and the "tail" or postabdomen with six cylindrical articles. The sixth segment of the tail is bulbous and is called the telson; it leads into the curved terminal stinger with a very sharp tip. The telson contains two poison glands with two independent ducts, leading into the terminal stinger and opening each one by a slit a short distance from the end of the tip. The scorpion can move its stinger in almost any direction, except downward. For this reason the tail is usually curved over its back, ready for stinging, or to one side, when resting. The impossibility of moving the stinger downward was a very important discovery, made by us 15 years ago. This fact allowed us to manipulate the scorpion between the human fingers for electrical extraction of the venom.

II. MORPHOLOGY, CLASSIFICATION, AND DISTRIBUTION

A. Morphology

The main body of the scorpion is divided into trunk (truncus) and tail (cauda or postabdomen). The trunk is subdivided into cephalothorax and preabdomen (Fig. 1). On the cephalothorax are two larger median eyes, often situated over an ocular tubercle, and at each side, of the extreme front edge, from two to six lateral eyes, arranged in one or two groups. Often there are several granular keels, named from the front to back of the cephalothorax anterior, internal median, external median, and posterior keels, united in different ways, parallel, diverging or joining. When the carapace is deeply excised anteriorly, the two curves on each side of the front are called frontal lobes. On each side of the median eyes exist commonly a smooth surface, the "mirror," limited by a granular area.

The preabdomen has dorsally seven tergites, marked generally with a median longitudinal keel and often with lateral keels, completely or partly granulated or without granules. On the ventral side only the five last sternites are visible, each of the first four bearing a pair of respiratory spiracles. In front of the five sternites is the basal piece of the combs. The series of sclerites that form the front margin of a comb are called marginal lamellae, next to these there is a series of more or less rounded sclerites, the middle lamellae; the small, rounded sclerites between the middle area and the teeth, are the fulcra; the longer and slender appendages of the comb are the teeth

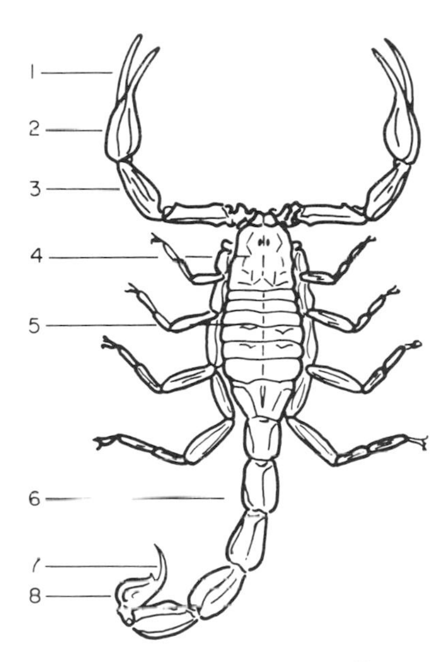

FIG. 1. *Tityus bahiensis:* 1, fingers; 2, hand; 3, pedipalp; 4, cephalothorax; 5, pre-abdomen; 6, tail; 7, stinger; 8, telson.

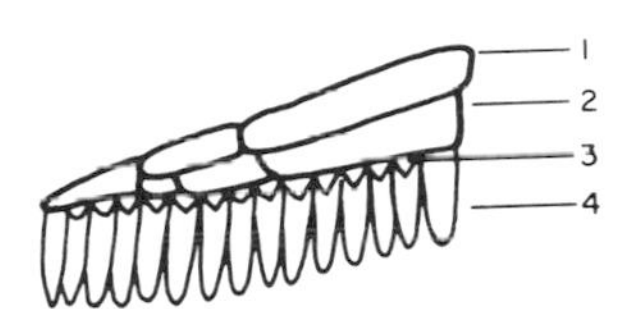

FIG. 2. *Tityus bahiensis:* comb. 1, marginal lamellae; 2, middle lamellae; 3, fulcra; 4, teeth.

(Fig. 2). In front of the combs, in the middle line, are the two genital sclerites (operculum genitale), and in front of these, the sternum. The sternum is situated just between the coxae of the last pair of legs. The external form of the sternum varies widely, from triangular to pentagonal or reduced to two small but very wide sclerites, offering important characters for the determination of the families (Fig. 3).

On the first five articles of the cauda there are often granulated or non-granulated keels. The segments are commonly mesally excavated above, with two dorsal median keels upon the dorsal side, and two ventral median keels on the ventral side of the same segment. Outside of these are the dorsal lateral keels and the ventral lateral keels. In addition to these principal keels —eight in number—a pair of "accessory keels" may be present.

FIG. 3. Different sterna, left to right: *Buthidae, Scorpionidae, Chactidae, Bothriuridae.* 1, sternum; 2, genital opercules.

The keel formation on the fifth segment is generally different. Commonly there is only one ventral median keel. The sixth segment or telson often bears a subaculear tooth or spine at the ventral base of the stinger.

The mouth appendages consist of a small pair of chelicerae and a large pair of pedipalps. The chelicerae or mandibulae are chelate, three-segmented, with a movable and an immobile finger, both forming a pincher. On the margins of the fingers one or more teeth may exist; they are very important for the classification of the families and the genera (Fig. 4). The pedipalps

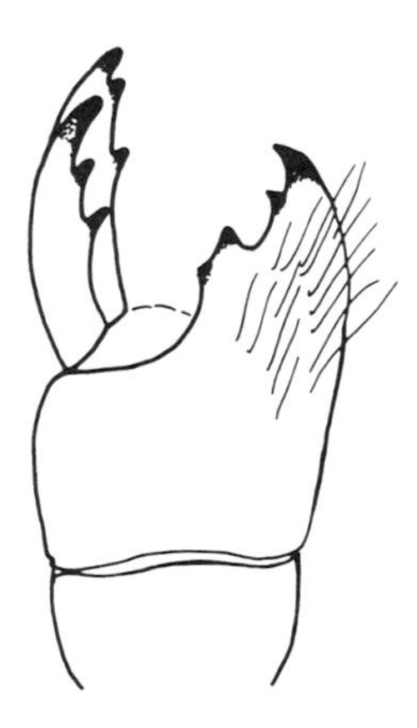

FIG. 4. *Tityus trinitatis*. Chelicerae, with the movable and the immobile finger.

are of leglike structure and end with powerful pinchers; the coxae form the maxillary lobes for crushing the prey, which is seized by the mandibulae; the trochanter is short and ringlike; the femur is long, cylindrical, rounded or with four keels; the tibia commonly bears large and pointed teeth; the large basal portion of the pincher is termed the hand with a fixed finger and with the movable finger, being the last segment of the pedipalp. Parts of the hand may be smooth and polished, weakly or strongly crested, with several longitudinal keels, furnished or not with granules, etc. Both fingers, when closed, may be in contact throughout or may bear a basal lobe. Very important for systematic studies is the aspect of the inner surface of the movable finger, furnished with rows of fine granular teeth from the base to the apex; a median series of teeth and lateral series may exist on each side; they consist of more scattered denticles, some of which are enlarged.

The four pairs of ambulatory legs show coxa, trochanter, femur, tibia and three tarsi. The coxae of the first two pairs bear each an endite, which is directed toward the mouth, forming the so-called maxillary lobes, which doubtless help to hold the prey; the endites of the first pair are more or less covered by those of the second pair. In several species of scorpions a stridulating apparatus is present on the coxae of the pedipalps and on those of the first pair of legs (*Pandinus* and *Heterometrus*). Very important for the separation of families are: (1) the presence or absence of a tarsal spur (tibial spur according to some authors) at the distal end of the first tarsus of the last pair of legs; (2) if the pedal spur on the arthrodial membrane connecting the second and the third tarsus of all legs is present only on the anterior side (one pedal spur only) or also at the posterior side (a pair of pedal spurs) (Fig. 5); some authors call it the tarsal spur; (3) sometimes the distal portion of the third tarsus is prolonged above the claws, forming a dorsal lobe; sometimes this segment is prolonged into a lobe on each side, forming the lateral lobes, rounded and covering the claws; and there may often exist a more or less clawlike plantula below, between the claws, the empodium.

The color of the body, the pedipalps and ambulatory legs are only of secondary importance for the classification of scorpions. Microcolor changes as an adaptation to different soils, woods, camps, on different ages of the same specimen. The formations on the trunk of keels, whether granulated or not, smooth and polished areas, and finely granular areas are distinguishing

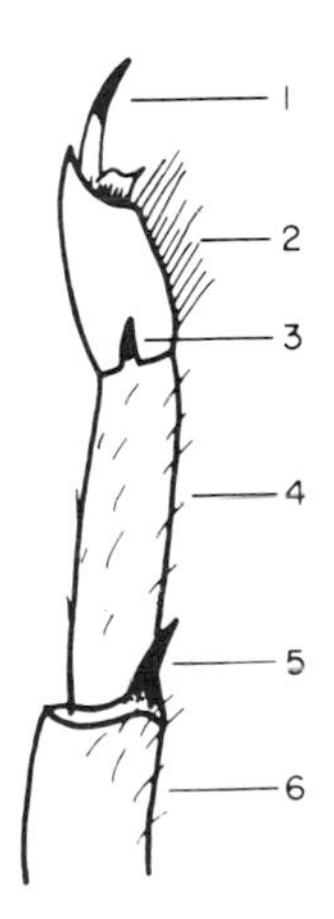

FIG. 5. Tarsi of an ambulatory leg of a scorpion. 1, claw; 2, tarsus III; 3, pedal spur; 4, tarsus II: 5; "tibial" or "tarsal spur"; 6, tarsus I.

characteristics. The number of trichobothriae, chiefly on the pectinal teeth and on the tibia and the pincers of the pedipalps, is important.

It is not always easy to distinguish the sexes. Vachon insists that in adult females after pairing a "spermatocleutrum" may be present, occluding the genital opercules; in some representatives of adult males two small, fingerlike processus may be present at the inner margin of the genital opercules. Secondary sexual characters are often present and may be useful to distinguish the sexes, such as different granulation, formation of keels only in one sex, much longer tails in the males as well as thicker or even smaller pincers, longer fingers, etc.

B. Classification of Scorpions

I. The Scorpions of the World

Key for Families (Figs. 3 and 5):

1
- Sternum longer than broad or as long as broad or only a little shorter than broad 2
- Sternum several times broader than long, often not visible, consisting of two small lamellae or transverse bars; middle lamellae of combs pearl shaped—Family Bothriuridae

2
- In the arthrodial membrane between the second and the third tarsus of the legs on the anterior and the posterior side with one or two tarsal spurs; sternum often narrowed in front or nearly so 3
- Legs furnished with a single tarsal spur on the anterior side of the arthrodial membrane connecting the second and the third tarsus; sternum pentagonal, the lateral margins parallel—Family Scorpionidae

3
- From three to five lateral eyes on each side, or more 4
- With only two, or rarely without, lateral eyes, often with a yellow spot behind the second 5

4
- Sternum narrowed in front, triangular, with anteriorly converging sides; in a few species it is pentagonal, but in this case there is a tarsal spur at the end of the first tarsus of the third and fourth pairs of legs; lateral eyes three to five in number; often a spine under the stinger—Family Buthidae
- Sternum with lateral margins parallel, generally broader than long, with a deep median furrow; middle sclerites of combs often pearl formed; three lateral eyes; without tarsal spur at the end of the first tarsus, but two simple, unbranched pedal spurs at the arthrodial membrane between the second and the first tarsus; without a spine under the stinger—Family Vejovidae

5 { Two lateral eyes with a yellow spot behind the second; movable finger of pedipalps with several rows of granules on top of each other; sternum nearly pentagonal with a median depression and a median furrow; without tarsal spur (but with two pedal spurs); movable finger of mandible with inferior teeth; pectines very short; leg claws not overlapped by the sides of the tarsus—Family Chaerilidae
Two lateral eyes or rarely no eyes, but always without yellow spot; movable finger of pedipalps with only one longitudinal row of teeth or with two very indistinct longitudinal rows; sternum pentagonal, generally as broad as long, without medial depression behind, but in front with two depressions "T"-like in form; with two pedal spurs, without tarsal spur; stigmata often circular; leg claws not overlapped; movable finger of mandible without teeth or only with one small inferior tooth —Family Chactidae

Key to Subfamilies:

Family Bothriuridae

Without subfamilies, with 7 genera, several subgenera, with about 50 species.

Family Scorpionidae

1 { With a distinct subaculear spine under the aculeus—Subfamily Diplocentrinae
No subaculear spine under the stinger 2

2 { Sides of the end of the third tarsus with rounded lobes, which form an acute angle with the dorsal lobe; with a rounded pincer 3
Lateral lobes of the tarsus not rounded, forming a right angle with the dorsal lobe; pincer flat, always with a distinct finger-keel 4

3 { Underside of all segments of the tail with only one median keel; movable finger of pedipalps with several more or less irregular granules, not in rows; with two lateral eyes—Subfamily Urodacinae
Underside of the tail segments with two median keels, only in the fifth segment with one median keel or without a distinct keel; movable finger of pedipalps with one or two indistinct longitudinal rows of granules; with three lateral eyes—Subfamily Scorpioninae

4 { Underside of all segments of the tail with only one median keel; underside of the third tarsus with two rows of about six setae each; telson of the male long, with two short processus on each side of the stinger —Subfamily Hemiscorpioninae
Underside of the segments 1 to 4 of the tail with two median keels; with only one median on the fifth article or the keel indistinct; telson in both sexes normal—Subfamily Ischnurinae

Subfamily Diplocentrinae with the following 3 genera: *Nebo* Simon 1878; *Oeclus* Simon 1880 (*O. purvesi* (Becker) 1880), and *Diplocentrus* Peters 1861 with 6 species: *D. antillanus* Pocock 1893, *D. grundlachi* Karsch 1880, *D.*

hasethi Krpln. 1896, *D. keyserlingi* Karsch 1880, *D. scaber* Poc. 1893, and *D. whitei* (Gervais) 1844.

Subfamily Urodacinae with only 1 genus and about 10 species: *U. abruptus* Poc. 1884, *U. armatus* Poc. 1888, *U. darwini* (Poc.) 1891, *U. excellens* Poc. 1888, *U. granifrons* Poc. 1898, *U. hoplurus* Poc. 1898, *U. novaehollandiae* Peters 1861, *U. planimanus* Poc. 1893, and *U. woodwardi* Poc. 1893.

Subfamily Scorpioninae with 4 genera and about 90 species. The genera and the most important species are: *Heterometrus* Hempr. & Ehrenb. 1828 with about 14 species: *H. longimanus* (Herbst) 1800—up to 15 cm long; *H. liphysa* (Thorell) 1886—14 cm long; *H. swammerdami* Simon 1836—up to 18 cm long; *H. fulvipes* (C. L. Koch) 1838—10 cm; *H. latimanus* (Poc.) 1894; *H. indus* (Geer) 1778—up to 15 cm long; *H. caesar* (C. L. Koch) 1842, *H. phipsoni* (Pocock) 1893—up to 14 cm long; *H. bengalensis* (C. L. Koch) 1842—13 cm; *H. liurus* (Poc.) 1897, *H. xanthopus* (Poc.) 1897, *H. cyaneus* (C. L. Koch) 1836—up to 13 cm; *H. scaber* (Thorell) 1872;

Pandinus Thorell 1877 with the following species: *P. meidensis* Karsch 1879, *P. exitialis* (Poc.) 1888—up to 14 cm long; *P. arabicus* (Krpln.) 1894, *P. pallidus* (Krpln.) 1894, *P. phillipsi* (Poc.) 1896, *P. colei* (Poc.) 1896, *P. bellicosus* (L. Koch) 1875, *P. cavimanus* (Poc.) 1888, *P. viatoris* (Poc.) 1890, *P. imperator* (C. L. Koch) 1842—one of the biggest scorpions of the world up to 180 mm in length, *P. dictator* (Poc.) 1898; *Scorpio* L. 1758: *S. maurus* L. 1758, *S. testaceus* (C. L. Koch) 1839, *S. boehmi* (Krpln.) 1896;

Opisthophthalmus C. L. Koch 1838, with about 35 species: *O. opinatus, nitidiceps, wahlbergi, schlechteri, ater, granicauda, carinatus, laticauda, pallipes, intermedius, fossor, capensis, fuscipes, granifrons, pictus, macer, austerus, calvus, gigas* Purcell 1898—with a length of about 140 mm, *karrooensis, flavescens, glabrifrons, pugnax, breviceps,* etc.

The subfamily Hemiscorpioninae includes only one genus with the species *Hemiscorpion lepturus* Peters 1861.

The subfamily *Ischnurinae* includes the genera:

1. *Hadogenes* Krpln. 1894 with the species: *H. trichiurus* (Gerv.) 1843—of about 130 to 180 mm long; *H. troglodytes* (Peters) 1861—male 180 mm; *H. paucidens* (Simon) 1887, *H. opisthacanthoides* Krpln. 1896.
2. *Opisthacanthus* Peters 1861, with the following species: *O. elatus* (Gerv.) 1805, *O. lecomtei* (Lucas) 1858, *O. africanus* Simon 1876, *O. madagascariensis* Krpln. 1894, *O. punctulatus* Poc. 1896, *O. asper* (Peters) 1861, *O. validus, O. rugulosus* Poc. 1896, and *O. cayaporum* Vellard 1932.

Family Buthidae

There are 4 subfamilies, in accordance with geographical distribution and with different morphological characters, as follows:

1. Movable finger of pedipalps with four to seven rows of granules, not on top of each other and disposed in a nearly longitudinal series; with only one and from one to two internal and external granules along the longitudinal rows; with a spine under the stinger—Subfamily Isometrinae
   Movable finger of pedipalps with more than seven rows of granules in oblique position 2
2. The oblique rows of granules not overlapping each other or nearly so; at the sides of each row there are only one or two isolated granules —subfamily Buthinae
   Oblique rows of teeth of the movable finger of pedipalps largely overlapping with isolated lateral tooth or they do not overlap very much, but in this case accessory rows of teeth exist at the sides of the central rows; without tarsal spur on the two last pairs of legs............ 3
3. Oblique rows of teeth of the movable finger of pedipalps with secondary rows of teeth at the sides—subfamily Rhopalurinae
   Oblique rows of teeth of the movable finger largely overlapping, with only isolated teeth at the sides, not forming secondary rows—subfamily Tityinae

The name Isometrinae seems to be more correct than Ananterinae, used by some authors, because the species *Isometrus maculatus* is much older than *Ananteris balzani.*

As the genus *Rhopalurus* is valid, extremely different from *Centruroides*, and much older than the last, it seems to be more logical to use the name Rhopalurinae in substitution for *Centruroidinae* or *Centrurinae* of some authors.

The subfamily Isometrinae includes the following 3 genera: *Isometrus* Hempr. & Ehrenb. 1828 with 10 species, *Ananteris* Thorell 1891 with 3 species, and *Ananteroides* Borelli 1911, with only 1 species; all of these of minor importance.

To the subfamily Buthinae belong the following genera:

1. *Cicileus* Vachon 1948 with the species *C. exilis* (Pallary) 1928;
2. *Buthiscus* Birula 1905 with the species *B. bicalcaratus* Birula 1905;
3. *Lissothus* Vachon 1948 with the species *L. bernardi* Vachon 1948;
4. *Butheoloides* Hirst 1925 with *B. maroccanus* Hirst 1925 and *B. milloti* Vachon 1948;
5. *Anoplobuthus* Caporiacco 1932 with *A. parvus* C. 1932;
6. *Androctonus* H. & E. 1829 with *A. aeneas* C. L. Koch 1839, *A. crassicauda* (Oliver) 1807, *A. mauretanicus* (Poc.) 1902, *A. sergenti* Vachon 1948, *A. hoggarensis* (Pallary) 1929, *A. australis* (Linnaeus) 1758, *A. amoreuxi* (Aud. & Sav) 1812;

7. *Buthacus* Birula 1908, with *B. foleyi* Vachon 1948, *B. villiersi* Vachon 1949, *B. arenicola* (Simon) 1885, *B. leptochelys* (H. & E.) 1829;
8. *Leiurus* (H. & E.) 1829, with *L. quinquestriatus* (H. & E.) 1829;
9. *Gompsobuthus* Vachon 1949, with *G. acutecarinatus* (Simon) 1882 and *w. werneri* (Birula) 1908;
10. *Orthochirus* Karsch 1891, with 5 species: *O. innesi* Simon, 1910;
11. *Buthotus* Vachon 1949, with *B. franzwerneri* (Birula) 1914, *B. judaicus* Simon 1872 and other species;
12. *Buthus* Leach 1815 with more than 20 species: *B. atlantis* Poc. 1889, *B. maroccanus* Birula 1903, *B. barbouri* Werner 1932, *B. occitanus* (Amoreux) 1789, etc.
13. *Microbuthus* Krpln. 1898, with *M. pusillus* Krpln. 1898, *M. litoralis* Birula 1905, and *M. fagei* Vachon 1949;
14. *Mesobuthus* Vachon 1950 with *M. eupeus* (C. L. Koch) 1839;
15. *Odontobuthus* Vachon 1950, with *O. doriae* (Thorell) 1877;
16. *Parabuthus* Poc. 1890, with from about 30 species, often very venomous, like *P. liosoma* (H. & E.) 1828, *P. granulatus* (H. & E.) 1828, *P. fulvipes* (Simon) 1887, *P. capensis* (H. & E.) 1828, *P. villosus* (Peters) 1863, *P. brevimanus* (Thor.) 1877, etc.
17. *Odonturus* Karsch 1879, with *O. dentatus* Karsch 1879 and *O. baroni* (Poc.) 1890;
18. *Butheolus* Simon 1883, with *B. melanurus* (Kessler) 1876, *B. pallidus* Poc. 1897, *B. bicolor* Poc. 1897, *B. thalassinus* Simon 1883, *B. ferrugineus* Krpln. 1898, etc.
19. *Neobuthus* Hirst 1911: *N. berberensis* Hirst 1911;
20. *Nanobuthus* Poc. 1895, with *N. andersoni* Poc. 1895;
21. *Charmus* Karsch 1879, with *Ch. laneus* Karsch 1895 and *Ch. indicus* Hirst 1911;
22. *Stenochirus* Karsch 1891, with *St. sarasinorum* Karsch 1891 and *St. politus* Poc. 1895;
23. *Isometroides* Keyserling 1884, with *I. vescus* (Karsch) 1880 and *I. angusticaudus* Keys. 1884;
24. *Lychas* Werner 1934, with from about 35 more or less good species;
25. *Pseudolychas* Krpln. 1912, with 2 species;
26. *Uroplectes* Peters 1861, with about 25 species;
27. *Karasbergia* Hewitt 1914, with *K. methueni* Hew. 1914;
28. *Psammobuthus* Birula 1911, with *Ps. zarudnyi* Bir. 1911;
29. *Liobuthus* Birula 1914, with *L. kessleri* Bir. 1914;
30. *Hemibuthus* Poc. 1900, with *H. crassimanus* Poc. 1900;
31. *Babycurus* Karsch 1886, with 6 species;
32. *Apistobuthus* Finnegan 1932, with *A. pterygocerius* F. 1932;
33. *Anomalobuthus* Krpln. 1900, with *A. rickmersi* Krpln. 1900.

The subfamily Rhopalurinae (Centrurinae Krpln. 1999; Werner 1934; Isometrinae Birula 1917, Mello-Leitão 1945) includes the genera *Centruroides* (*Centrurus* Krpln.) and *Rhopalurus*. Both may be distinguished as follows:

1. Tail, chiefly in males, not posteriorly expanded, longer in male than in female; sternum of first abdominal somite scarcely sulcate, without a distinct groove; the median rows of teeth on the movable finger of pedipalps not overlapping at the apices or nearly so and flanked on each side by rows of more scattered teeth, which occupy the interspaces between the large teeth. —genus *Centruroides* Marx 1889; genotype—*C. exilicauda* (Wood) 1863.

2. Tail, chiefly in males, posteriorly expanded, much thicker but scarcely longer in males than in females; first abdominal sternum strongly sulcate, marked with a pair of grooves; frequently there is a strong median lateral keel on the second and a weaker one on the third caudal segment; dentition of the finger of pedipalps resembling that of Centruroides—genus *Rhopalurus* Thorell 1876.

Centruroides with about 30 species. The most important are: *C. thorelli* (Krpln.) 1891, *C. elegans* (Thor.) 1877, *C. vittatus* (Say) 1821, *C. ornatus* Poc. 1902, *C. subgranosus* (Krpln.) 1898, *C. bertholdi* (Thor.) 1877, *C. nitidus* (Thor.) 1877, *C. flavopictus* (Poc.) 1898, *C. ochraceus* (Poc.) 1898, *C. margaritatus* (Gervais) 1841, *C. gracilis* (Latreille) 1804, *C. rubricauda* (Poc.) 1898, *C. bicolor* (Poc.) 1898, *C. limbatus* (Poc.) 1898, *C. nigrimanus* (Poc.) 1898, *C. nigrescens* (Poc.) 1898, *C. fulvipes* (Poc.) 1898, *C. danieli* (Prado & Rios Patiño) 1939, *C. exsul* (Meise) 1933, *C. granosus* (Thor.) 1877, *C. insulanus* (Thor.) 1877, *C. testaceus* (Geer) 1778, *C. sculpturatus* Ewing, *Centruroides gertschi* Stahnke 1949, *C. pantheriensis* Stahnke, etc.

Rhopalurus includes the species, *R. junceus* (Herbst) 1800, *R. acromelas* Lutz & Mello 1922, *R. agamemnon* (Koch) 1859, *R. borelli* Poc. 1902, *R. debilis* (C. L. Koch) 1841, *R. iglesiasi* Werner 1927, *R. intermedius* (Penther) 1913, *R. laticauda* Thor. 1876, *R. pintoi* Mello-Leitão 1933, *R. rochai* Borelli 1910, *R. stenochirus* (Penther) 1913.

The subfamily Tityinae includes the 2 genera *Zabius* Thorell 1894, with 2 species, and the very important genus *Tityus* C. L. Koch 1836, the genotypic species being *T. bahiensis* (Perty) 1830. The most important are: *Tityus bahiensis*, *T. serrulatus* Lutz & Mello 1922, *T. trivittatus* Krpln. 1898, *T. antillanus* (Thor.) 1877, *T. insignis* (Poc.) 1889, *T. androcottoides* (Karsch) 1879, *T. trinitatis* Poc. 1897, *T. pictus* Poc. 1893, *T. melanostictus* Poc. 1893, *T. clathratus* C. L. Koch 1845, *T. atriventer* Poc. 1897, *T. pusillus* Poc. 1893, *T. silvestris* Poc. 1897, *T. paraguayensis* Krpln. 1895, *T. colombianus* (Thor.) 1876, *T. paraensis* Krpln. 1896, *T. costatus* (Karsch) 1879, *T. stigmurus* (Thor.) 1877, *T. bolivianus* Krpln. 1895, *T. ecuadorensis* Krpln. 1896, *T. championi* Poc. 1898, *T. metuendus* Poc. 1897, *T. pachyurus* Poc. 1897, *T. timendus* Poc. 1898, *T. pugilator* Poc. 1898, *T. forcipula* (Gerv.) 1844, *T.*

macrochirus Poc. 1897, *T. cambridgei* Poc. 1897, *T. magnimanus* Poc. 1897, *T. discrepans* (Karsch) 1879, *T. obtusus* (Karsch) 1879, and several other partially described species.

Family Vejovidae

With 6 subfamilies, as follows:

1. Underside of caudal segments 1 to 4 with only 1 median keel—Subfamily Syntropinae, with only 1 genus: *Syntropis* Krpln. 1900, with only 1 species, *S. macrura* Krpln. 1900;
   Underside of caudal segments one to four with two median keels.... 2
2. Third tarsus with a distinct empodium between the claws 3
   Third tarsus without empodium; on the median line a fine papillary row below, distally bifurcate—Subfamily Caraboctoninae, with 2 genera: *Caraboctonus* Pocock 1893, with the species, *C. keyserlingi* Poc. 1893, and *Hadruroides* Poc. 1893, with 5 species of minor importance: *Hadruroides lunatus* (L. Koch) 1867 is the most frequent;
3. Intermediate lamellae of pectines not broken up into subequal subspherical sclerites, or when broken up in this manner only about 6 in number; fulcra small, triangular or even absent 4
   Intermediate lamellae of the pectines broken up into numerous (eight or more) subequal, rounded sclerites; fulcra spherical—Subfamily Vejovinae
4. Movable finger of the pedipalps with few transverse rows of teeth, overlapping each other; movable finger of mandible with a robust tooth beneath—Subfamily Jurinae, with the genus, *Jurus* Thorell 1877, with species, *J. dufoureius* (Brulle) 1832;
   Movable finger of pedipalps with only one longitudinal row or two parallel rows of teeth; movable finger of mandible with zero to small teeth beneath 5
5. Movable finger of the pedipalps with two indistinct parallel rows of teeth; with two external lateral granules; movable finger of mandible with four to six small teeth beneath—Subfamily Scorpiopsinae, with 2 genera: *Parascorpiops* Banks 1928, with the species, *P. montanus* Banks 1928, and *Scorpiops* Peters 1861, with 10 to 12 species of minor importance;
   Movable finger of the pedipalps with only one longitudinal row of teeth, with only one lateral granule on each side; movable finger of the mandible with zero to five teeth below—Subfamily Uroctoninae.

Subfamily Vejovinae

With the following 2 genera:

1. *Vejovis* C. L. Koch 1836 (*Vaejovis*), with the following important species: *V. mexicanus* C. L. Koch 1836, *V. carolinus* C. L. Koch 1843, *V. pusillus* Poc. 1898, *V. granulatus* Poc. 1898, *V. cristimanus* Poc. 1898, *V. sub-*

cristatus Poc. 1898, *V. nitidulus* C. L. Koch 1843, *V. spinigerus* (H. C. Wood) 1863, *V. boreus* Webster, *V. variegatus* Poc. 1898, *V. punctatus* Karsch 1879, *V. intrepidus* Thor. 1877, *Vejovis flavus* Banks;

2. *Hadrurus* Thorell 1877: *H. hirsutus* (H. C. Wood) 1863, etc.

Subfamily Uroctoninae

With the following genera:

1. *Physoctonus* Mello-Leitão 1934: *Ph. physurus*;
2. *Uroctonoides* Chamberlin 1920: *U. fractus*;
3. *Paruroctonus* Werner 1934: *P. gracilior*;
4. *Uroctonus* Thorell 1876: *U. mordax*;
5. *Anuroctonus* Poc. 1893: *A. phaeodactylus* (Wood) 1863.

Family Chaerilidae

With 2 genera: *Chaerilus* Simon 1877, with 11 species, and *Calchas* Birula 1899, with *C. nordmanni.*

Family Chactidae

1. Caudal segments from 1 to 4 furnished with a single inferior median keel; two indistinct rows of small teeth on the movable finger of the pedipalps; third tarsus below with a series of long setae; the whole underside granulate—Subfamily Megacorminae, with the genera: *Megacormus* (Megacormus) Karsch 1881, with the species, *M. granosus* (Gerv.) 1844, and the genus, *Plesiochactas* Poc. 1900, with the species, *Pl. dugesii* and *Pl. dilutus* Karsch 1881;
   Caudal segments 1 to 4 either keelless or with a pair of parallel inferior keels; lower border of the movable finger of the mandible without teeth or with one very small tooth; underside of body smooth.......... 2

2. Old world—subfamily Euscorpioninae, with genera, *Euscorpius* Thor. 1876—*E. italicus* (Herbst) 1800, *E. flavicaudis* (Geer) 1778, *E. carpathicus* (Linnaeus) 1763, *E. germanus* (C. L. Koch) 1836—and genus, *Belisarius* Simon 1879, with *B. xambeui*;
   New world—subfamily Chactinae, with the following genera and important species:

1. *Chactas* Gerv. 1844, with about 12 species;
2. *Broteas* C. L. Koch 1838, with about 5 species;
3. *Broteochactas* Poc. 1893, with about 7 species;
4. *Teuthraustes* Simon 1878, with 13 species;
5. *Chactopsis* Krpln. 1912, *Ch. isignis*;
6. *Parabroteas* Penther 1913, *P. montezuma*;
7. *Acanthothraustes* Mello-Leitão 1945, *A. brasiliensis*

2. The Neotropic, Dangerous Scorpions

The neotropic dangerous scorpions belong to Buthidae, Tityinae and Rhopalurinae. The following species may be considered as the most dangerous to men: *Tityus serrulatus, Tityus bahiensis, Tityus trinitatis* and *Tityus trivittatus.*

Tityus serrulatus

Female: Color of upper side of trunk yellowish or blackish brown on the cephalothorax and the six first tergites, the last tergite pale yellow, like the caudal segments, legs and pedipalps and mandibles; underside of the fifth caudal segment blackish from the middle to the posterior border; telson and pincers reddish, fingers and stinger blackish. Measurements—from about 70 mm of total length, carapace 30 and tail 40 mm. Terga and carapace densely and irregularly granular, with a distinct median keel; tergite 7 with a posterior median keel and posterior lateral keels; sternites finely granular, 1 and 2 with the posterior border smooth and shining, sternite 3 very smooth with a posterior triangle; sternite 4 with two keels at the posterior half, sternite 5 with five longitudinal keels, the median longer; pectinal teeth 18 to over 24; dorsal median keels of caudal segments with small granules, on segments 2 to 4 the posterior granules are bigger and very distinct, chiefly on the fourth segment.

Tityus bahiensis

With the same morphological characters as *T. serrulatus*, except the general color, which is uniformly brown (not yellow), legs reddish, tail reddish-brown, and the dorsal median keels of the tail, which bear only small granules, without bigger granules. Measurements of about 70 mm, the adult males 75 mm; pectinal teeth from 19 to 23.

Tityus trinitatis

Carapace and cauda yellowish-brown, often deeper brown, pedipalps and legs reddish-yellow, in juveniles often with blackish spots, finger deeper brown, sterna yellowish, finely punctured and densely granular, sternum 4 with median keel and two lateral keels, sternum 5 with four granulate keels; second segment of the tail with two ventral median keels completely separate, in the third segment united distad, in the fourth united at their posterior half. Cauda in the males seven times longer than the cephalothorax, cephalothorax much shorter than the third caudal segment; caudal segments between the keels granulate.

Tityus trivittatus

Longitudinal keels of the tibia of the pedipalps with pearllike small granules. Color: upper side of trunk yellow or yellowish-red, with three longitudinal black bands passing backwards, these black bands often more or less indistinct; fifth caudal segment yellowish-red; accessory keels on the second caudal segment only distad; median dorsal keel of caudal segments without bigger granules at the ends; pectinal teeth 19 to 23; measurements—from 50 to 60 mm.

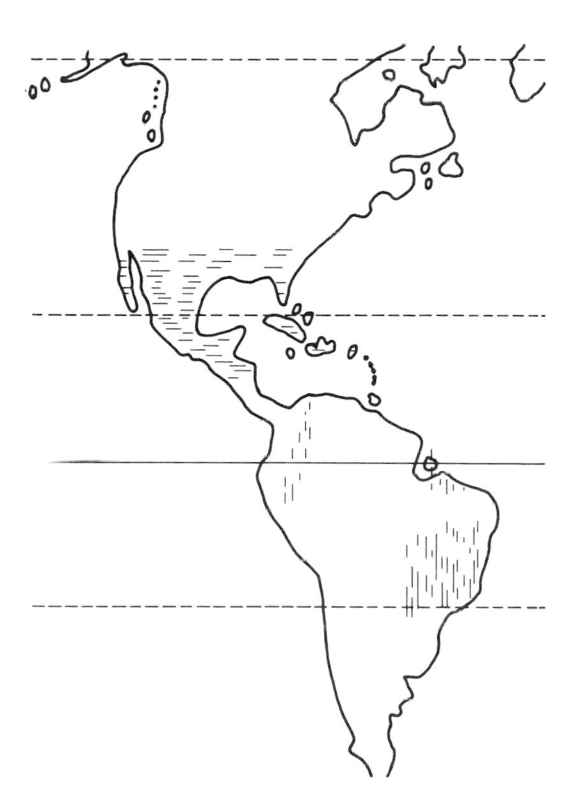

FIG. 6. Geographical distribution of dangerous scorpions in America: southern states of U.S.A., Mexico, and Central America—*Centruroides*; South America and Trinidad—*Tityus*.

C. Geographical Distribution

1. *Distribution of South American Dangerous Scorpions*

The South American dangerous scorpions belong to the genus *Tityus*, distributed from Mexico, the West Indies to South America, Argentina, Paraguay, Bolivia, Brazil, Peru, Ecuador, Colombia, the Guianas, Venezuela, Panama, Antilles: Puerto Rico, Trinidad, Sta. Lucia, St. Vincent, Grenada, etc.

Tityus serrulatus is found only in Brazil, in the states of Sergipe, Bahia, Minas Gerais, Epirito Santo, Rio de Janeiro, Goiás and São Paulo (see Fig. 6).

Tityus bahiensis—Brazil: states of Bahia, Espirito Santo, Minas Gerais,

Goiás, Rio de Janeiro, Guanabara, São Paulo, Santa Catarina, Mato Grosso; Argentina: Misiones, província de Buenos Aires.

Tityus trinitatis: Trinidad and Venezuela.

Tityus trivittatus: Argentina: Buenos Aires, Corrientes, Santa Fé; Paraguay, Uruguay; Brazil: Rio Grande do Sul, Mato Grosso, Goiás, Paraná, Rio de Janeiro, Minas Gerais (See Fig. 6).

2. *Distribution of Centruroides with Dangerous Representatives*

Panama, Guatemala, Curacao, Barbados, Mexico, Jamaica, Cuba, Haiti, Puerto Rico, St. Thomas, Nicaragua, Yucatan, U.S.A.: Arizona and parts of California (see Fig. 6).

3. *Distribution of Old World Dangerous Scorpions*

Distribution of Heterometrus—India.

Distribution of Pandinus—Africa and Arabia.

Distribution of Opistophthalmus—South Africa (see Fig. 7).

Distribution of Scorpio—North Africa (Fig. 7.).

Distribution of Hadogenes—South Africa and Madagascar (Fig. 7).

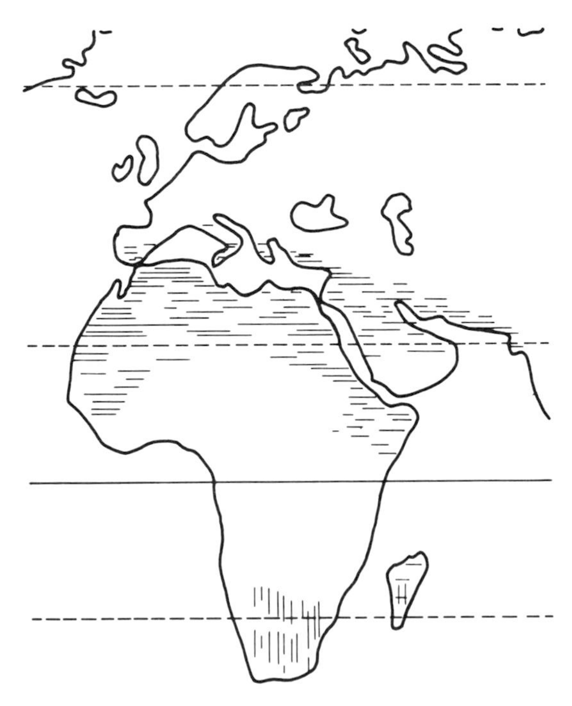

FIG. 7. Geographical distribution of dangerous scorpions of the Old World. South Africa—*Opisthophthalmus, Hadogenes* and *Parabuthus;* North Africa, Mediterranean region to Persia—*Leiurus, Buthacus, Buthotus, Buthus, Androctonus,* etc.

Distribution of Androctonus—From India and Persia to Atlantic coast of Morocco and Eastern Mediterranean region to Senegal and upper Egypt.

Distribution of Buthacus—from the Atlantic coasts to Palestine, Senegal, and Tunis.

Distribution of Leiurus—Syria, Palestine, Egypt, Yemen.

Distribution of Buthotus—Palestine, Syria, Algeria, Morocco.

Distribution of Buthus—North Africa, Egypt, Ethiopia, Somalia, parts of Palestine, Spain, France.

Distribution of Parabuthus—South Africa to Sudan.

4. *Distribution of Vejovis*—From British Columbia to Mexico.
5. *Distribution of Hadrurus*—Mexico and Southern States of U.S.A.
6. *Distribution of Euscorpius*—European Mediterranean region, Asia Minor, Caucasus, North Africa.

III. BIOLOGY OF SCORPIONS

Scorpions have been in existence since the Silurian Era, and the Carboniferous Era. Representatives of *Tityus* have been found in the Oligocene. The worldwide distribution of the representatives of Buthidae, wherever scorpions actually are found, prove that this family may be the most archaic of living scorpions.

A. The Life of Scorpions

1. Pairing

The mating habits of scorpions have been the subject of ardent controversies. According to Maccary (1810) the male of *Buthus occitanus* attacks the female from the front, turns her over on her back and remains for several minutes on her. Brongniart and Goubert, in 1891, observed a pair of scorpions clinging body to body with their interdigitated pectines. Fabre (1907) described the *"l'arbre droit,"* when the scorpions move forwards and backwards with their tails raised vertically, and the *"proménàde à deux"* as a prenuptial walk. Pawlowsky (1924) found in the vaginal opening of the fertilized females a brown, whitish shrunken mass, called by him spermatocleutrum. He suggested that this matter may be a part of a spermatophore excreted by the paraxial organ of the males, and concludes that the mating of scorpions may be similar to that observed in pseudoscorpions by Kew, in 1912. Piza, in several studies on the mating habits of *Tityus bahiensis* (1939, 1940, 1943), called the male organs hemipenis, fused to only one penis at the moment of mating and possibly recollected again by the male after pairing. Vachon (1952) noted also in African scorpions that the proximal ends of the so-called paraxial organs really meet together and suggested that the extrusion of both "hemiorgans" may result in the formation of a true penis. Southcott (1955) described the "sexual thrust" in *Urodacus* and reported that the male turns the female over on her back and mounts her body against body. Angermann, in the same year, described the mating and the emission of the spermatophore of *Euscorpio italicus*, and Bücherl, in 1955, of *Tityus bahiensis*, and in 1955, 1956, of *T. bahiensis* and *T. trivittatus*, as well as the full description of the male sexual apparatus and the formation of both hemispermatophores, the prenuptial play with two stages, the mating, the expulsion of the male spermatophore, which is fixed to the soil, just under the female, and the "promènade" of both over the spermatophore to introduce the spermatozoides into the female body. Zolessi (1956) observed the sexual behavior of *Bothriurus bonariensis*: "After a preliminary courtship carried out in the usual way, with the male holding the female by the chelae of the pedipalps, the male produces a spermatophore, which is glued to the soil by its pointed anterior end as soon as it is extruded; then the female is pulled to it by the male, and two small caudal projections of the spermatophore enter the female genital opening. The spermatophore is afterwards discarded, and the female frequently kills and eats her mate." The fertilization of the female scorpion occurs without actual copulation, the spermatozoa being transferred to the female by means of a chitinous spermatophore, fixed to the soil by the male. Shulow (1958) reported the prenuptial behavior, the structure of the spermatophore, and its expulsion in *Leiurus quinquestriatus* and *Buthotus judaicus*. In 1958 he extended the same observations to *Nebo hierochonticus*, with suggestions on the formation of the spermatophore and the spermatocleutrum. Alexander, in 1956 and 1957, described the same occurrence in *Opisthophthalmus latimanus* and Berg, a few years earlier, in *Centruroides*.

The sexual behavior in scorpions is as follows: The male scorpion crawls toward a virgin female of the same species and seizes the nearest part of her body with his pedipalps. Often several males seize the same female, repulsing each other with their stingers; finally he grasps the pedipalps of the female with the fingers; both partners go forward and backward in a space from about 20 to 40 cm, to find a proper place for mating. This walk I called the first stage of *promènade à deux*; it may continue for a few minutes or half an hour, with several moments of rest. When this proper place is found, the partners continue the walk *à deux*, forward and backward with the pectines in a vertical position: they clean the soil completely, so that even small granules of sand are "brushed" away. This is the second phase of the promenade. Then the expulsion of the male spermatophore sets in, being ejected in a few seconds towards the female; the larger front part of the spermatophore falls to the ground and is fixed to the soil, while the proximal end of this organ shows an elastic flagellum remaining attached to the genital portions of the male. The third stage of promenade sets in, both going cautiously forward and backward, a few centimeters only just over the spermatophore. The flagellum serves as a control of the posture of the spermatophore: the female, directed by the pedipalps of the male, is pulled over the more or less erected spermatophore so that her genital opening is situated exactly over the upper end of the spermatophore, where the openings for the spermatozoa are situated. The upper part of the spermatophore comes in direct contact with her genital aperture; the genital valves apparently exert an action of expression, or suction, or both. This last "spermatozoa-transmission-walk," observed by us over a period of several years and in several hundred cases in *Tityus trivittatus* and *T. bahiensis*, is executed by the pair for 5 to15 seconds only. Then the female, who has been passive from the beginning to this stage, becomes very active, freeing her pedipalps from the males' fingers and curving back her stinger. Immediately the male retreats, the final flagellum of the spermatophore extends from 10 to 20 cm and breaks, the male going away, while the female turns to her burrow. It is not true, of course, that the female kills the male. Within a week the same male can eject another and a week later a third spermatophore with sperm transmission to other females. After incision of the so-called paraxial organs, one on each side of the testes, one will find from 12 to 20 hemispermatophores, superposed in a vertical layer and with the larger ends directed toward the genital opening.

The same mating behavior, with insignificant variations, seems to be the rule for all scorpion families with separate sexes. But exceptions may exist and the sexual biology of scorpions may often be more complicated. In adult scorpions the sexes may be distinguished by secondary sexual characters. The males and females of *Tityus bahiensis* and *T. trivittatus* and several other species of the same genus may be easily distinguished. But in the most common and most poisonous *Tityus serrulatus*, of which about 60,000 have been in our laboratories, we never found one male. About 8,000 scorpion bodies have been dissected, and in all we found ovaries only. In 1962, Matthiesen described parthenogenesis in this species, observed in three individuals born in his laboratory. They became adults in 2 years and have had young.

2. Motherhood

The female genital system consists of an ovario-uterus, made by three longitudinal ovarial tubes, connected by three or four transverse tubes. Along these tubes may be seen the pearl-formed oocytes, found commonly on one side only. A certain ovarian asymmetry is common in scorpions. Each of the two lateral tubes leads into a relatively short oviduct. Each oviduct dilates in front, at the middle line, to form the seminal receptacle, and both receptacles lead into a vaginal part, which opens at the genital aperture. After mating, the spermatozoa, leaving the receptacles at the volition of the female, migrate along the oviducts and fertilize the oocytes. Each year from about 20 to 35 oocytes may be fertilized. It is possible also that two fertilizations and two births occur each year; 200 or even 350 oocytes may be commonly found in *Tityus bahiensis* and in *T. trivittatus*, studied by us in the years from 1953 to the present. This fact is important and leads to the conclusion that scorpions after maturity may live more than 5 to 8 years. The fertilized eggs develop in place, fixed inside special diverticula of whitish color and spherical form, the corresponding ovarian portion now being called the ovario-uterus. After successive cell divisions, a small embryo about 1 mm of size results, the follicle developing to a much larger, nearly spherical diverticulum. Each diverticulum is attached to the uterine tube by a single stalk. With the development of the embryo the diverticulum becomes of much larger size and forms the so-called incubation chamber, from which a whitish cylindrical cord, named the appendix issues distally. The appendix ends in a spherical knob. The internal passage, between the chamber and the stalk, is blocked by the so-called valvula diverticuli. The larger embryo remains in the chamber, curved and directed distally. Into its mouth a short nourishment-cord leads from the appendix. With the progress of pregnancy the chamber enlarges more and more, its external wall becoming thinner and more transparent, so that the whole yellowish embryo may be seen through the membrane after an incision in the body of the female. In a later stage the lumen of the stalk also will be occupied by the embryo, so that stalk and chamber become indistinct, only the knob being observable. In the final stage of the intrauterine development the embryo moves itself, becomes detached from the appendix, and moves into the uterine duct. The size of the incubation chamber is then greatly reduced, the distal knob of the appendix becomes brownish, hardened, and reduced in size to about 0.5 mm. It is discarded and often remains in the vaginal portion of the mother, as we have found several times in *Tityus serrulatus*. The embryos migrate into the ovarian-uterine tubes toward the vasa deferentia, vaginal portions, and force a small passage through the genital valves. Very often, after dissection of dead adult females of *T. bahiensis*, *T. serrulatus* and *T. trivittatus*, we have observed dead gigantic embryos, one in each mother, more than $1\frac{1}{2}$ cm in length and so wide that its birth must be considered impossible. Parturition lasts from several hours to one or more days, interrupted by large intervals of rest. The embryo emerges, more or less free of the embryonic envelope, crawls forward in the direction of the mouth parts of the mother, passes over the mother's mouth, where it is reversed (and often eaten) and migrates slowly onto the mother's back, where it stays in its proper "prefixed" place on one side of the first, second, third, tergite. When the birth is finished, often 20 or more young are on the mother's back, one by one, each one in its place. During all this time, from mating to parturition, the females are found in special burrows, undergroundly, with complete exclusion of daylight. One can consider scorpions as ovoviviparous or viviparous, as a strict separation is not possible.

3. Development

The young may be called larvae, because their external and internal morphological constitution is not complete; cephalothorax and preabdomen are large and thick with supplementary nourishment inside, while their tail is short, the coxae of the legs are swollen, the legs short and without spurs and claws, the stinger without venom openings, etc. Generally 1 week after birth the young change their entire skin, the ecdysis being complete and the mother often feeding on the exuviae. Then the young begin their independent life, changing skin four to five times in the first year and only from one to two times in the second year. The partial life story is known only for *Tityus bahiensis*, which develops to an adult in 3 years. We have observed the molting process several times: Before the rupture of the old cuticle, the scorpion stays almost motionless in a dark, protected place; its posterior half of the body, chiefly its tail, is raised, the tergites of the preabdomen arched; the chelicerae and pedipalps are moved actively forward and backward; finally the old cuticle ruptures at the front and side margins of the carapace; at first the chelicerae are withdrawn out of the exuvia, then the pedipalps and the legs, one by one; at last the cephalothorax, the preabdomen, and the tail became free after vigorous movements. The process is not rapid, often interrupted by intervals of rest, and takes place in the dark or during the night. Only after 1 to 3 days does the new cuticle acquire its final color. From eight to ten moltings may occur between the larval and the adult stages, lasting from 2 to $3\frac{1}{2}$ years. Matthiesen studied the life and development of *T. serrulatus*, R. San Martin of *Bothriurus*, and Shulow of several species native to Israel.

4. Means by Which Scorpions Obtain Their Prey

Scorpions live upon insects, arthropods, and even other scorpions. Small spiders, chiefly of the family of Lycosidae, cockroaches, and crickets, are their common food. They refuse older dead material, but accept fresh pieces of "tarantulas." They can go without food for several months, but they must always have fresh water. Scorpions are nocturnal hunters. They do not see very well and may come into contact with the body of the prey occasionally. They rapidly grasp it firmly with their pincers and keep it distant for a short time; their chelicerae immediately begin a forward and backward movement and then the prey is securely held by the chelicerae alone and the scorpion runs in the direction of a dark place. Usually they do not bring the sting into action unless the captive resists; in this case, they hold the prey with the fingers of the pedipalps far away from the body, curve their tail, and sting one or several times in the softer parts of the abdomen; the prey immediately dies and is eaten. Hard, chitinous pieces are discarded. A scorpion may drag large prey, such as a whole walking leg of a mygalomorph spider, up to 8 cm. When several scorpions are in the same cage, each scorpion carries its prey to a quiet corner of the cage or as far away from its cage mates as possible; several times two or three scorpions fight or eat the same prey. When several hundred scorpions are in the same cage, as is the case in Butantan, where about 15,000 live scorpions of *Tityus serrulatus* are kept every year, cannibalism may occur. The victims are usually much smaller in size or handicapped in some way, as when they are molting. We often observed, chiefly when food was very scarce, that two females with young on their backs would stand side by side, each one feeding on the young from the back of the other.

The scorpions drink large amounts of water daily, chiefly when they cannot obtain sufficient liquid moisture from their prey. Often nearly a thousand scorpions of the species *Tityus serrulatus*, which are captured around Nova Era in the state of Minas Gerais, arrive at Butantan, Brazil, after being caged for 2 weeks. We always pour water into the cages when they arrive. The scorpions begin to drink immediately; they lift the posterior part of body with the back legs and bend the front legs so that their mouths come in direct contact with water. They may drink for several minutes.

5. The Domiciles of Scorpions

One may distinguish between forest, field and desert, or semidesert scorpions. The first ones have a dark color from brown to dark brown; the field scorpions may be darker also, or yellowish-brown with few darker or clearer spots, and the desert or semidesert scorpions commonly show a yellowish color. The forest scorpions, like *Tityus cambridgei* of the Amazon, live under bark, trees, roots. Dendricole scorpions of smaller size also exist in South America. The field scorpions prefer damp places, in cane fields (Trinidad, Cuba, etc.) and banana plantations. They are often active burrowers, their burrowing being affected by the quality of the soil. They may also be found along certain rivers, chiefly under stones near the margins and very often under the hills of termites, outside, from 30 to 50 cm deep. The separation of habitats of forest and field scorpions is not rigorous. Even the forest scorpions can live perfectly well under ant or termite hills and under stones. The desert or semidesert scorpions show very widely varying biotopes and adaptation habits. Commonly they are perfect burrowers, with domiciles from 5 to 10 cm wide and often from 20 to 50 cm in length and depth. To avoid the abrupt variations of the desert climate, with its cool nights and very warm and dry days, they construct in the domiciles two chambers, an upper, "air-chamber" and a lower "domicile" chamber, their burrow being covered above by a large stone. Varying their position from the underside of the stone to the air chamber or the domicile chamber in the depth, they are well insulated against desert conditions.

It is curious that only the desert or semidesert scorpions all over the world, in Mexico, Brazil, the Sahara regions, in Palestine, Yemen, South Africa, Syria, Libya and Arabia are the most dangerous to man, and are representatives of the same family, the Buthidae. It may be curious also that several representatives of this group accept what I have called "domiciliar habits." They are found in hill regions and valleys, under stones, rocks, logs, loose bark of trees and around human habitations in gardens, old lumber, boxes, bricks, rays. They also live in old buildings, garages, cellars, washhouses, old chimneys no longer in use, and chiefly under houses. Often whole cities may be invaded, like Ouro Preto and Belo Horizonte, Nova Era, Passagem and other cities in the State of Minas Gerais or Ribeirão Preto, in the State of São Paulo, as well as Aparecida in the valley of the Paraíba river; scorpions may be found in cement blocks, boards, boxes, sacks, bricks, sand boxes, eucalyptus trees, porches, chicken houses, floors, behind consoles, kitchens, under wash cloths, couches, beds, in old mattresses, in cellars, in shoes, in clothing and even on piles of newspapers, stored in darker places. In these cities even the modern, new dwellings are not immune against scorpion invasion, especially when old dwellings are near. It is a fact that scorpion districts exist. In the years 1949 to 1963 about 138,500 scorpions were captured and sent to Butantan from the cities of Ribeirão Preto (about 2,000), Belo Horizonte (about 10,000), Nova Era (about 126,000) and Aparecida (about 500). All of the specimens were *Tityus serrulatus*. During the same period of time 38,744 *Tityus bahiensis* were captured, chiefly in the city of Ouro Preto, and sent to Butantan. During the eradication program with contact insecticides, such as DDT and BHC, several pails full of dead scorpions have been swept from the streets of Ribeirão Preto and Belo Horizonte. These invasions can be attributed to the climate, soil, and long-settled conditions which favored enormously the propagation and multiplication of scorpions of both species.

B. The Venom Apparatus of Scorpions

The venom apparatus of all scorpions consists of two venom glands, each one with its proper venom canal, which leads to the pit of the stinger and ends in an elliptical small opening at the outer side of the stinger. Macroscopic descriptions of the glands have been given by Meckel and Dufour. Blanchard reported that the canals could be fused into one canal only with two openings. This was contradicted by J. Miller and L. Dufour. Histological sections of the glands have been made by Joyeux-Laffuie in *B. occitanus*; they show that the venom-excreting epithelium is of prismatic aspect, a high cylindrical epithelium. Pawlowsky, in 1927, made comparative histological studies in several scorpions of different families: in *Euscorpius carpathicus* the two glands are elliptical, completely separated, each one with its proper canal. The basal portion of this canal does not show a proper muscularis and the lumen of the canal shows a chitinous sheath, followed by a simple layer of epithelial cells without excreting character. The venom glands are situated, on their upper side, directly under the cuticular epidermis, but their inferior walls are formed by a powerful muscularis with cross-striated fibriles. The venom glands are saclike. To the basement membrane is attached the glandular epithelium of high columnar character with very small and rounded nuclei, situated near the basement membrane. The venom elaborated in the cells is granular; small and large droplets also are found. Near the basement membrane and attached to it are small epithelial cells.

In the representatives of the Old World Buthinae, the venom glands show the same structural character as those of *Euscorpius*, with only insignificant alterations.

We have dissected several glands of *Tityus serrulatus* and *Tityus bahiensis* and stained the sections with hematoxylin-eosin; longitudinal, tangential, and cross sections have been made. In the venom glands of *Tityus* three elements can be distinguished: the external muscularis, the intermediate basement membrane, and the venom-excreting epithelial cell layer.

The external muscularis is double, forming a double layer, separated by a membrane of connective tissue. Both are striated. The muscularis covers the whole venom gland from the base of the canal to the apex. The external layer is circular, three or four muscle bundles going around the whole gland, while the internal layer is longitudinal, showing about 50 muscle bundles. The nuclei are more or less central in position, the muscle fibriles making a perfect cross striation.

The basement membrane forms a very thin but continuous layer inside the muscularis.

The venom-excreting cells form a simple epithelium. Two kinds of epithelial cells are present: the first forms a thin layer of small, subcuboidal cells, attached to the basement membrane, with elliptic nuclei in a rest stage. These cells will substitute the venom elaborating cells. The second forms the actually functional epithelial cells, also attached with their basal part at the basement membrane, between the substituting cells. These epithelial cells are of a high columnar type, cell walls, nucleus, cytoplasm, and venom granules being very visible. The venom elaborated forms at first small fine granules; with the agglutination of a granular portion, thicker droplets are formed at the middle of the cell body, the cell itself being enlarged. In the final stage of venom production, thick droplets break off of the apical cell wall, the droplets being extruded into the central lumen.

The venom-producing cells in scorpions are of the apocrine type, the same cell producing venom as long as its nucleus is active and cytoplasm exists; then the cell will degenerate, being replaced by a basal cuboidal cell.

The two, long, cylindrical venom canals of both glands are slightly curved, show a very small central lumen, a chitinous internal layer, simple low, cuboidal cells, and a thin basement membrane, without a proper muscularis.

When a scorpion stings, he ejects the venom at his volition. The venom ejection is violent and rapid, conditioned to the voluntary contraction of both muscle layers of the gland, and is favored especially by the rapid movements of the curved tail.

IV. EXTRACTION AND TOXICITY OF SCORPION VENOMS

A. Extraction of Venoms

The simplest method for obtaining scorpion venoms is to kill the scorpions, cut up the telsa, crush it, make a maceration in physiological saline solution or in distilled water and purify the solution.

The second method to obtain directly the crude venom consists in a proper manual technique, described by us in 1953.

Pure venoms can be obtained also with electric stimulus, first described by us in 1953, by provoking the contraction of the venom gland muscularis with a high frequency electric current—about a 5 V current obtained from a transformer or a battery, and redescribed by Lissitzky and Mirander, in 1956 and many others. From 1951 until June of 1953 we have collected the first 309 mg of dry *Tityus serrulatus* venom, obtained after 4,105 electric extractions of several hundred specimens and 697 mg of dry venom of *Tityus bahiensis* after 6,148 extractions from 1,112 individuals. Table I shows the number of scorpions and the quantities of dry venoms obtained with electrical stimulus.

The first drops of venom are commonly limpid and transparent, the following being more and more opalescent and viscous. The fresh, natural venom has a pH from neutral to alkaline, while immature venom is acid to litmus. Both have a foamy whitish aspect, with astringent properties and without any taste on the tongue. The vacuum-dried venom is of whitish-grayish color, hygroscopic, and may be stored perfectly in a vacuum dessicator, over calcium chloride, at room temperature and in the dark or in a small văcuum glass, without apparent change of activity after 3 or 5 months.

TABLE I

NUMBER OF SCORPIONS AND QUANTITIES OF DRY VENOM OBTAINED

Year	*T. serrulatus*[a] (no. of animals)	Dry venom (mg)	*T. bahiensis*[b] (no. of animals)	Dry venom (mg)
1953	3,216	275	11,988	1,000
1954	2,814	355	2,375	320
1955	1,231	720	4,052	1,170
1956	396	770	2,770	2,500
1957	5,713	2,230	6,569	3,690
1958	11,637	10,870	5,072	2,041
1959	6,060	5,490	1,075	450
1960	16,486	14,780	9,222	4,024
1961	32,862	25,000	2,450	2,200
1962	38,625	13,875	132	365
1963	22,245	13,440	285	89
Total	141,285	87,805	45,990	17,849

[a] Mean yield per individual was 0.62 mg.
[b] Mean yield per individual was 0.39 mg.

The first electric extractions give a higher quantity of venom. After 10 or 15 extractions, repeated with the same individuals every 3 or 4 weeks, the venom quantities decrease until they reach about 0.03 to 0.04 mg per specimen. Several scorpions have much more venom than 0.62, or 0.39 mg. About 2 mg per specimen may be the maximum of venom that a *Tityus* scorpion can inject in a victim.

B. Toxicity of Some Scorpions

According to the effects produced by the venom of scorpions upon people, two groups may be recognized: Those whose venom causes mainly a local action, rapid and transient and those whose venom produces not only local but primarily a systemic general action. The local reaction of some representatives of *Heterometrus, Pandinus, Scorpio, Rhopalurus, Opistacanthus, Orthochirus, Buthotus, Caraboctonus, Hadruroides, Vejovis, Megacormus, Teuthraustes, Acanthothraustes,* may consist of a sharp burning sensation around the place of the sting, which may last from a few minutes up to several hours. Others, like several species of *Bothriurus, Ananteris, Isometrus, Zabius, Jurus* and several species of *Vejovis,* may produce local burning for a few minutes only, while a third group, like species of *Diplocentrus, Vejovis, Tityus,* several representatives of Buthinae of the old and the new world, may cause not only local pains but also pronounced but temporary local swelling. In other scorpions, often of the same genus as those listed, venom may cause intensive local pain, highly discolored swelling and local burning sensation. Inflammation and pains may show an irradiating character. An individual stung on a finger may feel pain in the whole arm. However, under normal conditions, all these local symptoms are transient, without more serious danger.

Several species of *Opistophthalmus, Hadogenes* and *Parabuthus* in South Africa, of *Androctonus, Buthacus, Buthus* in North Africa, Palestine and Mediterranean subregions, *Centruroides* in some southern states of the U.S.A., Mexico and Central America, and *Tityus* in Brazil, Paraguay, Bolivia, have a more powerful venom which may cause in adults a very painful situation and often a pronounced local swelling of irradiating character, with more or less noticeable systemic effects of transient character. Even in these cases there may not exist serious danger of death, unless the victim is a very tiny child.

On the other hand, the stings of *Tityus serrulatus, T. trinitatis, T. bahiensis,* partly also *T. trivittatus* of South America, *Centruroides noxius, C. limpidus* and *C. suffusus* of Mexico, the Arizona scorpion, *C. sculpturatus,* the Texas scorpions, *C. vittatus* and *C. gertschi,* the South African scorpions, *Parabuthus granulatus, P. capensis,* and perhaps species of *Hadogenes* and *Opistophthalmus,* the Moroccan, Algerian, Libyan, Egyptian, Palestinian, Syrian, Arabian, and Persian scorpions, *Androctonus aeneas, A. amoreuxi, A. australis, A. crassicauda, A. maroccanus* and *Buthus occitanus, Buthotus judaicus, Leiurus quinquestriatus,* and *Buthacus arenicola*—produce a more or less intensive local pain, rarely with insignificant swelling. But the effects are principally systemic. The venoms may be considered as powerful neurotoxins with convulsant action. Death due to "scorpionism" all over the world is caused almost entirely by these species. Wilson (1904) recorded many deaths in Sudan; according to Sitt (1923), the mortality among young children stung by *L. quinquestriatus* is about 50%; Waterman (1957) states that in Trinidad the case mortality rate under 5 years is 25%; the same rate for children is found in the statistics of the Municipal Hospital in Ribeirão Preto, in 1952, with *T. serrulatus*; Magalhães (1935) quoted 874 case observations in Belo Horizonte with 100 deaths; the famous Durango scorpion (*Centruroides*) of Mexico was responsible for 1,608 deaths between 1890 and 1926 in the city of Durango with a total population of about 40,000 people; many deaths occur yearly in Egypt.

It seems to us to be correct to state that in Brazil the most venomous scorpion, *Tityus serrulatus*, may cause the death of 0.8 to 1.4% of adults, 3 to 5% of school children, and 15 to 20% of very young children. The affected cities are around Belo Horizonte, Nova Era, Aparecida do Norte, around Ribeirão Preto and other older cities in the States of Minas Gerais and Goiás and São Paulo, where this species is present. We determined, in 1953 and 1955 the LD_{50} of *T. serrulatus* venom per gram white mouse, as follows, including *T. bahiensis* venom for comparison:

T. serrulatus	intravenously	0.75 μg.
	intramuscularly	1.50 μg.
T. bahiensis	intravenously	4.25 μg.
	intramuscularly	9.35 μg.

Weissman and Shulow found a 50% LD of *Buthotus judaicus* per gram mouse, subcutaneously to be 8.5 μg.

The comparative potency of the venoms of various scorpion species, according to the LD_{50} reported by different authors—subcutaneously—per 15 gm of white mouse are more or less as follows (in milligrams):

South America	
Tityus serrulatus	0.022 mg
Tityus bahiensis	0.140 mg
Tityus trinitatis	0.030 mg
Mexico	
Centruroides limpidus	0.075 mg
Old World	
Leiurus quinquestriatus	0.005 mg
Buthacus arenicola	0.052 mg
Androctonus australis	0.091 mg
Buthus occitanus	0.115 mg
Buthotus judaicus	0.127 mg
South Africa	
Parabuthus sp.	0.5 to 2.0 mg
Opistophthalmus sp.	10.0 to 15.0 mg
Hadogenes sp.	30.0 to 40.0 mg

Symptoms in White Mice. In mice, injected with a solution of pure, dried scorpion venom, the following symptoms were commonly observed: immediately after injection the mice react by jumping and leaping in the cage, licking the place where the venom was inoculated. Within a few minutes begins hypersalivation, the animals now remain stationary; respiration becomes irregular; abundant sweat at the neck occurs; deep gasping, respiratory movements exist sporadically. In 20 to 30 minutes later the mice lie prone, motionless, with their hind legs paralyzed; respiration becomes more and more rapid, irregular, and asphyxiation processes set in; a powerful convulsive spasm, passing from the tail along the whole body leads to death. All mice dying from scorpion venoms show an extreme rigidity of the body as the result of the violent generalized muscular contraction. Large amounts of venoms, given intravenously, kill all mice immediately with asphyxia and tetanism. By injecting quantities of venoms a little lower than the LD_{50}, intravenously or intramuscularly, one may observe the progress of intoxication with abundant sweat, in all cases of *T. serrulatus* or *T. bahiensis* venoms, paralysis, respiratory failures, and so on; 2 hours after the intravenous venom, from about 12 to 15 hours after the subcutaneous injections, the mice begin to show recovery and 1 day later they are perfect again. Scorpions venoms act very rapidly and are destroyed and eliminated quickly. Postmortem examinations of mice killed reveals no macroscopic lesions.

For more elucidative and detailed information about dangerous scorpions in the Old World, see the chapter of L. Balozet; for biochemistry and pharmacology of the mexican dangerous scorpions of *Centruroides* several studies have been published by E. C. Del Pozo in 1948, and 1968, by E. C. Del Pozo and G. Anguiano in 1947, by E. C. Del Pozo, G. Anguiano and Gonzalez, in 1944. Biochemical and pharmacological aspects of the south american *Tityus serrulatus* venom have been investigated by C. R. Diniz, M. V. Gomez, A. Antonio, and A. P. Corrado, by J. M. Gonçalves, by Oswaldo Vital Brazil, and others.

C. Treatment of Scorpion Poisoning

The scorpions, dangerous for humans, belong—with a few exceptions—to the genera *Centruroides, Tityus, Androctonus, Buthacus, Leiurus, Buthotus, Buthus,* and *Parabuthus.* Their geographical distribution has been described above. They can be present in some places in large amounts and in the immediate neighborhood of populated settlements and may even appear in the dwellings as undesired "domestic animals." The estimate of 5,000 accidents a year due to scorpion stings is probably not too high. In adults there is a fatality rate of 0.25–1.80% which can rise to 15–25% in small children up to 5 years. These children and domestic personnel are most frequently exposed to the danger.

All severe cases of poisoning, particularly in small children *must be treated* with antiscorpion serum. The following rules for the serum therapy are of *vital* importance.

1. Specific or polyvalent antiscorpion serum has to be given as early as possible. Even an interval of 2 hours after the accident can make the success of treatment questionable.

2. The total dose of the serum has to be applied at once, one half intravenously, the other half by the subcutaneous or intramuscular route.

3. The serum dose must be large enough to neutralize at least 2 mg dry scorpion venom, which may amount to 5–10 ampules of serum.

Tourniquet, cryotherapy, acupuncture around the lesion, and suction are recommended as auxiliary methods if no serum is available, but they do not replace the serum treatment. Local analgesics and other forms of supporting therapy, frequently also antibiotics, should be used. For more details see chapter 54 by L. Balozet.

REFERENCES

Adam, K. R. and Weiss, Ch. (1966/67). *Mem. Inst. Butantan* (*Sao Paulo*) **33**, 603–614.
Amoreux, M. (1789). *J. Phys.* **35**, 9–16.
Birula, A. (1903). *Bull. Acad. Imp. St. Petersbourg* **19**, No. 3, 104–113.
Birula, A. (1905). *Zool. Anz.* **29**, No. 14, 445–450 and 621–624.
Birula, A. (1927). *Deut. Akad. Wiss. Wien* **101**, 79–88.
Blanchard, E. (1851–1864). *Arachnides, Paris,* pp. 1–232.
Brazil, V. (1966/67). *Mem. Inst. Butantan* (*Sao Paulo*) **33**, 610–612.
Bücherl, W. (1953). *Mem. Inst. Butantan* (*Sao Paulo*) **25**, No. 1, 53–83.
Bücherl, W. (1955). *Arzneimittel-Forsch.* **5**, No. 2, 68–72; *Cien. Cult.* (*Sao Paulo*) **7**, 3.
Bücherl, W. (1955–1956). *Mem. Inst. Butantan* (*Sao Paulo*) **27**, 121–155.
Caporiacco, L. (1937). *Atti. Soc. Ital. Sci. Nat. Museo Civico Storia Nat. Milano* **76,** 355–362.
Chaix, E. (1940). Thesis, Alger.
Charnot, A., and Faure, L. (1934). *Bull. Inst. Hyg. Maroc.* **4**, 1–72.
Christensen, P. A. (1966). *Mem. Inst. Butantan* (*Sao Paulo*) **33**, 245–250.
Corrado, A. P., Antonio, A., and Diniz, C. R. (1966/68). *Mem. Inst. Butantan* (*Sao Paulo*) **33**, 957–960.
de Mello, C. O. (1925). *Mem. Inst. Oswaldo Cruz* **17**, No. 2,
Del Pozo, E., and Anguiano, G. (1947). *Rev. Inst. Salubrid. Enfermedades Trop.* (*Mex.*) **8**, 231–263.
Del Pozo, E. (1948a). *Gaz. Med. Mex.* **78**, 387–397.
Del Pozo, E. (1948b). *Brit. J. Pharmacol.* **3**, 219–222.
Del Pozo, E., Anguiano, G., and Gonzalez, Q. J. (1944). *Rev. Inst. Salubrid. Enfermedades Trop.* (*Mex.*) **5**, 227–240.
Del Pozo, E. (1966/67). *Mem. Inst. Butantan* (*Sao Paulo*) **33**, 615–626.
Deoras, P. J. (1966/67). *Mem. Inst. Butantan* (*Sao Paulo*) **33**, 767–774.
Diniz, C. R., and Gonçalves, J. M. (1956). *In* "Venoms, American Association for the Advancement of Science," p. 131.
Diniz, C. R., and Gonçalves, J. M. (1960), *Biochim. Biophys. Acta.* **41**, p. 470.
Diniz, C. R., *et al.* (1966/67). *Mem. Inst. Butantan* (*Sao Paulo*) **33**, 453–456.
Fabre, J. H. (1907). "Delagrave," pp. 1–374. Paris.
Gomez, M. V., and Diniz, C. R. (1966/68). *Mem. Inst. Butantan* (*Sao Paulo*) **33**, 899–902.
Grasset, E., Schaafsma, H., and Hodson, J. A. (1946). *Trans. Roy. Soc. Trop. Med. Hyg.* **39**, No. 5, 397.
Guyon, D. (1842) *Compt. Rend.* 232; **34**, 404–407 (1852); **59**, 16–22 (1864).
Hemprich, F. G., and Ehrenberg, C. G. (1829). *Ges. Nat. Freunde Verh.* **1**, 348–262.
Hirst, S. (1925). *Ann. Mag. Nat. Hist.* **15**, 414–416.
Karsch, F. (1881). *Berlin Entomol. Z.* **25**, 89–91.
Koch, C. L. (1839). *Arachniden, Nurnberg* **5**, 1–158.
Kraepelin, K. (1898). *Mitt. Naturhist. Museum Hamburg* **15**, 1–6.
Kraepelin, K. (1899). *Tierreich* **8**, 1–265.
Krag, P., and Weiss, B. (1966). *Mem. Inst. Butantan* (*Sao Paulo*) **33**, 251–280.
Leach, W. E. (1815). *Trans. Linnean Soc. London* **11**, 390–301.
Linnaeus, C. (1758), "*Systema Naturae,*" 10th ed., pp. 624–625.

Lucas, H. (1849). *Imprimerie Nation, Paris* pp. 271–273.
Magalhães, O. (1935). *Ann. Fac. Med. Belo Horizonte* **4**, No. 1,
Magalhães, O. (1946). *Mem. Inst. Oswaldo Cruz* **44**, No. 3, 425–439.
Matthiesen, F. A. (1962). *Evolution* **16**, No. 2, 255.
Maurano, H. R. (1951). Thesis, Rio de Janeiro.
Pallary, P. (1937). *Arch. Inst. Pasteur d'Algerie* **15**, No. 1, 97–101.
Pawlowsky, E. N. (1924). *Trav. Soc. Naturalistes Leningrad* **53**, 17–86.
Pawlowsky, E. N. (1926). *Ann. Mus. Zool. Acad. Leningrad* **26**, 137–205.
Petrunkevitch, A. (1913). *Trans. Conn. Acad. Arts Sci.* **18**, 1–137.
Pflugfelder, O. (1930). *Z. Wiss. Zool.* **137**, 1–23.
Phisalix, M. (1922). "Scorpions," Vol. 1, pp. 208–217 and 234–255. Masson, Paris.
Piza, S. T. (1939). *J. Agron. Piracicaba* **2**, No. 1, 49–59; and **2**, No. 4, 285–287.
Piza, S. T. (1940). *Rev. Agr. Piracicaba* **15**, No. 5 and 6, 215–218.
Pocock, R. I. (1889). *Ann. Mag. Nat. Hist.* **6**, 334–351.
Pocock, R. I. (1902). *Ann. Mag. Nat. Hist.* **10**, 364–380.
Rochat, C. *et al.* (1966/67). *Mem. Inst. Butantan. (Sao Paulo)* **33**, 447–452.
Rosin, E., and Shulow, A. (1963). *Proc. Zool. Soc. London* **140**, No. 4, 547–556.
Schulze, W. (1927). *Philippine J. Sci.* **33**, 375–390.
Sergent, E. (1938–1945). *Arch. Inst. Pasteur Algerie* **16**, 23.
Sergent, E. (1948). *Arch. Inst. Pasteur Algerie* **16**, No. 1, 21–24.
Sergent, E. (1949). *Arch. Inst. Pasteur Algerie* **16**, No. 1, 31–34.
Shulov, A. (1966). *Mem. Inst. Butantan (Sao Paulo)* **33**, 93–100.
Shulow, A. (1955). *J. Med. Assoc. Israel*, **49**, No. 6, 131–133.
Shulow, A. (1958). *Proc. 10th Intern. Congr. Entomol. Montreal*, 1956 Vol. 1, 877–880.
Shulow, A., and Amitai, P. (1959). *Bull. Res. Council* Israel Bl, 41.
Shulow, A., and Amitai, P. (1958). *Arch. Inst. Pasteur Algerie* **36**, No. 3, 354–369.
Shulow, A., and Amitai, P. (1959). *Arch. Inst. Pasteur Algerie* **37**, No. 1, 218–225.
Simon, E. (1878). *Ann. Soc. Entomol. France* **8**, 158–160.
Simon, E. (1911). *Bull. Soc. Entomol. Egypt* **2**, 57–87.
Stahnke, H. (1956). "Scorpions," Poisonous Animal Res. Lab. Arizona State College, Tempe, Arizona.
Thompson, J. L. C. (1958). *Entomologist's Monthly Mag.* **94**, 229–230.
Vachon, M. (1952). *Arch. Inst. Pasteur Algerie.*
Watermann, J. A. (1939). *Trans. Roy. Soc. Trop. Med. Hyg.* **33**, No. 1, 113–118.
Watermann, J. A. (1951). *Caribbean Med. J.* **12**, No. 5, 3–15.
Watermann, J. A. (1957). *Caribbean Med. J.* **19**, Nos. 1 and 2, 113–128.
Weissmann, A., and Shulow, A. (1959). *Arch. Inst. Pasteur Algerie* **37**, No. 1, 202–217.
Werner, F. (1935). *Bronn's Klassen* **5**, No. 4,
Whittemore, F. W., Keegan, H. L., and Borowitz, J. L. (1961). *Bull. World Health Organ.* **25**, 185–188.
Zolessi, L. C. (1956). *Fac. Argon. Montevideo* **35**,

Chapter 56

*Scorpionism in the Old World**

LUCIEN BALOZET

INSTITUT PASTEUR D'ALGÉRIE, ALGIERS, ALGERIA†

I. History 349
II. Geographic Distribution of Dangerous Scorpions 351
III. The Venom 352
A. Collection 352
B. Chemical Properties 354
C. Enzymes 357
D. Other Biological Features 358
IV. Experimental Study of Toxicity 359
V. Envenomation in Man 360
VI. Pathological Physiology 362
A. Action on Muscles 363
B. Respiratory Effects 364
C. Vascular Action 364
D. Other Reactions 364
VII. Treatment 365
A. Specific 365
B. Nonspecific 367
References 368

I. HISTORY

The scorpion, by the terror which it inspires as well as the strangeness of its form, has struck the imagination of the peoples of the Orient and of the Mediterranean since ancient times and has been mythicized in the legend of Mithras. When Mithras sacrificed the bull, the first being created whose blood was to fertilize the Universe, a scorpion was sent by Ahriman, the Genie of Evil, to destroy the sources of life by attacking the testicles of the

* Translated by Wiltrand Segel.

† Present address: "Le Rouvre," 43, Traverse du Commandeur, Marseille 13°, France.

animal. Scorpions are prominent in monuments pertaining to the mysteries of Mithras, whose cult was prolonged until the second and the third centuries of our era, especially in North Africa. In ancient Egypt, also, scorpions were frequently represented on monuments. They are mentioned in the Ebers Papyrus and in several passages in the Book of the Dead. Numerous incantations had the reputation of protecting against the sting of the scorpion.

In Greek mythology the legend of Orion is well known. Orion, son of Zeus, was a great hunter. Artemis, whom he dared defy, produced a scorpion which stung and killed him. Then Zeus turned Orion into the constellation that brings rain and storms.

The Talmud and the Bible frequently mentioned the scorpion as a repugnant and formidable animal.

A curious transposition of the legend of Cleopatra was created at Cherchell (the ancient Cæsarea) on the Algerian coast, where important Berber and Roman ruins are to be found. There was a palace in Cæsarea about which Qazouini, a Berber of the 3rd century, told the following tale: This palace had been built by a king for his son, whom the astrologers predicted would one day be killed by the sting of a scorpion. The palace was constructed of stone to keep out scorpions, for these animals were said to be unable to reproduce in such surroundings or to find their way in because of the polished surface of the marble and columns. However, one day a basket of grapes containing a scorpion was delivered. The young prince, desiring fruit, was stung and died. This tale is connected with that of the death of Cleopatra of Egypt, victim of the bite of an asp, in the fact that the king Juba II, whose capital was Cæsarea, had married Cleopatra Selena, daughter of Cleopatra of Egypt (Dor, 1895).

Many other Greek and Latin classics mention scorpions. Aristotle wrote that their harmfulness was not the same in all lands: their sting was inoffensive in some lands, but always fatal in others (especially in Caria, south of Anatolia). Most authors shared this opinion, and thought Africa to be the country of origin of scorpions. References which are beyond the scope of this paper can be found in the Encyclopedia of Pauly (1929). It is to the Latins that we owe the still cited proverb: *in cauda venenum.*

In the 12th century Moses Maimonides, a Jew converted to Islam, mentioned the ineffectiveness of bezoar and certain stones to which popular belief attributed (and still attributes) a therapeutic value for the treatment of scorpion stings (see Maimonides, circa 1180). An Arab historian of the 14th century, Ibn Khaldun, recounts that Sidi Okba fixed the place where the city Kairouan should be built, in a marsh infested with ferocious and venomous animals. He then invoked, with the aid of the 18 companions of the Prophet who accompanied him, Allah the All Powerful for protection. Sidi Okbas's wishes were always granted, and for 40 years no serpents or scorpions were seen in Ifrikya (see Ibn Khaldun, circa 1400). The Arab Léon l'Africain (1550) mentions many localities in which the scorpions' sting is rapidly fatal and without remedy.

In the religious art of the 14th, 15th, and 16th centuries, the scorpion is often used as an emblem of the Jewish people to symbolize perfidy (Bulard, 1935). A painting by Fra Angelico, in the museum of San Marco in Florence, shows Christ carrying the Cross, surrounded by 3 Roman soldiers wearing tunics ornamented with scorpions.

The belief in fables and ancient proverbs is still great in the primitive populations of North Africa. A study of these legends is found in a work by Pallary (1936). The first investigation of scorpions, free of popular beliefs, superstitions, and myths, and founded rather on observation and experimentation was performed by Francesco Redi (1668), considered one of the first followers of the method of Descartes. Redi studied the scorpions of Italy, which he declared not venomous, the scorpions of Egypt, and specially those of Tunisia, whose sting, he declared often lethal. The illustrations in Redi's work show without doubt that the Tunisian scorpions studied were *Androctonus australis*.

Until the beginning of our century, there have been very few investigations of poisoning by scorpions. De Maupertuis (1731) experimented on dogs and chickens, Maccary (1810) tried the stings of *Buthus occitanus* on himself; Bert (1865, 1885), Jousset de Bellesme (1872, 1874), Valentin (1876), and Joyeux-Laffuie (1882, 1883) also contributed to the literature.

II. GEOGRAPHIC DISTRIBUTION OF DANGEROUS SCORPIONS

Despite the great number of species of scorpions, only a few are considered truly dangerous. In the countries on the north coast of the Mediterranean, Spain, the south of France, Italy, Greece, the Balkans, and the Greek Isles, the scorpions of the genus *Buthus* and, especially, *B. occitanus* are the most venomous, but are not really dangerous. The genus *Euscorpius* is almost innocuous.

The countries on the south coast of the Mediterranean harbor venomous species. In northwest Africa, 5 species are dangerous: *Androctonus australis* (L.), *Buthus occitanus* (Am.) whose toxicity appears greater than in Europe, *Buthacus arenicola* (E. Sim.), *Androctonus amoreuxi* (Aud. and Sav.), *A. æneas* (C. L. Koch). Among these species, *A. australis* is responsible for 80% of the accidents and 95% of the deaths. The geography of scorpionism of this region corresponds to the habitat of this species: in Algeria, north of the Sahara and south of the High Plateaus; in Morocco, south and east of the Atlas; in Tunisia, the southern half or two-thirds of the country. One must add *A. hoggarensis* (Pallary) which is found in the Sahara mountains at Hoggar and Tassili n'Hajer; *Leiurus quinquestriatus* (H. and E.) which is found southeast of the Sahara and in the territory of the Niger and of the Chad; and *A. amoreuxi* which is found in Mauritania.

In Libya (Tripolitan and Cirenaica) are found *A. australis* and *Leiurus quinquestriatus*. This formidable species is common in Egypt, where it is associated with *A. amoreuxi* and *A. crassicauda*; this association also occurs in the Sudan (northern and central parts). In Abyssinia and Erythria are found species which are slightly dangerous: *Parabuthus liosoma* (H. and E.) and *Pandinus meidensis* Karsch. The African *Pandinus*, like the *Heterometrus* and *Palamneus* of southern Asia, is only slightly venomous in spite of its large size.

We find *Leiurus quinquestriatus* again in the countries of the Middle East: Arabia, Jordan, Israel, Syria, and Turkey. A venomous *Androctonus* (a little less venomous than *A. australis*), *A. crassicauda* (Olivier), is found in Turkey, Lebanon, Iran, and along the western coast and islands of the Persian Gulf; *Buthus judaicus* found in Israel.

In eastern India, *Buthus famulus* is of medical interest. Dangerous scorpions are not found in Ceylon, Malaysia, or the southeastern Asiatic countries. *Buthus martensis*, whose rather active toxin has been studied by Mori (1917,1919), Kaku and Kaku (1950), and Kaku (1954), is found in Korea and in Manchuria.

In southern Africa, *Parabuthus transvalicus, P. triradulatus*, 2 species of *Opisthophtalmus*, and one of *Hadogenes* have been studied by Grasset *et al.* (1946) and were used in the preparation of a serum. The venom of *Parabuthus* possesses a moderate toxicity; that of the other species is less active.

III. THE VENOM

A. Collection

The simplest process for obtaining scorpion venom is maceration of the crushed telsons in distilled water or in physiological saline solution. The telsons are removed from living scorpions at the point of their articulation with the last caudal segment. They are washed rapidly in water, dried in a towel, and then placed in a dessicator over $CaCl_2$ at 2–4°C for at least 30 days. They are then preserved in evacuated ampuls at 2–4°C.

As has been claimed by de Magalhães (1938), the toxicity of the telsons is less stable than that of pure venom. This is illustrated by the following figures for the LD_{50} in mice (Balozet, 1956); new telsons: 1/66 of the telson; 6-year-old telsons, 1/30 of the telson; 12- and 13-year-old telsons, 1/25 of the telson.

To obtain a solution of venom, a number of telsons are ground, and the powder added, with glass beads, to a measured quantity of physiological saline solution (for example, 1 ml/telson). The mixture is refrigerated, with periodic agitation, for 12 to 24 hours and then centrifuged. The solution contains soluble tissue products in addition to the venom. This method is convenient for the preparation of toxin used in the production of serum by hyperimmunization of horses. The toxin unit can be evaluated only as the number of telsons or as a fraction of a telson. It is imprecise, and the presence of other substances in the venom extract is often troublesome. This extraction method was first employed by Jousset de Bellesme (1872) and then adopted by many investigators for routine use, mainly in the production of serum.

Pure venom, or of most of the secretion of the venom gland, can be obtained by provoking contraction of the muscles of the gland with high frequency electrical current (a Ruhmkorff coil charged by a 6-V current obtained from a battery or transformer). One electrode is fixed on a clamp which holds the scorpion at the 2nd or 4th ring of the tail, the other electrode is brought into contact with the telson. The droplets which appear at the orifice of the stinger are collected in a Petri dish or the tip of a pipette. The first drops are limpid, but the following ones become more and more opalescent and viscous. This opalescence is caused by granules which disappear in the presence of acetic acid or potassium hydroxide. The toxicity of the first few drops is greater than that of the following (C. Phisalix and de Varigny, 1896). M. Phisalix (1922) explains this difference as follows: the venomous granules secreted by the glandular cells probably react chemically with substances in the glandular liquid to produce the final venom.

Grasset *et al.* (1946) extracted the venom of *Parabuthus* by aspirating the tip of the cut telson by a capillary pipette. Another method, used by Lissitzky *et al.* (1956), involved the technique of Maurano (1915): The scorpion is held by the end of the tail with forceps, and venom is obtained by manual excitation of the cephalothorax. The toxicity of the venom obtained by this method is 10 times greater than that of the venom collected by electrical stimulation. This is so, according to these authors, because the secretion obtained by manual excitation contains mainly toxins, while electrical tetanization of the muscles of the glandular apparatus provokes the emission of nontoxic substances which dilute the toxin. It is also possible that some of these nontoxic substances, in particular mucoproteins, partially neutralize the toxic substances.

Finally, in species in which the chitin of the telsons is sufficiently supple, venom can be obtained by simple compression of the telsons with forceps (Kraus, 1930; Grasset *et al.*, 1946, for *Hadogenes* and *Opisthophtalmus*).

The collection of venom by electrical or manual stimulation permits the use of the live scorpions for further collections. To avoid cannibalism, it is expedient to avoid placing too large a number of scorpions in the same cage. Food for the scorpions can be easily supplied from a colony of *Gryllus domesticus*. These very prolific insects are easily bred in the laboratory. The best yield of venom is obtained by making a collection every 2 weeks from well-nourished scorpions (Balozet, 1964).

The quantity of venom obtained by electrical stimulation varies with the species. For *Androctonus australis* the average yield is 1.4 mg (0.8–1.8) of dried venom per collection; for *Buthacus arenicola*, 0.7 mg; and for *Buthus occitanus*, 0.29 mg. Scorpions kept in activity produce less venom than wild ones.

Venom, when dried, is a gray powder, with the exact shade depending on the species. When refrigerated in dry ampuls, the venom is potent for many years.

It is difficult to compare the yields of toxin obtained by the 2 main methods (maceration of ground telsons *versus* dried venom obtained by manual or electrical stimulation). The toxic substances of the telsons are present in the venom-producing glands in the form of liquid protids and as granules in the cytoplasm of the epithelial cells. The total amount of these toxic substances is greater than the quantity obtained by electrical stimulation. The residual toxicity of the venom apparatus after the extraction of the venom, is approximately one-fourth the total toxicity (gland plus venom). A full telson contains 66 LD_{50} units while an empty telson contains 15 LD_{50} units (a difference of 51 LD_{50} units). The venom obtained, 1.5 mg, represents only about 16 times the LD_{50} (Balozet, 1956). M. Phisalix (1922) noted that the electrical stimulation yielded one-fourth to one-fifth the total amount of venom contained in the telson. This great difference probably arises from a loss of venom during extraction by electrical stimulation and perhaps also from a reduction in toxicity of the venom obtained by this method (see above: Lissitzky *et al.*, 1956; the comparison between venom obtained by electrical and manual stimulation).

In experimental envenomation there are differences when the venom is given by injection of macerated telsons, by the injection of a solution of dried venom, or by an experimental sting. In the 2 latter techniques death occurs more rapidly than in the former (Shulov, 1939; Schöttler, 1954).

B. Chemical Properties

The work of Wilson (1904, 1921) established that the 2 proteins of Egyptian scorpion venom, obtained by maceration of the telsons, coagulate at 56° and 74°C, respectively, while other substances present in the venom (mucin, nucleoprotein) can be precipitated by acetic acid in the cold.

Dried venom is not completely soluble in distilled water, physiological saline solution, or glycerin. The insoluble part, which can be separated by centrifugation, has the consistency of mucus. After washing in distilled water or saline solution, this insoluble part loses all toxicity. It appears to be a mucoprotein, but its exact composition has not been studied.

The soluble component of the dried venom contains all the toxin. This active part is insoluble in neutral solvents, such as methyl or ethyl alcohol, ether, chloroform, acetone, benzene, xylene, or oils. The dissolved toxin loses only part of its activity when it is heated to 100°C for 15 to 30 minutes. It is precipitated by a saturated solution of ammonium sulfate or alcohol. It loses its activity in the presence of ammonia or a small amount of iodine. It is not retained by Chamberland or Berkefeld filters of medium porosity, and it is only partially dialyzable through collodion or cellophane membranes.

The elemental composition of the venom of *Leiurus quinquestriatus* of Syria (lethal to mice at a dose of 0.8 μg/gm, is, according to Tetsch and Wolff (1937), as follows: C, 43.6%; H, 6.8%; N, 13.6%; S, 3.8%. Kaku (1954) found an elemental composition for the venom of *Buthus martensis* that is significantly different in sulfur content: C, 45.58%; H, 5.83%; N, 15.21%; S, 28.8–29.2%.

The methods of purification described confirm the stability and the resistance of the toxin to relatively severe chemical treatments; a short summary and references are available (Balozet, 1960).

Electrophoresis and chromatography have permitted great progress in the analysis of venoms. Paper electrophoresis has been employed for the venom from *Centruroides* (Bolivar and Ornelas, 1949; Bolivar and Rodriguez, 1953), *Tityus* (Diniz and Gonçalves, 1956; Fischer and Bohn, 1957), *Leiurus quinquestriatus* and *Buthotus minax* (Adam and Weiss, 1958, 1959), *Buthus judaicus* (Weissman *et al.*, 1958), *Androctonus australis* (Balozet, 1960), and *A. australis* and *B. occitanus* (Miranda *et al.*, 1961), The protein fractions thus revealed are few: 8 in the venom of *Centruroides*; 6 or 7 in that of *Tityus*; 6 in the venom of *B. judaicus*; and 4 principal ones, 3 of which are positively charged at pH 8.5 to 8.8; 1 negatively charged, and 1 or 2 minor negatively charged fractions, in the venom of *A. australis*. The toxin is associated with 1 or 2 positively charged fractions. This has been confirmed by Miranda *et al.* (1961) in the course of successive operations for the isolation of scorpamines by electrophoresis, which further established the perfect electrophoretic homogeneity of these protein fractions.

The precipitation of venom with a homologous antivenin serum in an agar medium shows 4 lines representing 4 protein fractions of *A. australis* (Balozet, 1960). Such techniques have contributed interesting details about the chemical composition of the venom and the nature of the toxin.

Fischer and Bohn (1957) found that venom is composed mainly of proteins. There are no free amino acids except for 1 or 2% lysine. Histamine is absent. The concentration of phosphate ions is low. The total amount of protein in fresh *Buthus judaicus* venom is 8.7 gm/100 ml, representing 50% of the dry weight of the venom. Lipoproteins are absent, but small amounts of glycoproteins are present.

The results obtained by Miranda (1964), Miranda and Lissitzky (1958, 1961), and Miranda *et al.* (1960, 1962, 1964a,b,c) are especially interesting. These authors, working with the North African scorpions *Androctonus australis* and *Buthus occitanus* have successfully purified the neurotoxins in the venom of these 2 species. They used 3 types of biological material: telson, venom obtained by electrical stimulation, and venom obtained by manual stimulation.

Their purification of the venom involved 5 steps: the first 2 followed the traditional methods already used by investigators working with other species of scorpions (Wilson, Mohamed, and Anguiano). In contrast, the last 3 steps used modern techniques of ion-exchange chromatography and filtration through a molecular sieve (Sephadex-dextran gel). The final procedure was chromatographic separation on an Amberlite carboxylic resin (CG-50) followed by a filtration through Sephadex G-25 and a second chromatographic separation on Amberlite. The pH and molarity of the ammonium acetate buffer used for the 2 separations on Amberlite were not the same. The filtration through the Sephadex gel is done according to a special technique called "reversible retention" (Miranda *et al.*, 1962). For the purification of venoms from material much richer in toxins than telsons, the purification procedure only consists of 3 stages: extraction in pure water, filtration on Sephadex G-25, and chromatography on Amberlite CG-50.

The toxic products isolated at the end of the purification are homogeneous upon chromatography on Amberlite, paper-electrophoresis, and ultracentrifugation. There are 2 toxins called scorpamines which share almost equally the total toxicity obtained when purification is complete. These substances seem to be basic proteins of low molecular weight. Their pH_i is high (above pH 8). Their sedimentation constant is low (mean, 1.2). Their molecular weight is between 10,000 and 18,000, depending on the species of scorpion and the type of toxin. Their extinction coefficient ($E_{1\,cm}^{1\%}$) is about 20. The 2 toxins have different amino acid compositions, however, they have, in common, a high proportion of cystine and basic amino acids (with lysine predominating among these) (Miranda, 1964).

In the case of toxin II of *A. australis*, maceration in 0.4 M acetic acid leads to the dissociation of the toxin into two proteins; only one accounts for the total activity. The authors' explanation for this is that the active constituents of scorpion venoms are able to form complexes, stable under certain conditions, which dissociate during chromatography in systems containing dilute acetic acid (Rochat, 1964).

These authors, using chromatographic methods (employing DEAE-

Sephadex A-50 and CM-Sephadex C-50) and molecular filtration with LKB Recychrom were able to purify 12 scorpion neurotoxins from the species of North Africa: *Androctonus australis hector*, *Buthus occitanus tunetanus*, and *Leiurus quinquestriatus quinquestriatus* (C. Rochat *et al.*, 1970; Miranda *et al.*, 1970). The method applied to the venom of *Naja haje* leads to the purification of the neurotoxins of this snake (Miranda *et al.*, 1970). This last fact strongly supports the authors' opinion that the method they worked out can be used for the purification of all basic animal neurotoxins.

In 1967, H. Rochat *et al.* showed that a common sequence exists in scorpion neurotoxins and they suggested that these toxins make up a further development of their work, leading to the determination of the primary structure of toxin I of *A. australis hector* and N-terminal amino acid sequences of 3 other scorpion neurotoxins (toxin II of *A. australis hector*, toxin I of *B. occitanus tunetanus*, and toxin II of *L. quinquestriatus quinquestriatus*) confirmed their hypothesis (H. Rochat, 1969; H. Rochat *et al.*, 1970).

The presence of 5-hydroxytryptamine in the venom of *Tityus* has been suggested by Diniz and Gonçalves (1956). Paper chromatography and fluorometry permitted Adam and Weiss (1956, 1958) to establish the existence of 5-hydroxytryptamine in the venom of *L. quinquestriatus* and to estimate its concentration as 2–4.5 μg/mg venom. 5-Hydroxytryptamine exists in many animal venoms, especially the venoms of wasps and bees. A histological study of the glands by enterochromaffin reactions shows that the 5-hydroxytryptamine is localized in the intracellular granules of the venom glands (Adam and Weiss, 1959a,b). Its action on isolated muscle has been studied by the same authors (1959c).

The 5-hydroxytryptamine, which is present in greater amount in scorpion venoms than in other animal venoms, plays no role in actual intoxication. It was found that treatment of a solution of venom with an equal volume of 95% acetone removes 5-hydroxytryptamine but does not perceptibly alter the neurotoxicity of the residue. In contrast, chymotrypsin destroys the neurotoxin but not the substance which causes the violent local pain, i.e., 5-hydroxytryptamine (Adam and Weiss, 1956, 1959a).

C. Enzymes

1. Lecithinase

The hemolytic properties of venoms of *Ophidians* (with the exception of *Dendroaspis angusticeps*) are the result of the action of phosphatidases on the lecithin of plasma or on the vitellin of chicken egg to form lysolecithin. Scorpion venoms generally contain few or no hemolysins although there are some exceptions. The venom of the following species is devoid of hemolysin: *Leiurus quinquestriatus* (Todd, 1909; Houssay, 1919; Balozet, 1962), *Buthus occitanus, Androctonus australis* (Balozet, 1951, 1962), *A. amoreuxi*,

A. aeneas, *A. aeneas liouvillei*, *Buthacus arenicola* (Balozet, 1962), *Tityus serrulatus* (Houssay, 1919; Balozet, 1962), and *Buthotus judaicus* (Weissmann and Shulov, 1959). The hemolytic properties of the venom of South African scorpions are very weak and are almost nonexistent in *Parabuthus* and *Hadogenes* (Emdin, 1933; Grasset *et al.*, 1946). The venom of *Buthus martensis* is hemolytic (Kubota, 1918; Mori, 1919). The venom of *Scorpio maurus* shows hemolytic properties comparable to those of the venoms of *Ophidians* (Levy, 1924; Balozet, 1951, 1962). Similarly the venom of *Heterometrus* of India, studied by Caius and Mhaskar (1932) is strongly hemolytic. These 2 genera, *Heterometrus* and *Scorpio*, are closely related zoologically, both belonging to the family Scorpionidæ.

Lecithinase also decreases the sedimentation rate of red blood cells and brings about the formation of the spherogene factor in the venom of *Scorpio maurus*. Other scorpion venoms are without this activity (Balozet, 1962).

Certain discrepancies between the results of different investigations should be noted. Moḥamed *et al.* (1953, 1953a) found a very strong hemolytic activity in the venom of *L. quinquestriatus* and de Magalhães (1928, 1938) found a similar activity in the venom of *Tityus*. Different experimental conditions may be responsible for these discrepancies; for example, the red blood cells used were from different species (horse, guinea pig, rabbit, sheep, etc.), another is the origin of the scorpion. The venom of *Buthacus arenicola* may have hemolytic properties (Balozet, 1953), or it may not (Balozet, 1962).

2. Hyaluronidase

Tarabini-Castellani (1938) detected the presence of a spreading factor in the venom of *B. bicolor* (= *A. australis*) and of *Euscorpius italicus*. The factor was more abundant in the latter even though this species is supposedly innocuous. Favilli (1956) confirmed spreading factor activity in *E. italicus*. Jaques (1956) made the same observation for the venoms of *B. occitanus* (France) and *S. maurus* (Algeria). The hyaluronidase activity of these 2 scorpions is fairly high. The enzyme is inhibited by many mammalian sera (MacClean, 1942).

3. Other Enzymes

L. quinquestriatus is devoid of any proteolytic action (Houssay, 1919). The same is true of the venoms of *A. australis*, *A. amoreuxi*, *B. occitanus*, *B. judaicus*, and *S. maurus*. The venom of *L. quinquestriatus* does not act on coagulated milk or starch (Houssay, 1919).

D. Other Biological Features

The venom of *Centruroides sculpturatus* does not induce an allergic response in guinea pigs (Doro *et al.*, 1957). This is probably true of all scorpion venom. Pavan (1952) found in the venom gland of *Euscorpius italicus* antibiotic properties against *Staphylococcus aureus, Vibrio comma,* and *Eberthella typhosa.*

The venom of *L. quinquestriatus* is devoid of all blood coagulating activity (Todd, 1909; Houssay, 1919), as are the venoms of *A. australis, A. amoreuxi, B. occitanus* (Balozet, 1952), and *B. judaicus* (Weissman and Shulov, 1959). The venom of *Buthacus arenicola* shows a slight coagulating activity (Balozet, 1953); while that of *Scorpio maurus* acts as a weak anticoagulant (Balozet, 1951).

IV. EXPERIMENTAL STUDY OF TOXICITY

The sensitivity of various animal species to scorpion venom is extremely variable. Todd (1909) observed variations even among different breeds of dogs. Certain animals are very resistant to scorpion venom (Todd, 1909). Cats, porcupines, hedgehogs, chickens, turkeys, and toads lack sensitivity to the venom according to Sergent (1939, 1948). Frogs, according to C. Phisalix and de Varigny (1896) and Sergent (1939), are almost completely resistant. The scorpion can be killed by his own venom, but relatively high doses must be used (C. Phisalix and de Varigny, 1896; Shulov, 1955).

In laboratory white mice, the variations in sensitivity can be very important. Miranda (1964), using *A. australis* venom, obtained an LD_{50} value of 70 μg in mice of an unspecified strain, and a value of 17.2 μg in male Swiss mice from controlled breeding colonies of the Centre National de la Recherche Scientifique. This consideration, among others, makes it difficult to compare toxicity figures published by different investigators, because they used different types of mice.

Another very important factor in the degree of sensitivity is the age of the test animals. Young animals are much more sensitive than adults. It is necessary, for example, when measuring the LD_{50} of a venom in mice, to employ only adult males of known weight within a specified range (18–20 gm for the Algerian strain of mice). For very precise measurements, young male adults of nearly the same age must be used.

The variations in toxicity for the venom of scorpions, as for the venom of serpents, depends on the animal's geographical origin. Thus, for the venom of *A. australis*, one lot of venom from scorpions of the Bou Saada region had an LD_{50} titer of 91 μg for 20-gm mice. Venom of the same species from Chellala-Laghouat titered 62.5, 62.2, and 67 μg, and the venom from El-Oued, 51.5 and 67 μg. The venom used by Lissitzky *et al.* (1956) from Mecheria titered 70 μg. The toxicity of venom from Algerian *A. australis* may therefore vary by a factor of 2 according to the scorpions place of origin. The following are some LD_{50} evaluations of toxicity for other species:

B. arenicola, 52 μg; Algerian *B. occitanus*, 115 μg (Balozet, 1956). For *L. quinquestriatus* the numbers vary considerably: Tetsch and Wolff (1937) report 0.8 μg/gm mouse (= 16 μg for a 20-gm mouse); Shulov *et al.* (1957) 5.09 μg; and Adam and Weiss (1959a) 6.5 mg/kg or 130 μg for a 20-gm mouse. The toxicity of *Buthus judaicus* is 127 μg (Weissmann and Shulov, 1959); that of *A. crassicauda* is comparable (Tulgat, 1960). The toxicity of *B. famulus* seems less (Deoras, 1961), the same as that of the South African scorpions (Grasset *et al.*, 1946). According to Mori (1917–1919) and Kaku (1954), the venom of the Korean and Manchurian scorpion *B. martensis* is highly toxic.

All the parenteral routes of injection have been used. The intravenous route yields the most severe reaction. The subcutaneous, intramuscular, and intraperitoneal routes give equivalent results, the one to the others. The intraperitoneal route seems the most convenient and gives the most constant results. The venom is inactive when deposited on mucous membranes and in the digestive tract.

The symptoms of envenomation, in mice, appear almost immediately after injection and are manifested initially by cries, great agitation, vertical leaps, great aggressiveness, and violent battles. This period of excitation is followed by a period of inactivity during which time the mice are doubled up, with their hair bristling, and an accelerated respiration. Salivation and lacrimation occur, with bleeding from the mouth and eyes. New phases of agitation may be repeated. Then, especially when the dose is lethal, paralysis and contractions appear: the posterior limbs are taut and extended back. Respiration becomes short and superficial and death follows. The time lapse between injection and the death of the animal varies from 2–3 minutes to 3 hours. This period of time, however, is not inversely proportional to the dose.

In other experimental animals envenomation follows an analogous course. In the rat, the guinea pig, and the rabbit, the period before death is prolonged, but does not exceed 6–7 hours. Tears mixed with blood are more frequent in the rat. In the horse, nonlethal doses of venom provoke salivation, abundant sweating, and exaggerated intestinal peristalsis which manifests itself in violent colic and diarrhea.

In all species, if death is not too rapid, hyperglycemia, glycosuria, hypertension, and cardiac arhythmia is observed. No lesion, other than congestion at the site of injection (subcutaneous) or peritoneal bleeding (intraperitoneal), is seen at autopsy.

V. ENVENOMATION IN MAN

Scorpion stings always occur under accidental circumstances, more often at night than during the day, and more often when the weather is stormy, the temperature elevated, and the wind hot. They occur in houses or in tents, in palmeries or gardens during the course of agricultural work, or on trips. Women are somewhat less frequently attacked because of their more sedentary habits. In children, although the number of stings is not perceptibly higher than in adults, the consequences are much graver. The parts of the body most frequently stung are the limbs or, sometimes, the head and neck of sleeping persons. The latter case is more frequent in infants and young children.

The sting of a scorpion causes an extremely severe local pain, somewhat like a burn. After a variable amount of time, the pain lessens and is replaced by a prickling sensation and a horripilation which spreads around the stung region. A relative insensitiveness follows.

General symptoms follow more or less rapidly: the sick person becomes agitated, sometimes impressively so, and this agitation yields with difficulty to sedatives and tranquilizers. Children cry. Muscle tone increases and sometimes contractures result, affecting the limbs, the abdominal muscles, or the pharynx. They can be generalized (opisthotonos). Very young children often have convulsions. Priapism is frequent. Tendon reflexes are often abolished, plantar reflexes are diminished or suppressed, and often the corneal reflex disappears. Consciousness is retained at the beginning of envenomation, but during the course of its evolution a sensation of anguish may appear, followed by a certain degree of obnubilation. Syncope is rather frequent.

Salivation and perspiration are greatly increased. The eyes tear, the pupils are mydriatic, and visual acuity is impaired. Frequently there is exophthalmia. The pulse rate is greatly increased, becoming progressively arhythmic. The arterial pressure is first increased and diminishes later; there can be either hyper- or hypotension.

The temperature is unstable, sometimes higher and sometimes lower than normal, and this frequently alternates at short intervals. Hypothermia is always a sign of aggravation.

Respiratory movements are irregular. They are diminished in amplitude and their rhythm is altered; they are sometimes of the Cheyne-Stokes type. Death is always due to respiratory paralysis and may occur after sudden apnea. Auscultation reveals a humid rattle due to a bronchial hypersecretion which gradually impedes the respiratory apparatus and may provoke an attack of coughing.

Some victims sneeze spasmodically. Sneezing is often produced in laboratory personnel who grind dry telsons without covering their nostrils with a humid gauze mask. This may be caused by an irritating substance which occurs throughout the scorpion, since manipulation of scorpions, without handling the venom, or simply inhaling the air over the scorpion cages causes sneezing. No desensitivation occurs. After more than 10 years of scorpion handling, sensitivity is just as great.

Vomiting is the first sign that the nerve centers have been attacked by the toxin; the prognosis is poor in such cases. The vomitus may contain some blood. Intestinal peristalsis is exaggerated and sometimes results in diarrhea and melena.

Sometimes oliguria, sometimes polyuria and involuntary micturition and hematuria are observed. Glycosuria is not always present, but hyperglycemia is typical. The level of sodium in the plasma is lowered and that of potassium increased, the concentrations in milliequivalents being modified as a result.

Sergent (1942a) established, after many observations, that the first symptoms other than local pain, appear from 5 minutes to 24 hours after the sting and in the majority of cases, between 20 minutes and 4 hours. The interval between the sting and death varies from several minutes to 30 hours, on the average, between 2 and 20 hours.

Table I, compiled from data in the documents of the Service de Santé des Territoires du Sud (Algeria) from 1942 to 1958, illustrates the lethality of scorpion stings in that country (almost all stings are due to *A. australis*), and the difference in sensitivity according to age (Balozet, 1964).

TABLE I

IMPORTANCE OF SCORPION STINGS ACCORDING TO AGE

	No. of stings	No. of deaths	Percent lethality
Infants	411	32	7.78
Children	5211	191	3.66
Adults	13130	145	1.10
Aged	1412	18	1.27
	20164	386	Mean: 1.90

Finally, an important characteristic of poisoning from scorpion stings should be mentioned: the sudden aggravation of the condition of the victims after amelioration of symptoms and an apparent recovery. This is marked by the sudden reappearance of grave respiratory trouble. This was mentioned by Barros (1938), and Sergent (1942b) reported 17 relapsed cases of which 13 were lethal. Close surveillance of cases must continue for at least 12 hours after the complete disappearance of the symptoms, to guard against such a relapse.

VI. PATHOLOGICAL PHYSIOLOGY

The mechanism of poisoning by scorpion venom is simple. It consists, primarily, of neurotoxic action on the nerve centers, and secondarily of local action of 5-hydroxytryptamine. It can be distinguished from poisoning by serpent venoms which possess several enzymes and thus produce a more complex effect.

The symptoms of scorpion poisoning in man and experimental animals shows that the action of the neurotoxin resembles that of parasympatheticomimetic substances (muscarine, acetylcholine, pilocarpine, and eserine) and sympatheticomimetic compounds (adrenaline). The best analogy is with the action of nicotinic compounds which produce bradycardia and hypotension following tachycardia and hypertension, stimulation of the adrenal medulla to produce adrenaline and hyperglycemia, polypnea, peristalsis, and mydriasis.

Most of the effects of the toxin can be explained by its very marked selective action on the sympathetic and parasympathetic autonomic centers in the hypothalamus. Excitation of these areas produces hypertension, tachycardia, mydriasis, sweating, hyperglycemia, and gastrointestinal inhibition, all reactions of the sympathetic nervous system. If the anterior region of the hypothalamus is stimulated, the reactions are of the parasympathetic type: contraction of the bladder, increased gastrointestinal motility, bradycardia, and vasodilatation. Stimulation of certain other hypothalamic areas results in secretion of adrenaline.

The neurotoxic nature of scorpion venom was apparent to the first investigators: Bert (1885), Talaat (1904), Wilson (1904, 1921), Todd (1909), and Houssay (1919). Progress in the understanding of the poisoning mechanism followed that of the physiology of the nervous system. The work of the Brazilian and Mexican authors are, in that regard, particularly important, and it is not possible in this chapter to study the physiological effects of scorpion toxin without referring to the works of these American authors.

In 1928, de Magalhães demonstrated that the venom acted selectively on the cerebrospinal and sympathetic nerves. He later (1938) specified that the venom acted on the vagus nerve and sympathetic centers of the medulla.

The works of Del Pozo (*cf.* especially 1948, 1956), of Anguiano, and their collaborators (*cf.* especially Anguiano and Adriano, 1959, where the related references can be found) are particularly important.

A. Action on Muscles

The venom provokes fibrillar and fascicular contractions which increase and become clonic. This activity results from a central action on spinal motor neurons. The venom does not act on peripheral nerves or denervated muscle. When electrical or acetylcholine stimulation is applied to a muscle under the influence of venom, larger and longer contractions are obtained; a series of shocks provokes tetany-like responses. When a blocking dose of curare is given, scorpion venom produces a quick and complete decurarization.

These actions suggest an eserine-like anticholinesterase activity, but the measurement of this activity compared with that of eserine on frog abdominal muscles did not confirm an inhibition of cholinesterase by the venom. Other differences have been found between the action of venom and eserine on isolated guinea pig intestine. The venom response was not abolished by atropine, whereas the eserine response was. The effect of scorpion venom on muscle results from its direct action on motor end plates. Very large doses of venom produce a blocking action on the muscular response to nerve stimulation which is not abolished by neostigmine or curare.

Diniz and Gonçalves (1956) demonstrated that the action of venom on muscles was inhibited by atropine and by cocaine, but not by antihistamines. This distinguishes it from the action of histamine and acetylcholine which is not inhibited by cocaine. It is analogous to the action of 5-hydroxytryptamine.

B. Respiratory Effects

The effect of scorpion venom on respiration and the resulting paralysis are due to its action on the respiratory centers in the brain stem. Animals can be kept alive by artificial respiration or by rhythmical stimulation of the phrenic nerve. The respiratory paralysis caused by the venom is not due to a curarization effect, bronchiolar constriction, asphyxia by bronchial secretions, or laryngeal spasm.

C. Vascular Actions

The initial rise of blood pressure following the injection of venom results from a stimulation of the spinal nerves and the resultant increase in adrenaline secretion.

D. Other Reactions

Hyperglycemia, sweating, salivation, and mydriasis are all due to the action of the venom on the vegetative nerve centers.

Egyptian authors have reached the same conclusions: that the venom has a selective action on the sympathetic and parasympathetic branches of the autonomic nervous system. They searched for a neutralizing effect among sympatheticolytic drugs (ergotine, ergotamine) and parasympatheticolytic drugs (atropine) (Mohamed, 1950). Barsoun *et al.* (1954), also confirmed the results of the Brazilian and Mexican authors. Samaan and Ibrahim (1959) proved experimentally that salivary hypersecretion caused by the venom was not due to stimulation of the glands, or the parasympathetic postganglionic end organs but, rather, to a central nervous system stimulation and also, in part, to enhancement of the response of the peripheral neurosecretory salivary mechanisms to chorda tympani impulses.

What is the mode of action of venom on the nerve centers and, especially, on those centers of the hypothalamus? Does poisoning selectively alter the cells, or does it affect the chemical mediators of the synapses such as adrenaline (for the sympathetic system) and acetylcholine (for the sympathetic, parasympathetic, and the neuromuscular systems)? These questions remain unanswered. We know, however, that scorpion venom does not act on monoamine oxidase which could neutralize adrenaline (Anguiano *et al.*, 1957) or on cholinesterase which neutralizes acetylcholine (Del Pozo and Derbez, 1949).

VII. TREATMENT

A. Specific

Antiscorpion serum was first produced in the old world by Todd in Cairo in 1909. Later it was produced in Johannesburg, Algiers (Sergent, 1936, 1938), Tunis (Renoux and Juminer, 1958), Ankara (Tulgat, 1960), and Bombay (Deoras, 1961).

In the Pasteur Institute in Algiers, horses are immunized by injection, twice weekly at first, then once a week, and finally twice a month with increasing quantities of a ground telson preparation from *Androctonus australis*. Scorpion venom is a poor antigen, and immunization requires at least 8 months, and involves the use of from 400 to 500 telsons. We begin with 0.5 telson and increase the dose to 25 telson, a quantity which cannot be exceeded, since larger quantities provoke severe reactions. Penicillin and streptomycin are added to the saline solution used as a diluent. Not all horses produce serum of satisfactory titer. Immunization, moreover, is never complete and certain physiological reactions to the venom are never suppressed, even in horses immunized for several years. These reactions are, principally, local pain at the site of injections (which is always acute and must be alleviated with novocaine) and action on smooth muscle. Thus, 15 to 30 minutes after the injection, the horses experience violent colic accompanied by diarrhea and profuse sweating.

When tested by the method of Ipsen, the neutralizing power of the serum averages 1–1.2 mg/ml of serum. *Androctonus australis* serum exerts a weaker neutralizing action on the venom of other North African species: 0.4 mg venom/ml of serum for *B. arenicola* and *B. occitanus* venoms.

Scorpion antivenins are rather specific. The serum from Algiers affords no protection against venom of *Tityus*, and the *anti-Tityus* serum from Butantan (São Paulo) has a little neutralizing power against *A. australis* venom (Bücherl and Balozet, 1960).

Moreover, Whittmore *et al.* (1961) have studied the paraspecific action of antiscorpion sera prepared in Algiers, Ankara, Johannesburg, Mexico, and Sao Paulo. The serum anti-*A. crassicauda* from Ankara had the widest paraspecific action. It neutralized *A. australis* venom better than the homologous serum. It was as effective as the *A. australis* serum in neutralizing venom of *B. occitanus* and venoms of the South American species. *Tityus serrulatus* and *T. bahiensis*. The antivenins from *A. australis* and *A. crassicauda* gave some protection against the venom of *Centruroides sculpturatus* from Arizona, U.S.A. The *A. crassicauda* venom, therefore, has more important and wider antigenic properties than all other scorpion venoms. However, the 4 heterologous antivenins gave only slight protection against the venom of the Turkish *A. crassicauda*.

The neutralizing power of a serum is measured by injecting a mixture of the serum and venom prepared a certain time before the injection. Its therapeutic activity is something else again, and is important medically. One can measure it by calculating the LD_{50} for several groups of mice which, after an injection of venom, receive a fixed quantity of serum after variable amount of time. The LD_{50} decreases progressively as a function of the time elapsed between the injection of venom and that of the serum, and attains, at the limit, the LD_{50} of venom alone (without serum). This limit marks the moment from which the serum is without action because the venom has already produced irreversible lesions. The antiviper serum from the Pasteur Institute of Algiers has therapeutic activity for a little more than 3 hours after envenomation with the venom of *Cerastes cerastes*, and a little more than 2 hours with the venom of *Vipera lebetina* (Balozet, 1959). The following experiment conducted with the antiscorpion serum of Algiers (neutralizing power, 1 mg/ml) and the venom of *A. australis*, yielded the following results: LD_{50} of venom alone, without serum, 57 μg; venom injected simultaneously but at a different site, with 0.5 ml of serum: 109 μg; serum 5 minutes after the venom: 70 μg; 10 minutes after: 60 μg; 15 minutes after: 57 μg. The therapeutic activity of serum, in mice, is zero when the serum is injected 10 to 15 minutes after the venom (Balozet, 1964). This time of 10 to 15 minutes, for mice, is no doubt longer for man. However, the rapidity of the appearance of grave symptoms in certain cases is well known, as is the fact that central nervous system lesions are reversed with difficulty. Sergent (1943b) concluded, after some medical observations, that the serum began to lose its effectiveness when administered 4 hours after the sting. Effectiveness diminishes rapidly as the time between the sting and serotherapy increases. The serum must be injected as soon as possible, within the hour or within 2 hours after the sting, with 60 to 80 ml injected intramuscularly, even in infants. Hyaluronidase increases the diffusion of the antivenin (Boquet, 1956). Nonspecific therapeutic medication, in addition to the serotherapy, is useful, especially if medical intervention is late.

B. Nonspecific

The first aid recommended in the case of a scorpion sting is the same as that recommended for snake bites: a tourniquet applied a little above the sting, a light incision, and sucking. This can be accompanied by cryotherapy (L.C.-treatment; Stahnke, 1953). The effectiveness of the latter is uncertain and disputable. It seems that, if it is employed immediately after the sting, the absorption of the venom and its diffusion throughout the organism may be slowed down, thus facilitating the neutralizing action of antiscorpion serum which should be injected as soon as possible. The slackening of diffusion of the venom is better if the patient is placed immediately and completely at rest.

The destruction of the venom in the vicinity of the sting by cauterization, the use of potassium permanganate, hypochlorites, ammonia, or phenol is useless. Christensen and Widdicombe (1950) tested the local infiltration of a carbolic soap aqueous 5% solution; some favorable results in guinea pigs injected with the venom of *Parabuthus* were obtained. Calcium gluconate, procaine, and antihistamines have no effect. The injection of large volumes (200 to 500 ml) of a physiological saline solution is recommended (Sergent, 1943a) because the solution dilutes the venom transported by the blood and counterbalances sodium elimination. Glucose solution should not be employed as it increases hyperglycemia.

Hassan and Mohamed (1940) prescribed the combination of a sympathetic drug (ergotoxine) with a parasympathetic drug (atropine) as an antidote in experimental envenomation with the venom of *L. quinquestriatus*. Mohamed (1942) used this combination for the immunization of animals producing serum. Mohamed and el Karemi (1953) replaced atropin by bellafoline (Sandoz) and ergotoxine by ergotamine tartrate for treatment in man. Rohayem (1953) recommends the use of a synthetic sympathetic drug, phentolamine (regitine or C 7337 Ciba) in place of ergotoxine, while Barsoun *et al.* (1954) recommend a dehydrated alkaloid of rye ergot, hydergine.

This combination of sympathetic and parasympathetic drugs is perfectly logical, from the physiological point of view, as compensation for the sympathetico- and parasympatheticomimetic action of the venom. But even these can only diminish the symptoms of poisoning. There is no neutralization of toxin and no true repair of neural lesions. The protective effect of the atropine-ergotamine combination could not be confirmed in the experiments of Anguiano and Cordova (1946).

Mohamed *et al.* (1954b) advise the use of corticoids. Those substances effect a retention of sodium, chlorides, and water, and favor the elimination of potassium. Thus they oppose certain actions of the venom.

Other drugs may be used in symptomatic therapy. These are the analepticcardiovascular drugs camphor and nicethamide. Sparteine increases the effects of adrenaline and induces syncope and therefore must not be used. Tranquilizers such as chlorpromazine are recommended for their ganglioplegic and adrenolytic effect.

Barbiturates, morphine, and morphine derivatives are contraindicated as they have an inhibitory effect on the bulbar respiratory centers.

Artificial respiration can be a great help when respiratory paralysis appears probable. The techniques of refrigeration and hibernation may be useful. Several trials of these techniques have been made in Algeria, but the results are inconclusive.

Reliable statistics cannot be established in a comparison of the effectiveness of different therapeutic treatments. It is certain that rapidity of medical intervention plays the most important role, that serotherapy must be begun as soon as possible, that medical therapy must be initiated and accompany the evolution of symptoms in the victim, and that the patient must be kept under continual medical surveillance.

REFERENCES

Aristotle, VII, 28, § 2.
Adam, K. R., and Weiss, C. (1956). *Nature* **178**, 421–422.
Adam, K. R., and Weiss, C. (1958). *J. Exptl. Biol.* **35**, 39–42.
Adam, K. R., and Weiss, C. (1959a). *Z. Tropenmed. Parasitol.* **10**, 334–339.
Adam, K. R., and Weiss, C. (1959b). *Nature* **183**, 1398–1399.
Adam, K. R., and Weiss, C. (1959c). *Brit. J. Pharmacol.* **14**, 334–339.
Anguiano, G. L., and Adriano, J. A. (1959). *Rev. Fac. Med. Mex.* **1**, 237–245.
Anguiano, G. L. and Cordova, C. A. (1946). *Bol. Inst. Estud. Med. Biol.* (*Mex.*) **4**, 95–100.
Anguiano, G. L., Montano, C. R., and Adriano, J. A. (1957). *Bol. Inst. Estud. Med. Bil.* (*Mex.*) **15**, 97–100.
Balozet, L. (1951). *Arch. Inst. Pasteur Algerie*, 29, 200–207.
Balozet, L. (1952). *Arch. Inst. Pasteur Algerie*, **30**, 1–10.
Balozet, L. (1953). *Arch. Inst. Pasteur Algerie*, **31**, 400–410.
Balozet, L. (1956). *In* "Venoms," Publ. No. 44, pp. 141–144. Am. Assoc. Advance. Sci., Washington, D.C.
Balozet, L. (1959). *Arch. Inst. Pasteur Algerie*, **37**, 387–400.
Balozet, L. (1960). *Arch. Inst. Pasteur Algerie*, **38**, 465–471.
Balozet, L. (1962). *Arch. Inst. Pasteur Algerie*, **40**, 149–178.
Balozet, L. (1964). *Bull. Soc. Pathol. Exotique* **57**, 33–38.
Barros, E. F. (1938). *Mem. Inst. Biol. Ezequiel Dias* (*Belo Horizonte*) **2**, 101–287; *O Hospital* (*Rio de Janeiro*) **14**, 317 and 581.
Barsoun, G. S., Nabawy, M., and Salama, S. (1954). *J. Egypt. Med. Assoc.* **87**, 857–894.
Bert, P. (1865). *Compt. Rend. Soc. Biol.* **17**. 136–139.
Bert, P. (1885). *Compt. Rend. Soc. Biol.* **37**, 574.
Bolivar, J. I., and Ornelas, V. (1949). *Ciencia* (*Mex.*) **9**, 207–208.
Bolivar, J. I., and Rodriguez, G. (1953). *Ciencia* (*Mex.*) **12**, 277–283.

Boquet, P. (1956). *In* "Venoms," Publ. No. 44, pp. 387–391. Am. Assoc. Advance. Sci., Washington, D.C.
Bücherl, W., and Balozet, L. (1960). Unpublished data.
Bulard, M. (1935), *In* "Le scorpion symbole du peuple juif dans l'art religieux des XIVe, XVe et XVIe siècles" (E. de Bocard, ed.). Paris.
Caius, J. F., and Mhaskar, K. S. (1932). *Indian Med. Res. Mem.* **24**, 1–102.
Christensen, P. A., and Widdicombe, M. (1950). *Lancet* 259, 774.
Del Pozo, E. C. (1948). *Gace. med. Mex.* **78**, 387–397.
Del Pozo, E. C. (1956). *In* "Venoms," Publ. No. 44, pp. 123–129. Am. Assoc. Advance. Sci., Washington, D.C.
Del Pozo, E. C., and Derbez, J. (1949). *Rev. Inst. Salubridad Enfermedades Trop.* (*Mex.*) **10**, 203–213.
de Magalhães, O. (1928). *Mem. Inst. Oswaldo Cruz* **21**, 5–159.
de Magalhães, O. (1938). *J. Trop. Med. Hyg.* **41**, 393–399.
de Maupertuis, P. L. M. (1731). *Mem. Acad. Roy. Sci., Paris* p. 223.
Deoras, P. J. (1961). *Probe* (*Calcutta*) **1**, 45–54.
Diniz, C. R., and Gonçalves, J. M. (1956), *In* "Venoms," Publ. No. 44, pp. 131–139. Am. Assoc. Advance Sci., Washington, D.C.
Doerr, R. (1947–1950). "Die Immunitätsforschung," 6 Vols. Springer, Vienna.
Dor, F. (1895). *In* "Cherchell et la commune mixte de Gouraya" (A. Manguin, ed.). Blida, Algérie.
Doro, D. S., Ornelas, E. S., and Johnson, R. M. (1957). *J. Allergy* **28**, 540–541.
Emdin, W. (1933). *Arch. Intern. Pharmacodyn.* **45**, 241–250.
Favilli, G. (1956). *In* "Venoms," Publ. No. 44, pp. 281–289. Am. Assoc. Advance. Sci., Washington, D.C.
Fischer, F. G., and Bohn, H. (1957). *Z. Physiol. Chem.* **306**, 269–272.
Grasset, E., Schaafsma, A., and Hodgson, J. A. (1946). *Trans. Roy. Soc. trop. Med. Hyg.* **39**, 397–421.
Hassan, A., and Mohamed, A. H. (1940). *Lancet* **238**, 1001–1002.
Houssay, B. A. (1919). *J. Physiol. Pathol. Gen.* **18**, 305–317.
Ibn Khaldun (about 1400) "Histoire des Berbères." Trad. baron De Slane, Paul Genthner édit. Paris, 1925. Vol. I, p. 327.
Jaques, R. (1956). *In* "Venoms," Publ. No. 44, pp. 291–293. Am. Assoc. Advance. Sci., Washington, D.C.
Jousset de Bellesme, (1872). *Rev. Mag. Zool.* **23**, 150–156.
Jousset de Bellesme, (1874). *Ann. Sci. Nat., Zool.* **19**, 24.
Joyeux-Laffuie, J. (1882). *Compt. Rend. Acad. Sci.* **95**, 866–869.
Joyeux-Laffuie, J. (1883). Thèse de Médecine, Paris.
Kaiser, E., and Michl, H. (1958). "Die Biochemie der tierischen Gifte." F. Deuticke, Vienna.
Kaku, T. (1954). *Kumamoto Pharm. Bull.* **1**, 21–6.
Kaku, T., and Kaku, I. (1950). *J. Pharm. Soc. Japan* **70**, 83–89.
Kraus, R. (1930). "Serumtherapie der Vergiftungen durch tierische Giftes," 3rd ed., Vol. III–1. Fischer, Jena.
Kubota, S. (1918). *Pharmacol. Exptl. Therap.* **11**, 379–388.
Léon l'Africain (1550). "Description de l'Afrique."
Lévy, R. (1924). *Compt. Rend. Acad. Sci.* **179**, 1093.
Lissitzky, S., Miranda, F., Etzenperger, P., and Mercier, J. (1956). *Compt. Rend. Soc. Biol.* **150**, 741–743.
Maccary, A. (1810). "Mémoire sur le scorpion qui se trouve sur la montagne de Cette." Gabon, Paris.

Mac Clean D. (1942). *J. Path. Bact.* **54**, 284.
Maimonides, Moses (circa 1180). "Specimen dieteticum." Trad. allemande de V. Winternitz, Vienna, 1873.
Maurano, H. R. (1915). Thèse, Rio-de-Janeiro. Collecteanea de trabalhos 1901–1917. Inst. Butantan, São Paulo, Brazil.
Miranda, F. (1964a). Thèse de doctorat en Sciences, Marseille.
Miranda, F. (1964b). Personal Communication.
Miranda, F., and Lissitzky, S. (1958), *Biochim. Biophys. Acta* **30**, 217–218.
Miranda, F., and Lissitzky, S. (1961). *Nature* **190**, 443–444.
Miranda, F., Rochat, H., and Lissitzky, S. (1960). *Bull. Soc. Chim. Biol.* **42**, 379–391.
Miranda, F., Rochat, H. and Lissitzky, S. (1961). *Bull. Soc. Chim. Biol.* **43**, 945–952.
Miranda, F., Rochat, H., and Lissitzky, S. (1962). *J. Chromatog.* **7**, 142.
Miranda, F., Rochat, H., and Lissitzky, S. (1964a). *Toxicon* **2**, 51–69.
Miranda, F., Rochat, H., and Lissitzky, S. (1964b). *Toxicon* **2**, 113–121.
Miranda, F., Rochat, H., and Lissitzky, S. (1964c). *Toxicon* **2**, 123–138.
Miranda, F., Kupeyan, C., Rochat, H. Rochat, C., and Lissitzky, S. (1970). *2nd Int. Symp. Animal Plant Toxins, Tel Aviv, Israel.*
Mohamed, A. H. (1942). *Lancet* **243**, 364–365.
Mohamed, A. H. (1950). *Nature* **166**, 734–735.
Mohamed, A. H., and el Karemi, M., (1953). *J. Trop. Med. Hyg.* **56**, 58–59.
Mohamed, A. H., Rohayem, H., and Zaky, O. (1953). *J. Trop. Med. Hyg.* **56**, 242–244.
Mohamed, A. H., Rohayem, H., and Zaky, O. (1954a). *Nature* **173**, 170.
Mohamed, A. H., Rohayem, H., and Zaky, O. (1954b). *J. Trop. Med. Hyg.* **57**, 85–87.
Mori, H. (1917). *Chosen Igaku-Kwai Zasshi* **16**, 46–51.
Mori, H. (1919). *Japan. Med. Lit.* **4**, 10.
Pallary, P. (1936). *Rev. Africaine* pp. 368–369.
Pauly, A. (1929). "Realencyclopädie der classischen Altertumwissenschaft," Ser. 2, Vol. 3. J. B. Metzlerche, Stuttgart.
Pavan, M. (1952). *Boll. Inst. Sieroterap. Milan.* **31**, 232.
Phisalix, C., and de Varigny, H. (1896). *Bull. Museum Hist. Nat.* **2**, 67–73.
Phisalix, M. (1922). "Animaux venimeux et venins," 2 Vols. Masson, Paris.
Redi, F. (1668). "Esperienze intorno alla generazione degl' insetti." Florence.
Renoux, G., and Juminer, B. (1958). *Arch. Inst. Pasteur Tunis* **35**, 365–376.
Rochat, C., Rochat, H., Miranda, F., and Lissitzky, S. (1967). *Biochemistry* **6**, 578–585.
Rochat, H. (1964). Thèse de doctorat en Pharmacie, Marseille.
Rochat, H. (1969). Thèse de doctorat en Sciences, Marseille.
Rochat, H., Rochat, C., Miranda, F., and Lissitzky, S. (1967). *Biochem. Biophys. Res. Commun.* **29**, 107–112.,
Rochat, H., Rochat, C., Kupeyan, C., Lissitzky, S., Miranda, F., and Edman, P. (1970). *2nd Int. Symp. Animal Plant Toxins, Tel Aviv, Israel.*
Rohayem, H. (1953). *J. Trop. Med. Hyg.* **56**, 150–158.
Samaan, A., and Ibrahim, H. H. (1959). *Arch. Intern. Pharmacodyn.* **122**, 249–264.
Schöttler, W. H. A. (1954). *Am. J. Trop. Med. Hyg.* **3**, 172–178.
Sergent, E. (1936). *Compt. Rend.* **202**, 989.
Sergent, E. (1938). *Arch. Inst. Pasteur Algerie* **16**, 257.
Sergent, E. (1939). *Arch. Inst. Pasteur Algerie* **17**, 625–632.
Sergent, E. (1942a). *Arch. Inst. Pasteur Algerie* **20**, 130–134.
Sergent, E. (1942b). *Arch. Inst. Pasteur Algerie* **20**, 357–358.
Sergent, E. (1943a). *Arch. Inst. Pasteur Algerie* **21**, 24–27.
Sergent, E. (1943b). *Arch. Inst. Pasteur Algerie* **21**, 263–267.

Sergent, E. (1948). *Arch. Inst. Pasteur Algerie* **26**, 253.
Shulov, A. (1939). *Trans. Roy. Soc. Trop. Med. Hyg.* **33**, 353.
Shulov, A. (1955). *Harefuah* **49**, 131–133.
Shulov, A., Weissmann, A., and Ginsburg, H. (1957). *Harefuah* **53**, 309–312.
Stahnke, H. L. (1953). *Am. J. Trop. Med. Hyg.* **2**, 142–143.
Talaat, M. (1904). *1st Cong. Égypt. Méd.*
Tarabini-Castellani, G. (1938). *Arch. Ital. Med. Sper.* **2**, 969–978.
Tetsch, C., and Wolff, K. (1937). *Biochem. Z.* **290**, 394–397.
Todd, C. (1909). *J. Hyg.* **9**, 69–85.
Tulgat, T. (1960). *Turk Ijiyen Tecrubi Biyol. Dergisi* **20**, 201–203.
Valentin, (1876). *Z. Biol.* **12**, 170–179.
Weissmann, A., Shulov, A., and Shafrir, E. (1958). *Experientia* **14**, 175–176.
Weissmann, A., and Shulov, A. (1959). *Arch. Pasteur Algerie* **37**, 202–217.
Whittmore, F. W., Keegan, H. L., and Borowitz, J. L. (1961). *Bull, World Health Org.* **25**, 185–188.
Wilson, W. H. (1904). *Records Egypt. Govt. School Med.* **2**, 9–44.
Wilson, W. H. (1921). *Bull. Inst. Egypte* **3**, 67–73.

VENOMOUS MOLLUSKS

Chapter 57

Mollusks—Classification, Distribution, Venom Apparatus and Venoms, Symptomatology of Stings*

DONALD F. McMICHAEL†

CURATOR OF MOLLUSKS, THE AUSTRALIAN MUSEUM, SYDNEY

I. Introduction 373
II. Poison Cone Shells 374
A. Classification 374
B. Distribution 381
C. Venom Apparatus and Venom 381
D. Symptomatology of Stings 383
III. Octopuses 384
A. Classification 384
B. Distribution 388
C. Venom Apparatus and Venom 388
D. Symptomatology of Stings 390
References 391

I. INTRODUCTION

The active venomous mollusks fall into two of the major classes of the phylum Mollusca. These are the Gastropoda and the Cephalopoda. There are no known venomous species of the other four classes, Monoplacophora, Amphineura, Bivalvia, and Scaphopoda. Among the Gastropoda, several families of carnivorous marine snails are known to produce toxic venoms which are used in the capture of prey. The four families Mitridae, Terebridae, Turridae, and Conidae are grouped in the order Toxoglossa because of similarities in the structure of their venom apparatus, but only the latter are known to be actively venomous to man, a number of species having been responsible for inflicting poisonous stings. Among the Cephalopoda, which

* Received for publication March, 1964.
† Present address: National Parks and Wildlife Service, 189 Kent St., Sydney, Australia.

are all active carnivores, a venom apparatus which is used in the capture of prey is usually present. On a number of occasions poisonous bites have been inflicted on human beings, but all of these were caused by octopuses, which fall into the order Octopoda.

II. POISON CONE SHELLS

A. Classification

The family Conidae includes a large number of species, conservatively estimated at several hundred. Various attempts have been made to subdivide the group into both genera and subgenera, but none of these has been completely successful. For the present purpose, the subgeneric groupings can be disregarded and all species can be referred to a single genus, *Conus*. Within this genus, a number of species have been recorded as causing poisonous stings, though only a few of these records are based on firsthand observations. In some cases the identification of the species can be accepted without question, as the shells responsible were determined by competent malacologists. In other cases, however, the venomous nature of the species has been reported from hearsay or secondhand information, and it is quite possible that there have been mistakes in identification. The cases of *Conus* stingings which have been recorded in the scientific literature are listed in Table I. Much of this information has been taken from the papers of Clench (1946) and Kohn (1958, 1963), but some additional data have been included which are based on newspaper accounts of recent cases.

The known venomous species can be summarized as follows:

1. *Conus geographus* Linné. Definitely responsible for at least four, possibly five, fatal stingings and three others of some severity. It may be regarded as the most dangerous species.

2. *Conus textile* Linné. Apparently responsible for four fairly severe stingings. Some doubt has been cast on the venomous nature of this species by Endean and Rudkin (1963) whose experiments on venom extracts from *C. textile* revealed that they were not toxic for laboratory mice, guinea pigs, or blennies, whereas the same extracts were toxic for worms and mollusks, the normal prey of the species. Endean and Rudkin concluded that "the effects of the venom on vertebrates appear to be restricted to tissue damage at the site of injection and it is unlikely that the venoms would be capable of causing injury to man." These results are in conflict with those obtained by Kohn, Saunders, and Wiener (1960) who found that in some cases at least, laboratory mice were killed by *Conus textile* venom duct extracts. The four recorded cases of stinging by this species include three which can be regarded as doubtfully identified [those of Macgillivray (1860) and Crosse and Marie (1874)] since it appears that in none of these cases was the shell responsible

actually seen by an observer of known competence. However, in the case reported by Kohn, which occurred in Hawaii in 1956, the victim was a shell collector, who should have had no difficulty in identifying the species responsible. One case of stinging by "*Conus textile*" often quoted was originally reported by the Rev. W. Wyatt Gill (1876, pp. 274–275) who included a figure of the shell responsible. This proves that the species was in fact *Conus geographus*, as has been confirmed recently by S. P. Dance in Kohn (1963).

3. *Conus omaria* Linné. Definitely responsible for one severe case and must be regarded as very dangerous.

4. *Conus tulipa* Linné. Definitely responsible for both mild and moderately severe stingings.

5. *Conus aulicus* Linné. Definitely responsible for one mild stinging.

6. *Conus obscurus* Sowerby. Definitely responsible for three mild stingings.

7. *Conus nanus* Sowerby. Definitely responsible for one mild stinging.

8. *Conus pulicarius* Hwass. Definitely responsible for one mild stinging.

9. *Conus lividus* Hwass. Doubtfully responsible for one fairly severe stinging.

10. *Conus imperialis* Linné. Reputedly responsible for one mild stinging.

11. *Conus quercinus* Solander. Apparently responsible for one mild stinging.

12. *Conus litteratus* Linné. Definitely responsible for one mild stinging.

13. *Conus catus* Hwass. Definitely responsible for one mild stinging.

14. *Conus marmoreus* Linné. Reputed to sting (Montrouzier, 1877) but no actual cases reported.

15. *Conus striatus* Linné. Reputed to sting (Hiyama, 1943) but no actual cases reported.

These species are all illustrated in Fig. 1.

Various authors (Kohn, 1959a,b, Endean & Rudkin, 1963) have shown that cone shells fall into three groups as far as their food preferences are concerned. Some species feed only on fishes; some feed only on gastropod mollusks, while a large group of species feed on "worms" (including polychaetes, echiuroids, and some enteropneusts). The experiments of Kohn *et al.* (1960) demonstrate that for the most part the species which feed on fishes have venoms which are highly toxic for vertebrates whereas those which feed on mollusks and worms have venoms which affect vertebrates only mildly. Of the species known to have been responsible for stinging man, *Conus catus*, *Conus geographus*, *Conus obscurus*, and *Conus tulipa* are piscivorous. Other known piscivorous species are *Conus magus*, *striatus*, and *stercusmuscarium*. Species which are known to be molluskivorous include *Conus capitaneus*, *coronatus*, *lividus*, *marmoreus*, *pennaceus*, *quercinus*, *textile*. Species which are known to be vermivorous include *Conus abbreviatus*,

TABLE I

RECORDED CASES OF STINGINGS BY CONE SHELLS

Name	Place and date	Species	Identification	Symptoms	Reference
Native of Matupi	New Britain about 1884	*C. geographus*	Not positive	None described, but small mark at site of bite.	Hinde and Cox, (1884)
Mrs. B.	Levuka, Fiji, 1901	*C. geographus*	Positive (by Charles Hedley from a similar shell)	Numbness at bite, loss of speech, paralysis of voluntary muscles, confusion, eye trouble, no cardiac or respiratory trouble. Recovery after about 24 hours.	Corney (1902), Cleland (1912), Hallen (1914)
M. F. de Lafontaine Aged 32	Ile-aux-Cerfs Seychelles	*C. geographus*	Fairly positive	Sharp sting, burning sensation, spreading numbness, nausea, sight impaired, dizziness, paralysis, difficulty in speech; pulse, respiration both normal.	Hermitte (1946)
Native of Mare	Mare, Loyalty Islands	*C. geographus*	Said to be *C. textile*, but shell figure is clearly *C. geographus*	Painful sensation running up arm to shoulder. Great pain, agony, body swelling, death.	W. Wyatt Gill (1876)
Vu Nhu Ngos (Vietnamese) Aged 23	Vila, New Hebrides, Dec. 1962	*C. geographus*	Unknown	Bitten on abdomen. Acute pains in stomach, recovered after 24 hours' hospitalization.	*Pacific Islands Monthly* July, 1963
Charles Garbutt Aged 27	Hayman Is., Queensland June 27, 1935	*C. geographus*	Positive, shell in Queensland Museum	Slight numbness, spreading to lips, failing eyesight, double vision, drowsiness, paralysis of legs, coma, death.	Flecker (1936)
Name unknown Aged 32	Loc. uncertain, June 29, 1935	*C. geographus*	Unknown	Great pain, temperature rose very high, breathing became difficult, fingertips became purple, unconscious, died within hours.	Yasio (1939)

Theophile Gnai of Koumac	Pouebo, New Caledonia, Oct. 11, 1960	*C. geographus*	Fairly positive	Died within 2 hours.	*France Australe* of 28.10.'60. *Bulletin du Commerce,* Noumea. 5.11.'60, and *Pacific Islands Monthly,* Dec. 1960
Pomo Dassa Poapit Aged 9	Poindimié, East New Caledonia, September, 1963	*C. geographus*	Unknown	No details. Death resulted.	Sydney *Morning Herald* 21/IX/1963
Niuenham (native)	Aneiteum, New Hebrides, June 9, 1860	"Intrag" *C. textile*	Identity reported by Macgillivray, so fairly positive	Numbness of whole arm, agonizing pain, (possibly from tight bandage). Recovery after incisions and release of bandage.	McGillivray (1860)
Unnamed woman	Aneiteum, New Hebrides, May 14, 1859	*C. textile*	Doubtful. Stinging took place 14 days before observations	No early symptoms recorded. Tight bandaging and incisions, possibly caused gangrene and death.	McGillivray (1860)
Native of Pouebo	Pouebo, New Caledonia, before 1874	*C. textile*	Doubtful, not seen by any scientist	Swelling, hand and arm, pain, swelling persisted for some time.	Crosse and Marie (1874)
Adult male	Nanakuli, Oahu, Hawaii, July 5, 1956	*C. textile*	Positive by shell collector	Immediate shortness of breath, severe headaches, extreme nausea, stomach cramps, recovered in 24 hours, except for local pain.	Kohn (1958)
Sir Edward Belcher	Meyo, Moluccas, 1844	*C. aulicus*	Fairly positive	Sensation of burning phosphorus under skin.	A. Adams (1848)
A. Garrett	Tuamotus, before 1874	*C. tulipa*	Positive, by Garrett, a conchologist	Pain like a wasp sting.	A. Garrett (1874)

TABLE I (*Cont'd.*)

Name	Place and Date	Species	Identification	Symptoms	Reference
R. Pahl	Samoa, Dec. 1, 1961	*C. tulipa*	Positive by shell collector	Numb finger, numb arm, chest constricted. Antihistamine injection relieved constriction; symptoms persisted for months.	*Pacific Islands Monthly* for July 1963 from R. Wright in *South Pacific Bulletin* January 1963 Kohn (1963)
Adult male	Kwajalein, Marshall Ids. April, 1954	*C. tulipa*	Positive by shell collector	Pain like bee sting lasting 5–10 minutes.	Kohn (1958)
Native girl Aged 8	Manus Is., August 27, 1964	*C. omaria*	Positive, by Kleckham, shell collector	Slurred speech, shallow breathing, palsy, absence of reflexes, increased heart rate, no temp. elevation, paralysis of respiratory muscles, necessitating artificial respiration. Black spot at site of bite.	Petrauskas (1955)
Adult male	Kahuluu, Hawaii, April, 1947 and Nanakuli, Oahu, April, 1956	*C. obscurus*	Positive, by shell collector	Sharp stinging, edema, redness, swelling, lasted 1 hour, no aftereffects.	Kohn (1958)
Dr. J. Kiser	Waianae, Hawaii, Spring,	*C. obscurus*	Positive, by shell collector	Pain similar to bee sting, immediate blanching, then throbbing and reddening. Still sore, puffy and red after several hours, and slight soreness next day.	*Hawaiian Shell News*, 10,(11), 3
Adult female	Waianae, Oahu, June, 1955	*C. nanus*	Positive by collector.	Intense stinging localized at puncture site. Red spot lasted at least 18 months.	Kohn (1958)

Adult male	Nanakuli, Oahu, Nov. 13, 1955	*C. pulicarius*	Apparently positive by collector	Stinging sensation, slight bleeding.	Kohn (1958)
Adult male	Guam, April, 1957	*C. lividus*	Doubtful, shell not seen as far as known	Severe pain throughout entire arm. No other details.	Kohn (1958)
Adult female	Hope Is., Queensland, 1948	Unidentified, one of several in pocket		Sting, severe headache, plus inflammation.	Kohn (1958)
Child female	Beau Vallon Mahé, Seychelles	*C. imperialis*	?	Severe pain at wound, diminished during 1 hour, fully recovered next day.	Kohn (1958)
Adult female	Oahu, Hawaii, March 22, 1959	*C. quercinus*			*Hawaiian Shell News* 7, (6),60
Adult male	Eniwetok, September, 1959	*C. litteratus*		Like a bee sting, went away after 5 minutes.	Kohn (1963)
Adult female	Guam, date unknown	*C. catus*		Sharp pain, followed by local ache, no worse than wasp sting in a couple of hours.	Kohn (1963) *Hawaiian Shell News* 9,(2),4
Native girl Aged 9	Tanga Is., off New Ireland, New Guinea, November 15, 1963	Unknown		No details. Death resulted.	Sydney *Morning Herald* and Sydney *Daily Telegraph* 16/XI/1963
Native boy Aged 7	Tanga Is., New Guinea, 1963	Unknown		No details. Recovered.	Sydney *Morning Herald* and Sydney *Daily Telegraph* 16/XI/1963

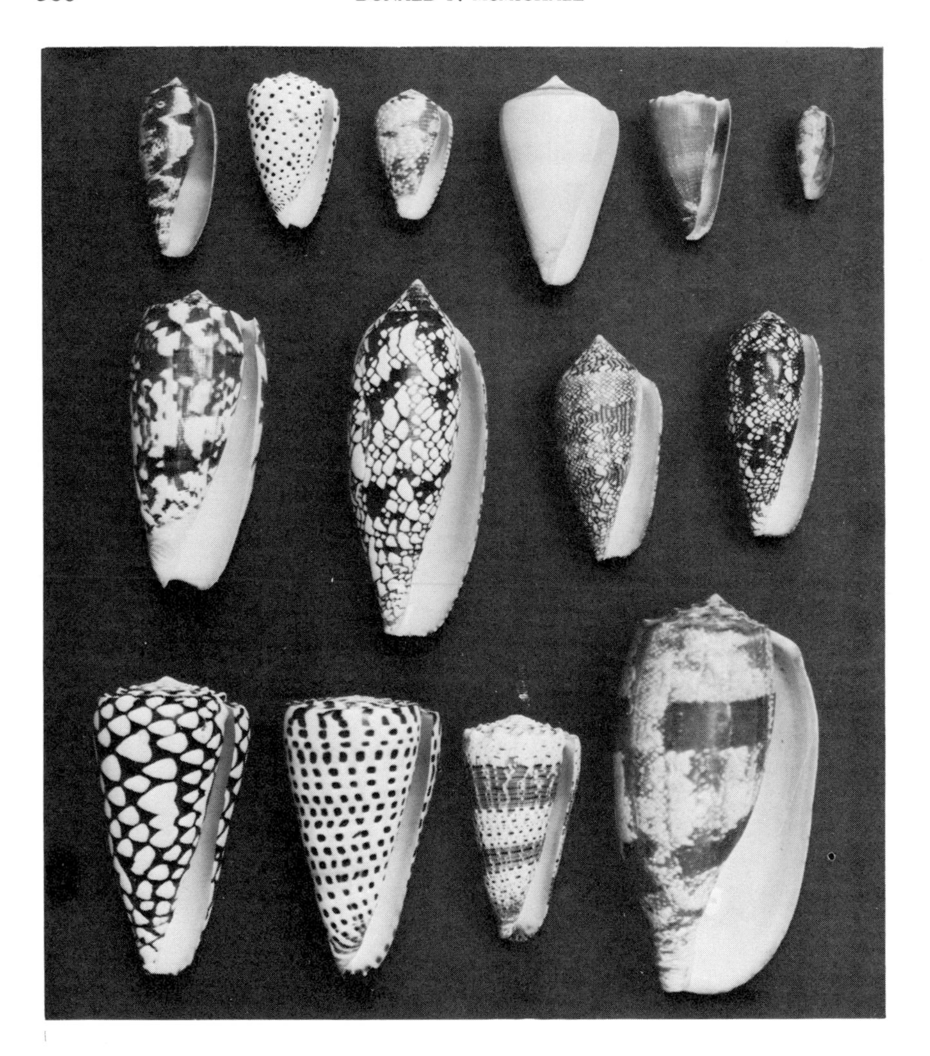

FIG. 1. Venomous cone shells. Top row, left to right: *Conus tulipa, C. pulicarius, C. catus, C. quercinus, C. lividus, C. obscurus*. Center row, left to right: *Conus striatus, C. aulicus, C. textile, C. omaria*. Bottom row, left to right: *Conus marmoreus, C. litteratus, C. imperialis, C. geographus*.

ceylanensis, *chaldeus*, *clavus*, *coronatus*, *distans*, *ebraeus*, *emaciatus*, *figulinus*, *flavidus*, *glans*, *imperialis*, *leopardus*, *lividus*, *miles*, *pertusus*, *pulicarius*, *quercinus*, *rattus*, *sponsalis*, *tiaratus*, *vexillum*, *virgo*, and *vitulinus*.

All piscivorous species should be regarded as highly dangerous and apparently molluskivorous and vermivorous species may cause serious harm to susceptible individuals.

B. Distribution

In common with most cone shells and indeed the majority of tropical Indo-Pacific mollusca, the species which are known to have inflicted poisonous stings on man are all widespread and are found practically throughout the tropical Indian and Pacific Oceans.

C. Venom Apparatus and Venom

The venom apparatus in cone shells consists of a single poison gland and duct, a storage sac in which are housed a number of modified radular teeth and into which the duct of the poison gland opens, and a retractible proboscis which carries the radular teeth forward and thrusts or shoots them into the prey. The essential features of this venom apparatus are shown in Fig. 2.

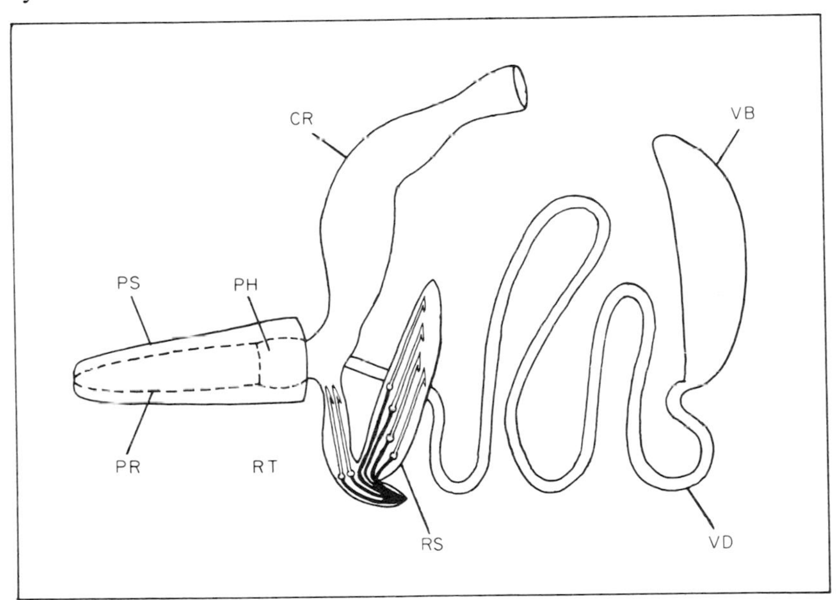

FIG. 2. Venom apparatus of *Conus*. PS, Proboscis sheath; PR, proboscis; PH, pharynx; RS, radular sac; RT, radular teeth; CR, crop; VD, venom duct; VB, venom bulb.

The large, muscular structure which has been known as the poison gland, or venom bulb, is a sausage-shaped organ lying transversely just under the epidermis in front of the visceral mass. The organ is largely muscular, and although it was originally regarded as a poison gland (Bergh, Shaw, Clench), later workers have postulated that it is not concerned with the secretion of venom, but rather with storage and ejaculation (Hermitte, Bouvier, Hinegardner). Hinegardner suggested that the venom was secreted by the long thin, coiled tube leading from the venom bulb to the pharynx. This duct, known as the poison or venom duct, is highly convoluted and has been measured as 130 mm long in a specimen of *Conus textile* (Hinegardner, 1958), and more than five times the length of the shell in *Conus marmoreus* (Kohn *et al.*, 1960). It consists of outer muscular and connective tissue layers, and an inner layer of radially arranged columnar epithelial cells, which have been described as containing granular material. However, Endean (personal communication) has cast some doubt on the histological studies made previously. He believes that venom is secreted in the venom bulb and that the long duct is principally concerned with storage. Whatever the source, it is still certain that the venom is concentrated in the duct and passes from it to the pharynx. The venom duct opens just posterior to the muscular anterior portion of the alimentary canal known as the pharynx, in front of which the extensible proboscis is located. The radular sac opens into the pharynx just anterior to the opening of the venom duct. The radular sac is a modification of that found in most gastropod mollusks, and consists of a somewhat Y-shaped organ of considerable size, which lies immediately in front of the venom bulb. The sac consists of two portions, a long arm in which the radular teeth are believed to develop, and a short arm in which the teeth appear to be held ready for use. The method by which the teeth are transferred from the long arm to the short arm, and then to the proboscis is not known. However, it seems likely that the retraction of the proboscis exerts pressure on the pharynx and this may assist the transfer of the radular teeth to a position where they can be picked up by the proboscis. The teeth are elongate, very fine, arrowlike structures, consisting of a single sheet of chitinous material, rolled up to form a hollow dart. The barbed head of the tooth is formed by indentations in the edge of the chitinous sheet. While lying in the radular sac, the teeth are arranged in pairs and are anchored by ligamental structures to the angle of the sac. The teeth are considered to be homologous with the lateral teeth of the radula; the highly venomous species have teeth as much as 8 mm or more in length.

Presumably the venom is introduced into the hollow tube of the tooth when it lies in the muscular pharynx prior to being carried forward by the proboscis. The precise method by which the tooth is thrust into the prey varies from species to species. Kohn (1956) has shown that in some piscivorous species, the tooth is retained until the tip of the proboscis comes into contact with the fish prey, when it is thrust with great rapidity into the victim. In the case of *Conus striatus*, the tooth is retained by the proboscis, which contracts around the terminal knob and is used to pull the paralyzed prey toward the mouth. In other species, the tooth is released until the prey is paralyzed and the cone subsequently moves up to consume it. Some observers have described the tooth as being shot out from the proboscis like a dart, though such observations have usually been made when a living cone shell has been handled out of water. However, it seems possible that in some cases at least, the *Conus* animal is capable of firing the teeth at its prey from a short distance.

The nature of the venom produced by species of *Conus* is unknown. Endean (personal communication) reports that the venom of *C. textile* consists of spheroidal bodies, 5.7 μ long and 2.0 μ wide, which contain protein-bound amino groups in a polysaccharide-containing fluid. Endean has found that amebocytes play a major role in venom formation. Pharmacologically, the venom of *Conus geographus* caused muscle paralysis, but this did not appear to be by blocking transmission at the neuromuscular junctions (Whyte and Endean, 1962). On the contrary, the venom acted directly on the muscles, and Endean and Rudkin (1963) supposed that the prime cause of death of laboratory mice after injection with venom duct extracts was paralysis of the respiratory musculature.

D. Symptomatology of Stings

The stingings which have been recorded have seldom been observed by trained medical persons. However, there have been a few cases where doctors have been in attendance from a short time after the event, and hence detailed observations on symptoms and treatments were obtained. In general the recorded symptoms fall into two distinct categories: mild and severe. In the cases of stings by nine species (*C. aulicus*, *C. catus*, *C. imperialis*, *C. litteratus*, *C. marmoreus*, *C. nanus*, *C. obscurus*, *C. pulicarius*, and *C. tulipa*) the victims mostly described the symptoms as a sharp stinging, variously described as similar to a bee or wasp sting, sometimes with numbness and usually with redness at the site of the bite which persisted for some time. Residual numbness and muscular soreness was noted in some cases, and in others secondary symptoms included edema, swelling, headache, chest constriction, and in one case, slight bleeding.

In the cases of stings by three species (*C. geographus*, *C. textile*, and *C. omaria*) the symptoms were all severe, sometimes resulting in death. In the case reported by Dr. A. H. Hallen, Resident Medical Officer at Fiji in 1901, the symptoms may be summarized as follows: sharp sting and numbness, bluish discoloration at puncture, paralysis of arm, spreading to all parts of body, and loss of sensation, speech difficulty, throat muscles affected, difficulty in swallowing; recovery began after about 6 hours, paralysis diminished over 24 hours, numbness of finger and residual pain lasted considerably longer. In the case of Charles Garbutt, the symptoms were generally similar, except that the sight was affected after about 20 minutes, and the patient became unconscious after 60 minutes, dying about 5 hours after being stung. The stinging of M. de Lafontaine in the Seychelles, reported by Hermitte (1946), followed a course almost identical with that described by Dr. Hallen. A medical officer, Dr. L. Petrauskas, was present soon after the stinging of a native girl at Manus Island in 1954. The symptoms in this case, which were caused by *Conus omaria*, were generally similar to those described above, except that an early symptom was shallow breathing, and respiratory muscle paralysis became so severe that artificial respiration by an endotracheal tube was necessary, which was maintained for 4 hours, with the administration of oxygen.

III. OCTOPUSES

A. Classification

The octopuses which have been reported as inflicting poisonous bites have seldom been identified as to species with any accuracy. In most cases, they have been small specimens which were not easily identified, or else the individual animal responsible was not collected by the victim. All reported cases of octopus bite are listed in Table II. The names which have been applied to those species available for identification are as follows: *Octopus* c.f. *apollyon* Berry (two cases); *Octopus* sp., possibly *rubescens* Berry; *Octopus fitchi* Berry; *Octopus rugosus* Bosc, and *Hapalochlaena maculosa* (Hoyle). Of these identifications, only *Octopus fitchi* Berry and *Hapalochlaena maculosa* (Hoyle) can be regarded as positive. The small specimens identified as *O.* c.f. *apollyon* may be correctly identified with the largest of all octopuses, especially as the determinations were due to S. S. Berry. The identification of *O. rubescens* Berry as responsible for a bite at Monterey Bay, California, is speculative only. Flecker and Cotton (1955) identified as *Octopus rugosus* Bosc an octopus reputed to be identical with one which caused the death of an Australian skindiver at Darwin in 1954. The specimen was illustrated by a photograph of the preserved animal, and it was described as "iridescent blue" when first produced and about 6 inches long, but subsequently it became a dull blue. Lane (1957) quotes Pickford as saying that the name *Octopus rugosus* is not valid and has been used indiscriminately, for a variety of specimens belonging to other species. Although I have not examined the specimen,* the illustration shows a short-armed species with a rather large web, and the description of the color as iridescent blue suggests that the animal was a specimen of *Hapalochlaena*, which is characterized by all three characters mentioned. Although both *H. lunulata* (Q. and G.) and *H. maculosa* (Hoyle) typically have iridescent blue circles on the body, at least one form regarded by Robson as referable to *H. maculosa* had the pigment uniformly distributed (Robson, 1929, p. 213).

* Identification as *H. lunulata* (Q. and G.) subsequently confirmed [see McMichael, D. (1964). *J. Malac. Soc. Aust.* **1** (8), 23–24].

TABLE II

RECORDED BITES BY OCTOPUS

Name	Place and date	Species	Identification	Symptoms	Reference
Donald A. Simpson	San Francisco, Calif., October 1947	*Octopus* c.f. *apollyon* Berry	Apparently by expert	Sharp scratch, two small skin punctures on back of hand. Profuse bleeding; after a few minutes, tingling spreading to entire hand; later swelling, which persisted after tingling subsided. Heat, redness and itching, which gradually subsided over a 4 week period. No paralysis or sensory loss.	Halstead (1949)
Brother of Harold Fujii	Pearl Harbor, Oahu, Hawaii, Summer, 1927	Unknown	—	Copious bleeding; dizzy spells during next 2 days. No nausea, paralysis, or sensory loss. After 2 days, swelling and reddening which gradually subsided. Healed in about 1 month.	Halstead (1948)
Mrs. Giles W. Mead	Monterey Bay, June/July, 1948	? *Octopus rubescens* Berry	Speculative	Bite like mild bee sting, with moderate bleeding. No pain, no other aftereffects.	Berry and Halstead (1954)
Fred Herms, Jr.	San Francisco, Calif., March 25, 1949	*Octopus* c.f. *apollyon Berry*	Apparently by expert	Two small wounds which bled excessively for 20 minutes. Tingling and pulsating sensation, confined to area of wounds. Pain, comparable to bee sting but more intense, continuing for 1 hour. Swelling, edema, gradual subsidence over 10 days, the wound being sensitive.	Berry and Halstead (1954)
John E. Fitch	San Felipe, Baja, California, April, 1950	*Octopus fitchi* Berry	Positive, by author of species	Sharp stinging pain like a bee or wasp sting. Dark spot with white center; swelling about wound. Only one drop of blood appeared; pain subsided in half-hour, swelling persisted for some hours.	Berry and Halstead (1954)

TABLE II (*Cont'd.*)

Name	Place and Date	Species	Identification	Symptoms	Reference
Harvey Bullis	Biscayne Key, Fla., May, 1950	Unknown	—	Sharp pain, which persisted for several minutes. Small red dot appeared, without noticeable bleeding. Pain disappeared within an hour or two; red dot turned bluish. Itching developed after 5 or 6 days, which persisted for about 1 month.	Berry and Halstead (1954)
Clifton S. Weaver	Oahu, Hawaii. Two unspecified dates	Unidentified	—	Painful sting felt on each occasion.	Weaver (1963)
Unknown	Admiralty Bay, Cook Strait, New Zealand, 1904	Unknown	—	Agonizing bite, immediate swelling which subsided within 3 days. Left side slightly affected. Little pain after actual bite, but a feeling of numbness followed.	McMichael (1958)
E. P. Henn	Wollongong N.S.W., May, 1950	*Hapalochlaena maculosa* (Hoyle)	Almost positive from description	No bite felt. Numbness of mouth and tongue, failing eyesight, weakness in legs, vomiting within 10 minutes, thickening of speech, inability to swallow. No loss of hearing, some mental confusion. Loss of coordination of limbs, paralysis of legs. Gradual recovery during next few days, persistent soreness and swelling of hand, and itchiness which persisted for 4 days.	McMichael (1957)

K. Wyborn	Lake Macquarie, N.S.W., April, 1961	*Hapalochlaena maculosa* (Hoyle)	Almost positive from description	Apparently no painful bite. Gradual onset of nausea, vomiting, loss of coordination of limbs, thickening of speech, difficulty in breathing, skin cold and clammy, tongue coated with white mucus, spasmodic jerking of limbs. Speech and use of arms recovered after 8 hours. Could not walk properly for several days.	McMichael (1951)
Arthur Thompson	Beaumaris Beach near Melbourne, Vic. December, (1962)	*Hapalochlaena maculosa* (Hoyle)	Positive from press photos	No bite felt. Dry feeling in mouth, tingling, numbness spreading to all parts of body, vomiting. Tiny spot of blood on back of hand. Complete paralysis, quite unable to speak. Recovered from paralysis after 6 hours, discharged from hospital after 24 hours. At no time did victim lose his awareness.	*Mal. Soc. Australia. Newsletter* No. 40. (January, 1963), Sydney *Morning Herald* December 26, 27, 1962
D. Baker	Cowes, Victoria, circa January, 1961	*Hapalochlaena maculosa* (Hoyle)	Authority unknown, probably from description	Numbness, illness (=vomiting) and paralysis. Recovery after 24 hours in hospital.	*Mal. Soc. Australia. Newsletter* No. 40 (January, 1963)
Kirke Dyson-Holland	East Point, near Darwin, N. T. September, 1954	"*Octopus rugosus* Bosc" [=*Hapalochlaena lunulata* (Q. and G.)]	Almost positive	No bite felt. Dry feeling in mouth, difficulty in swallowing, small puncture wound and trickle of blood, vomiting, rapid deterioration, loss of coordination, failure of limbs, semiconsciousness, respiration ceased, cyanosis, death within 2 hours of being bitten.	Flecker and Cotton (1955), Lane (1957), McMichael (1964)

Probably a large number of species will prove to be capable of inflicting poisonous bites, but at this stage there are no known criteria by which potentially venomous species can be recognized. All the species listed above belong to the subfamily Octopodinae, but the classification at the generic level may need revision.

B. Distribution

All the venomous bites by octopuses recorded to date have occurred in only six localities—California; Florida; Hawaii; Darwin, Northern Australia; Central New South Wales; and Victoria. This may reflect the fact that the principal interest in bites from these mollusks has been in the United States and Australia. It is probable that other cases have occurred elsewhere which have not been recorded in scientific literature. Lane (1957) gives no other specific cases, though he records that fishermen from Madras, southern India, fear a small octopus, which they claim is capable of inflicting a poisonous bite; Lane does not indicate the identity of the species concerned.

In general, it should be assumed that all octopuses are capable of biting if sufficiently aroused and that there is a high probability that many such bites would be venomous, since octopuses are carnivorous and apparently all possess poison glands which produce a venom to help in capturing their prey. Although there are many instances of octopuses having been handled freely without any bites being inflicted, it should not be assumed that these species cannot bite.

C. Venom Apparatus and Venom

The venom apparatus of *Octopus fitchi* Berry was described by Berry and Halstead (1954) and is the only venom apparatus yet described from species known to inflict poisonous bites. [Flecker and Cotton (1955) diagrammed the venom apparatus of an octopus which is probably a copy of that described for *Eledone* by Isgrove (1909).] However, as all octopuses seem to have essentially similar anatomy, the following generalized description may be given. A pair of sharp, horny, biting jaws, similar in shape to the beaks of a parrot, but with the lower beak projecting in front of the upper, surround the mouth and pharynx; these are embedded in the strong buccal musculature. The esophagus leads from the pharynx to the crop and stomach. A pair of anterior "salivary" glands are situated on the posterior dorsolateral areas of the buccal mass, to which they are bound by fibrous connective tissue. The ducts of these glands arise in the center of the glands and pass directly into the buccal mass to open into the pharynx. A pair of posterior "salivary" glands is situated on either side of the esophagus, between the brain and the hepatopancreas. A duct arises from each gland and passes toward the midline where the ducts from the two glands fuse into a single common salivary duct, which runs forward ventral to the esophagus, passes into the buccal mass and opens into the pharynx. A fifth, sublingual "salivary" gland is located ventral to the radula in the buccal mass. It should be remembered that neither the anterior nor the posterior pair of glands is known to secrete "saliva." All are considered to produce a variety of compounds including the poisons.

The venom apparatus of *Hapalochlaena maculosa* has not been examined previously. Figure 3 shows its structure in a small male specimen from

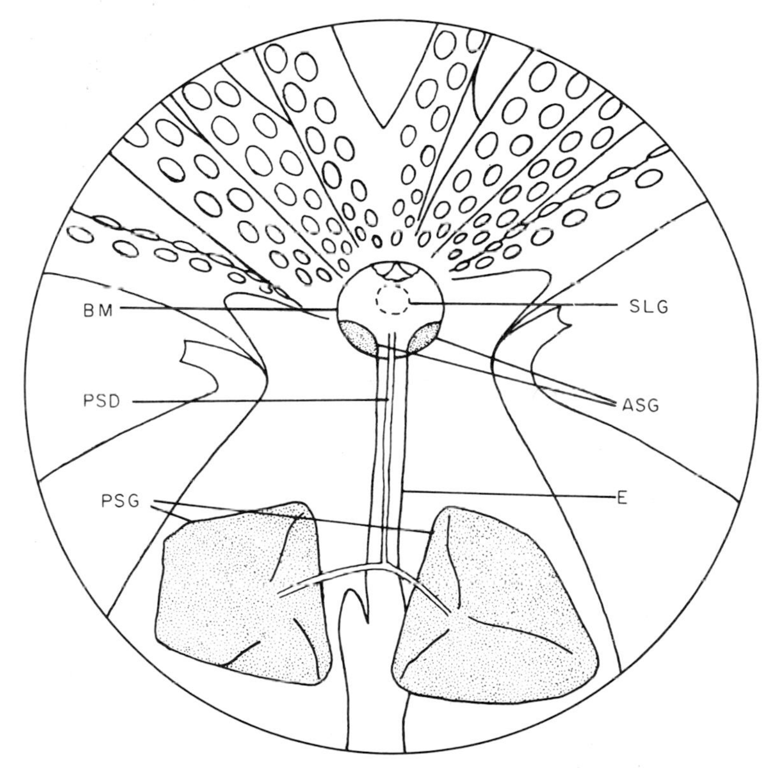

FIG. 3. Venom apparatus of *Hapalochlaena maculosa* (Hoyle), dissected from the ventral surface. BM, Buccal mass; SLG, position of sublingual gland; PSD, posterior salivary gland duct; ASG, anterior salivary glands; E, esophagus; PSG, posterior salivary glands.

Warriewood Beach, near Sydney, New South Wales. The essential features agree with those described for *Eledone* and *Octopus fitchi*, though the posterior pair of glands is relatively very large, almost twice the size of the buccal mass, whereas in the *Eledone* they are a little smaller than the buccal mass. The brilliant coloration adopted by species of *Hapalochlaena* when disturbed (iridescent blue rings and darker patches and bands on a yellow body) may represent a warning coloration associated with its extreme toxicity.

The nature of the venom produced by those octopuses responsible for poisonous bites on human beings is not yet known. Rouville demonstrated many years ago (1910b) that the extract of the salivary glands of *Octopus vulgaris* was toxic to mammals, and the existence of substances toxic to crabs in the posterior salivary glands was demonstrated as long ago as 1888. Ghiretti has recently investigated the nature of the crab-paralyzing poisons of cephalopods, and has isolated a protein from the posterior salivary glands of the cuttlefish *Sepia officinalis*, which when injected into crabs produced the characteristic paralysis caused by injection of the saliva. This protein has been named "cephalotoxin," and has also been extracted from the posterior salivary glands of two species of *Octopus* (Ghiretti, 1959). Tyramine was first isolated from the posterior salivary glands of *Octopus* by Henze (1913) and is thought to be the toxic agent in the venom (Ghiretti, 1959). It is also important in the control of chromatophore action which gives the animal its remarkable ability to change color. The anterior salivary glands also secrete toxic substances which are, however, less potent than those of the posterior salivary glands (Rouville, 1910a). The nature of the secretion produced by the sublingual gland has not been investigated to my knowledge. Romanini (1952) has shown that both the anterior and posterior glands contain hyaluronidase which may act as a spreading agent for the venom.

D. Symptomatology of Stings

The symptoms of octopus bite fall into two very distinct categories. The first of these includes all the bites from localities outside Australia; these have been caused by a number of species, of which only two have been identified with any certainty (*O.* cf. *apollyon* and *O. fitchi*). All these bites have resulted in relatively mild symptoms, ranging from a sharp bite resembling a bee-sting with no subsequent pain or aftereffects, to those in which the bite was followed by copious bleeding, dizziness, swelling and redness, and edema of the affected limb, tingling and throbbing, pain and itching. In general, there was no paralysis or sensory loss, though the case reported from New Zealand mentioned numbness. The initial symptoms of pain and tingling generally disappeared after about 1 hour, whereas the swelling and redness and accompanying itching persisted for varying periods, ranging from "some hours" to "almost a year" but generally for a number of days.

The cases of octopus bite recorded from Australia are possibly all attributable to species of *Hapalochlaena*, and all can be described as causing serious symptoms, in contrast to the relatively mild cases occurring elsewhere. The most curious feature, observed in each case recorded in detail, is the complete absence of initial pain at the time of the bite. Indeed, in most of the cases the fact that the victim had been bitten was not realized at the time. The first symptoms are generally dryness of the mouth, numbness of the lips and tongue, nausea and difficulty in swallowing. The symptoms followed in rapid succession, including numbness spreading to all parts of the body, vomiting, failing eyesight, difficulty or thickening of speech, loss of muscular coordination, leading to complete paralysis of voluntary muscles. In most cases, there was no loss of hearing or mental awareness and in most instances, there was no respiratory distress, though in two cases this was noted, one terminating fatally. The site of the bite usually was revealed by a slight trickle of blood, but there was no copious bleeding, and the wound disappeared quickly, though in one case swelling, soreness, and itchiness of the hand persisted for some days. In all but one of the cases, there was a partial recovery after 6–8 hours, with complete recovery in 1–4 days.

The similarity of these symptoms to those described in cone shell stingings suggest that similar toxic substances are involved, and the paralytic effect may be due to similar modes of action. In such cases, the advantages of artificial respiration until the initial symptoms begin to pass off seem clearly indicated.

REFERENCES

Adams, A. (1848). *In* "Narrative of the Voyage of H.M.S. Samarang," Vol. 2, pp. 356–357. Reeve, Benham & Reeve, London.

Berry, S. S. and B. W. Halstead (1954). Octopus Bites—A Second Report. *Leafl. in Malacol.*, **1**, 59–65, fig. I.

Bergh, R. (1895). Beiträge zur Kenntnis der Coniden. *Nova Acta Leop. Deutsch. Akad. Naturforsch.*, **65**, 67–214, pls. 1–13.

Bouvier, E.-L. (1887). Système nerveux, morphologie générale et classification des Gastéropodes prosobranches. *Ann. Sci. Nat.*, (7), *Zool.*, **3**, 1–510, pls. 1–19.

Cleland, J. B. (1912). Injuries and Diseases of Man in Australia Attributable to Animals (except insects). *Aust. Med. Gaz.*, **32**, 269–274.

Clench, W. J. (1946). The Poison Cone Shell. *Occ. Pap. Mollusks Harv.*, **1**, 49–80, figs. 1–5.

Clench, W. J., and Y. Kondo (1943). The Poison Cone Shell. *Amer. J. Trop. Med.*, **23**, 105–120.

Corney, B. G. (1902). Poisonous Molluscs. *Nature, Lond.*, **65**, 198–199.

Crosse, H., and E. Marie (1874). Catalogue des Cônes de la Nouvelle—Caledonie et des îles qui en dependent. *J. Conchyliol.*, **22**, 333–359.

Endean, R., and Clare Rudkin (1963). Studies of the Venoms of Some Conidae. *Toxicon*, **1**, 49–64.

Flecker, H. (1936). Cone Shell Mollusc Poisoning, With Report of a Fatal Case. *Med. J. Aust.*, April 4, 1936, pp. 464–466, figs. 1, 2.

Flecker, H., and B. C. Cotton (1955). Fatal Bite from Octopus. *Med. J. Aust.*, August 27, 1955, pp. 329–332, figs. 1, 2.

Garrett, A. (1878). Annotated Catalogue of the Species of Conus collected in the South Sea Islands. (*Quart.*) *J. Conchol.*, **1**, 365.

Ghiretti, F. (1959). Cephalotoxin; the Crab-paralysing Agent of the Posterior Salivary Glands of Cephalopods. *Nature, Lond.*, **183**, 1192–1193.

Gill, W. Wyatt (1876). "Life in the Southern Isles: or Scenes and Incidents in the South Pacific and New Guinea," pp. 274–275. Religious Tract Society, London.

Hallen, A. H. (1914). Poisoning by the Bite of *Conus geographus*. *Nautilus*, **27**, 117–120.

Halstead, B. W. (1949). Octopus Bites in Human Beings. *Leafl. in Malacol.*, **1**, 17–22.

Henze, M. (1913). *p*-Oxythenylophylamin das Speicheldrüsenteist der Cephalopoden. *Z. Physiol. Chem.* **87**, 51.

Hermitte, L. C. D. (1964). Venomous Marine Molluscs of the genus *Conus*. *Trans. R. Soc. Trop. Med. Hyg.*, **39**, 485–512, pls. 1, 3, 4, 5.

Hinde, B., and J. C. Cox (1884). Poisonous Effects of the Bite Inflicted by *Conus geographicus* (sic). *Proc. Linn. Soc. N.S.W.*, **9**, 944–946.

Hinegardner, Ralph T. (1958). The Venom Apparatus of the Cone Shell. *Hawaii Med. J.*, **17**, 533–536, figs. 1–6.

Hiyama, T. (1943). Report of an Investigation of Poisonous Fishes of South Seas. *Nissan Fisheries Exp. Stat., Odawara Branch. Spec. Pub.*, 137 pp., pls. 1–29. English Transl. by W. G. van Campen, (1950). *In* "Poisonous Fishes of the South Seas." *Spec. Sci. Rept. Fisheries, U.S. Dept. Interior, Fish & Wildlife Service*, No. 25, 188 pp.

Isgrove, A. (1909). *Eledone. L.M.B.C. Memoir*, **18**, pp. 1–105, pls. 1–10.

Kohn, A. J. (1956). Piscivorous Gastropods of the Genus *Conus*. *Proc. Nat. Acad. Sci., Wash.*, **42**, 168–171, figs. 1–7.

Kohn, A. J. (1958). Recent Cases of Human Injury Due to Venomous Marine Snails of the Genus *Conus*. *Hawaii Med. J.*, **17**, 528–532, fig. 1.

Kohn, A. J. (1959a). The Ecology of *Conus* in Hawaii. *Ecol. Monogr.*, **29**, 47–90. figs. 1–30.

Kohn, A. J. (1959b). Ecological Notes on *Conus* (Mollusca: Gastropoda) in the Trincomalee Region of Ceylon. *Ann. Mag. Nat. Hist.*, (13), **2**, 309–320, figs. 1–3.

Kohn, A. J., P. R. Saunders, and S. Wiener (1960). Preliminary Studies on the Venom of the Marine Snail *Conus*. *Ann. N.Y. Acad. Sci.*, **90**, 706–725, figs. 1–7.

Kohn, A. J. (1963). Venomous Marine Snails of the Genus *Conus*. *In* "Venomous and Poisonous Animals and Noxious Plants of the Pacific Area," pp. 83–96. Pergamon Press, Oxford.

Lane, F. W. (1957). *Kingdom of the Octopus*. Chapter II. Jarrolds, London.

Macgillivray, J. (1860). Zoological Notes from Aneiteum, New Hebrides. *Zoologist*, **18**, 7136–7138.

McMichael, D. F. (1957). Poisonous Bites by Octopus. *Proc. Roy. Zool. Soc. N.S.W.*, 1955–56, 110–111.

McMichael, D. F. (1958). Another Octopus Bite. *Proc. Roy. Zool. Soc. N.S.W.*, 1956–57, 92.

McMichael, D. F. (1963). Dangerous Marine Molluscs. *In* "The Hazards of Dangerous Marine Creatures," pp. 74–80. Publ. by the Committee of the 1960 Internat. Convent on Life Saving Techniques.

McMichael, D. F. (1964). The Identity of the Venomous Octopus responsible for a fatal bite at Darwin, Northern Territory. *J. Malac. Soc. Aust.* **1**, (8), 23–24.

Montrouzier, X. (1877). (Letter to the Editor). *J. Conchyliol.*, **25**, 99.

Petrauskas, L. E. (1953). A Case of Cone Shell Poisoning by "Bite" in Manus Island. *Papua & New Guinea Med. J.*, **1**, 67–68.

Robson, G. C. (1929). "A Monograph of the Recent Cephalopoda. Pt. 1: Octopodinae." British Museum (Nat. Hist.), London.

Romanini, M. G. (1952). Osservazioni sulle ialuronidasi delle ghiandole salivari anteriori and posteriori degli Octopodi. *Pub. Staz. Zool. Napoli*, **23**, 251–270.

Rouville, E. de (1910a). Études physiologiques sur les glandes salivaires des Cephalopodes, et en particulier sur la toxicité de leurs extraits. *C. R. Soc. Biol. Paris*, **68**, 834–836.

Rouville, E. de (1910b). Sur la toxicité des extraits de glandes salivaires des Cephalopodes pour les Mammifères. *C. R. Soc. Biol. Paris,* **68**, 878–880.

Shaw, H. O. N. (1914). On the anatomy of *Conus tulipa,* Linn., and *Conus textile,* Linn. *Quart. J. Micr. Sci.,* **60**, 1–60.

Weaver, C. S. (1963). Tiny Hawaiian Octopus Has Toxic Bite. *Hawaiian Shell News,* **11,** No. 8, 7.

Whyte, J. M., and R. Endean (1962). Pharmacological Investigation of the Venoms of the Marine Snails *Conus textile* and *Conus geographus. Toxicon,* **1**. 25–31, figs. 1–3.

Yasiro, H. (1939). Fatal Bite of *Conus geographus. Venus,* **9**, 165–166.

VENOMOUS COELENTERATES, ECHINODERMS, AND ANNELIDS

Chapter 58

Venomous Coelenterates: Hydroids, Jellyfishes, Corals, and Sea Anemones

BRUCE W. HALSTEAD

WORLD LIFE RESEARCH INSTITUTE, COLTON, CALIFORNIA

I. Introduction 395
  A. Hydrozoa 396
  B. Scyphozoa 397
  C. Anthozoa 397
II. List of Representative Venomous Coelenterates 398
  A. Hydrozoa 398
  B. Scyphozoa 401
  C. Anthozoa 406
III. Nature of Coelenterate Venoms, Medical Aspects 407
  A. Mechanism of Intoxication 407
  B. Venom Apparatus 409
  C. Chemistry of Coelenterate Venoms 410
  D. Pharmacology of Coelenterate Venoms 411
  E. Pathology 412
  F. Clinical Characteristics 413
  G. Treatment 414
  H. Prevention 415
References 416

I. INTRODUCTION

The coelenterates or cnidarians are simple tentacle-bearing metazoans having primary radial, biradial, or radio-bilateral symmetry, composed essentially of two epithelial layers with nematocysts, and but one internal cavity, the gastrovascular cavity or coelenteron, which opens only by the mouth. The group is further characterized by showing a remarkable degree of polymorphism. A single species may present a variety of forms of either the sessile polyp or free-swimming medusoid type. Most coelenterates are attached or sedentary for at least a part of their life span.

The phylum Coelenterata is generally divided into three classes, the Hydrozoa, Scyphozoa, and Anthozoa.

A. Hydrozoa

To this class belong the hydroids which are commonly found growing in plumelike tufts on rocks, seaweeds, and pilings. Small medusae are budded from these branching polyps. The order Siphonophora includes the venomous *Physalia*, which is a planktonic colonial medusa, exhibiting maximum polymorphic development.

Hydroids are generally found attached to a substratum in shallow waters, from low tide level down to depths of 1000 m or more. The ecological conditions in which hydroids are found are extremely variable, fluctuating according to the species. Some species have a wide distribution, whereas others are restricted to certain definite latitudes. Hydroids are generally more abundant in temperate and cold zones. The form of the colony may be radically altered by environmental factors. Colonies are usually small or moderate in size, but some species may attain a length of 2 m. Because of their sessile habits, commensalism with other animals occurs frequently. Hydroids may be attached to pilings, rafts, shells, rocks, or algae, and because of their fine mosslike growth are sometimes mistaken for seaweed. The nematocysts, nettle or stinging cells, are restricted to the tentacles.

The order Hydrocorallina, or hydroid corals, of which *Millepora* is the best known genus, are widely distributed throughout tropical seas in shallow water down to depths of 30 m. Hydroid corals are an important constituent in the formation of coral reefs. They may form upright, clavate, bladelike, or branching calcareous growths, or encrustations over other corals and objects. They vary in color from white to yellow-green, and because of their variable appearance, are sometimes difficult to recognize. According to Hyman (1940), millepores have two or three types of powerful nematocysts which are located on the polyps, polyp bases, and in the general coenosarc.

The order Siphonophora, of which *Physalia* is a pertinent example, are highly polymorphic, free-swimming or floating colonies composed of several types of polypoid or medusoid individuals attached to a floating stem. Siphonophores are pelagic animals, inhabiting the surface of the sea. Their movements are largely dependent on currents, wind, and tides. They are widely distributed as a group, but are most abundant in warm waters. The float or pneumatophore of *Physalia* is greatly enlarged, representing an inverted modified, medusan bell, whereas the remainder of the coenosarc is correspondingly reduced. *Physalia* may attain a large size with a float 10–30 cm in length. From the underside of the float hang gastrozooids, dactylozooids, and the reproductive gonodendra with their gonophores or budding medusoids. The female gonophores are medusoid and may swim free, but the male reproductive zooid remains attached. The gastrozooids or feeding polyps are without tentacles. Some of the tentacled dactylozooids are small, whereas several of the large dactylozooids are equipped with very elongate "fishing" tentacles. These elongate tentacles are equipped with batteries of powerful nematocysts capable of producing a painful sting, and may extend down into the water for a distance of 9 m or more, presenting a dangerous menace to bathers in areas inhabited by these creatures. Considering the large number of fishing filaments on each *Physalia*, this presents a very formidable venom apparatus.

B. Scyphozoa

This class includes the larger medusae having eight notches in the margin of the bell. All Scyphozoans—jellyfishes—are marine and the majority are pelagic. Most jellyfishes are free-swimming in the adult stage. Because of their relatively weak swimming ability, currents, tides, and wind greatly influence their movements. Scyphozoans are widely distributed throughout all seas. Many medusae demonstrate sensitivity to light intensity by surfacing during the morning and later afternoons, and descending during the midday and in the darkness, whereas others react in just the opposite manner. They usually descend during periods of stormy weather. Swimming is accomplished by rhythmic pulsations of the bell, and is used primarily to determine the vertical rather than the horizontal progress of the animal. Jellyfishes display a remarkable ability to withstand considerable temperature and salinity changes. All jellyfishes are carnivorous in their feeding habits, and some of the larger species are capable of capturing and devouring large crustaceans and fishes. A wide variety of sizes, shapes, and colors are displayed by jellyfishes. Many of them are semitransparent or glassy in appearance, often having brilliantly colored gonads, tentacles, or radial canals. They may vary in size from a few millimeters in some species to more than 2 m across the bell with tentacles more than 36 m in length, as in *Cyanea capillata*.

C. Anthozoa

These are the sea anemones, corals, and alcyonarians. The medusa stage is absent in this group, and many of the polyps are colonial. The corals are notable for their precipitation of calcareous skeletal structures and their role as reef builders. Important venomous examples are the anemones, *Sagartia*, *Actinia*, and *Anemonia*.

The nematocyst apparatus, which is one of the morphological characteristics of all coelenterates, makes these animals of significance to venomologists. However, the majority of coelenterates are not harmful to man or are able to inflict only minor irritations.

Actiniarians, or sea anemones, are one of the most abundant of seashore animals. It is estimated that there are about 1000 species. Their bathymetric range extends from the tidal zone to depths of more than 2900 fathoms. They abound in warm tropical waters, but some species inhabit arctic seas. Anemones range in size from a few millimeters to a half meter or more in diameter. Most species are sessile and attached to objects of various kinds but are nevertheless able to creep about to some extent. When they are covered, by water and undisturbed, the body and tentacles are expanded, and because of their variety of colors they frequently have a flowerlike appearance. If the animal is irritated or the water recedes, the tentacles may be rapidly invaginated and the body contracted. The food of anemones consists of mollusks, crustaceans, other invertebrates, and fishes. Most anemones have a short cylindrical body and a flat oral disc margined with a variable number of tentacles around a slitlike mouth. The base or pedal disc of the anemone serves for attachment to objects. The mouth is connected to the enteron by a tubelike gullet. The body is internally divided into radial compartments by septa, in multiples of six. The free inner margins of the septa within the enteron may bear convoluted septal filaments, which are continued as threadlike acontia, bearing nematocysts. On occasion the acontia may be extended through pores in the body wall or through the mouth to aid in subduing prey. Nematocysts are also situated on the marginal tentacles.

II. LIST OF REPRESENTATIVE VENOMOUS COELENTERATES

A. Hydrozoa

MILLEPORIDAE—stinging coral, false coral, fire coral

Millepora alcicornis Linnaeus [Fig. 1(a,b)]

Range: Tropical Pacific, Indian Ocean, and Caribbean Sea.

OLINDIADIDAE

Olindioides formosa Goto

Range: Tokyo Bay, southern Japan, and northwest Pacific.

PHYSALIIDAE

Physalia physalis (Linnaeus)—Atlantic Portuguese man-o-war

Range: Tropical Atlantic, occasionally as far north as the Bay of Fundy and the Hebrides; Mediterranean.

Physalia utriculus (La Martinière)—Pacific Portuguese man-o-war (Fig. 2)

Range: Indo-Pacific, as far north as southern Japan, Hawaii.

PLUMULARIIDAE

Aglaophenia cupressina Lamouroux

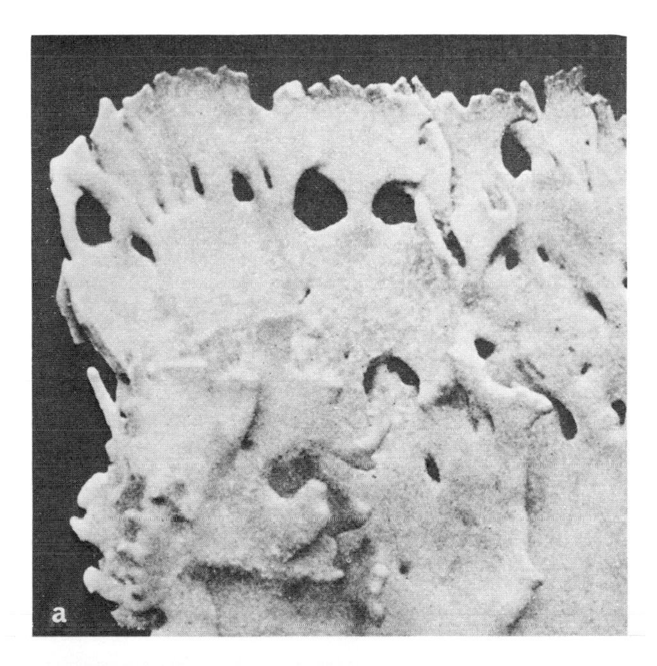

FIG. 1. (a) *Millepora alcicornis* Linnaeus. Colony with fused branches showing platelike growth. (Courtesy of Boschma.) (b) Reef showing growth of *Millepora* sp., fire coral.

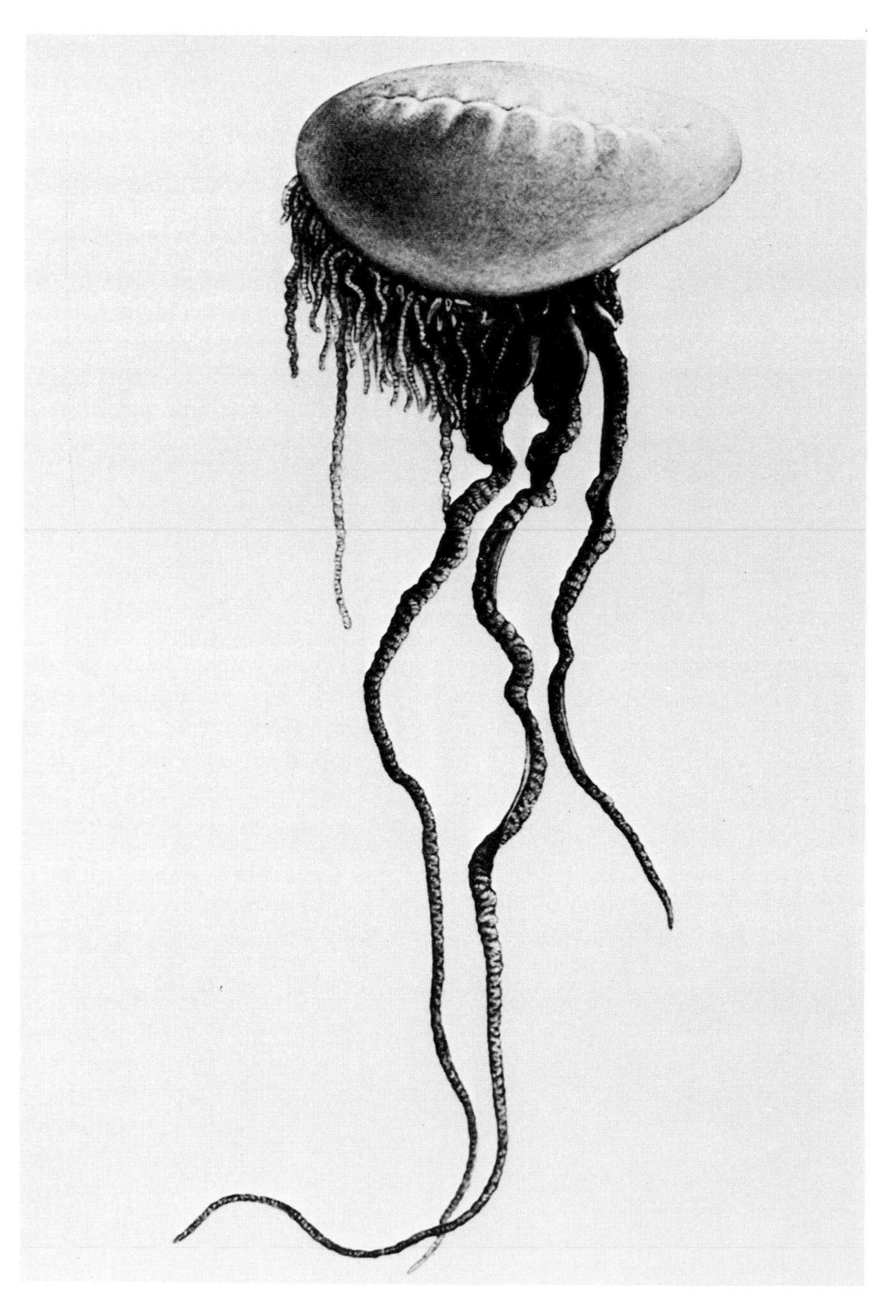

FIG. 2. *Physalia utriculus* (La Martinière). Pacific Portuguese-man-o-war. (M. Shirao.)

Range: Indian Ocean; Zanzibar to North Australia and the Philippines.

Lytocarpus philippinus (Kirchenpauer)—feather hydroid

Range: Tropical and subtropical Pacific, Indian, and Atlantic Oceans; Mediterranean.

B. Scyphozoa

CARYBDEIDAE—sea wasps

Carybdea alata Reynaud (Fig. 3)

Range: Tropical Pacific, Atlantic, and Indian Oceans.

Carybdea marsupialis (Linnaeus)

Range: Atlantic; West Indies to Bermuda Islands, West Africa, and Portugal; Mediterranean.

Carybdea rastoni Haacke

Range: Tropical Pacific; Malayan Archipelago to South Australia, Marquesas Islands, Hawaiian Islands, and southern Japan.

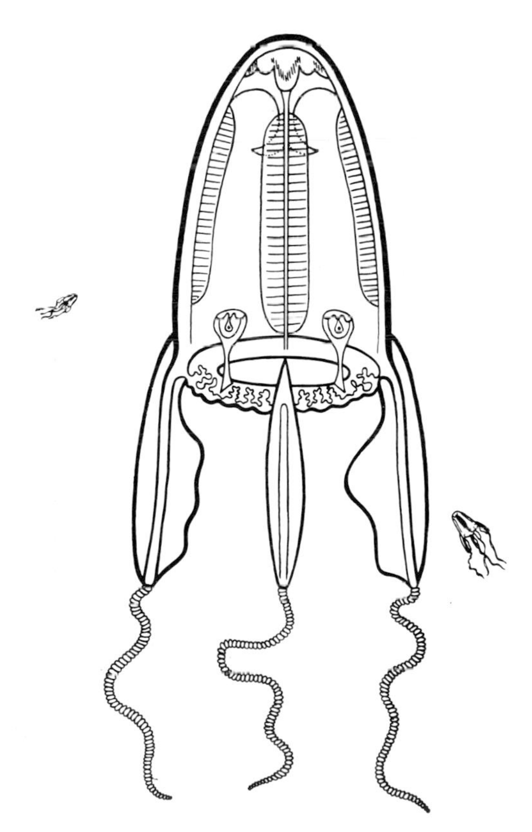

FIG. 3. *Carybdea alata* Reynaud. Sea wasp. (R. Kreuzinger.)

CATOSTYLIDAE—sea blubbers

Catostylus mosaicus (Quoy and Gaimard)

Range: Brisbane to Melbourne, Australia, New Guinea, Philippines.

CHIRODROPIDAE—sea wasps

Chironex fleckeri Southcott (Fig. 4)—sea wasp

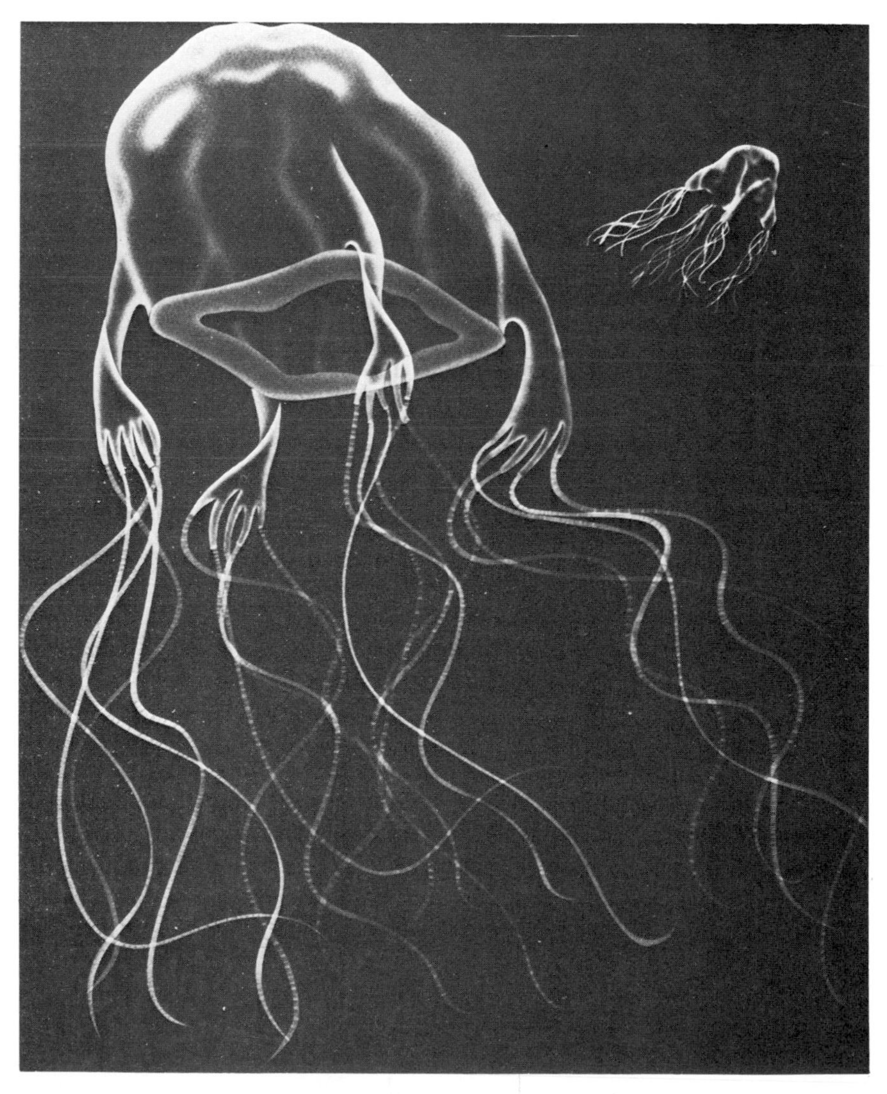

FIG. 4. *Chironex fleckeri* Southcott (after Mayer). Semidiagrammatic. (R. Kreuzinger.) Sea wasp.

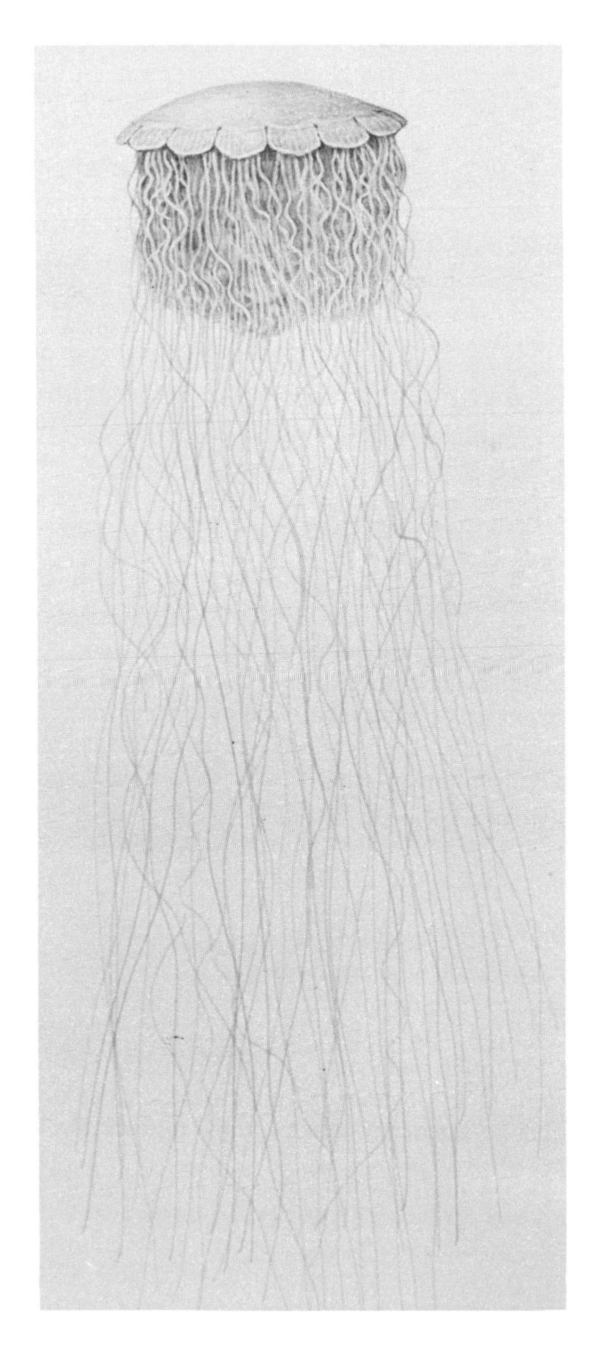

FIG. 5. *Cyanea capillata* (Linnaeus). Sea blubber. (M. Shirao.)

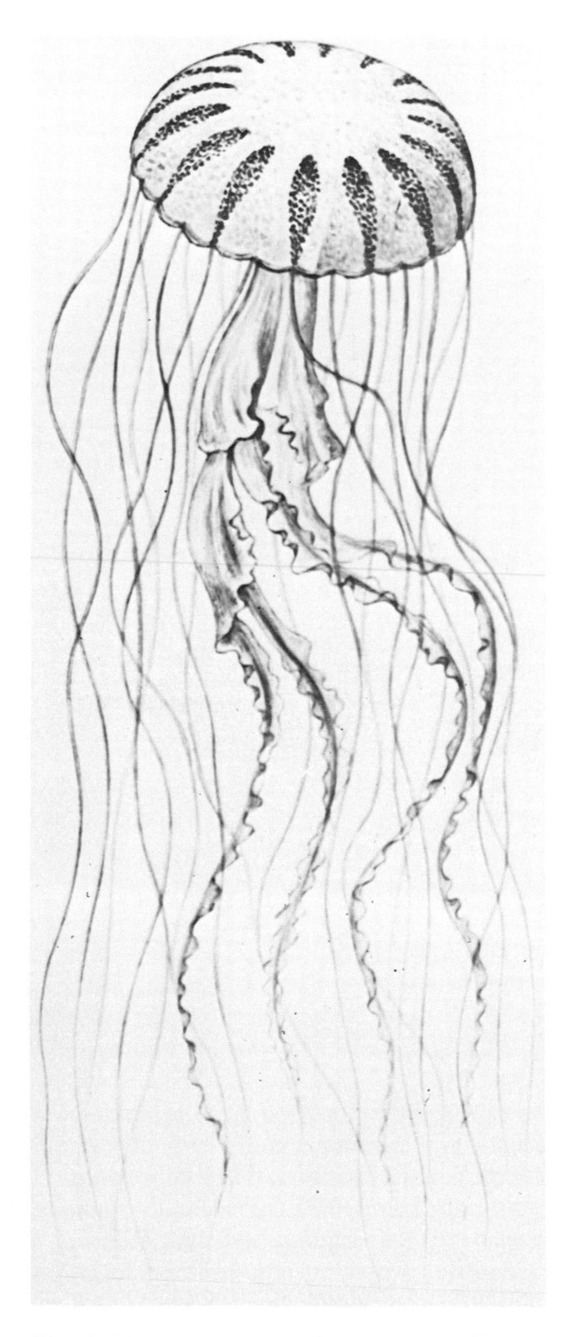

FIG. 6. *Chrysaora quinquecirrha* (Desor). (M. Shirao.)

Range: Northern Territory and North Queensland, Australia.

Chiropsalmus quadrigatus Haeckel

Range: Northern Australia, Philippines, Indian Ocean.

CYANEIDAE—sea blubber, hairy stinger, sea nettle, lion's mane

Cyanea capillata (Linnaeus) (Fig. 5)

Range: North Atlantic and Pacific; southern coast of New England to the Arctic Ocean; France to northern Russia; Baltic Sea; Alaska to Puget Sound, Japan, and China.

PELAGIIDAE

Chrysaora quinquecirrha (Desor)—sea nettle (Fig. 6)

Range: Azores and Woods Hole, New England, to the tropics; West Africa, Indian Ocean, western Pacific from Malayan Archipelago to Japan; Philippines.

Pelagia noctiluca (Forskål)—sea blubbers

Range: Warmer parts of Atlantic, Pacific, and Indian Oceans; West Indies, Florida, to New England.

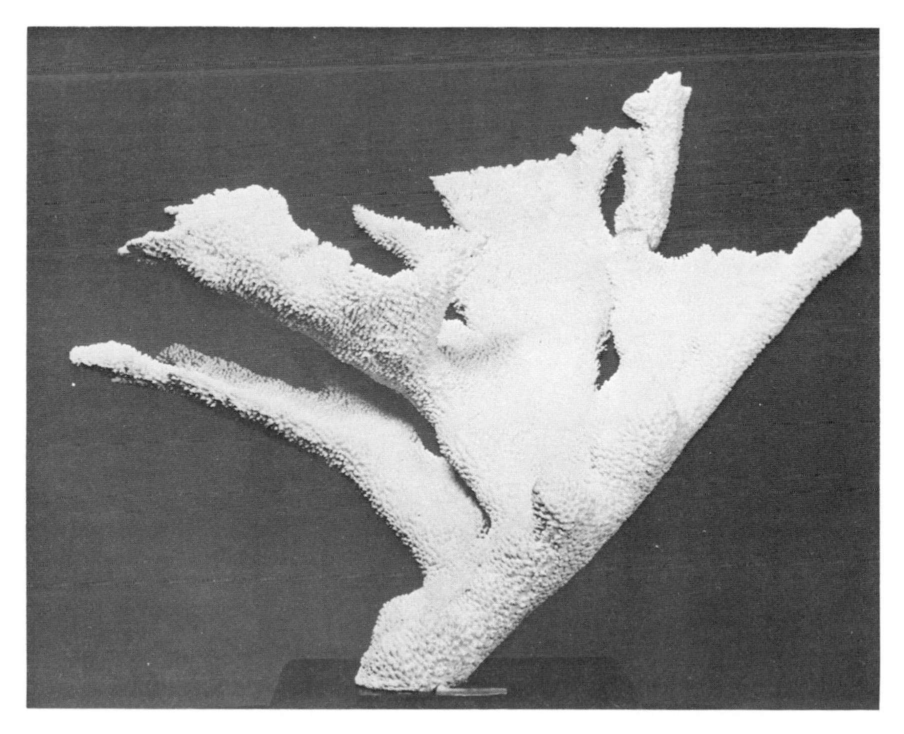

FIG. 7. *Acropora palmata* (Lamarck). (Courtesy of Smithsonian Institution.)

C. Anthozoa

ACROPORIDAE—elk horn coral
Acropora palmata (Lamarck) (Fig. 7)

Range: Florida Keys, Bahamas, and West Indies.
ACTINIIDAE—sea anemone
Actinia equina Linnaeus (Fig. 8)

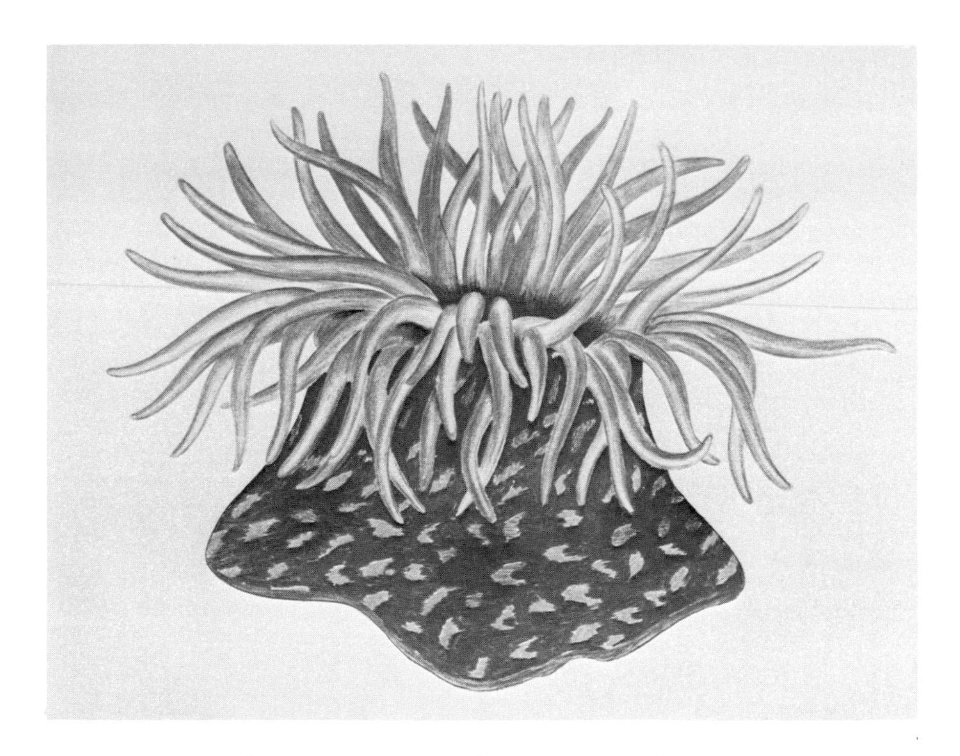

FIG. 8. *Actinia equina* Linnaeus. (R. Kreuzinger.)

Range: Eastern Atlantic from Arctic Ocean to Gulf of Guinea, Mediterranean, Black Sea, Sea of Azov.
Anemonia sulcata (Pennant)
Range: Eastern Atlantic, from Norway and Scotland, to the Canaries, Mediterranean.
SAGARTIIDAE—rosy anemone
Sagartia elegans (Dalyell) (Fig. 9)
Range: Iceland to Atlantic coast of France, Mediterranean Sea and coast of Africa.

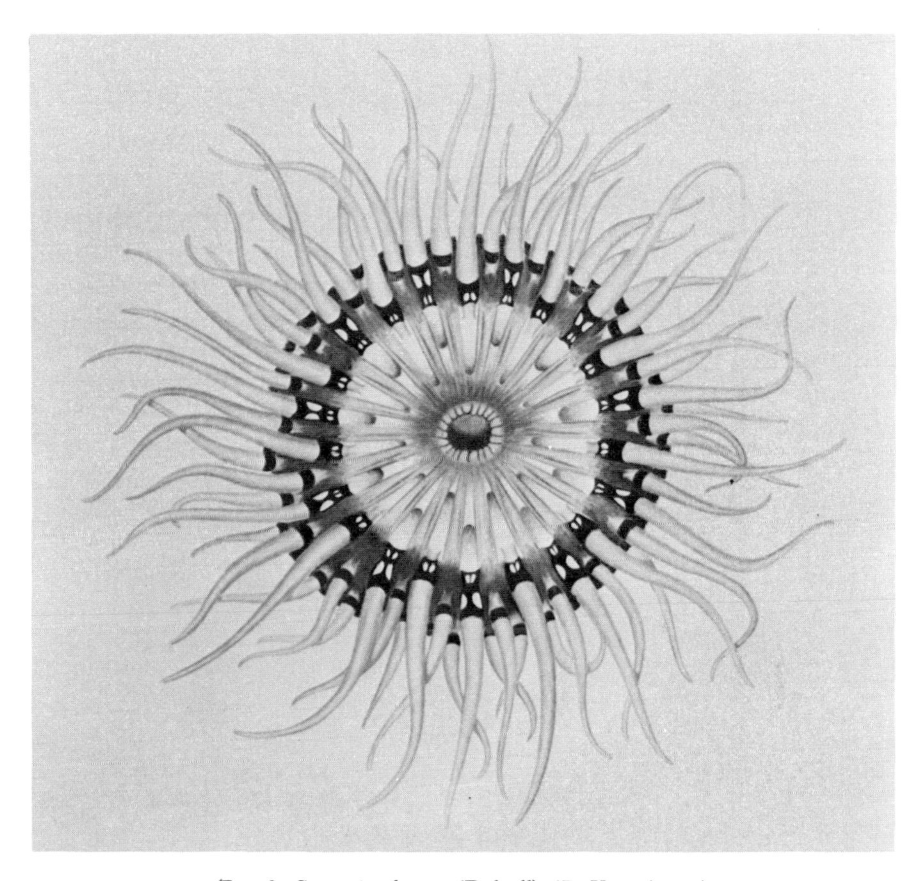

FIG. 9. *Sagartia elegans* (Dalyell). (R. Kreuzinger.)

III. NATURE OF COELENTERATE VENOMS, MEDICAL ASPECTS

A. Mechanism of Intoxication

Most coelenterates inflict their injurious effects upon man by their nematocyst apparatus. The venom is conveyed from the capsule of the nematocyst through the tubule into the tissues of the victim. Theoretically, any coelenterate equipped with a nematocyst apparatus is a potential stinger. The injurious effects may range from a mild dermatitis to almost instant death. The severity of the stinging is modified by the species of coelenterate, the type of nematocyst it possesses, the penetrating power of the nematocyst, the area of exposed skin of the victim, and the sensitivity of the person to the venom.

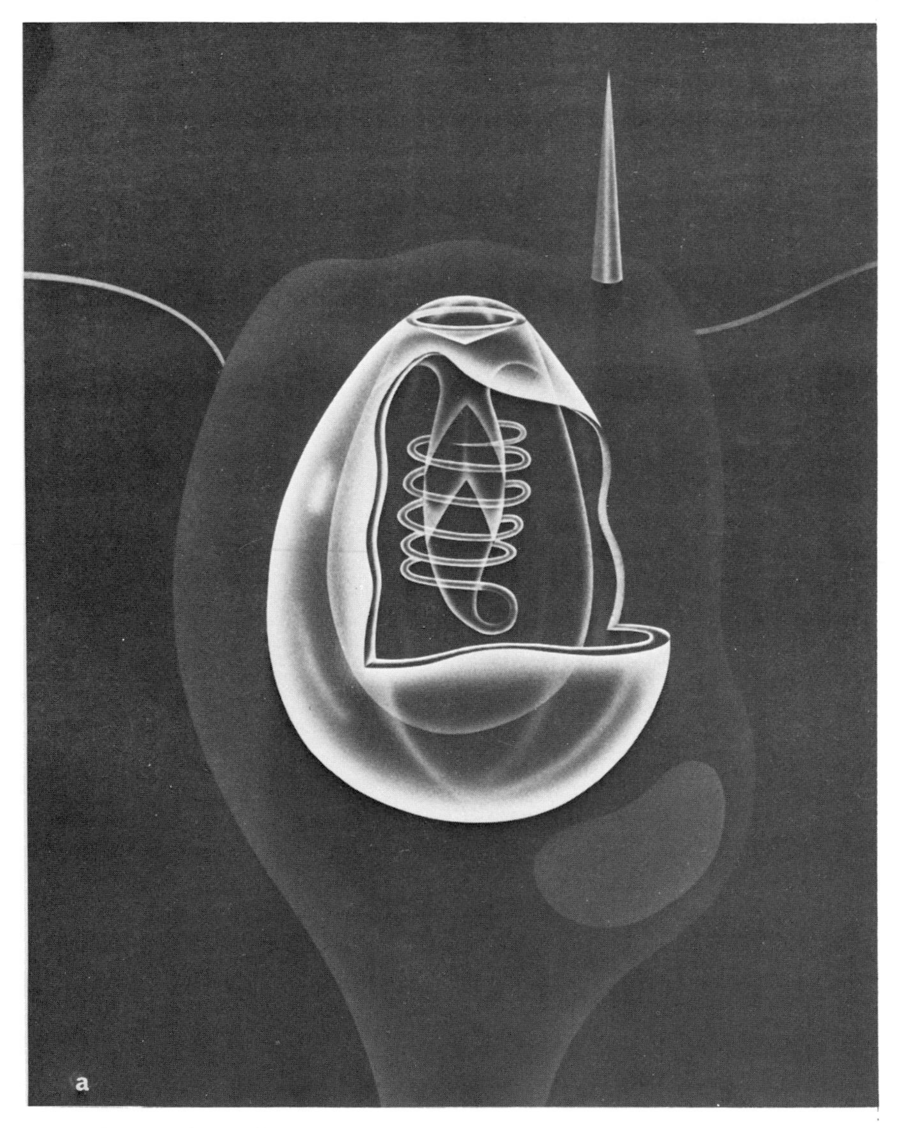

FIG. 10(a). Undischarged nematocyst of a coelenterate. (R. Kreuzinger.)

Some of the sea anemones are poisonous to eat, particularly when raw. It is not known whether oral actinian intoxications are caused by their nematocyst poisons, or are the result of other noxious chemical substances contained in the tissues of their tentacles.

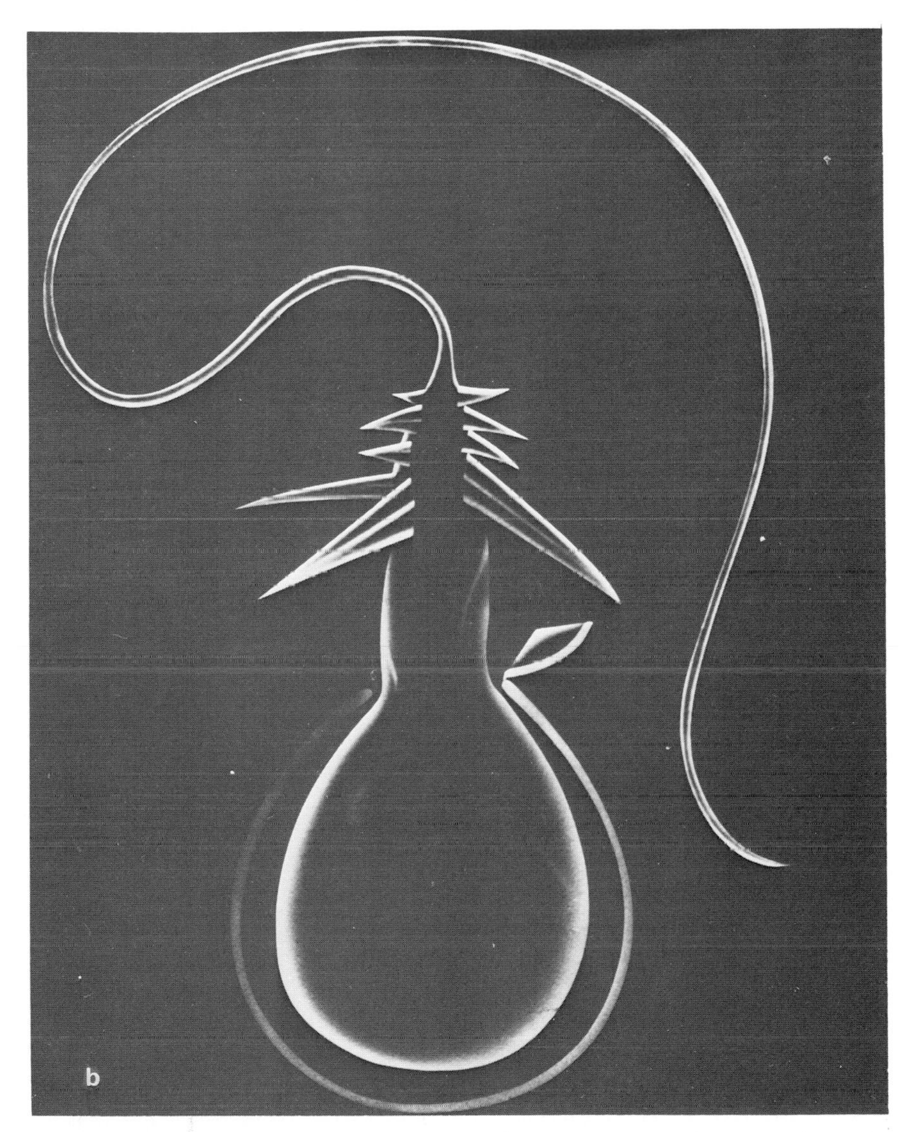

FIG. 10(b). Discharged nematocyst. Note that the operculum has opened and the coiled thread tube has everted. Semidiagrammatic. (R. Kreuzinger.)

B. Venom Apparatus

The venom apparatus of coelenterates consists of the nematocysts, which are largely located on their tentacles. Nematocysts are also called cnidocysts, stinging cells, or nettle cells.

Nematocysts are usually most abundant on tentacles, grouped on protuberances, or circular or spiral ridges. They are also found in the epidermis of the oral region, and internal tentaclelike structures, i.e., the gastric filaments, the septal filaments, and the acontia. Nematocysts initially develop inside interstitial cells, termed cnidoblasts or nematocytes. The nematocyst usually develops at a site some distance from the region in which it is finally utilized. The basal enlargements of the tentacles of medusae are regular developmental sites. The cnidoblasts containing developing nematocysts are transported by amoeboid movement through the body wall or by way of the gastrovascular cavity to their final destination in the ectodermal epithelium.

The structure of the nematocyst is remarkably complex considering their minute size, which ranges from 5 μg to 1.12 mm in length. The largest nematocyst known is found in the siphonophore *Halistemma*, which discharges a tube several millimeters in length. The essential features of the nematocyst apparatus are as follows [Fig. 10(a,b)]: The capsulelike nematocyst is contained within the outer cnidoblast, which is fixed in the epidermis by a slender stalk connecting with the mesoglea. Projecting at one point on the outer surface of the cnidoblast is the triggerlike cnidocil. Near the base of the cnidocil, in the periphery of the cnidoblast, there is usually found a circlet of stiff rods, which probably have a supporting function, and frequently a basketwork of sinuous fibrils, which extend down into the stalk of the cnidoblast. Contained within the fluid-filled capsule is the hollow, coiled, thread tube containing the folded spines. The opening through which the thread tube is everted is closed, prior to discharge, by a lidlike device called the operculum. The fluid within the capsule is the venom.

Stimulation of the cnidocil appears to produce a change in the capsular wall of the nematocyst, causing the operculum to spring open and the thread tube to evert. A variety of chemical substances, including food materials, are also known to stimulate discharge of the nematocyst.

C. Chemistry of Coelenterate Venoms

Most of the early research on the chemistry of coelenterate venoms has been reviewed by Courville, Halstead, and Hessel (1958) and Kaiser and Michl (1958). Aside from the reports on the pharmacology of coelenterate venoms, which also deal to a limited extent with the chemistry of coelenterate poisons, Lane and his co-workers have provided most of the more recent data on the chemical properties of *Physalia* poison (Lane and Dodge, 1958; Lane, 1960, 1961), which is one of the few coelenterate poisons that has been chemically evaluated to any extent.

The activity of *Physalia* toxin is decreased markedly by heating to 60°C. for 15 minutes, and by precipitation with acetone or by extraction with ether. The toxin is nondialyzable. Both before and after hydrolysis, it gives a positive ninhydrin test but negative Benedict's test. The toxin was found to be stable for at least 2 months when stored at $-5°C$. Precipitation of the toxin with alcohol, followed by redissolving in sea water, resulted in some loss of toxicity. Doses lethal for crabs did not cause death for at least 24 hours. However, the characteristic action on the legs was noted immediately. Adsorption on paper followed by elution released only the autotomy-producing activity.

The lethal dose for mice by a solution of the toxin containing 0.201% total nitrogen was 2.1 ml/kilo when injected subcutaneously, and 0.037 ml/kilo when injected intraperitoneally. Death seems to be due to respiratory failure. A dose of 0.5 ml of crude toxin containing 2.43 μg of nitrogen per milliliter was uniformly lethal when injected into the left ventral lymph sac of frogs.

The crude toxin is strongly biuret-positive. The Molish test is negative, suggesting the presence of little or no polysaccharide. The crude toxin withstands lyophilization without significant loss of toxicity. It is separable into component peptides by one-dimensional chromatography in 80% aqueous *n*-propanol or by electrophoretic methods. The crude toxin isolated by lyophilization assays between 15 and 16% nitrogen.

D. Pharmacology of Coelenterate Venoms

Work on the general pharmacological properties of coelenterate poisons is of recent origin and represents the findings of only a few investigators. Most of the studies have been conducted on invertebrate organ systems; relatively little is known of the effects on mammals. The principal actions of coelenterate poisons appear to relate to pain production, paralysis, and the urticarialike effects on the skin.

It has been suggested by Welsh (1956, 1961) that pain production from coelenterate venom is probably due to the direct effect of 5-hydroxytryptamine, a known potent pain-producer and histamine-release, on the pain receptor organs in the skin. The presence of 5-hydroxytryptamine in coelenterate tentacles has been well established by the investigations of Welsh (1956, 1961), Mathias, Ross, and Schachter (1957, 1958), Welsh and Prock (1959), and Welsh and Moorhead (1959).

In his investigations on the effects of tentacular extracts from *Physalia*, and the sea anemones, *Condylactis* and *Aiptasia*, on the fiddler crab, *Uca mordax*, Welsh (1956) found that the tendency to drop or autotomize legs was reduced by the injections of tentacular extracts, and that the crabs exhibited signs of paralysis. The results led to the conclusion that a paralyzing agent of some type was present in the tentacular extract. Studies by Ackermann, Holtz, and Reinwein (1923) have shown that tetramethylammonium chloride, or tetramine, is a normal constituent of coelenterates and is also a paralysant having curarelike properties in vertebrates.

The common clinical findings of wheals, redness, painful burning sensation, urticarial eruptions, etc., in coelenterate stings indicate the presence of histamine or a histamine-releaser substance. Uvnäs' (1960) investigations of the pharmacological properties of extracts prepared from the tentacles of *Cyanea capillata* revealed no histamine, but they did indicate the presence of a histamine-releasing agent. Similar conclusions have been presented by Welsh (1961).

E. Pathology

Very little has been reported in the literature on the autopsy findings in fatal cases of coelenterate stings. Wade (1928) reported on a single fatal case of jellyfish poisoning in the Philippine Islands. The identity of the jellyfish was unknown. The significant findings consisted of acute congestion of the viscera, serous edema of the lungs with diapedesis, acute toxic nephritis with diapedesis, persistent thymus, and moderate lymphoid hyperplasia of the pharynx, duodenum, ileum, and spleen. There was no evidence of an anaphylactic bronchospasm. Scott (1921) states that intense congestion of meningeal and cerebral vessels, and engorgement of the right ventricle have been observed in fatal jellyfish stings. Acute visceral congestion in man and experimental studies on dogs have also been reported by Nobre (1928) and Berntrop (1934). Pawlowsky and Stein (1929) obtained skin biopsies from persons exposed to the tentacles of *Actinia equina*. Pathological findings consisted of acute inflammatory changes of the skin.

Kingston and Southcott (1960) reported on the autopsy findings of two fatal cases in Australia believed to have been caused by a species of cubomedusa. One of the victims was a female, aged 11, and the other a male, aged 38, but the findings were similar in both instances. The lungs and air passages were filled with large quantities of frothy mucus. The abdominal organs, kidneys, and brain were congested, but otherwise normal. The heart was normal. In the case of the male, there was evidence of minute hemorrhages into the substance of the cerebral hemispheres. The skin of the legs of the child showed linear reddish-brown streaks about 6 mm in width where the jellyfish tentacles had adhered. It was estimated that the child died in ½ to 10 minutes, and the adult, in less than 35 minutes. The child had been wading in about 2½ feet of water about 15 yards from shore, in the vicinity of Tully, North Queensland; and the adult had been standing in about 2 feet of water, near Townsville, Queensland.

Small strips of affected skin, which included areas of whealing, were excised from the legs of the victim. Paraffin sections were prepared, stained with eosin, and other routine staining techniques. Microscopic examination revealed numerous nematocysts on the surface of the skin, the threads of which had penetrated through to the dermis. The stratum corneum was edematous and its layers separated. The Malphighian layer was thinned and the cells degenerate, having pycnotic nuclei. No pathological changes were observed in the deep layers. The surface nematocysts were demonstrable with suitable illumination and standard reticulum-staining techniques. The nematocysts were identified as microbasic mastigophores, unusually large, having a distinctive elongate capsule; and they were identical in appearance to those obtained from the tentacles of the cubomedusae, *Chironex fleckeri* and *Chiropsalmus quadrigatus*. Although the coroner's report indicated that the child had died of anaphylactic shock and pulmonary edema, Kingston and Southcott were unable to find any histological evidence of anaphylaxis; they believed that death was a direct result of the jellyfish venom.

F. Clinical Characteristics

The symptomatology of coelenterate stings varies according to the species, the site of the sting, and the person. In general, the symptoms produced by hydroids and hydroid corals are primarily local skin irritations. Sea anemones and true corals produce a similar reaction, but may be accompanied by general symptoms. Ulcerations of the skin may be quite severe in lesions produced by *Sagartia* in the sponge fishermen's disease. Symptoms resulting from siphonophores and scyphozoans vary greatly. The sting of most scyphozoans is too mild to be noticeable, whereas *Cyanea, Catostylus, Dactylometra, Physalia*, and others are capable of inflicting very painful local and generalized symptoms. *Physalia* is reputed to cause death in debilitated persons. *Chironex* is probably the most venomous marine organism known, and may produce death within 3 to 8 minutes in humans (Light, 1914a,b; Wade, 1928; McNeill and Pope, 1943a,b; McNeill, 1945; Flecker, 1952, Southcott, 1952, 1956, 1958a,b, 1959, 1960, 1963).

Contact with the tentacles of coelenterates results in symptoms ranging from an immediate mild prickly or stinging sensation like that of a nettle sting, to an intense burning, throbbing, or shooting pain which may render the victim unconscious. In some cases the pain is restricted to an area within the immediate vicinity of the contact or it may radiate to the groin, abdomen, or armpit, depending upon the initial site of the lesion, or it may become somewhat generalized. The localized pains may be followed by a feeling of numbness, or sometimes hyperesthesia. The area coming in contact with the tentacles usually becomes erythematous, followed by a pronounced urticaria, blistering, swelling, and petechial hemorrhages. In the syndrome known as the sponge fishermen's disease, *Sagartia* may produce a rash that becomes cyanotic, and is followed by multiple abscesses, necrosis, sloughing of the tissues, and ulceration. Jellyfishes particularly may cause redness and flushing of the face, increased perspiration, lacrymation, coughing, sneezing, and rhinitis. An asthmalike condition has been reported (Anon., 1924) as a result of inhaling dust containing dried jellyfish tentacles.

The above-mentioned symptoms may be accompanied by headache, malaise, primary shock, collapse, faintness, pallor, weakness, cyanosis, nervousness, hysteria, chills, and fever. In severe cases there may be muscular cramps, abdominal rigidity, diminished tactile and temperature sensation, nausea, vomiting, severe backache, aphonia, frothing at the mouth, sensation of constriction of the throat, respiratory distress, paralyses, delirium, convulsions, opisthotonus, cardiac standstill, and death. The recovery period varies from a few hours to several weeks.

According to Richet (1905), Scott (1921), Berntrop (1934), Giunio (1948), and others, recurrent sting from coelenterates may result in developing a sensitivity to the toxin, and a subsequent sting may produce a fatal anaphylactic reaction in the victim. Richet (1902, 1993, 1904a,b, 1905) and Portier and Richet (1902a,b) experimentally demonstrated anaphylaxis with the use of *Actinia* venom on dogs, and Binet, Burstein, and Lemaire (1951) have produced anaphylaxis with *Physalia* venom in dogs and guinea pigs.

The so-called "status thymolymphaticus" appears to be present in individuals especially susceptible to stress. Persons having this condition are probably more susceptible to fatalities resulting from coelenterate stings than normal individuals. This situation probably accounts for the high incidence of "status thymolymphaticus" among coelenterate fatalities (Wade, 1928; Berntrop, 1934). Another important factor in the causation of jellyfish-sting fatalities is primary shock and subsequent drowning before the victim can be rescued.

The clinical characteristics of coelenterate stings have been discussed by Zervos (1903, 1934, 1938), Cleland (1912, 1916, 1932), Light (1914a,b), Schwartz and Tulipan (1939), Levin and Behrman (1941), Southcott (1952, 1956, 1958a,b, 1959, 1960, 1962,) Pope (1953), Fish and Cobb (1954), Barnes (1960), and Kingston and Southcott (1960).

G. Treatment

A variety of measures have been suggested for the therapeutic management of coelenterate stings. Treatment must generally be directed toward accomplishing three primary objectives: relieving pain, alleviating neurotoxic effects, and controlling primary shock. Morphine sulfate has been found most effective in relieving pain. Intravenous injections of calcium gluconate have been recommended by Stuart and Slagle (1943), Essex (1945) and De Oreo (1946) for the prompt relief of muscular spasms. According to Waite

(1951), oral antihistaminics and topical cream afford alleviation of urticarial lesions and symptoms. Dilute ammonia hydroxide, sodium bicarbonate, olive oil, sugar, ethyl alcohol, and other types of soothing lotions are sometimes of value (Graham and O'Roke, 1943). Artificial respiration, cardiac and respiratory stimulants, and other forms of supportive measures may be required. There are no known specific antidotes for coelenterate venoms (see also Southcott, 1963).

H. Prevention

Particular care should be exercised by bathers swimming in areas in which dangerous coelenterates, and especially the cubomedusae (*Chiropsalmus*, *Chironex*, *Carybdea*) and *Physalia* are known to exist. It should be kept in mind that the tentacles of some species may trail a great distance from the body of the animal; consequently, they should be given a wide berth. Tight-fitting long woolen underwear or rubber skin diving suits are useful in affording protection against attacks from these creatures. Jellyfishes washed up on the beach, even though appearing dead, may be quite capable of inflicting serious stings. The tentacles of some jellyfishes may cling to the skin. Care should be exercised in the removal of the tentacles, or additional stings will be received. Use a towel, rag, seaweed, stick, or a handful of sand. Swimming soon after a storm in tropical waters in which large numbers of jellyfish were previously present may result in multiple severe stings from remnants of damaged tentacles floating in the water. Upon being stung the victim should make every effort to get out of the water as soon as possible because of the danger of drowning. Dilute ammonia and alcohol should be applied, and other therapeutic measures should be instituted as promptly as possible.

Since coral cuts are frequently involved with coral stings, a few remarks regarding this lesion are pertinent to the general subject at hand. Anyone handling or wading around corals soon suffers from their sharp traumagenic exoskeletons. A mere scratch, left untreated under adverse living conditions, within a few days may become an ulcer with a septic sloughing base surrounded by a painful zone of erythema. In general, the severity of the symptoms appears to be far in excess of the visible lesions present, and is probably due to a combination of factors: laceration of tissues, introduction of foreign substances into the wound, i.e., minute bits of calcium carbonate from the animal's exoskeleton and possibly nematocyst venom, secondary bacterial infection, and adverse climatic and living conditions. Treatment consists of prompt cleaning of the wound, removal of foreign particles, debridement if necessary, and application of antiseptic agents. In severe cases it may be necessary to give the patient bed rest with elevation of the limb, kaolin poultices, magnesium sulfate in glycerin solution dressings, and antibiotics

(Paradice, 1924; Byrne, 1924; Preston, 1950). Levin and Behrman (1941) found roentgen therapy useful in resolving a chronic coral ulcer with keloidal formation.

REFERENCES

Ackerman, D., Holtz, F., and Reinwein, H. (1923). *Z. Biol.* **79**, 113–120.
Anonymous. (1924). *U.S. Nav. Med. Bull.* **20**, 466–467.
Barnes, J. H. (1960). *Med. J. Australia.* **2**, 993–999.
Berntrop, J. C. (1934). *Ned. Tijdschr. Geneesk.* **78**, 2084–2089.
Binet, L., Burnstein, M., and Lemaire, R. (1951). *Compt. Rend. Acad. Sci.* **233**, 565–566.
Byrne, K. (1924). *Med. J. Australia.* **2**, 649–650.
Cleland, J. B. (1912). *Australasian Med. Gaz.* **32**, 269–272.
Cleland, J. B. (1916). *6th Rept. Govt. Bureau Microbiol., Dept. Public Health, N.S.W.*, p. 266–276.
Cleland, J. B. (1932). *Med. J. Australia.* **4** 157–166.
Courville, D. A., Halstead, B. W., and Hessel, D. W. (1958). *Chem. Rev.* **58**, 235–248.
De Oreo, G. A. (1946). *Arch Dermatol. Syphilol.* **54**, 637–649.
Essex, H. E. (1945). *Physiol. Rev.* **25**, 148–170.
Fish, C. J., and Cobb, M. C. (1954). *U.S. Fish Wildlife Serv., Res. Rept. No.* **36**, p. 17-20.
Flecker, H. (1952). *Med. J. Australia.* **1**, 35–38.
Giunio, P. (1948). *Higijena (Belgrade).* **1**, 282–318.
Graham, S. A., and O'Roke, E. C. (1943). *In* "On Your Own; How to Take Care of Yourself in Wild Country," p. 56–73. Univ. Minnesota Press, Minneapolis.
Hyman, L. H. (1940). "The Invertebrates: Protozoa through Ctenophora." McGraw-Hill Book Co., New York.
Kaiser, E., and Michl, H. (1958). "Die Biochemie der Tierischen Gifte." Franz Deuticke, Wein.
Kingston, C. W., and Southcott, R. V. (1960). *Trans. Roy. Trop. Med. Hyg.* **54**, 373–384.
Lane, C. E. (1960). *Ann. N.Y. Acad. Sci.* **90**, 742–750.
Lane, C. E. (1961). *In* "The Biology of Hydra" (H. M. Lenhoff and W. F. Loomis, eds.), p. 169–178. Univ. Miami Press, Coral Gables.
Lane, C. E., and Dodge, E. (1958). *Biol. Bull.* **115**, 219–226.
Levin, O. L., and Behrman, H. T. (1941). *Arch. Dermatol. Syphilol.* **44**, 600–603.
Light, S. F. (1914a). *Philippine J. Sci.* **9**, 195–231.
Light, S. F. (1914b). *Philippine J. Sci.* **9**, 291–295.
McNeill, F. A. (1945). *Med. J. Australia* **2**, 29.
McNeill, F. A., and Pope, E. C. (1943a). *Australian J. Sci.* **5**, 188–191.
McNeill, F. A., and Pope, E. C. (1943b). *Australian Mus. Mag.* **8**, 127–131.
Mathias, A. P., Ross, D. M., and Schachter, M. (1957). *Nature* **180**, 658–659.
Mathias, A. P., Ross, D. M., and Schachter, M. (1958). *J. Physiol.* **142**, 56–57.
Nobre, A. F. (1928). *Inst. Zool. Univ. Porto* **1**, 1–8.
Paradice, W. E. (1924). *Med. J. Australia* **2**, 650–652.
Pawlowsky, E. N., and Stein, A. K. (1929).
Pope, E. C. (1953). *Australian Mus. Mag.* **11**, 111–115.
Portier, P., and Richet, C. (1902a). *Compt. Rend. Acad. Sci.* **134**, 247–248.
Portier, P., and Richet, C. (1902b). *Compt. Rend. Acad. Sci.* **54**, 170–172.
Preston, F. S. (1950). *Brit. Med. J.* **1**, 642–644.

Richet, C. (1902). *Compt. Rend. Soc. Biol.* **54**, 837–838.
Richet, C. (1903). *Compt. Rend. Soc. Biol.* **55**, 707–710.
Richet, C. (1904a). *Compt. Rend. Soc. Biol.* **56**, 302–303.
Richet, C. (1904b). *Compt. Rend. Soc. Biol.* **56**, 775–777.
Richet, C. (1905a). *Compt. Rend. Soc. Biol.* **58**, 109–112.
Richet, C. (1905b). *Compt. Rend. Soc. Biol.* **58**, 112–115.
Schwartz, L., and Tulipan, L. (1939). *In* "Occupational Disease of the Skin," p. 603–605. Lea & Febiger, Philadelphia.
Scott, H. H. (1921). *In* "The Practice of Medicine in the Tropics" (W. Byam and R. G. Archibald, eds.), Vol. I, p. 790–798. H. Frowde & Hodder & Stoughton, London.
Southcott, R. V. (1952). *Med. J. Australia* **1**, 272–273.
Southcott, R. V. (1956). *Australian J. Marine Freshwater Res.* **7**, 254–280.
Southcott, R. V. (1958a). *S. Australian Nat.* **32**, 53–59.
Southcott, R. V. (1958b). *Discovery* **19**, 282–285.
Southcott, R. V. (1959). *Military Med.* **124**, 569–579.
Southcott, R. V. (1960). *Good Health* **113**, 18–23.
Southcott, R. V. (1963). *In* "Venomous and Poisonous Animals and Noxious Plants of the Pacific Region" (H. L. Keegan and W. V. MacFarlane, eds.), p. 46–65. Pergamon Press, Oxford.
Stuart, M. A., and Slagle, T. D. (1943). *U.S. Nav. Med. Bull.* **41**, 497–501.
Uvnäs, B. (1960). *Ann. N.Y. Acad. Sci.* **90**, 751–759.
Wade, H. W. (1928). *Am. J. Trop. Med.* **8**, 233–241.
Waite, C. L. (1951). *U.S. Armed Forces Med. J.* **2**, 1325–1326.
Welsh, J. H. (1956). *Deep Sea Res., Suppl.* **3**, 287–297.
Welsh, J. H. (1961). *In* "The Biology of Hydra" (H. M. Lenhoff and W. F. Loomis, eds.), p. 179–186. Univ. Miami Press, Coral Gables.
Welsh, J. H., and Moorhead, M. (1959). *Science* **129**, 1491–1492.
Welsh, J. H., and Prock, P. B. (1958). *Biol. Bull.* **115**, 551–561.
Zervos, S. G. (1903). *Semaine Méd.* **25**, 208–209.
Zervos, S. G. (1934). *Paris Méd.* **93**, 89–97.
Zervos, S. G. (1938). *Bull. Acad. Méd.* **119**, 379–395.

Chapter 59

Venomous Echinoderms and Annelids: Starfishes, Sea Urchins, Sea Cucumbers, and Segmented Worms

BRUCE W. HALSTEAD

WORLD LIFE RESEARCH INSTITUTE, COLTON, CALIFORNIA

I. Echinoderms 419
A. Asteroidea—Venomous Starfishes 420
B. Echinoidea—Venomous Sea Urchins 424
II. Venomous Annelids 433
References 440

I. ECHINODERMS

Echinoderms are characterized by having radial symmetry, with the body usually consisting of five radii around an oral-aboral axis, comprised of calcareous plates which form a more or less rigid skeleton, or with plates and spicules embedded in the body wall. Spines and pedicellariae are present in the asteroids and echinoids, but are absent in some of the others. The coelom is complex and includes a water vascular system with tube feet. The digestive tract may be with or without an anus. The sexes are usually separate. Echinoderms, with the exception of a few planktonic holothurians, are all benthic marine organisms.

The phylum Echinodermata contains only two classes that are of interest to the venomologist, and they are the Asteroidea (starfishes) and the Echinoidea (sea urchins and their relatives). Several of the other classes contain poisonous species, but none of the other groups contain members known to possess a true venom apparatus.

A. Asteroidea—Venomous Starfishes

The class Asteroidea is represented by a single family of venomous starfishes, the Acanthasteridae, which contain at least two species, *Acanthaster planci* (Linnaeus) (Fig. 1), an Indo-Pacific form that ranges from Polynesia to the Red Sea, and *A. ellisi* (Gray), found in the Gulf of California, Galapagos, and Clarion Islands. Despite the relatively large size of this starfish (diameter up to 60 cm or more), it is a relatively inconspicuous reef inhabitant —its color is brown, or gray to red and blends closely with the surrounding

FIG. 1. *Acanthaster planci* (Linnaeus). Venomous starfish. Taken at Heron Island, Great Barrier Reef, Australia. (Courtesy of K. Gillett.)

corals which it generally feeds upon. Little is known about the biology of *A. planci*, and even less is known regarding *A. ellisi*. The long pungent spines that cover the aboral surface of this starfish make it a real hazard to anyone who comes in contact with it.

1. Mechanism of Envenomation

Wounds are inflicted by means of the long, stinging, spines situated on the aboral surface of the starfishes *A. planci* and *A. ellisi*. The spines are enveloped by an integumentary sheath which secretes a venom. Wounds are usually contracted as a result of either stepping upon the spines, or careless handling of the starfish.

2. Venom Apparatus

Acanthaster planci attains a large size, up to 60 cm or more in diameter, and possesses 13 to 16 rays or arms. The outer surface of the entire body is covered by a series of large, pungent spines, which are completely enveloped in a thin layer of rugose integument. The spines may measure up to 6 cm or more in length, and tend to be somewhat less friable than those of equivalent size in sea urchins.

Histological sections reveal that the spines of *A. planci* are covered by an integument consisting of an outer cuticle, an epidermis, and an underlying dermis, which surrounds the endoskeletal calcareous supportive structure of the spine [Fig. 2(a-c)]. The epidermal cells are tall, flagellated, and apparently of pseudostratified columnar epithelial type. The epidermal cells are interspersed with neurosensory cells, pigment cells, and two types of glandular cells: acidophilic cells having a densely granular cytoplasm, and a basophilic cell having a light-staining, vacuolated or reticulated cytoplasm, resembling a sebaceous type of cell. The acidophilic cells are peripheral in their distribution whereas the basophilic cells are for the most part situated adjacent to the basement membrane. There are observed at periodic intervals along the outer surface of the epidermal layer indentations or small accumulations of acidophilic glandular cells, which lie wholly within the epithelium and contain their own small lumen. They have the appearance of a typical intra-epithelial gland [Fig. 2(c)]. It is believed that the venom is produced by these acidophilic cells. The basophilic cells are probably mucus-producing cells. However, the exact site of venom production remains to be determined.

3. Chemistry of Starfish Venom

Unknown.

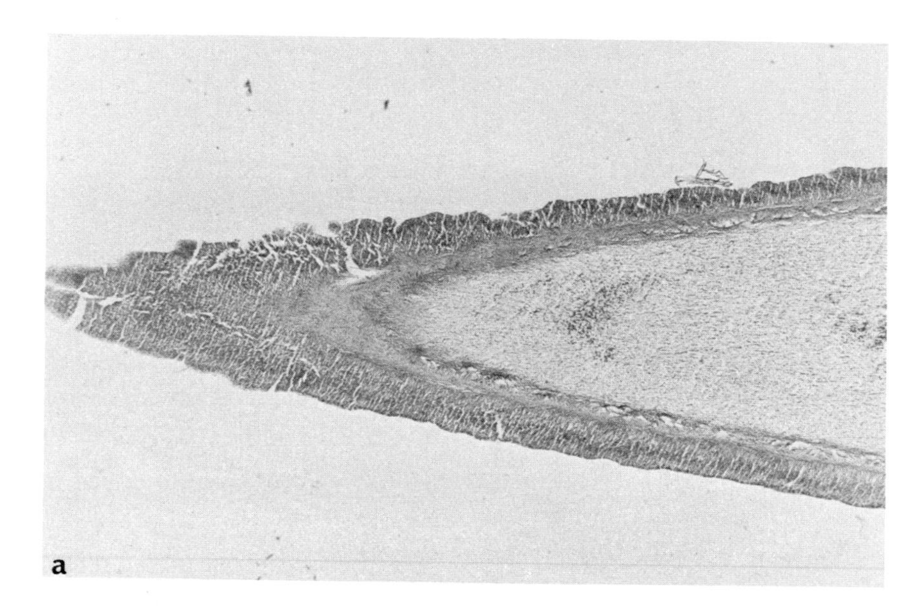

FIG. 2a.

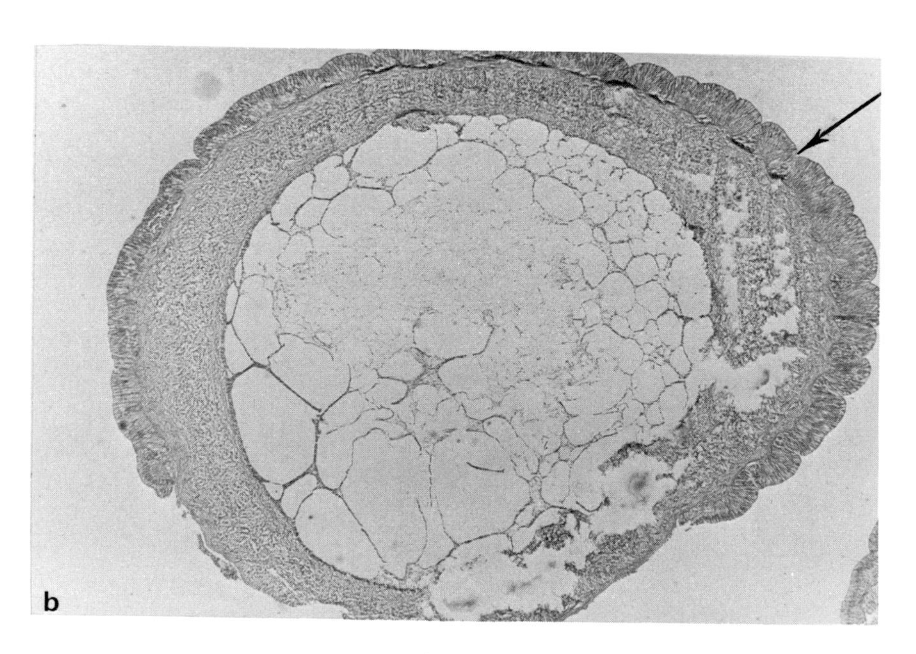

FIG. 2b.

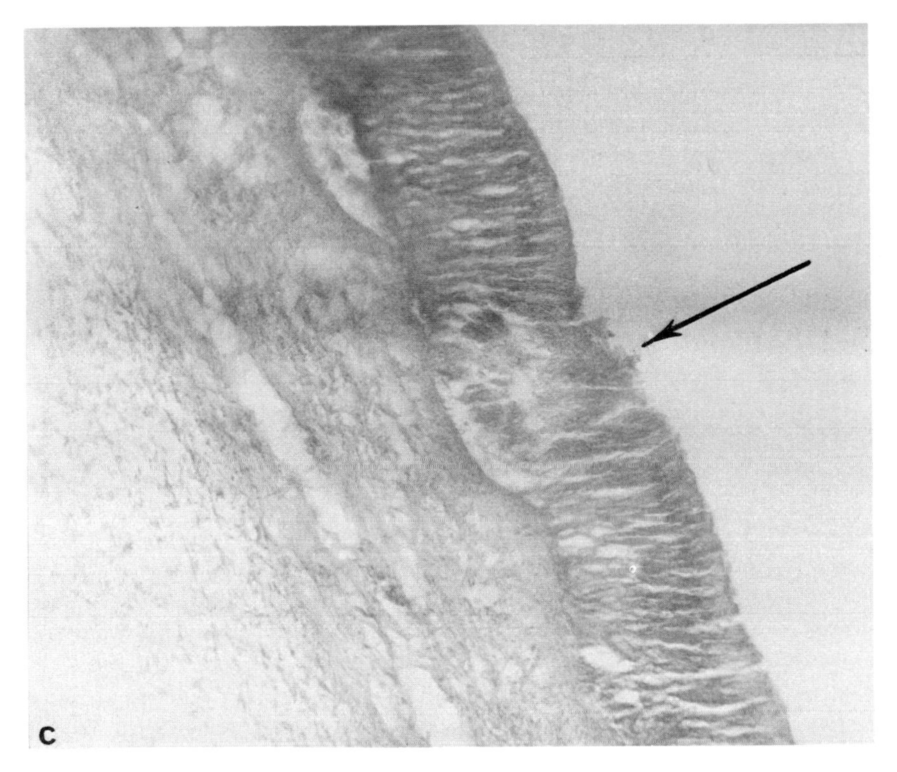

FIG. 2c.

FIG. 2(a). Photomicrograph of a longitudinal section of the tip of an aboral spine of *Acanthaster planci*. The epidermis, dermis, and underlying endoskeletal supportive structure of the spine are clearly delineated. × 110. (b) Cross section of the same spine. Arrow is pointing to the indentation of the epidermal layer at which point is located an intra-epithelial venom gland. × 110. (c) Enlargement showing, at tip of arrow, a typical intra-epithelial gland comprised of acidophilic glandular cells which are believed to secrete the venom. × 260. (E. Rich.)

4. Pharmacology of Starfish Venom

Unknown.

5. Pathology

Unknown.

6. Clinical Characteristics

Contact with the venomous spines of *Acanthaster planci* may cause an extremely painful wound, redness, swelling, numbness, and paralysis (Fish and Cobb, 1954; Pope, 1963). Little else is known about the clinical characteristics of venomous starfish stings.

7. Treatment

The treatment of venomous starfish stings should be handled in the same manner as fish stings (see Vol. II, p. 625).

8. Prevention

The handling of venomous starfish should be done with considerable care since their long sharp spines can readily penetrate the average glove. Work in areas inhabited by venomous starfish should be conducted with caution since the starfish are quite inconspicuous in their native surroundings

B. Echinoidea—Venomous Sea Urchins

The class Echinoidea, which includes the sea urchins and their relatives, contains the majority of venomous species of echinoderms. Sea urchins are free-living echinoderms, having a globular, egg-shaped, or flattened body. The viscera are enclosed within a hard shell or test, formed by regularly arranged plates, carrying spines articulating with tubercles on the test. Between the spines three-jawed pedicellariae are situated, which are of direct concern to the venomologist and will subsequently be described in greater detail. The mouth is situated on the lower, or oral surface, turns downward, and is surrounded by five strong teeth incorporated in a complex structure termed "Aristotle's lantern." Sea urchins move about slowly by means of the spines on the oral side of the test. Some forms have the ability to burrow into rocks, whereas others cover themselves with shells, sand, and bits of debris. Some urchins are nocturnal in their habits, hiding under rocks during the day and coming out to feed at night. Echinoids tend to be omnivorous in their feeding habits, ingesting algae, mollusks, foraminifera, and various other types of benthic organisms.

1. List of Representative Venomous Sea Urchins

ARBACIIDAE

Arbacia lixula (Linnaeus)

Range: Mediterranean Sea, west coast of Africa, Canaries, Madeira, Azores, and Brazil.

DIADEMATIDAE

Diadema antillarum Philippi

Range: West Indies.

FIG. 3. *Diadema setosum* (Leske). Long-spined or black sea urchin. The spines of this sea urchin can inflict very painful stings. (Photo by Ron Church, courtesy of *Skin Diver Magazine*.)

Diadema setosum (Leske) (Fig. 3)

Range: Indo-Pacific area, from East Africa to Polynesia, China, Japan, Bonin Islands.

Echinothrix calamaris (Pallas)

Range: Indo-Pacific area, from East Africa to Polynesia, Red Sea, Australia, and Japan.

ECHINIDAE

Psammechinus microtuberculatus (Blainville)

Range: Mediterranean and Adriatic Seas.

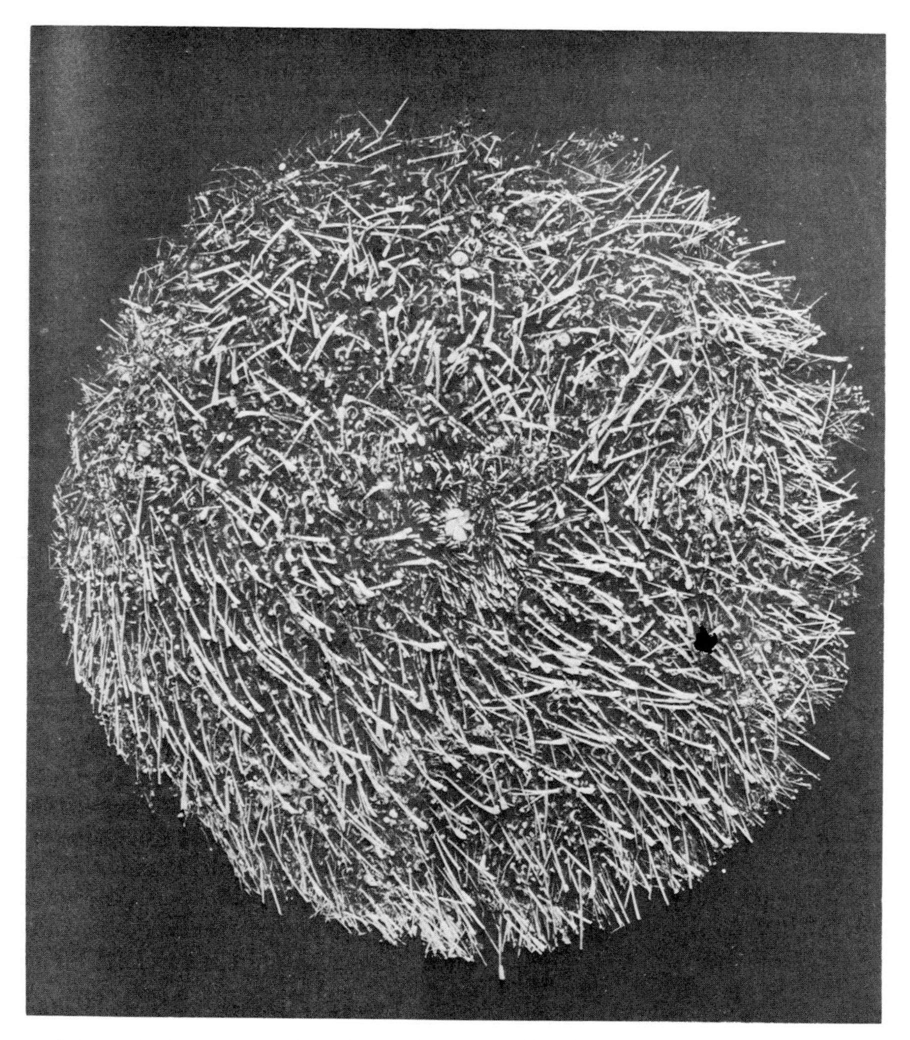

FIG. 4. *Asthenosoma varium* Grube. Leather urchin. The venomous globiferous pedicellariae of this sea urchin can inflict painful stings. Oral side. (From Mortensen.)

ECHINOTHURIDAE

Aesthenosoma ijimai Yoshiwara

Range: Southern Japan to Moluccan Sea.

Asthenosoma varium Grube (Fig. 4)

Range: Indonesia, Indian Ocean, and Gulf of Suez.

TOXOPNEUSTIDAE

Sphaerechinus granularis (Lamarck)

Range: Mediterranean Sea, eastern Atlantic Ocean, from the Channel Islands to Cape Verde, Canaries and the Azores.

Toxopneustes elegans Döderlein

Range: Japan.

Toxopneustes pileolus (Lamarck) (Fig. 5)

Range: Indo-Pacific area, from Melanesia to East Africa, and Japan.

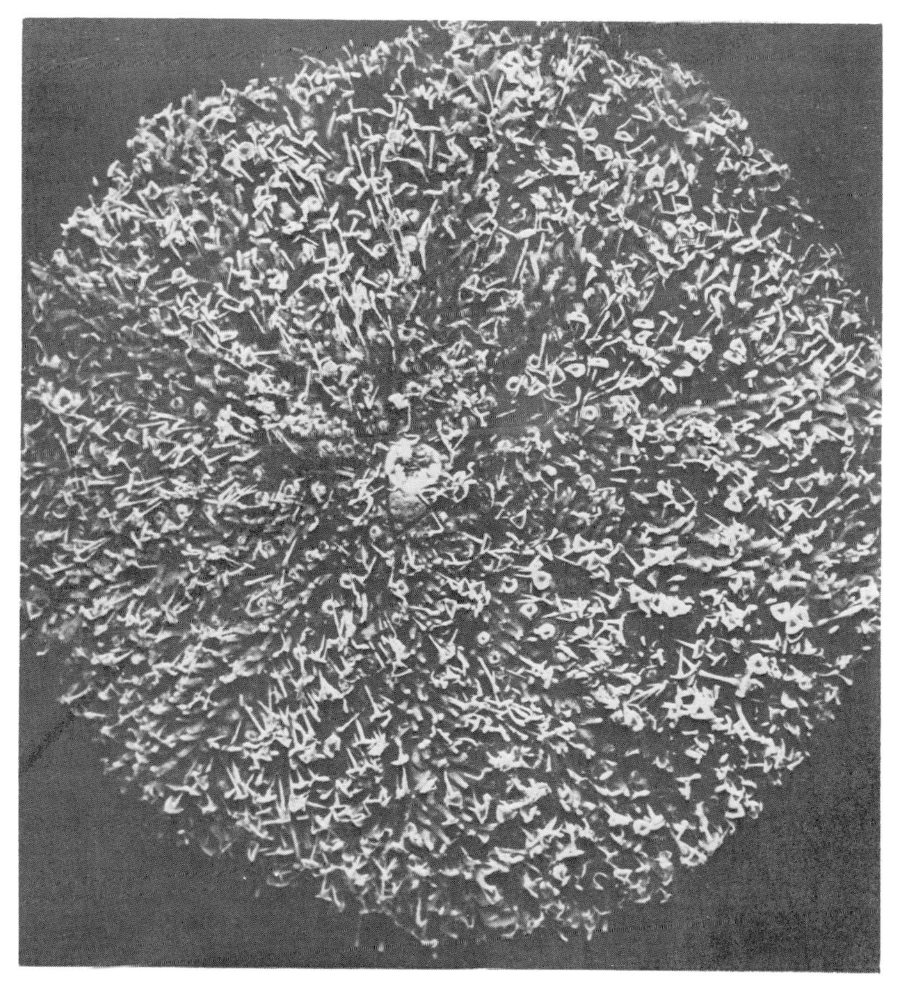

FIG. 5. *Toxopneustes pileolus* (Lamarck). A sting from the pedicellariae of this sea urchin may result in intense pain, numbness, paralysis, and death. The large venomous globiferous pedicellariae are seen projecting beyond the spines as rounded light-colored objects. Oral side. (From Mortensen.)

2. Mechanism of Envenomation

Venomous echinoids inflict their stings by means of spines, pedicellariae, or both.

3. Venom Apparatus

a. Spines. The spines of sea urchins vary greatly from group to group. In most instances the spines are solid, have blunt, rounded tips, and do not constitute a venom organ. However, members of the families Echinothuridae and Diadematidae possess long, slender, hollow, sharp spines, which are extremely dangerous to handle. The acute tip and the retrorse spinules, which are present in some instances, allow the spines to penetrate deep into the flesh, and because of their extreme brittleness, they break off readily in the wound. The spines in *Diadema* are very long, frequently attaining a length of 25 cm or more. The aboral spines of *Asthenosoma* are particularly suited as a venom organ because of their unique structure. Mortensen (1935) states "The smaller spines in *Asthenosoma varium* are developed into special poison-organs, carrying a single large gland, the point of the spine acting like the poison fang of a snake" (Fig. 6). The secondary spines of other echinothurids are also known to terminate in a sharp point, which is enclosed in a venom gland.

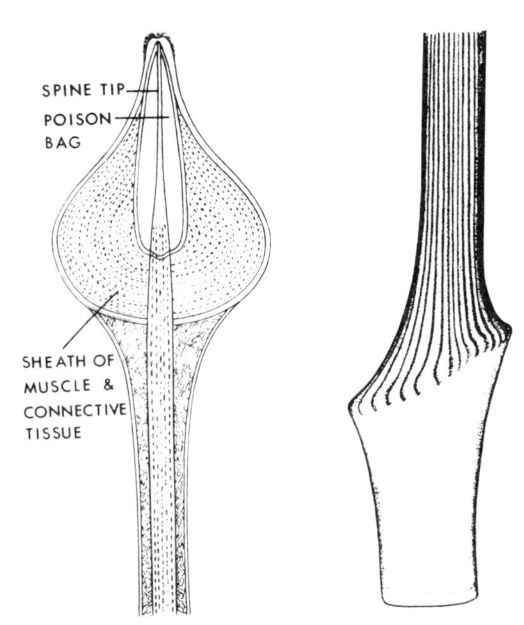

FIG. 6. A secondary aboral spine of *Asthenosoma vaium*. Note the venom or "poison bag" surrounding the tip of the spine. (R. Kreuzinger, after Mortensen.)

b. Pedicellariae. These are small, delicate, seizing organs, which are found scattered among the spines of the test peculiar to the echinoderm classes Echinoidea and Asteroidea. They are comprised of two parts, a terminal, swollen, conical head, which is armed with a set of calcareous pincerlike valves or jaws, and a supporting stalk. The head is attached to the stalk either directly by the muscles, or by a long flexible neck. The head may consist of two or four, but more commonly three, calcareous valves or jaws. On the inner side of each valve is found a small elevation provided with fine sensory hairs. Contact with these sensory hairs causes the valves to close instantly. The pedicellariae of each sea urchin species has its own characteristic skeleton. The skeletal elements of the stalk are formed by a calcareous rod composed of loose filaments united by five crosspieces. This rod is generally dilated at its base into a larger cylindrical structure, which is connected by soft parts to a small tubercle on the shell. The length of the stalk varies according to the species and type of pedicellaria. The soft parts of the pedicellariae consist of an outer epithelium, an underlying layer of connective tissue, muscles, venom, and mucous glands.

Five different types of pedicellariae are generally recognized, namely, globiferous, tridentate, triphyllous, ophiocephalous, and dactylous. Since it is the globiferous type that is usually the venomous type of concern to the

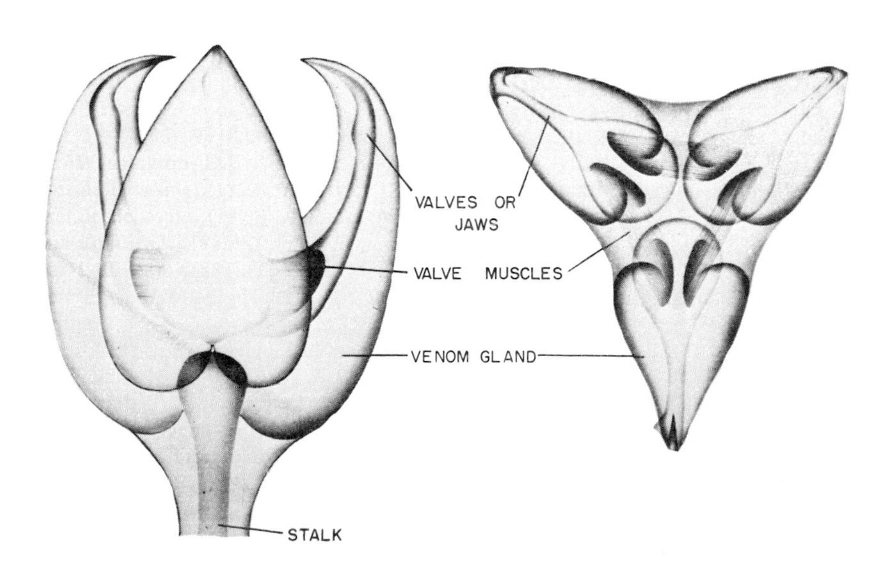

FIG. 7. Showing the structure of a typical venomous globiferous pedicellaria. (R. Kreuzinger, after Mortensen.)

clinician, only the globiferous pedicellariae will be described in detail.

The head of large globiferous pedicellariae may reach a size of 2 mm by 3 mm, with an overall height of 10 mm or more. The head is globular in shape and comprised of three calcareous valves (Fig. 7). Each valve terminates in a sharp incurved spine, or fang. Attached to each valve is an adductor muscle, which closes the valve, and an abductor muscle, which opens the valve. There are also three flexor muscles which join the tip of the calcareous pedicellarial stalk to the base of the valves that are used to flex the head of the pedicellaria. The outer surface of each valve is covered by a large venom gland which in many sea urchins has two efferent venom ducts that empty in the vicinity of a small toothlike projection on the terminal fang of the valve. The cells within the venom glands appear to undergo dissolution, indicating a holocrine type of secretion. The opening and closing of the globiferous pedicellariae is believed to be triggered by a sensorial or "nervous papillae" located on the inner aspect of each valve. These organs are comprised of sensitive cells furnished with a rigid bristle, and are intercalated with supporting cells. Nerves are said to extend from these papillae to the muscles. Stimulation of these bristles causes the adductor muscles to contract, and the pedicellaria thereby inflicts its sting with release of the venom into the wound. Venomous globiferous pedicellariae have been reported in the families Arbaciidae, Echinidae, Spatangidae, Strongylocentrotidae, Toxopneustidae, Temnopleuridae, and possibly others.

One of the primary functions of pedicellariae is that of defense. When the sea urchin is at rest in calm water, the valves are generally extended, moving slowly about, awaiting their prey. When a foreign body comes in contact with them, it is immediately seized. The pedicellariae do not release their hold as long as the object moves, and if it is too strong to be held, the pedicellariae may be torn from the test, but continue to seize the object. Detached pedicellariae may remain active for several hours after separation from the sea urchin. Since they continue to inject venom into the wound, it is important that they be removed immediately.

4. Pathology

Unknown.

5. Pharmacology of Sea Urchin Venom

Laboratory data on the toxicology and pharmacology of sea urchin venom is quite meager. Uexküll (1899) tested pedicellarial venom by stinging various marine animals with the pedicellarae of *Sphaerechinus granularis*. Stings were fatal to marine snails and eels. Octopuses became hyperactive when stung. A single pedicellarial bite was able to stop a frog's heart, and if the bite was inflicted on a frog's spinal cord, it resulted in generalized convulsions.

Henri and Kayalof (1906), and Kayalof (1906) macerated the globiferous, tridentate, and ophicephalous pedicellariae of *Paracentrotus lividus, Arbacia lixula, Sphaerechinus granularis*, and *Spatangus purpureus*. The pedicellariae were removed from live sea urchins, macerated in sea water, and the solution injected into the visceral cavity of crabs, sea cucumbers, starfishes, frogs, and lizards, and intravenously into octopuses and rabbits. They found that only sea cucumbers, starfishes, and frogs were not affected by the venom, but the other animals were paralyzed and finally died.

Fujiwara (1935) tested the toxicity of the venom of *Toxopneustes pileolus* by stinging the shaved abdomens of mice with the globiferous pedicellariae of living sea urchins. Stings by seven or eight pedicellariae resulted in respiratory distress, apparently the result of muscular paralysis, and a drop in body temperature in the mouse. Similar results were obtained from saline extracts of the pedicellariae. Fujiwara later stung himself with eight pedicellariae, which resulted in severe pain, respiratory distress, giddiness, paralysis of the lips, tongue, and eyelids, relaxation of the muscles of the arms and legs, difficulty in speech, and loss of control of his facial muscles. He recovered after about 6 hours.

Okada (1955), and Okada, Hashimoto, and Miyauchi (1955) reported the pedicellarial venom of *Toxopneustes pileolus* to have a cardio-inhibitory effect on oyster heart.

Mendes, Abbud, and Umiji (1963) prepared saline extracts from homogenates of globiferous pedicellariae of *Lytechinus variegatus* and tested them on cholinergic effector systems, namely, guinea pig ileum, rat uterus, amphibian heart, longitudinal muscle of holothurians, the protractor muscle of a sea urchin lantern, and the blood pressure of dogs. They concluded that the venom contained a dialyzable acetycholinelike substance.

6. Chemistry of Sea Urchin Venom

Unknown.

7. Clinical Characteristics

The hollow, elongate, fluid-filled spines of echinothurid and diadematid sea urchins are particularly dangerous to handle. Because of their sharp, needlelike points, they are able to penetrate the flesh with ease, producing an immediate and intense burning sensation. Upon penetration of the spines, a violet-colored fluid is released that causes discoloration of the wound. The intense pain is soon followed by redness, swelling, and aching sensations. Partial motor paralysis of the legs, slight anesthesia, edema of the face, and irregularities of the pulse have also been reported (Pugh, 1913). According to

Earle (1941), secondary infection is a frequent complication with some species. The pain usually subsides after several hours, but the discoloration may continue for 3 or 4 days. There is a divergence of opinion as to whether or not the spines actually contain a venom. However, the clinical characteristics indicate that a poison of some type is present, since the pain far exceeds that produced by mere mechanical injury. According to Mortensen (1935), there seems to be no doubt of the presence of venom in the spines of *Asthenosoma varium*. The symptomatology of stings from the spines of sea urchins has been discussed by Fonssagrives (1877), Verrill (1907), Cleland (1912), Pugh (1913), Paradice (1924), Kopstein (1926), Levrat (1927), Whitley and Boardman (1929), Mortensen (1935), Tweedie (1941), Earle (1941), Taft (1945), Phillips and Brady (1953), and Fish and Cobb (1954).

Although numerous writers refer to the dangers of pedicellarial stings, there is very little worthwhile clinical data. Oshima (1915) and Fujiwara (1935) state that a sting from the pedicellariae of *Toxopneustes pileolus* results in immediate, intense, radiating pain, faintness, numbness, generalized paralysis, aphonia, respiratory distress, and death. The pain may diminish after about 15 minutes, and completely disappear within an hour, but paralysis may continue for 6 hours or longer. This subject has been discussed briefly by Mortensen (1943), Clark (1950), and Fish and Cobb (1954).

8. Treatment

Insofar as the venom is concerned, sea urchin stings should be handled in a manner similar to venomous fish stings (see Vol. II, p. 625). However, pedicellariae should be removed from the wound promptly. Pedicellariae are frequently detached from the parent animal, and may continue to be active for several hours.

The extreme brittleness and retrorse barbs of some sea urchin spines present an added mechanical problem. Cleland (1912), Earle (1940), and others are of the opinion that some sea urchin spines are readily absorbed, and need not be removed. Absorption of the spines is said to be complete within 24 to 28 hours. However, Earle (1941) later pointed out that the spines of *Diadema setosum* are not readily absorbed, and may be observed in the wound by roentgenological examination many months later. It is recommended that the spines of this species be removed surgically.

9. Prevention

No sea urchin having elongate, needlelike spines should be handled. Moreover, leather and canvas gloves or shoes do not afford protection. The spines of *Diadema* and others are able to penetrate leather and canvas with ease. Care should be taken in handling any tropical species of short-spined sea urchin without gloves because of the pedicellariae. Members of the family Toxopneustidae are especially dangerous, and should be avoided.

II. VENOMOUS ANNELIDS

The phylum Annelida, or segmented worms, are organisms having an elongate, usually segmented body with paired setae. The outer covering of the body consists of a thin, nonchitinous cuticle. The digestive system is tubular, coelom large, and the vascular system is closed. Some species have well-developed chitinous jaws. Annelids are cosmopolitan in distribution and are found in marine, freshwater, and terrestrial environments. Marine annelids also have a wide distribution bathometrically, and are for the most part benthic, either creeping or burrowing in mud, pilings, corals, or under rocks. Some are sedentary in their habits, building calcareous or fibrous tubes, whereas a few are pelagic. It is estimated that there are more than 6200 species of annelids. Very little is known regarding the toxicity of marine annelids. Although there are five different classes of the phylum Annelida, species of concern to the venomologist are all members of the class Polychaeta, the marine polychaetes.

The polychaetes are divided into two major groups, the Errantia, which include most of the free-moving kinds, and the Sedentaria, or tube-dwelling and burrow-inhabiting species. The polychaetes that have been incriminated as toxic are largely errant forms.

Polychaetes have cylindrical bodies and are metameric, having numerous somites, each bearing a fleshy paddlelike appendage, or parapodia, that bears many setae. The head region has tentacles. There is no clitellum. Sexes are usually separate. There are no permanent gonads and fertilization is commonly external. Polychaetes have a trochophore larval stage, and there is a sexual budding in some species.

Most polychaetes are free-living; a few are ectoparasitic. They have a bathymetric range from the tide line to depths of more than 13,000 feet. A few species are pelagic. Several of the polychaetes inhabit fresh water. Polychaetes are largely carnivorous. Some of the burrowing worms feed on bottom detritus, whereas the tube dwellers subsist on plankton. Generally, polychaetes spend their existence crawling under rocks, burrowing in the sand or mud, in and around the base of algal growths, or construct tubes, which they may leave at intervals in search of food. The majority of polychaetes range in size from 5 cm to 10 cm. However, some of the syllids are only 2 mm in length, whereas the giant Australian species, *Onuphis teres* and *Eunice aphroditois*, may attain a meter or more in length. These two polychaetes have powerful chitinous jaws and are able to inflict a nasty bite.

The blood worm, *Glycera dibranchiata,* has venom glands associated with its jaws and can inflict a painful bite.

Many polychaetes are very beautiful, colored red, pink, or green, or display a combination of colors and iridescence.

I. List of Representative Venomous Annelids

AMPHINOMIDAE—Bristle worms

Chloeia flava Pallas. Sea mouse or bristle worm

Range: Indo-Pacific, Japan, China Sea, and Indian Ocean.

Chloeia viridis Schmarda. Sea mouse or bristle worm.

Range: West Indies, Gulf of California, Mexico south to Panama, Galapagos Islands, tropical America, both sides.

Eurythoë brasiliensis Hansen. Bristle worm.

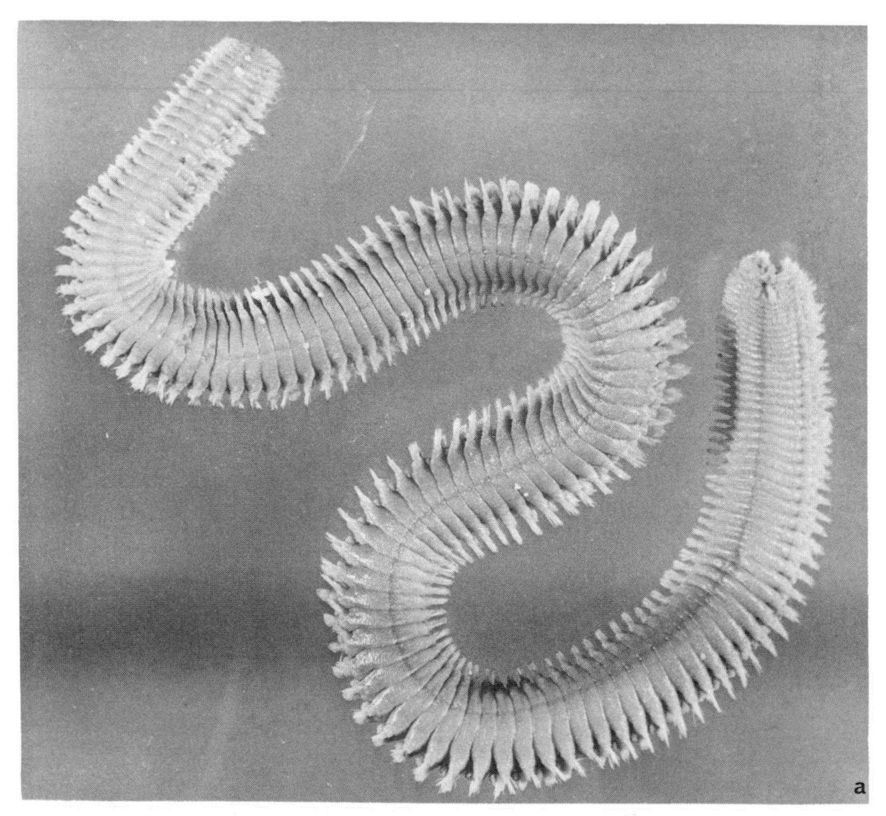

FIG. 8(a). *Eurythoë complanata* (Pallas). The bristle worms, which are relatively common in many tropical marine areas, can inflict painful stings with their bristlelike setae. The setae in this specimen are retracted, but they can be readily extended.

FIG. 8(b). *E. complanata* with the setae extended. The setae are extremely sharp and can readily penetrate the average glove. (Courtesy of Smithsonian Institution.)

Range: Brazil.

Eurythoë complanata Pallas. Bristle worm. [Fig. 8(a–c)]

Range: Circumtropical.

Hermodice carunculata Pallas. Bristle worm. (Fig. 9)

Range: Florida and the West Indies.

GLYCERIDAE

Glycera dibranchiata Ehlers. Blood worm. [Fig. 10(a–c)]

Range: North Carolina to northeast Canada.

Glycera ovigera Schmarda

Range: New Zealand.

2. Mechanism of Envenomation

Injurious effects from marine annelids are generally the result of careless handling. Bristle worms, *Chloeia, Eurythoë,* and *Hermodice*, inflict their injuries by means of their pungent parapodial bristlelike setae. *Glycera dibranchiata, G. ovigera,* and related species are capable of inflicting painful bites by means of their chitinous fangs or jaws, which are associated with venom glands. The fangs of *Glycera* are situated at the extreme tip of the

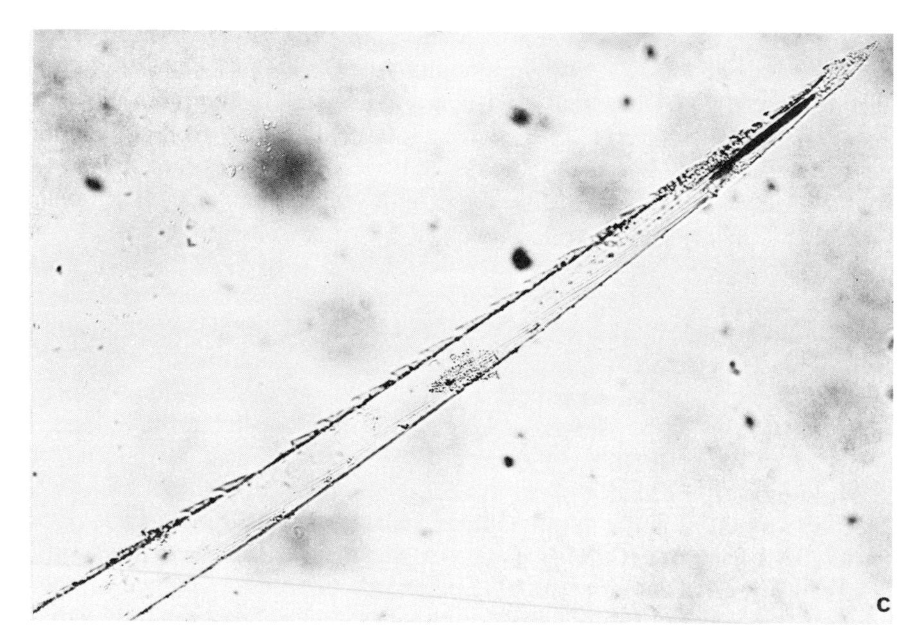

FIG. 8(c). Photomicrograph of a seta from *E. complanata*. The seta are hollow and are believed to be filled with venom. × 540. (N. Hastings.)

everted proboscis and can be everted for striking or retracted with extreme rapidity.

3. Venom Apparatus

The venom apparatus of polychaetes is comprised of setae, in the case of *Chloeia, Eurythoë, Hermodice*, and related species, or venomous jaws, as in the case of *Glycera* and its relatives.

a. Setae. These are elongate pungent chitinous bristles which project from the parapodia. The parapodia are a pair of lateral appendages which extend from each of the body segments. The structure appears as a more or less laterally compressed fleshy projection of the body wall. Each parapodium is biramous and consists of a dorsal portion, the notopodium, and a ventral part, the neuropodium. Each division of the parapodium is supported internally by one or more chitinous rods, termed acicula, to which are attached some of the parapodial muscles. Each of the distal ends of the two parapodial divisions are invaginated to form a setal sac or pocket in which are situated the projecting setae. Each seta is secreted by a single cell at the base of the setal sac. Generally the setae of polychaetes project some distance beyond the end of the parapodium. However, *Eurythoë* and *Hermodice* have the ability to retract or extend their setae to quite a remarkable degree.

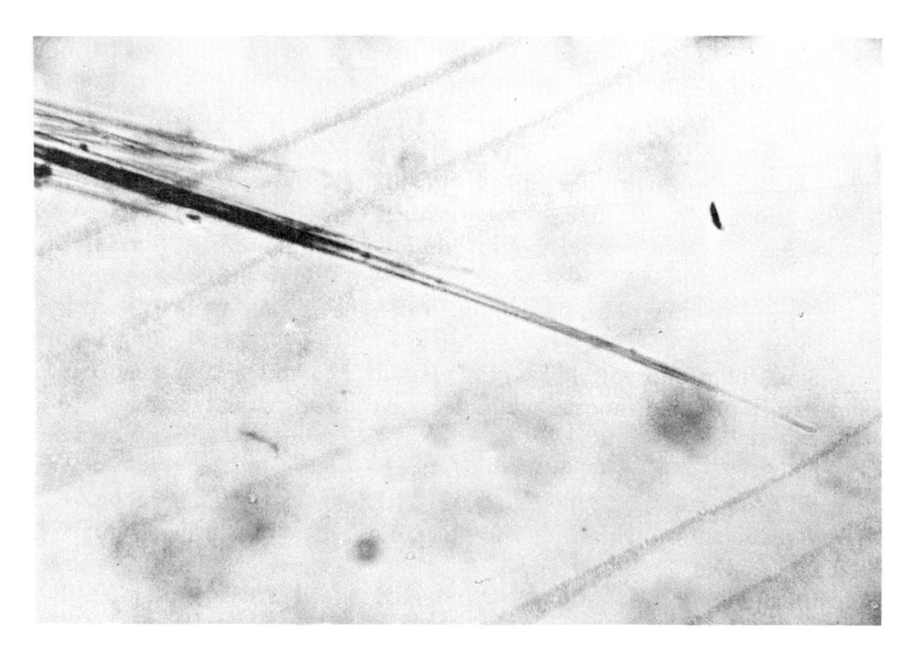

FIG. 9. Setae of the bristle worm *Hermodice carunculata* Pallas. × 540. (N. Hastings.)

When the living worm is at rest the setae appear to be quite short and barely in evidence, but when irritated the setae are rapidly extended and the worm appears as a mass of bristles.

The severity of symptoms reported in some of the clinical accounts lend credence to the belief that both *Eurythoë* and *Hermodice* possess venomous setae. The setae of both *E. complanata* and *H. caruncutata* appear to be hollow, and at times seem to be filled with fluid. A seta of *E. complanata* has a series of retrorse spinules along the shaft, whereas the seta of *H. carunculata* is without spinules and has a needlelike appearance. Final determination of whether glandular elements are present or not awaits further research. Thus far, no one appears to have described any venom glands.

b. Jaws. The jaws of *Glycera* consist of a long tubular proboscis which can be extended to about one-fifth the length of the body. When the proboscis is retracted, it occupies approximately the first twenty body segments. The pharynx is located behind the proboscis and bears four jaws or fangs, which are arranged equidistantly around the pharyngeal wall. Attached to, and immediately following the pharynx, is an S-shaped esophagus. The proboscis and associated portions of the digestive tract lie free, unattached, in the coelom. An explosive extension of the pharynx results from the sudden

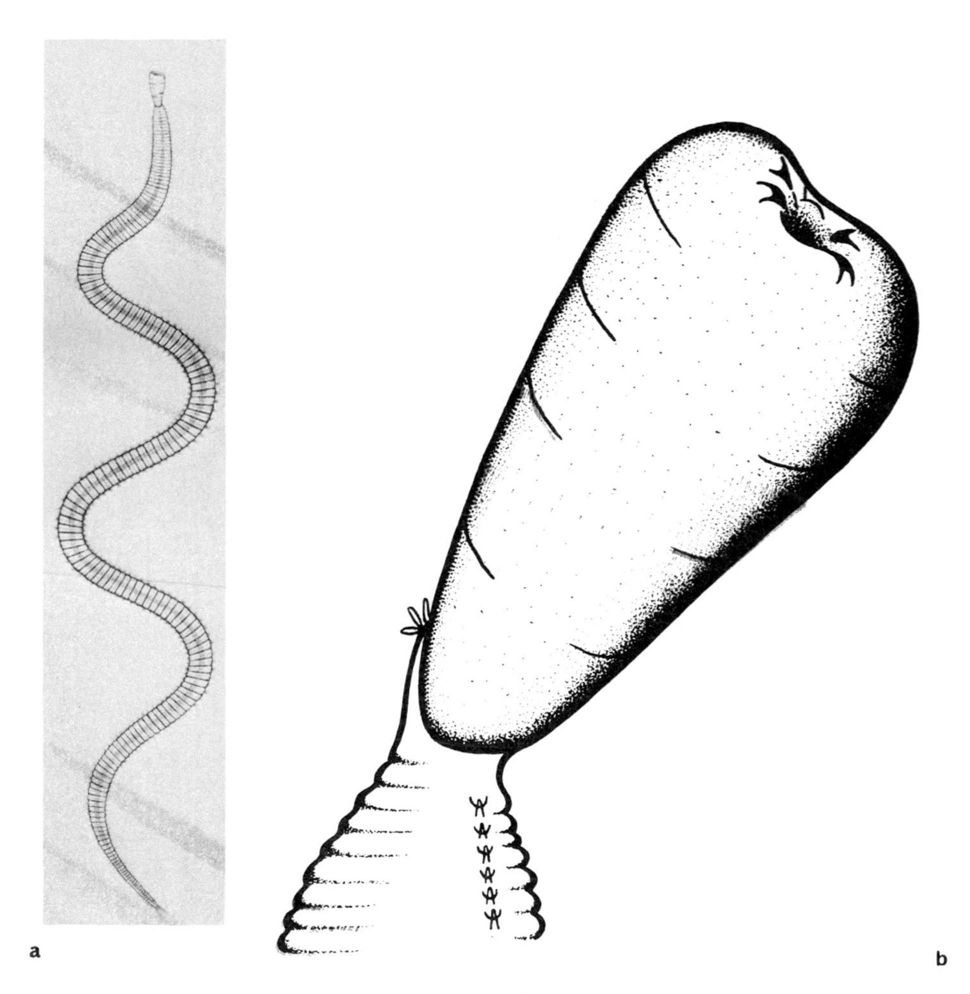

FIGS. 10a and 10b.

contraction of the longitudinal muscles, which slide the pharynx forward and straighten out the esophagus. The four fangs appear at the tip of the proboscis when the pharynx is in the extended position. Each jaw appears as a curved fang, to which is attached the venom gland. The duct and body of the gland lie within the pharyngeal wall and can be observed only by means of dissection. The gland empties its contents through the long slender duct which has its opening at the base of the fang.

4. Chemistry of Annelid Venom

Unknown.

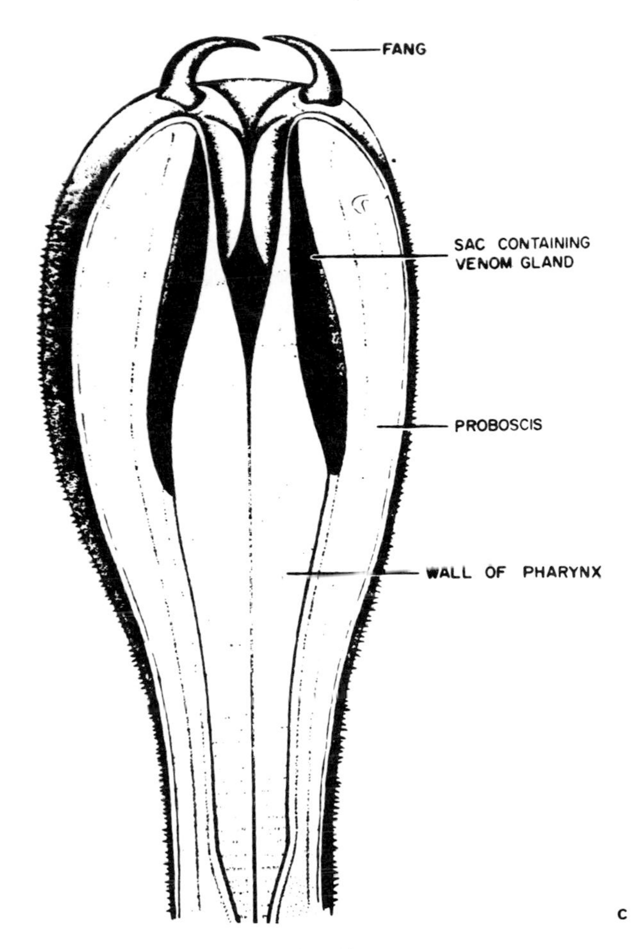

Fig. 10c

FIG. 10(a). *Glycera dibranchiata* Ehlers. The blood worm can inflict a painful bite with the use of the venomous fangs that are located at the anterior tip of the extended proboscis. (b) Enlarged view of the proboscis showing the fangs. (c) Longitudinal section of the proboscis of *G. dibranchiata* showing the anatomical relationships of the fangs and venom glands. (R. Kreuzinger, after Ehlers and others.)

5. Pharmacology of Annelid Venom

Unknown.

6. Pathology

Unknown.

7. Clinical Characteristics

Bristle worm (*Chloeia, Eurythoë, Hermodice*) stings may result in an intense inflammatory reaction of the skin, consisting of redness, swelling, burning sensation, numbness, and itching. According to Mullin (1923), *Hermodice caruncolata* is able to inflict a "paralyzing effect" with its setae. Contacts with the setae of bristle worms have been likened to handling the spines of prickly pear cactus or nettle stings. Severe complications may result, causing secondary infections, gangrene, and loss of the affected part. The clinical effects of bristle worm stings have also been discussed by Baird (1864), Paradice (1924), Levrat (1927), Roughley (1940, 1947), Halstead (1956, 1959), Pope (1947, 1963), LeMare (1952), Phillips and Brady (1953), and Gillett and McNeill (1962).

The glycerids are believed to secrete a venom at the time of biting. The bite has been compared to a bee sting, and results in swelling, redness, and pain. Two to four minute oval lesions are present at the site where the jaws have pierced the skin. There may be blanching of the skin immediately surrounding the lesions from the effects of the venom. With the exception of severe itching, recovery is generally uneventful and occurs in a few days.

8. Treatment

The treatment of bristle worm stings and glycerid worm bites is largely symptomatic. There are no specific antidotes. Secondary infections may occur which require antibiotic therapy. The setae of bristle worms can best be removed from the skin with adhesive tape; ammonia or alcohol may be applied to the area to alleviate discomfort.

9. Prevention

Care should be taken in handling glycerid and bristle worms. Heavy rubber gloves should be worn.

REFERENCES

Clark, A. H. (1950). *Bull. Raffles Mus., Singapore.* **22**, 53–67.

Cleland, J. B. (1912). *Australasian Med. Gaz.* **32**, 269–272.

Earle, K. V. (1940). *Trans Roy. Soc. Trop. Med. Hyg.* **33**, 447–452.

Earle, K. V. (1941). *Med. J. Australia.* **2**, 265–266.

Fish, C. J., and Cobb. M. C. (1954). *U.S. Fish Wildlife Serv., Res. Rept. No.* **36**, p. 21–23.

Fonssagrives, J. B. (1877). "Traite d'Hygiene Navale." Bailliere et Fils, Paris.

Fujiwara, T. (1935). *Annotationes Zool Japon.* **15**, 62–67.

Gillett, K., and McNeill, F. (1962). "The Great Barrier Reef and Adjacent Isles." Coral Press Pty., Sydney.

Halstead, B. W. (1956). *In* "Venoms" (E. E. Buckley and N. Porges, eds.), p. 9–27. Publ. No. 44, Am. Assoc. Adv. Sci., Washington.

Halstead, B. W. (1959). "Dangerous Marine Animals." Cornell Maritime Press, Cambridge.

Henri, V., and Kayalof, E. (1906).
Kayalof, E. (1906).
Kopstein, F. (1926).
LeMare, D. W. (1952). *Med. J. Malaya.* **7**, 1–8.
Levrat, E. (1927). *Toulouse Med.* **28**, 194–202.
Mendes, E. G., Abbud, L., and Umiji, S. (1963). *Science* **139**, 408–409.
Mortensen, T. (1935). "A Monograph of the Echinoidea." Vol. II. Carlsberg Fund, C.A. Reitzel, Copenhagen.
Mortensen, T. (1943). "A Monograph of the Echinoidea." Vol. III. Part 3. Carlsberg Fund, C.A. Reitzel, Copenhagen.
Mullin, C. A. (1923). *Univ. Iowa Studies Nat. Hist.* **10**, 39–45.
Okada, K. (1955). *Records Oceanogr. Works Japan.* **2**, 49–52.
Okada, K., Hashimoto, T., and Miyauchi, Y. (1955). *Bull. Marine Biol. Sta. Asamushi, Tohoku Univ.* **7**, 133–140.
Oshima, H. (1915). *Dobutsugaku Zasshi.* **27**, 605.
Paradice, W. E. (1924). *Med. J. Australia.* **2**, 650–652.
Phillips, C., and Brady, W. H. (1953). "Sea Pests." Univ. Miami Press, Miami.
Pope, E. C. (1947). *Australian Mus. Mag.* **9**, 164–168.
Pope, E. C. (1963a). *In* "Proceedings First International Convention on Life Saving Techniques" (J. W. Evans, chmn.), Part III, p. 91–102. Bull. Post Grad. Comm. Med., Univ. Sydney
Pope, E. C. (1963b). *Personal Communication*, July 25, 1963.
Pugh, W. S (1913). *U.S. Nav. Med. Bull.* **7**, 254–255.
Roughley, T. C. (1940). *Natl. Geogr. Mag.* **77**, 823–850.
Roughley, T. C. (1947). "Wonders of the Great Barrier Reef." Scribner's, New York.
Taft, C. H. (1945). *Texas Rept. Biol. Med.* **3**, 339–352.
Tweedie, M. W. (1941). "Poisonous Animals of Malaya." Malaya Publishing House, Singapore.
Uexküll, J. von. (1896). *Z. Biol.* **34**, 298–339.
Verrill, A. E. (1907). *Trans Connecticut Acad. Arts Sci.* **12**, 204–323.
Whitley, G. P., and Boardman, W. (1929). *Australian Mus. Mag.* **3**, 330–336.

Chapter 60

Animal Venoms in Therapy

D. DE KLOBUSITZKY

SÃO PAULO, BRAZIL

I. Introduction 443
II. Snake Venoms 444
A. Analgesic Applications 445
B. Hemostatic Applications 452
C. Other Therapeutic Applications 455
III. Toad Venoms 458
IV. Spider Venoms 461
V. Bee Venom 461
A. Pharmaceutical Preparations 462
B. Antirheumatic and Antiarthritic Applications 463
C. Other Therapeutical Applications 465
VI. Venom Combinations 466
VII. Appendix: Snake Venoms as Anticoagulants 467
References 470

I. INTRODUCTION

That animal venoms were used in the "medicine" of prehistoric peoples must certainly be admitted; we have reliable evidence of their use among the ancient civilized peoples as well as at the time of iatrochemistry. It should be mentioned that precisely the most active of these venoms, namely, the snake venoms, found their way into medicine relatively late. The *Theriak,* invented by Mithridates VI, king of Pontus, and improved by Andromachus, Nero's physician, ordinarily contained only viper flesh as a characteristic component. Thus also the viper broths, viper ointments, viper pastilles, viper salts, and other remedies made from venomous snakes, always on sale in medieval pharmacies, contained no venom because, before the preparation, the heads of the animals were cut off. It was the Florentine physician and poet Francesco Redi (1626–1697) who first described the venom apparatus and pointed out that only extracts from the head could contain venom (1664). It would, however, lead us too far afield to discuss either the use of these natural products in those times or the part they play in the popular medicine of different countries. The primary purpose of this study is to outline the present state of the art, and, in view of this delimitation, this chapter will only report the therapeutic uses in modern medicine of snake, toad, spider, and bee venoms.

II. SNAKE VENOMS

If one does not take into account the statement of the Italian physician P. Bergamo (1835), who claimed to have successfully used viper (*Vipera berus*) venom against rabies, the Texan physician L. E. Self (1908) must then be considered the founder of the systematic and more or less critically checked treatment by snake venoms. He reported that after 15 years the continually recurring epileptic attacks of a 35-year-old man stopped after he had been bitten by a diamond-back rattlesnake (*Crotalus adamanteus*). As Self only reported this case after a 2-year control, his communication was met with interest. Another North American, Spangler (1910), who had previously used Crotalin, a commercial preparation made from the venom of the above-mentioned snake, in his laryngological practice, resorted to it for the therapy of genuine epilepsy and was soon able to report treatment of 11 patients with apparent success. He had observed his patients for 14 months and ascertained in all of them a considerable improvement of general condition, as well as the nearly complete cessation of attacks.

The study of snake venoms as remedies received a strong impulse from Monaelesser, who heard in New York, in 1929, that the pains of a leper in Cuba had almost completely ceased after he had been accidently bitten by a venomous spider (of the family Lycosidae). Monaelesser communicated this information to Calmette and Taguet. Thereupon Calmette, who was familiar with the paralyzing action of certain animal venoms on the sensory nerves, persuaded the two clinicians (Taguet and Monaelesser 1933), to test the pain-killing action of the venom of the Indian cobra (*Naja naja*) in cancer patients. He and his collaborators also examined the action of this venom on rat carcinoma.

Calmette's reports in turn induced many researchers in Europe and elsewhere to turn their attention to the therapeutic value of snake venoms. In the decades that followed, these venoms were used to treat genuine epilepsy, inoperable malignant tumors, neuralgia, hemophilia, hemorrhagic diathesis, various internal bleedings, metrorrhagias, menorrhagias, edema, and trachoma.

The therapeutic results obtained narrowed the field of application over the years until there remained as main applications relief from pain and stimulation of blood clotting.

A. Analgesic Applications

Malignant tumors make up by far the greatest number of cases in which snake venoms have been used to relieve pain.

The venoms were used in their natural state, strongly diluted, in formaldehyde solution as anatoxins. Many authors made experiments with fractions obtained by different processes and sold as commercial products, such as Cobratoxin in France, and Trachozid* in Austria.

The first report, on 20 cases, was made by Laignel-Lavastine and Koressios (1933) at the meeting of the Société Médicale des Hôpitaux de Paris on February 24, 1933. It was soon followed by a second one, covering 115 cases treated by Monaelesser and Taguet, that was presented to the Paris Medical Academy by Taguet (1933) on March 14 of the same year. The former report described six cases at length. All the patients suffered from inoperable and very painful tumors. The treatment was carried out by giving raw cobra venom intramuscularly in a dose of 0.1 mg in 2 ml of distilled water. In some patients, however, nosebleeds or bulbar troubles were noted. In general, a single dose was enough to obtain full analgesia for several days. By repeating the injection, the authors succeeded in suppressing the pain for 8–10 days. The cases observed by Monaelesser and Taguet concerned mostly carcinomas of the digestive tract (33 cases), of the breast (25 cases), and of the female genitals (20 cases). The only requirement made for the choice of patients was the presence of pain. Among the patients treated there were very serious cases with advanced cachexia, as well as those who had been operated upon shortly before or who had been recently treated by radiotherapy, etc. At the beginning the dosage of the venom (venoms of *N. naja* and *C. adamanteus*) was empirical, insofar as 2–3 drops of a very thin glycerin solution were given every 3, 5, or 10 days, and so on. Later, aqueous solutions were used, and the amounts of venom determined in mouse units. A mouse unit was considered to be 0.01–0.02 mg of venom, and 2.5 units were given as the initial dose. The authors increased these doses gradually to 20 units. Relief from pain could be observed in all patients.

Taguet (1933a,b) reported to the Paris Medical Association in 1933 that *N. naja* venom, free from hemorrhagins and hemolysins, administered to cancer patients in the increasing mouse-unit doses described by Monaelesser and Koressios, not only made morphine superfluous, but was also

* This preparation, developed by Brecher (1934), contained venom of the European viper (*V. berus*) and of the honeybee (*Apis mellifera*).

much superior to it since an injection every third day was quite sufficient.

If one analyzes the results given in past clinical publications on the relief of cancer pains, one comes to the conclusion that no uniform opinion has been formed on the matter.

Rossi (1933), without mentioning the venom used, described two cases. In one of them, he observed that the patient was rid of his pains for 10 days with only one injection; in the second case, the result was nil (cystic carcinoma). Calderón (1934) deduced from eight cases which he treated by increasing amounts of up to 30 rat units (1 such unit is equal to 0.075 mg) of the venom of the *Bothrops alternatus* ("urutu" in Brazil, "vibora de la cruz" in Latin America), that there really is relief from pain, but that it lasts only a few hours.

A conspectus published by Brazil (1934) describes 25 cases that various Brazilian clinicians treated with venom of *Crotalus durissus terrificus* in formaldehyde solution. On account of the very high neurotoxic action of this venom, the initial dose was very small (0.001 mg) and was increased only slowly, and at the most to 0.02 mg. The results published are very favorable, since in 22 patients good relief from pain was obtained. Reports about treatments described before 1934 were published almost simultaneously by Ravina (1934) and by Orticoni (1934).

Laignel-Lavastine and Koressios (1934) reported, in a second lecture to the Paris Medical Association, on 60 cases treated by *N. naja* venom, 25 of which they described at length. They found that the pain-killing action was lasting and better than that of morphine. They mentioned cases in which the patients had needed up to 0.02 gm of morphine daily, while afterward complete relief from pain was obtained with only 10 mouse units of venom daily. They emphasized the necessity of establishing the efficient dose for each patient individually. Interesting in their explanations is the assertion that the venom leads to regression of edemas, which substantially contributes in many cases to the relief from pain. The edema regression is put down to a vasodilatatory action of the venom on an extensive capillary plexus. This action has been carefully studied in animal experiments by Gautrelet *et al.* (1934), as by Laignel-Lavastine and associates (1934).

In 1935 Koressios summarized in a book the observations he had gathered in the use of *N. naja* venom. He thought that only unfiltered venom should be used, as the filter holds back the vasodilatory substances, which for the above-described reason is undesirable. In this publication, 25 later cases, gathered in 1933 and 1934, are also fully described. They are divided into three groups, according to the results obtained, namely "successful cases" (15 patients), "slight improvement" (two patients), and "tenacious cases" (eight patients). Thus relief from pain was obtained in 70% of the cases.

Meneghetti (1934) described four carcinoma cases in which he was able to obtain good relief from pain by means of the venom of a species of *Bothrops*. An injection of 0.1 mg of venom was usually enough to free the patients from their pains for two days.

Köbler (1934) treated two cases with *N. naja* venom and 26 by *Vipera ammodytes* venom. The former did really produce relief from pain in one patient, while in the second one the expected action did not occur. The viper venom actually caused more or less painful edemas and swellings at the injection spot, but they disappeared spontaneously within a few days and, as regards the relief from pain, produced a salutary effect in all of the cases.

Castro Araujo (1935) treated eight cancer cases with the venom of *C. durissus terrificus* and observed assuagement or cessation of the pains in all cases.

Barú (1935) described the treatment of a carcinoma mammae by the venom of *C. durissus terrificus*. The patient received 455 pigeon units in 11 weeks, that is, 0,455 mg of venom. The pain-killing action was satisfactory, though the patient's reaction to morphine was not tested.

Castro Escalada (1935) published a book on ophidiotherapy, in which he compiled 27 cases observed by him with results obtained by other, mainly South American, clinicians. As did most of the other South American clinicians, he used a combined procedure that consisted of alternate doses of the venom of the *C. d. terrificus* and a venom mixture containing gland secretions of *Bothrops alternatus*, *B. cotiara*, *B. jararaca*, and *B. jararacussu*. In 22 cases, he noted satisfactory relief from pain.

The results obtained by other clinicians given in Castro Escalada's book are not as good but still can be considered as favorable. Relief from pain was obtained in about 50% of the cases. There are two cases mentioned there (p. 66), observed by Calcano, that reacted excellently to *Micrurus* venom. Anyhow, these are the only observations we have found in the literature on the therapeutic use of this elapid venom. Castro Escalada was still more optimistic than Koressios in his appraisal of the value of snake venoms as analgesics, since he quotes 91% effectiveness.

Kirschen (1936) treated 23 cancer patients, among whom were 10 with inoperable stomach carcinoma. In his estimation, the analgesic action of the venom was better than that of morphine.

D. I. Macht (1936), who treated 150 inoperable cases (among them 39 carcinomas of the female genitals, 20 carcinomata mammae, 13 carcinomata recti, 13 carcinomas of the jaw, 7 carcinomas vesic. urin.) by various *Naja* venoms, stated 65% favorable results and confirmed this in his later publication (1938). He saw the advantage of snake venoms over morphine and the other analgesics in that the former do not lead to addiction and do not influence the mental faculties.

Penido Monteiro (1936) described a patient with melanosarcoma pharyngis, under observation for many months, who reacted very well to the venom of *C. durissus terrificus*. The doses had to be increased continuously and later it was necessary to give an injection daily.

Nekula (1936) used the *N. naja* venom with good result in 10 gynecologic carcinoma cases. The initial dose of 2.5 mouse units was increased tenfold in the course of time.

Not all the results of that time, however, were as favorable as the ones reported so far. Lavedan (1935), who had treated 53 patients by means of increasing amounts of *N. naja* venom, did not consider snake venom superior to morphine. His communication deserves special attention because he treated his cases regularly and observed them throughout four months at least. Most of the tumors were in the following locations: female genitals (14 patients), mamma (10 patients), lingua (6 patients). esophagus (3 patients), rectum (3 patients). In addition, 6 cases did not receive preliminary treatment; the venom was administered after operations or radiotherapy, or after treatments combining both. An analgesic action was obtained only after several injections and only in 10% of the cases. Lavedan mentioned, for example, two patients (carcinoma linguae, sarcoma orbitae, respectively) to whom the venom was administered throughout 6 and 8 weeks respectively, without the least result. As regards the influence on the general state, Lavedan declared that the improvement often observed at the beginning is only of short duration and is first and foremost psychologically conditioned. Also, he was not able to ascertain a favorable action on the evolution of the tumors, although in 18 patients the intratumoral form of administration was applied.

Gueytat (1934) likewise arrived at a negative result, and so did Seiler. They, however, tried the treatment with *N. naja* venom in only two (epithelioma pharyngis, ca. colli uteri) and three (two bronchial carcinomas and one prostatic carcinoma) cases, respectively.

Des Lignieris and Grasset (1936) reported later on 16 patients, 10 of which were treated by the venom of *Naja flava* and 6 by a mixture composed of the venoms of *N. flava* and *Bitis arietans*. The pure *Naja* venom was administered in separate doses of 0.1 to 0.5 mg; the *Bitis* venom, in separate doses up to 1.0 mg. Even as regards relief from pain, the results were not better than those of Lavedan (1935). The authors were actually able to notice an improvement in some patients, but it was always transitory. For some patients they had to fall back on morphine.

Van Esveld (1936) reported that the Cobratoxin he had prepared at the Netherland Cancer Institute in Amsterdam, was tested in 15 carcinoma cases. He pointed out that the results obtained both with the Dutch preparation and the French one remained well below the expectations aroused by the work of the French.

Since then the use of snake venoms for relieving cancer pains certainly has declined very much; at least it is the conclusion one comes to from the small number of publications on the matter during the past 25 years.

As many South American clinicians, Ferreira da Silva (1945) used the venom of the *C. durissus terrificus*. His initial dose was 0.01 mg (10 pigeon units). In the course of the treatment this dose was increased to 0.08 mg. He treated 17 cases and obtained very good results in relieving pain in 6 cases, that is, 35%.

Seliger (1951) used *N. naja* venom (Cobratoxin) in combination with X-ray therapy in five patients with recurring cancer. He not only praised the analgesic action of the venom, but also stressed that it exerts a favorable action on cachexia.

Blasiu (1945) treated 26 carcinoma patients by a preparation called Cobralgin, made from *N. naja* venom. In 17 patients exemption from pain was obtained, in 5 there was an alleviation, and in 4 only an improvement of short duration (respectively 65, 19, and 16%). He found out that this treatment acted favorably on the patient's psyche. He could not observe any reduction of the tumors.

For completeness, two studies that have not been accessible to the author, even in reviews, must still be mentioned: the publication of de Leon (1946), and that of Hills and Firor (1952).

Except for the 625 cases, recapitulated above and more or less lengthily described in the world literature accessible to the author, the outcome as regards relief from pain must be qualified as not favorable. Apart from the scanty good results bearing on the regression of tumors and the favorable results obtained only by Castro Escalada, it turns out that good analgesia was obtained in only 57% of all the cases. If one tries to explain the cessation of all interest in the treatment of cancer pains by snake venoms, one must, besides the results just shown, consider the morphine derivatives and substitute preparations that have been synthesized in the past decades. Among them, there are those that fully possess the analgesic action of morphine, without causing in many patients secondary effects (e.g., vomiting) that made it impossible to administer morphine to them and, therefore, forced the clinicians to fall back upon other means, among which were snake venoms.

Another vast field where snake venoms were put to use for treatment is that of relief from pain in general. Snake venoms have been used in attempts to eliminate or at least to alleviate pains of all kinds and of different origins, such as in persistent neuralgia, arthritis, rheumatism, sciatica, tabes dorsalis, lepra, and so on. The publications that deal with this field and seemed important to the author are reviewed here.

In Koressios' above-mentioned book (Koressios, 1935), nine trigeminal and facial neuralgias that were treated by the venom of *C. adamanteus*, already on the market before 1914 under the name Crotalin, as well as 11 tabes cases, observed by Negro in Turin, for which *N. naja* venom was used, are described. Koressios' conclusions on the fight against neuralgia tend to show that neither *Naja* venom nor that of the rattlesnake can alleviate serious attacks. On the other hand, he did observe good results in more than one patient.

Brazil (1934) reports on three patients suffering from neuralgias of different origins who were successfully treated by the venom of the South American rattlesnake (*C. durissus terrificus*).

Burkardt (1935) administered to 64 rheumatics increasing amounts of a mixture prepared from the venoms of *Vipera aspis* and *V. ammodytes*, and the results, not including unspecific articular rheumatisms, were very satisfactory. In arthritis go. or tbc. he was able to obtain a decrease of the swelling in a short time. This author mentions in his communication 14 cases of Gilbert-Querato and Polalen, who also were satisfied with the analgesic action obtained by venom therapy. As Hanzséros informed to the author, (1934-1936) only the first 1–2 injections of the *B. jararaca* were effective in ischiatic patients and he was not able to obtain any further results by continuing the treatment.

Ferri (1935) obtained good results in cases of neuralgia and arthritis, and a good analgesic effect in tumor pain treated with the venom of *V. ammodytes*. These results induced Pepeu (1938) to test the same venom in "many" ischias, trigeminal neuralgia, and chronic arthritis cases. He declared his results were good. The treatment of four asthmatics was also very satisfactory.

Brünner-Ornstein (1937) treated 14 especially obstinate trigeminal neuralgia cases by Cobratoxin with satisfactory results; six patients remained completely free from pain for 3–4 months.

Rottmann (1948) likewise used Cobratoxin to fight symptoms of sensory irritation. He reported satisfactory results. The venom combined with iodine and bismuth improved the condition of the spinal fluid and blood in cases that had resisted the usual therapy.

Hourand (1938) used viper venom (from 0.01 to 3.5 rat units) with good results in relief from pain in subacute and chronic arthritis, sciatica, lumbago, and arthropatia senilis cases.

Löwenstein (1938) employed Cobratoxin for polyarthritis, arthritis deformans of the spinal column, and against arthroses of the hip and shoulder joints. He observed relief from pain lasting 48 hours, but in no case was there either complete cessation of pain or a favorable influence on the course of the disease.

Steinbrocker *et al.* (1940) noted good results in 50% of the cases of arthralgias treated with *Naja naja* venom (administered in 5 mouse unit doses). Some patients received only physiological sodium chloride injections and they also claimed some remission of pain. The same venom was employed by Gürteler in cases of herpes zoster combined with neuralgia. He recorded only aggravation of the condition.

Behrmann (1940) noted very good relief from pain in glossopharyngeal and trigeminal neuralgia treated by viper venom. Horányi-Hechst (1941) reported long-lasting relief from pain in 36% of the cases treated with Cobratoxin for lancinating pains and gastric crises of tabetic origin. Schübel (1942) obtained lasting relief from pain in the most serious tabetic crises with the same venom preparation in 44% of his patients. Albee and Maier (1943) treated 56 arthritis cases by *N. naja* venom. "A marked salutary effect" was obtained in 46 patients.

Grundmann (1950) used a preparation called Vipericin. Each ampule of this product contains 5 mouse units of a mixture consisting of the venoms of *Cerastes cornutus*, *Vipera berus*, and *Lachesis muta*, to which Novalgin and caffeine sodium salicylate are also added. Having applied it to 54 patients with arthralgia, myalgia, and neuralgia, he declared 31 cases as cured, 14 as improved, and 8 as not affected. He noted no result in a tabes dorsalis case. Rudat (1952) treated 62 painful migraine and migrainoid conditions by Crotalus toxin, Epileptasid, with considerable improvement in 50% of the cases.

Meiselas and Schlecker (1957) used a drug called Nyloxin in the treatment of arthritis. It contains cobra venom and formic and silicic acids. They divided the patients into three groups: group A (15 patients) was treated with silicic acid only, group B (14 patients) was treated with Nyloxin, and group C (16 patients) was treated with formic and silicic acids. Their results are found in the accompanying tabulations.

Effect	Number of patients		
	Group A	Group B	Group C
Some relief	7	6	7
Considerable relief	3	3	3
No relief	5	5	6

Williams (1960) treated seven trigeminal neuralgia patients with a *Naja naja* venom preparation containing procaine. The success was complete in all cases; no patient had a relapse.

Other communications have been published on the same disease by Billiotet (1947), Taren (1953), Jackman (1954), Bryson (1954), and Lumpkin and Firor (1954), but they cannot be discussed here as they were not available to the author.

From the many experiments performed on especially rheumatic patients, it can be established without doubt that none of the snake venoms used in this group of diseases exerts a specific healing effect, and that the relief from pain obtained is merely due to their paralyzing action. That such an action actually exists has often been proved. To mention only recent studies, we may point out the results of the animal experiments carried out by Kellaway (1934), the publication of D. Macht and Macht (1939), and the studies mentioned by Boquet (1948) on this matter.

The scanty favorable observations made on lepers by Chowhan and Chopra, (1938) are also only with regard to relief from pain; there is no question of a specific action. Moreover, the results were of so little significance that the literature of the past 30 years does not mention treatment.

B. Hemostatic Applications

The action of several snake venoms in accelerating considerably blood coagulation, which was described for the first time by F. Fontana in 1767, was first applied therapeutically by American physicians.

The first applications were performed by Taylor (1929) as well as by Stockton and Franklin (1931), and were based, it is interesting to note, on the opposite action of the venoms, namely, large amounts of venom produced an anticlotting effect. On this basis each of these authors gave a snake venom immune serum in a case of purpura haemorrhagica. The favorable results observed were obtained by administering great amounts of serum, and were thus due to protein therapy and not to any specific action of the immune serum.

A treatment of the hemorrhages by the venoms themselves was described for the first time by Peck (1932) and Frank (1932). They started on the basis of the observations on rats by Peck and Sobotka (1931) to the effect that these animals did not react with the Schwartzman phenomenon when they were injected with 0.1 mg of the venom of the water mocassin (*Agkistrodon piscivorus*). Both authors used very dilute solutions of the native venom of the above-mentioned snake in the treatment of menorrhagias. On the basis of their favorable results, additional North American and European physicians began clinical experimentation in the treatment of hemorrhagic disorders of the most varied nature by using water mocassin venom as well as the venoms of *Vipera russelli, Notechis scutatus, Cerastes cornutus, Naja naja, Bothrops atrox,* and *B. jararaca* (Baker and Gibson, 1936; Hance, 1937; Hargreaves, 1943; Koressios, 1935; Macfarlane and Barnett, 1934; Peck *et al.,* 1935; Peck and Goldberger, 1933; Peck and Rosenthal, 1935; Rosenfeld and Lenke, 1935; Sergent, 1937). The Council of Pharmacy and Chemistry, on the basis of the good results obtained by North American observers, expressed, as far back as 30 years ago, a favorable opinion of the therapeutic value of the clotting properties of snake venom (1941, 1944).

In order to be able to put on the market easily controllable preparations with predictable uniform effects, the Preparation Board of the Brazilian

Academy of Pharmacy for the First Panamerican Pharmaceutic Congress, held in December 1948, in Havana, Cuba, honored this author by requesting him to report on this subject and to prepare recommendations of standards for pharmaceutical preparations of this type to be submitted to the Congress. The author's suggestions (1948, 1949) had reference chiefly to the following qualifications:

1. Establishment of an easily determinable coagulant unit for animal tests. Because the clotting unit suggested by Rosenfeld and Lenke (1935) proved to be too large, the proposed unit was to be the smallest quantity of the coagulant substance which, dissolved in 1 ml of physiological saline solution, causes firm (hard) clotting within from 8 to 10 minutes at 20° to 22°C in 5 ml of horse blood containing 0.134% sodium oxalate. It was decisive in the choice of the unit that the unit quantity be equivalent to a therapeutic dose for adults.

The proposed *in vivo* test would consist of injecting intravenously into the animal (rabbit or pigeon), the blood-clotting period of which had been previously determined, 1/70 ml per kilogram of body weight of the preparation and determining the coagulation times of blood samples taken 20 minutes and 24 hours later. Both samples must show curtailment of coagulation time to approximately 1/3 of the original value.

2. Establishment of criteria for the nontoxic nature of the preparation. As a test of nontoxicity, it was proposed that 1.5 ml of the product containing 1 coagulation unit per milliliter be injected intravenously into a pigeon weighing 300–350 gm.; no symptoms of poisoning such as vomiting or vertigo should be observed.

3. Establishment of a time limit within which the preparations will maintain their full potency.

4. The establishment of procedures to eliminate heat-precipitable proteins, because of their allergic effect.

In addition to the earlier and almost exclusively North American publications, it has been notably Brazilian authors who have written about the clinical application of purified venoms in speeding up coagulation and about the results obtained (Azevedo, 1946; Barbosa, 1944; Buchert, 1945; Freire, 1943; Klobusitzky, 1938; Moacyr, 1939; Vaz and Pereira, 1944; and others). There are also publications giving results obtained both in Europe and later in America (Aubry and Causse, 1936; Breda *et al.*, 1951; Davin *et al.*, 1937; Grundmann, 1950; Hanut, 1936; Kauer *et al.*, 1943; Kleinmann *et al.*, 1945; D. M. Macht, 1943; Malmierca, 1947; Mauro, 1950; Schmidt, 1952; Stöger, 1944; Croxatto and Croxatto, 1944; Page and De Bere, 1943; Githens, 1939; Hadley, 1940; Alexander, 1941; Wazen and Moura, 1942; Ghezzi and Giovanni, 1946; Clauser, 1950; Schmidt, 1952; and others).

Nowadays the venoms of *Bothrops atrox, B. jararaca*, and *Vipera russelli* are mainly used for therapeutic purposes. Many authors have ascertained the actual coagulation-accelerating action of the above mentioned venoms, and of the preponderant majority of the pharmaceutical preparations made from them, through laboratory tests carried out both *in vitro* and *in vivo*. To mention only some of the more recent studies on this matter, the publications of Fichmann and Henriques (1962), Henriques *et al.* (1960), Hohnen (1957), Hubert (1958), de Klobusitzky (1958), Kornalik (1960), Sanchez Fayos *et al.* (1960), Velázquez *et al.* (1960), and the papers of Zürn (1957, 1959) should be indicated.

In the past 25 years about 60 clinical studies on the use and the therapeutic value of these venom preparations in the most varied branches of medicine, such as surgery (Bruck, 1957; Bruck, and Salem 1954; Visetti, 1958, Wilms, 1960), urology (Burke *et al.*, 1960; Durante *et al.*, 1962; Kürn, 1955; Larget-Hue, 1963), gynecology and obstetrics (Baumgartner and Typl, 1956; Bleech, 1961; Cazzola, 1957; Cravarezza and Stura, 1955; Heiss, 1954; Picha *et al.*, 1957; Larralde and Labat, 1965; Iriart, 1965), opthalmology (Calvi-Zametti and Minini, 1956; Giffo, 1949a,b; Marucci, 1958; Valvo, 1968), otorhino-laryngology (Di Giunta and Cali, 1960; Glaninger, 1955; Gossens and Durr, 1953; Osako *et al.*, 1964; Ottoboni and Ciurlo, 1957; Sacristan Alonso and Gomez Lopez, 1960; Scola, 1959; Sugano and Maruyama, 1962; Gossare, 1967; Plath and Salchi, 1968), odontology and maxillofacial surgery (Dix and Sellach, 1958; Törteli and Hattyasy, 1965; Wagner and Hubbel, 1958), internal medicine (Frandoli, 1962; Giardiello *et al.*, 1959; Hohnen, 1957; Kock, 1957; Kuthy *et al.*, 1960; Larget-Hue, 1963; Lee *et al.*, 1955; Mac-Lagan *et al.*, 1958; Malmierca, 1947; Schoger, 1960; Stacher and Böhnel, 1959; Stefan, 1960; Steinhoff, 1960), haematology (Henderson and Gray, 1959; Holzknecht, 1958; Horvath and Ludwigh, 1962), phtisiology (Gürtler, 1964; Kuthy *et al.*, 1960; Schill, 1956), dermatology (Bertaggia and Cadrobbi, 1970), and in pediatrics (Keuth, 1959), have been published.

Also, various studies reporting the use of snake venoms as hemostatics have been published in the Soviet Union (Iskhski, 1959; Kots and Barkagan, 1957; etc.). Only the report of Iskhski has been accessible to the author in a review. This otorhinologist treated the postoperative bleeding following tonsillectomy in 108 patients by tampons impregnated with a native snake venom in a 1:10,000 dilution (the snake species is not indicated). He reports noticeable reduction of postoperative bleeding, secondary hemorrhages less frequent than usual, reduced postoperative pain, and the appearance of fibrin exudates in a much finer form that disappeared much sooner than usual.

The clinical results obtained confirm the supposition, based on laboratory tests, that ferments acting as thrombin (in the case of the *Bothrops* species), or as thromboplastin (when the venom of *Vipera russelli* is used) are concerned. From the clinical results covering over 3000 registered cases it can be inferred that the preparations made from the above-mentioned venoms considerably reduce the operative and postoperative bleeding and favorably influence the pathologically prolonged bleeding and coagulation times, when these are caused by prothrombin deficiency, deficiency in the accelerators of the transformation of prothrombin into thrombin, or deficiency of factor V or factor VII. In connection with the latter, it should be mentioned that Hougie *et al.* (1957) experimented with Russell's viper venom in dilutions of 1:10,000 and 1:20,000 in six cases of proconvertin deficiency, and they obtained favorable effects in four cases. The ferments and the drugs prepared from them are ineffective in cases of essential hemophilia, fibrigenopenia, or afibrinogenemia. In cases of essential hemophilia, beside the specific treatment, the part of adjuvans falls to them as they can shorten the coagulation time after restoration of the normal coagulation process. They are likewise ineffective in cases in which the bleeding is caused by extensive destructive lesions or, in women, by hormonal disturbances.

Recently several papers have been published concerning hemostatic use of Reptilase, a thrombin-like enzyme preparation derived from *Bothrops atrox* venom* Ohlenschlaeger and Schwalbe (1966) have treated with Reptilase 91 patients who had prothrombin deficiency caused by liver disease, symptomatic thrombocytopenies, various forms of purpure, and postoperative bleeding. They found the drug effective in all cases in which fibrinogen deficiency was not present. Noguera Toledo *et al.* (1965) had used Reptilase in aerosol form with "optimal" results in the treatment of 42 patients with bleeding in the respiratory region, among them seven cases with infiltrated pulmonary tuberculosis. Stroeder and Ambs (1967) had the opinion that Reptilase occasionally influences favorably certain types of hemophilia bleeding. Utz (1968) reported on treatment of von Willebrand-Juergens disease and Keuth (1968) on bleeding of premature infants by the named drug with satisfactory results.

Considering the unquestionable coagulant action of snake venoms used for therapeutic purposes, and the favorable results obtained by clinicians with these preparations, it is difficult to explain how Blombäck *et al.* (1957) in laboratory tests and Kock (1957) in operations did not gain positive results.

C. Other Therapeutic Applications

Modern medicine has pioneered the use of snake venom in the treatment of genuine epilepsy.

A year after the publication of Spangler's first results, mentioned above in the introduction (1910). Fackenheim (1911) reported satisfactory results he obtained in five patients with treatment by Crotalin. He used increasing

* It is produced by Pentapharm A. G. Basle (Switzerland).

doses (0.3–1.3 mg) in the form of injections and found that the disease is favorably influenced, specially in its initial stages.

Spangler (1911) later described 36 more cases in which he also observed favorable results. In a later communication he reported 18 patients treated with success (1913).

The theory by which Spangler tried to explain the favorable therapeutic action of snake venom is interesting, even if only from an historical point of view. It was based, on the one hand, on his observation that the blood of patients who had had Crotalin injections showed a very protracted coagulation time, and, on the other hand, on the fact that he knew of no case of epilepsy paired with hemophilia. He therefore thought that hemophiliacs could never get epilepsy and attributed the good results obtained with Crotalin to the anticoagulant effect of the venom, that is, to an artificially induced condition similar to hemophilia (Spangler, 1912). It is worthwhile mentioning that for some time even the well-known English epilepsy investigator Turner (1910) accepted this idea.

To explain and complete Spangler's observation, it should be mentioned that de Klobusitzky and König (1936) have shown that the anticoagulant action of large doses of a coagulant snake venom is only transitory, and that after 24 hours it also has a coagulant effect. Noc (1904) was the first to show that the snake venoms that in low concentration are coagulants prevent coagulation in higher concentrations. He did not know, however, that this action is only transitory.

Paula Souza (1914) was the first to object to Spangler's audacious theory by pointing out that hemophilia is such a rare disease compared to epilepsy that to claim a connection between them would be absolutely inadmissable. He himself treated seven patients with the venom of the *Crotalus d. terrificus*. He used the venom in two ways; to four patients he gave increasing weekly doses of four 0.1 to 10 mg per injection; to the others he gave only one large dose of 7.5, 8.0, or 8.5 mg. All of the patients had been under observation for 2 years before the venom therapy, and afterwards were observed for a whole year. The treatment with venom did not bring any improvement to any of the patients.

The results of Yawger (1914) were likewise unfavorable. This clinician treated six patients with increasing doses (from 0.32 to 2.6 mg) of Crotalin with the following results: the condition of two patients remained unchanged and two others got worse; the course of treatment had to be interrupted in one because of symptoms of poisoning, and, finally, one died.

Thom (1914), who treated 58 cases with Crotalin, summarized his experiences as follows: in 10% the number of attacks increased, in 35% there was a noticeable aggravation of the condition, and in 40% there was no change. Thom therefore considered the value of Crotalin as very questionable. Zalla

(1914) also expressed the same opinion, while Crotalin therapy is considered by Joedicke (1913) as totally ineffective. Still more pessimistic in their judgment were Jenkin and Pendleton (1913), as well as Anderson (1914). The former even considered Crotalin dangerous, because they never could obtain improvement in their patients through its use, but, on the contrary, some of their patients died from poisoning. Schübel (1942) also agrees with them, and he strictly refused to treat epilepsy with snake venom.

Opposed to these absolutely negative results stand 11 cases studied by Calmette and Mézie (1914), who, during a 3-year period of observation, were able to observe a 3.3–77.1% decrease in the number of attacks in 10 patients due to the use of Crotalin (0.3–1.5 mg per dose). In one patient, though, the frequency of the attacks increased by nearly 100% (before treatment 132, during treatment 216 attacks per year). These authors ascribe a double action to Crotalin. One consists in the sedation of the nervous system, the other in an increase in blood viscosity.

The results of Bouché and Hustin (1922) are even less favorable. They saw in the snake venom not a specific, but only a foreign albumine. These authors used, with 97 epileptics, snake venom, horse serum, or powdered erythrocytes prepared from the patient's own blood. These three methods were in part combined, and in part applied separately. Of the 97 patients, 49 did not show any change, in 6 a clear deterioration was registered, 14 improved transitorily, and 28 for a long time. Among the latter, there were 16 who remained free of attacks for months.

The best results of all have been registered by Fackenheim (1926) and Lindenberg (1953). The former treated 50 patients by snake venom and observed them during 10 years. Thirty of them were able to ply their trades "nearly without troubles." His conclusion is that by means of Crotalin, epilepsy can be controlled. Lindenberg tested on a large number of patients a preparation called Epileptasid that contains 0.1 mg of the venom of the South American *Crotalus durissus terrificus* and as preservative substance an orthooxybenzoic acid ester. All together 444 patients, 402 of whom suffered from traumatic epilepsy and 28 from genuine epilepsy (in 14 patients it was not possible to establish the genesis) were subjected to the venom treatment. The treatment consisted in subcutaneous injections given once a week for 35 weeks. The first dose was 0.02, the second 0.05, and the third 0.075 mg; the later doses were all 0.1 mg. His statistics show a 55.2% improvement, 15.1% good improvement, and 3.8% recovery. In 25.9% the disease continued unchanged. The author ascribes as a secondary advantage to his treatment a psyche-stimulating and euphoristic action. In the past 16 years no more studies have been published on this which seems to indicate that the treatment of epileptics by snake-venom preparations has been abandoned by clinicians, since it presented no advantages in any respect.

To conclude, it should also be said that Giffo (1949) has employed, or tried *Naja naja* venom in ophthalmology on a very large scale. He treated periorbital pains and sympathalgias by intracutaneous administration of Cobratoxin; on the other hand, he gave 2 mouse units of it, intramuscularly, or subcutaneously, twice a week to patients suffering from eyeground diseases, retinitis due to myopy, chorioretinopathy, glaucoma, Gräf's syndrome, intraocular bleedings. As he has compiled the relative literature in a book published the same year (Giffo, 1949), we consider unnecessary an enumeration of all publications before 1949. In his opinion the *Naja* venom is not specific, but it is, besides the etiological treatment, a "very precious" adjuvant. More recently, Oaks and Quinn (1954) as well as Clark *et al.* (1962) have reported on the treatment of eye diseases with snake venoms. The latter treated recurrent lesions of the cornea caused by herpes simplex virus with a neurotoxoid preparation. Their opinion coincides with that of Giffo.

The many contradictory opinions induced Bryson (1954) to add formic and silicic acids to cobra venom for the treatment of chronic arthritis, but without encouraging results.

Laterza and De Serio (1967) have employed the anticoagulant fraction of *Bothrops jararaca* venom in the treatment of chicken pox in 40 children (without declaration concerning method of fractionation or name of the drug). They observed the rapid cessation (in 1 to 3 days) of pruritus, healing in 38 cases within 4 to 7 days, and healing within 8 days in 2 cases. Nashbits (1969) has treated inflammatory diseases of female genitals with a Russian snake venom preparation (Virapin). The original of his publication was not accessible to this author.

Isolated experiments have been made with different snake venoms for the treatment of herpes simplex (Kelly, 1938), herpes s. recurrens (Sommerville, 1948), herpes zoster (Dennis *et al.*, 1949); Tan and Ines-Tan, 1947; Fisher, 1941), for different skin diseases (D. I. Macht, 1940), bronchial asthma (da Silva, 1948), angina pectoris (Bullrich and Ferrari, 1940; Riseman, 1945), rhinitis (Jeschek, 1938), eclampsia (Poulain, 1937), and, also urticaria (Cohen, 1951), but they are more of theoretical than practical interest.

III. TOAD VENOMS

When and where raw toad venom was first used for medicinal purposes escapes our knowledge. The ancient Egyptians often represented toads, especially on divining rods or in the form of amulets, probably as a frightening or protective figure against evil spirits (Egger, 1936), but there are no proofs that they also used their venom as medicine. We are equally ignorant of when the dried toad venom found its way into the therapeutic arsenal of the ancient Chinese (under the name Ch'an Su) and the ancient Japanese physicians (under the name Senso). In Europe, this venom appears for the first time in powder form under the designation *Bufones exsiccati* in the medicinal book of France, published in 1692. At the suggestion of Samuel Dale, it was recommended for the treatment of hydrops and in the following centuries it was used without interruption for this purpose, the only change being that, beside the powder form, it could also be found triturated with olive oil (Lémèry, 1716).

In modern medicine, attempts have been made to use the analgesic, heart-stimulating, and hemostatic action of the raw toad venoms therapeutically (that is, skin extracts, secretions of the big double glands, the so-called parotid glands, lying behind the head, as well as of the hindlegs glands).

The analgesic property of these venoms, which was discovered by Pierotti (1906) and more minutely studied by Kajimoto (1937), who found the anesthetic action of *Ch'an Su* considerably (up to five times) stronger than that of cocaine, was put to use in malignant tumors. The preparation had been put on the market as a subcutaneously injectable watery or oily solution (usually in 10% camphorated oil), titrated in mouse units. A mouse unit was considered to be the smallest amount of venom that could kill a 1-gm white mouse and was equivalent to about 1/25 of Cobratoxin mouse unit. The number of cancer cases in which this venom was used is, in comparison with the ones treated by snake venom, infinitesimal. To our knowledge, the course of the treatment has been described for only 12 such patients: Cornilleau reported four cases (1938), Sohier and Radaody-Ralarosy five cases (1940), Fiehrer one case (1940), Moch one case (1942), and Gascoin one case (1942). The first two and Moch agreed that relief from pain was obtained and that a general tonic and stimulating action, as well as the return of lost appetite, was observed. Fiehrer stresses further that both the bleeding caused by the tumors and the painful menometrorrhagias were favorably influenced. Gascoin treated a mediastinal tumor, the exact nature of which had not been established (Hodgkin's disease or lymphosarcoma?) in the same way with success for 6 months. (It was ascertained roentgenologically that the tumor had nearly completely disappeared and that the patient's general state had accordingly improved).

Gascoin had previously (1941–1942) made known the extraordinarily good action of this venom on a patient with Hodgkin's disease. He began the treatment by administering 1 ampule of the commercial preparation and later reduced this dose by half. In his estimation, the patient was completely cured and his opinion was reinforced by the specialists called in to check his results. This result, the only totally favorable one to be found in the special literature, led Denéchau (1942) to apply the toad venom treatment to two of his patients affected by the same disease. In one of them there was unquestionable and considerable improvement, but it proved however, to be only transitory. The second case was so aggravated by the venom therapy, that treatment had to be interrupted.

The heart-stimulating property of these venoms has been known for over a hundred years. Vulpian (1854) was the first to report on it (at the Paris Biological Association in 1853). The active substances were only identified, however, after Faust (1903) isolated his Bufotalin from the skin of the *Bufo b. bufo*. We know today that the steroids existing in all toad venoms and called bufogenines act as heart stimulants. Süttinger (1955a) tried a preparation called Bufomarin in heart diseases and their secondary complication (in a reduction of blood circulation dynamics). He treated per os 19 patients who had cardiac weakness, bronchial asthma or angina pectoris. He observed good results in six cases, including three of angina pectoris. In a later publication (1955b) he summarized his experiences by ascribing to the venom a compensation-maintaining but no decompensatory action. Consequently, toad venom therapy should only be used as an adjuvant for compensated hearts. For senile hearts, he rates the action higher, as he ascribes to the venom a coronary dilatator effect and, in the case of an old myocardium, an effect of awakening oxygen hunger. Since, however, the venom does not show any advantages at all over digitalis preparations and synthetic coronary dilators, and because the toxic dose cannot be clearly differentiated from the therapeutic one, toad venom therapy has been completely abandoned for heart diseases.

With regard to the hemostatic uses of toad venom, reports are scanty and restricted to only one field, gynecology. Fiehrer (1940), in addition to having used this venom on the cancer patient mentioned previously, has used it in 10 cases of menorrhagias and metrorrhagias, under that several of hormonal origin. He usually administered 8 mouse units daily; in especially serious cases he would give the same amount every 3 hours until the bleeding stopped. His good results deserve consideration insofar as the toad venom also proved to be efficient in bleeding caused by hormonal dysfunction, which, as is known, cannot be eliminated by snake venom ferments. In spite of this, no other reports of this possible use of toad venom have been published, at least, not to our knowledge. This certainly must be explained by improvements in the field of hormonotherapy. It should also be mentioned that toad venom seems to influence favorably the healing process (closing and cicatrization) in skin wounds. Machon (1944) reported that the Indians in Mexico use an ointment containing toad venom to hasten the closing of wounds caused in birds by plucking their feathers. This possibly has led Fischenko and Neiko (1961) to investigate the influence of this venom on the healing of experimental wounds. Unfortunately their results have not been available to the author, even in review form.

IV. SPIDER VENOMS

Chemical and pharmacological knowledge of the many spider venoms is limited. Only one, the raw venom of the European spider, *Latrodectus tredecimguttatus*, has been used for therapeutic purposes, and then only by a single group of clinicians.

Jelašić *et al.*, starting with the pharmacological properties of the venom ascertained by Maretić and Jelašić (1953) and later by Jelašić *et al.* (1954), got the idea of using it in diseases in which the local vasodilatation could have a pain-killing action and a favorable action on the local pathological process, as well as in rheumatic diseases. They have described the treatment by intracutaneous injection of four cases of periarthritis humeroscapularis, four chronic degenerative arthropathy, and three spondylarthrosis cases in which satisfactory relief from pain was obtained. In ischialgia, with herniation of the intervertebral disk, the result was questionable; only in one case did they observe an improvement lasting 3 months. In bronchial asthma, the improvement lasted 4 days; by repeating the treatment this result could again be obtained. The authors considered the action of the spider venom used very similar to that of bee venom, especially on the chronic pathological processes arising in the mesenchymal tissues.

V. BEE VENOM

The presence of the venom of the honeybee (*Apis mellifera*) in the therapeutic arsenal of various peoples can be traced back to ancient times. The Babylonians, Egyptians, Persians, Greeks, and Romans were beekeepers and they esteemed these hymenopteras not only for their honey, but because of the curative value in various illnesses which the people and physicians of the times ascribed to their use.

Alcoholic or aqueous bee extract (called variously bee broth or bee tea), bees triturated with honey, and ashes of bees were "medicines" known for a very long time when homeopathy began to make use of them.

Until about the middle of the nineteenth century, the whole animal was used; although it was known that the curative effect was connected with the venom, people still obviously found it easier to act according to the principle of *totum pro parte* than to collect the venom separately. An Austrian physician, F. Terč, introduced bee venom therapy into modern allopathy. He was familiar with the descriptions of many cures for gout and rheumatism that had been published in beekeeping magazines, and since he had had similar experiences in these areas, he decided to perform some controlled experiments. Both healthy persons and rheumatics were stung by bees at definite time intervals and the subsequent phenomena observed. In 7 years he had 173 persons stung by bees 39,000 times; after this, he handed his results over to the specialists (F. Terč, 1888). He reports that the rheumatic subjects reacted to the first sting as do immune persons, that is, the stung places did not swell. Reaction to stinging was observed only after several stings. Continued stinging caused a renewed absence of reaction, and during this period the rheumatism was cured. Terč designated this stage as the stage of obtained immunity.

Terč's report, though it described 23 cases in detail, did not meet with any consideration, principally because to let oneself be stung by bees as a treatment was rejected a priori by the specialists and because of the large differences in the number of stings necessary for treatment (Case No. 4, backache, required 8 stings, while No. 23, general rheumatic arthritis 1650 stings were needed). Keiter's report 16 years later (1904) did not do anything to change this attitude, although he reported favorable results with about 2000 rheumatic and arthritic subjects. Resumption of the use of bee venom took place only after pharmaceutical preparations (ointments and injectable products) had been made from it.

A. Pharmaceutical Preparations

The use of living bees for therapeutic purposes presented several undesirable concomitant circumstances. The stinging of the bees was painful, the measurement of the amount of venom applied was difficult, it was not always possible to get the bees, and many people were very sensitive to the proteins present in the venom; they showed allergic manifestations, accompanied by circulation disorders, headaches, attacks of dizziness, vomiting, etc., and also suffered from fever, somnolence, and polyuria. Langer, who was the first to work on the chemistry and on both the physiologic and toxicologic effects of bee venom (1897, 1899), was also the first to prepare injections from weakened but nevertheless native bee venom, which he used with success on children suffering from arthropathies (1915). The fact that this preparation still cause violent pain (the solution was injected subcutaneously) induced several researchers (Flury, 1920; Pollack, 1928; Kretschy, 1928; Forster, 1938; Tetsch and Wolf, 1936; Reinert, 1936), to produce from the native venom a purified sterilizable, standard pharmaceutical preparation that was also stable in the injectable form. The first of these remedies was put on the market under the name Apicosan.

In time, the injectable preparations were replaced in most countries by ointments containing bee venom. The first of these was produced by Forster (1938) and was named Forapinsalbe. Both the injections and the ointments contain the venom measured in mouse units. The smallest quantity of venom that, administered subcutaneously, killed a white mouse, weighing 13–15 gm, in 6 hours (about 0.12 mg) was established as the mouse unit.

The ointment can be introduced in different ways. There are ointments that contain very finely ground glass or other silicate that, when rubbed in with a special rubbing device, scarifies the upper skin layer and so considerably furthers the penetration of the venom. Ointments that do not contain a scarifying substance can be brought to resorption by rubbing, by iontophoresis alone, by iontophoresis combined with hyperemizating physiotherapy (Ferrière, 1937; Descoeudres and Wacker, 1938).

B. Antirheumatic and Antiarthritic Applications

The above-described experiences of the beekeepers, the publications of F. Terč (1888) and of Keiter (1904) reported there, in addition to R. Terč's account of 700 rheumatism cases with 82% recovery (1912), make one assume that this venom, when the production of pharmaceutical preparations had replaced the stinging method and, consequently, its application had spread among the physicians, was almost exclusively used in cases of "rheumatism," namely rheumatic diseases, of arthritis and other arthropathies.

The great enthusiasm of F. Terč, his son, R. Terč, and Keiter, each of whom reported almost 100% success, gave way to the restrained view that bee venom was not a wonder drug, treatment which hardly failed, but, even when the dosage was adjusted to each individual, the expected result was obtained only in 50–52% of the cases in which it was used.

The earliest reports, involving several hundred treatments, such as those of Grünewald (1934), Kübels (1923), Passow (1927), Wasserbrenner (1928), Kretschy (1928), Freund (1927), Krebs (1929), Löbl and Simo (1930), E. Lange, (1931), Zimmer (1931), Becker (1931), Nowotny (1932), Pollack (1932), Fehlow (1935), Burkardt (1935), and Kroner *et al.* (1938) are not discussed here, since they have already been discussed in the monographs of Beck (1935), Phisalix (1936), and Schwab (1934, 1937, 1938).

A brief discussion of some recent studies, as well as some earlier ones involving a large number of patients, follows. Pengler and Pribert (1935) reported 200 cases of myalgia, arthralgia, and neuralgia treated with bee venom containing histamine (1 part histamine to 5 parts bee venom). In only a few patients was this treatment completely ineffective. Baldermann (1936) treated approximately 24 patients afflicted with "rheumatism" by means of an injection preparation, and observed relief from pain. Frauchiger (1936). treated 66 cases: he obtained very good results in cases of chronic polyathritis but cases of localized arthroses exhibited little reaction to bee venom. Hofbauer (1936) related his experience with 68 patients, most of them rheumatics, including 24 chronic cases. Thirty-two patients recovered after a course of two treatments. Mackenna (1939) concluded that bee venom therapy provides satisfactory results, based on his work with 100 cases of rheumatism. Kirchner (1940) treated rheumatism of allergic origin with similar results.

Keller's four cases of malum coxae sen. (1936) deserve to be mentioned here, since through this treatment both remission of pain and long-lasting recovery of mobility were obtained. Ferrière (1937) used either bee venom alone or Paul's Cuti-Vakzin on his patients. The latter medication gave better results. Bornet (1937) and Bénitte (1938) have combined this therapy with natural warm radioactive baths so that the cause of their poor results (relief from pain, but not in all cases) cannot be established unquestionably. Attinger (1941) treated over 50 patients with subacute or chronic articular rheumatism deformations of varied types and origins. In almost all cases, he was able to ascertain a favorable effect from the bee venom preparation. Walch (1943) also reported good experience with lumbago, myalgia, epicondylitis, and sciatica. However, positive results were obtained in only 50–60% of the cases of neuritis, chronic rheumatism, and periarthritis scapulae, depending on the disease. Hesse (1949), reports several acute and chronic nonfebrile neuralgia, myalgia, arthritis, and spondylarthritis cases, in which he used 15–20 injections given every other day, without positive results. Lhermitte *et al.* (1954) generally observed good results from the injection therapy in chronic mono- and polyarthritis.

The last notable communications known to us are that of Steinhoff (1960), and Steigerwaldt *et al.* (1966). Steinhoff treated 70 patients (23 with muscular and articular rheumatism, 25 with athletic injuries, and 10 with pain in the limbs resulting from grippe), with ointment. He obtained freedom from complaints in 52% of the rheumatics and injured and an equal relief from pain in the other patients. Of six patients with a high degree of articular deformation, roentgenologically proved, two patients experienced freedom from pain of short duration, and four patients showed considerable improvement. Steigerwaldt and co-workers had treated arthritis in fifty patients and obtained 84% benefits.

In Soviet Russia, treatment by means of bee venom was begun about the time that it was being given up in other countries. The studies of Istomina and Konevetseva (1959) who used the venom against polyarthritis, of Petsulenko (1961) who used it against unspecified (rheumatoid) arthritis, and of Poriadin (1962) who used it against spondyloarthritis deformans date from the years 1959–1962. Prikhod'ko (1965) published treatments of lumbar sciatica by iontophoresis introduced bee venom. Since they are not known to the author, even in review, it is impossible for him to go into more detail about them. The development of physical and other therapeutic resources, the use of which seems to be simpler, and is in any case painless, has practically eliminated bee venom entirely from the present drug list, although its curative effect has been repeatedly established by objective research methods.

In the previously mentioned work of Burkardt (1935), a personal report of Körbler is cited which states that, after successful treatment with bee venom, improvement in bone structure could be ascertained by means of X-rays. Neumann and Stracke (1951) have been able to verify in rats, in which arthritis had been provoked by formaldehyde and which had been treated partially by pure bee venom with the addition of histamine, and partially by desoxycorticosterone-glycosides and ascorbic acid, that the bee venom "in appropriate dosage, possesses a statistically significant prophylactic and therapeutic action. The success of the bee venom treatment must not be ascribed either to the histamine content of the venom or the release of the histamine proper into the body." Fellinger *et al.* (1952) established once more by electrodermatography that this venom "causes a clear change of the tonicity condition toward normalization" in rheumatics.

C. Other Therapeutic Applications

It is a matter of experience in medicine that every completely new drug, soon after the first favorable news of its efficiency against a particular disease or group of diseases is published, is tried on a series of diseases. One must not be surprised that this fate also occurred with bee venom. At a time when the "stinging method" was still in use, Boinet (1923a,b), reported treatment of two lepers and two patients afflicted with lupus. The first leper, suffering from *L. maculo-tuberculosa*, received, during a 4-month period, stings on the lepromas in increasing daily doses (up to 120 per day). As a result, the author was able to record the retrogression of the lepromas, the return of skin sensibility (except on the left hand), and improvement in the general condition of the patient. The other leper received 3935 stings and felt so much better that he was able to return to his native country. He was able to control the two patients ill with lupus (a 50-year-old man and a 30-year-old woman) for over 10 years. The stings were applied to the skin lesions. The man received 1500 stings in 4½ months, and the woman received 4000 stings in 9 months. Both could be considered cured. At about the same time, Vigne and Bouyala (1923) reported treating a 45-year-old woman suffering from leg pains and a very bulky edema with 4450 stings given over a 9-month period. She regained the capacity to walk, and was able to return to work. The work of Strebel (1931) on the successful treatment of trachoma must likewise be mentioned here.

In addition, pains caused by malignant tumors (Yannovitch and Chabovitch, 1932; Natale, 1935), urticaria (Wolpe, 1937), trigeminal neuralgia (Bayne, 1948), hay fever (Haag and König, 1936; von der Trappen, 1938), trachoma (Bechner, 1934, 1935; Charamis, 1935; Lagomarsino, 1939), chilblains (Watson, 1941), *herpes corneae*, *keratitis* (Much, 1948), various eye diseases of an inflammatory nature (Fontana, 1941; Shershevskaya, 1949), and allergy (McLane, 1943), have been treated by pharmaceutical preparations containing bee venoms but not with very encouraging results.

The vasodilator action of this venom has been used by Skursky (1955) in the treatment of coronary circulation problems. For reasons easy to understand, no further therapeutic attempts have been made in this field.

Only brief mention of this subject is to be found in the literature of Soviet Russia. Vladimirova (1959) used bee venom on the pain syndrome, Voitik (1958), and also Alesker (1959), on internal diseases, Fishkov (1954, 1961) on surgical problems, and Liashenko *et al.* (1963) in gynecological diseases. Nothing can be reported on their results, as the corresponding periodicals have not been accessible to the author and he has not been able to find any reviews. It should, however, be mentioned that a patient of Karapata and Shumakow (1961) developed lung edema when his case of chronic nephritis was treated with bee venom.

VI. VENOM COMBINATIONS

As has been shown in the preceding sections, several clinicians have used venom mixtures (Castro Escalada, Des Ligneris and Grasset, Burkardt, and Grundmann) and these were also put on the market as pharmaceutical preparations, e.g., Vipericin. Only one such preparation, composed of venoms of different animal species and of other substances, has been produced that has been able to maintain itself up to the present time; every now and then reports have appeared in the clinical literature. This preparation came out in 1943 under the name Esberitox and it contained besides bee venom, *Crotalus* and *Lachesis* venoms (most probably it is the venom of a *Bothrops* species; earlier the *Bothrops* and *Lachesis* species were often confused) and extracts of *Echinacea,* and *Baptisia,* and *Thuja occidentalis.*

Esberitox has not been recommended or used specifically against a particular disease or disease group, but only as an adjuvant, mainly together with antibiotics or chemotherapy. Over 1000 cases treated by this means have been described in the literature and all authors (Pfannenstiel, 1952; Frenkler-Bleissner, 1956; Haevernick and Brandner, 1956; Hagenmüller, 1956; Johne and Sperr, 1956; Ströder, 1956; Wiese, 1956; Friede, 1957; Jornes, 1957; Ackermann, 1959, 1959; Schoger, 1960; Helbig, 1961; Graatz, 1962; etc.) stress the anti-infective property of the preparation, its capacity to increase the natural defensive powers of the organism. It shows special activity both as prophylactic and as therapeutic agent against upper respiratory infections. In general, it is considered to be an unspecific stimulant, the action of which manifests itself in a stimulative effect on the anterior hypophysial suprarenal cortex system. This can be inferred from the reduction of the eosinophile by 50% and more. It is mostly used in child and infant care to forestall catarrhal infections, due to colds or to shorten and to mitigate the course of such diseases.

VII. APPENDIX: SNAKE VENOMS AS ANTICOAGULANTS

The fact that some snake venoms have a strong anticoagulant property has been known for as long as their blood coagulant action. (The first known observations on the anticoagulant effect of snake venom are mentioned by Claude J. Geoffroy and Pierre Hunauld: Mém. dans lequel on examine si l'huile d'olive est un specifique contre la morsure des vipères. Amsterdam, 1741, pg. 282. See the more recent corresponding literature in Chaps. 6 and 7 in Volume I of this treaties). The therapeutic application of these venoms as anticoagulants in allopathy is, however, more recent, and presently is still in its elementary stage.

We can consider as the beginning of these clinical experiments those observations of Reid *et al.* (1963a,b) and Chan and Reid (1964),which they obtained on 250 patients who were bitten in 1960–1961 by *Agkistrodon rhodostoma* (Malayan pit vipers). Beside other symptoms, a clotting defect, consisting of complete defibrination, was the principal characteristic of systemic poisoning, which was present in 39% of the victims. In 15% of the cases an overt hemorrhagic syndrome was also observed. The coagulation deficiency is in many respects similar to other defibrination syndromes. A very interesting feature was the complete absence of unprovoked hemorrhages in these patients except in the area of necrosis at the site of the bite. The treatment of the described clotting defect by blood transfusions or by concentrated antivenin was very successful. Chan and his co-workers (1965) had demonstrated in later investigations that the clotting agent, which exists in the crude venom, had no effect on the antihemophylic globulin and it was not neutralized by formed fibrin. They have carried out experiments with the purified clotting enzyme (prepared by Esnouf and Tunnah, the method was published in 1967) and obtained very encouraging results concerning the possibility of its use in the prevention or treatment of thrombosis.

Somewhat later, Marsten and his co-workers (1966) studied the effect of the venom of *Agkistrodon rhodostoma* on experimental thrombosis of dogs. They obtained large red clots completely obstructing the inferior vena cava and were able to observe the appearance of hypofibrinogenemia and of accelerated fibrinolysis. The authors consider this venom an antithrombotic agent. Similar opinion has been expressed by Chan *et al.* also in their abovementioned paper.

Nearly simultaneously another group of investigators (Regoeczi and co-workers, 1966) verified that the purified clotting agent, if it is injected into animals, produces defibrination by continuously converting fibrinogen to fibrin in the circulation. Emboli were found in various organs of rabbits as a consequence of injection of the purified enzyme, but never in dogs or sheep.

In connection with these results, it must be mentioned that Djaldetti *et al.* (1966) found that the venom of *Echis coloratus* (an Arabian adder), which contains a strong procoagulant component, shows remarkable clot-dissolving activity. It was verified that this venom inoculated into dogs is also able to enhance dissolution of artificial pulmonary emboli.

However, clinical experiments have been carried out, in the meantime, only with the venom of Malayan pit viper. Esnouf and Tunnah (1967) succeeded in the purification of this crude venom, obtaining a coagulant enzyme free from the proteolytic and hemorrhagic fractions by chromatography on triethylaminoethyl-cellulose. This purified coagulant agent, which was named Arvin, had thrombin like activity, but it is certainly not identical with the thrombin; it is stable and unaffected by heparin.

Arvin is standardized for clinical use.* One Arvin unit contains 2 μg protein, being equivalent to 1 National Institute of Health unit of thrombin, using purified and standardized human fibrinogen as substrate.† One milliliter of the pharmaceutical product contains 75 units dissolved in sodium phosphate–sodium chloride buffer at pH 7.0. It keeps at 4°C for several months and 1 ml of specific antivenom against crude *Agkistrodon rhodostoma* venom, prepared by the Pasteur Institute in Paris, neutralized approximately 75 Arvin units.

The results of Chan (1968) on the modification of occlusive arterial thrombosis in the rat also should be mentioned. He has studied the effect of Malayan pit viper venom on the fate of experimental arterial thrombosis. The fibrinogenemia was obtained by daily injections of 500 μg venom. The results are summarized in Table I.

The first known clinical results were published simultaneously by Bell and co-workers (1968) and by Sharp and co-workers (1968)

Bell *et al.* treated seven patients with venom thrombosis, one with acute myocardial infarction, and one with atrial fibrillation and mitral stenosis. The drug was given intravenously, at a dose of 1 Arvin unit per kilogram body weight. The first dose was given in 100 ml of sterile 0.9% saline solution and infused slowly over a 6-hour period by means of a constant infusion pump. At the completion of the infusion, a second dose was injected in 20 ml

* It is prepared by Twyford Laboratories, Ltd.

† This unit is the quantity which clots 1 ml of this fibrinogen solution in 15 ± 0.5 seconds at 28° ± 1°C.

TABLE I

Days postoperative in control animals and days following cessation of treatment in treated animals	Percent occlusions		Percent red thrombus	
	Control	Treated	Control	Treated
2–3	63.2	36.8	47.4	16.8
4–5	76.9	36.4	69.2	9.1
7	100	50.0	100	33.3

of the same saline solution over 10–15 minutes repeated every 12 hours during the duration of the therapy. Forty-eight hours before discontinuation of Arvin, each patient was given oral anticoagulant, which was continued after discharge from the hospital. Complete resolution of the thrombus was observed in the patients who had venom thrombosis. Three patients had had the thrombosis for 5 weeks or more, and the authors interpret the results as follows: "the rapid clinical response in these patients was dramatic." The patients with infarction were treated also with phenindion, the result was recovery without any incident. The plasma-fibrinogen concentration fell rapidly to less than 40 mg/100 ml (the normal values range from 150–250 mg/100 ml) during the treatment and increased slowly after its termination.

Sharp and co-workers had treated 19 patients; of these, however, two (the cases 1 and 8 in their paper) could not be used for critical examination of the treatment by Arvin. Patient 1 died of cardiac arrest, patient 8 was submitted to an endarterectomy after which she (a woman, 65 years old) suffered excessive hemorrhage, which was treated by surgical intervention.

Arvin was given by continuous drip, single i.v. injections at suitable intervals, or, in one case only (with patient 19), by intramuscular injection. The initial dosage was 1–2 units per kilogram body weight and subsequent dosage was adjusted according to the fibrinogen level. The results of the treatment are reproduced in Table II. The authors are of the opinion that Arvin is worthy of more experimental trial.

The therapeutic (thrombolytic) action of Arvin rests, according to the opinion of Reid and Chan (1968), upon its highly specific effect of fibrinogen. Arvin, if used in suitable dosage, does not affect any of the other coagulation or hemostatic factors.

The manner and mechanism of action of Arvin require elucidation. The results of treatment known at present are few but are encouraging for further clinical and hematological studies.

TABLE II

Patient no.	Clinical indication for therapy	Injected Arvin units and duration of therapy (days)	Results of therapy
2	Suspected deep-vein thrombosis	150 (5)	Equivocal
3	Deep vein thrombosis	210 (6)	Satisfactory
4	Deep vein thrombosis	360 (6)	Satisfactory
5	Recurrent thrombosis or emboli brachial arteries	685 (9); 2150 (15)	Equivocal
6	Femoral thromboendarterectomy	175 (5)	Satisfactory
7	Femoral artery thrombosis, thromboendarterectomy	1800 (6; drip); 980 (14; injected)	Equivocal
9	Deep vein thrombosis, purpura gangrenosa	7350 (21)	Satisfactory
10	Painful Singer syndrome, digital vein thrombosis	1344 (12)	Satisfactory
11	Quadriplegia prophylaxis against deep vein thrombosis	1120 (5; drip); 1344 (12; injected)	Satisfactory
12	Bilateral iliac artery thrombosis, thrombectomy	2128 (19)	Unsatisfactory
13	Nephrotic syndrome, pulmonary emboli	1568 (7); 2338 (7)	Satisfactory; unsatisfactory[a]
14	Venous thrombosis	448 (2)	Equivocal
15	Deep vein thrombosis, pulmonary emboli	2400 (10; drip); 2880 (12; injected)	Satisfactory
16	Deep vein thrombosis, paradoxical brachial emboli, pulmonary emboli	5400 (15)	Satisfactory
17	Inoperable carcinoma of stomach, deep vein thrombosis	480 (2)	Satisfactory
18	Incipient bilateral gangrene of feet	960 (4)	Equivocal
19	Deep vein thrombosis left leg	580 (36 hours; intramuscular)	Satisfactory

[a] Appendicitis developed on the eleventh day, with resistance to therapy.

REFERENCES

Ackerman, J. (1959). *Deut. Med. J.* **10**, 161.

Albee, F. H., and Maier, E. (1943). *Med. Record* **156**, 217.

Alesker, E. M. (1959). *Tr. Leningr. Sanit. Gigien. Med. Inst.* **48**, 62.

Alexander, A. H. (1941). *Sci. Am.* **165**, 66.

Anderson, J. F. (1914). *J. Am. Med. Assoc.* **62**, 893.

Attinger, E. (1941). *Schweiz. Med. Wochschr.* **71**, 772.

Aubry, M., and Causse, J. (1936). *Presse Med.* **44**, 2091.
Azevedo, M. P. (1946). *Rev. Paulista Med.* **29**, 105.
Baker, G. A., and Gibson, P. C. (1936). *Lancet* **1**, 428.
Baldermann, G. (1936). *Med. Klin.* **32**, 1368.
Barbosa, A. (1944). *Rev. Brasil. Biol.* **4**, 367.
Barú, R. J. (1935). *Arch. Uruguay Med., Cir. Especial.* (*Montevideo*) **6**, 569.
Baumgartner, H., and Typl, H. (1956). *Wien. Med. Wochschr.* **106,** 1000.
Bayne, R. (1948). *Southern Surgeon* **14**, 209.
Becher, J. (1934). *Ars Med.* **24**, 286.
Beck, B. F. (1935). "Bee Venom Therapy. Bee Venom, its Nature and Effects on Arthritic and Rheumatoid Conditions." Appleton, New York.
Becker, S. (1931). *Therap. Gegenw.* **72**, 251.
Behrmann, W. (1940). *Deut. Med. Wochschr.* **66**, 817.
Bell, W. R., Pitney, W. R., and Goodwin, J. F. (1968). *Lancet* **11**, 490.
Bénitte, A. C. (1938). *Bull. Mens. Soc. Med. Mil. Franc.* **32**, 119.
Bergamo, P. (1835). Quoted by Sachs, (1837).
Bertaggia, A., and Cadrobbi, P. (1970). *G. Mal. Infett. Parasit.* **22**, 102.
Billiotet, J. C. (1947). *Rev. Med. Navale* **2**, 65.
Blasiu, A. P. (1954). *Praxis* (*Bern*) **43**, 836.
Bleech, D. M. (1961). *Bull. Univ. Miami School Med.* **15**, 11.
Blombäck, B., Blombäck, M., and Nilsson, I. M. (1957). *Thromb. Diath. Haemorrhag.* **1**, 76.
Blondeau, P. (1963). *Anesthesie Analgesie, Reanimation* **20**, 187.
Boinet, E. (1923a). *Marseille Med.* **60**, 793.
Boinet, E. (1923b). *Marseille Med.* **60**, 1193.
Boquet, P. (1948). "Venins de serpents et antivenins." Flammarion, Paris.
Bornet, J. (1937). Ph.D. Dissertation, Fac. Med., University of Paris.
Bouché, G., and Hustin, A. (1922). "Chocs thérapeutique contre chocs morbides." Masson, Paris.
Brazil, V. (1934). *Biol. Med.* **1**, 50.
Brecher, I. (1934). *Klin. Monatsbl. Augenheilk.* **93**, 327.
Breda, R., Bernardi, R., and Sala, F. (1951). *Giorn. Clin. Med.* (*Parma*) **32**, 882.
Bruck, H. (1957). *Wien. Klin. Wochschr.* **69**, 571.
Bruck, H., and Salem, G. (1954). *Wien. Klin. Wochschr.* **66**, 395.
Brünner-Ornstein, M. (1937). *Wien. Klin. Wochschr.* **50**, 127.
Bryson, K. D. (1954). *Am. Surgeon* **20,** 751.
Buchert, (1945). *Rev. Brasil. Med.* **2**, 189.
Bulbrich, R. A., and Ferrari, J. A. (1940). *Prensa Med. Arg.* **26**, 12.
Burkardt, A. (1935). *Deut. Med. Wochschr.* **61**, 1159.
Burke, D. E., Pogrund, R. S., and Clark, W. G. (1960). *Clin. Res.* **8**, 127.
Calderón, H. R. (1934). *Semana Med.* (*Buenos Aires*) **41**, 1109.
Calmette, A., and Mézie, A. (1914). *Compt. Rend.* **158**, 846.
Calvi-Zampetti, A., and Minini, F. (1956). *Osped. Maggiore* **44**, 510.
Castro Araujo, S. G. (1935). *Biol. Méd.* **1**, 13.
Castro Escalada, P. (1935). "Ofidioterapia." Vian y Zona, Buenos Aires.
Cazzola, D. (1957). *Ginecol. Obstet. Mex.* **12**, 11.
Chan, K. E. (1968). *Thromb. Diath. Haemorrhag.* **19**, 242.
Chan, K. E., and Reid, H. A. (1964). *Lancet* **1**, 361.
Chan, K. E., Rizza, C. R., and Henderson, M. P. (1965). *Brit. J. Haematol.* **11**, 646.
Charamis, I. (1935). *Klin. Monatsbl. Augenheilk.* **95**, 660.

Chowhan, J. S. (1941). *Antiseptic (Madras, India)* **38**, 1.
Chowhan, J. S., and Chopra, R. N. (1938). *Indian Med. Gaz.* **73**, 720.
Clark, W. B., Baldone, J. A., and Thomas, C. I. (1962). *Southern Med. J.* **55**, 947.
Clauser, F. (1937). *Clin. Ostet. Ginecol.* **39**, 289.
Clauser, F. (1950). *Ann. Ostet. Ginecol.* **72**, 699.
Cohen, V. L. (1951). *Ann. Allergy* **9**, 173.
Cornilleau, R. (1938a). *J. Sci. Med. Lille* **56**,
Cornilleau, R. (1938b). "La Bufothérapie." Vigné, Paris.
Council of Pharmacy and Chemistry (1941). *J. Am. Med. Assoc.* **116**, 2521.
Council of Pharmacy and Chemistry (1944). *J. Am. Med. Assoc.* **122**, 825.
Cravarezza, F., and Stura, L. (1955). *Minerva Ginecol.* **7**, 710.
Croxatto, M., and Croxatto, (1944). *Anales Soc. Biol. Bogota* **1**, 147.
Da Silva, D. F. (1948). *Hospital (Rio de Janeiro)* **33**, 553.
Davin, E. J., Spielmann, F., and Rosen, A. J. (1937). *Am. J. Obstet. Gynecol.* **33**, 463.
de Klobusitzky, D. (1938). *Anais Inst. Pinheiros (Sao Paulo)* **1**, (2), 3.
de Klobusitzky, D. (1948). *Bol. Acad. Nacl. Farm. (Rio de Janeiro)* **10**, 627.
de Klobusitzky, D. (1949). *Mem. 1st Congr. Panam. Farm.* p. 481.
de Klobusitzky, D. (1958). *Abstr. Commun., 4th Intern. Congr. Biochem., Vienna,* 1958, p. 164. Pergamon Press, Oxford.
de Klobusitzky, D., and König, P. (1936). *Arch. Exptl. Pathol. Pharmakol.* **182**, 577.
de Leon, W. (1946). *J. Philippine Med. Assoc.* **22**, 1.
Denéchau, D. (1942). *Sang* **15**, 98.
Dennis, C. C., Morgan, D. B., and Coombs, F. P. (1949). *J. Invest. Dermatol.* **12**, 153.
Descoeudres, P., and Wacker, T. (1938). *Schweiz. Med. Wochschr.* **68**, 179.
des Lignieris, M., and Grasset, E. (1936). *Am. J. Cancer* **26**, 512.
Di Giunta, E., and Cali, G. (1960). *Clin. Otorinolaring.* **12**, 58.
Dix, J., and Sellach, J. (1958). *Deut. Zahnarztebl.* **12**, 725.
Djaldetti, M., Sandbank, U., Dinstman, M., Bessler, H., Rosin, M., and de Vries, A. (1966). *Thromb. Diath. Haemorrhag.* **16**, 420.
Durante, I. J., Moutsos, A., Ambrose, R. B., Duncan, W. Y., Fleming, W., Zinsser, H. H., and Phillips, L. L. (1962). *Ann. Surg.* **156**, 781.
Egger, F. (1936). *Mitt. Geog.-Ethnol. Ges. Basel* **4**, 1.
Esnouf, M. P., and Tunnah, G. W. (1967). *Brit. J. Haematol.* **13**, 581.
Fackenheim-Cassel, S. (1911).
Fackenheim-Cassel, S. (1926). *Zentr. Neurol.* **44**, 778.
Faust, E. S. (1903).
Fehlow, W. (1934). Quoted by Schwab (1938).
Fehlow, W. (1935). *Z. Aerztl Fortbild.* **32**, 258.
Fellinger, K., Rummelhardt, K., Schirmer, K., and Schmid, T. (1952). *Wien. Klin. Wochschr.* **64**, 661.
Ferreira da Silva, D. (1945). *Biol. Med. (Niterói, Brazil)* **3**, 9.
Ferri, G. (1935). *Giorn. Ital. Anestesia Analgesia* **1**, 588.
Ferrière, F. (1937). *Rev. Med. Suisse Romande* **57**, 145.
Fichmann, M., and Henriques, O. B. (1962). *Arch. Biochem. Biophys.* **98**, 95.
Fiehrer, A. (1940). *Arch. Hospital.*
Fischenko, L. Y., and Neiko, E. M. (1961). *Biul. Eksperim. Biol. i Med.* **52**, 93.
Fisher, A. A. (1941). *Arch. Dermatol. Syphilis* **43**, 444.
Fishkov, E. L. (1954). *Klin. Med. (U.S.S.R.)* **32**, 20.
Flury, F. (1920). *Arch. Exp. Pathol. Pharmakol.* **85**, 319.
Fontana, F. (1767).

Fontana, G. (1941). *Rass. Ital. Ottalmol.* **10**, 116.
Forster, K. A. (1938). *Chem. Ztg.* **62**, 657.
Frandoli, G. (1962). *Riforma Med.* **76**, 737.
Frank, R. T. (1932). *Am. J. Obstetr. Gynecol.* **25**, 887.
Frauchiger, E. (1936). *Schweiz. Med. Wochschr.* **66**, 267.
Freire, R. (1943). *Rev. Brasil. Cirurg.* **12**, 197.
Frenkler-Bleissner, I. (1956).*Medizinische (Stuttgart)* p. 1719.
Freund, E. (1927). *Wien. Klin. Wochschr.* **40**, 1637.
Friede, K. H. (1957). *Deut. Med. J.* **8**, 416.
Gascoin, H. (1941). *Gaz. Med. France* **48**, 184.
Gascoin, H. (1942). *Sang* **15**, 103.
Gautrelet, J., Halpern, N., and Corteggiani, E. (1934). *Arch. Intern. Physiol.* **38**, 193.
Ghezzi, M., and Giovanni, F. (1946). *Minerva Med.* **1**, 283.
Giardiello, A., Boscaino, N., and Pennino, P. (1959). *Rass. Intern. Clin. Terap.* **39**, 462.
Giffo, E. (1949a). *Semaine Hop. Paris* **25**, 2420.
Giffo, E. (1949b). "Le venin de cobra en thérapeutique ophthalmologique." Delmas, Bordeaux.
Githins, T. S. (1939). *Clin. Med. Surg.* **46**, 167.
Glaninger, J. (1955). *Monatsschr. Ohrenheilk. Laryngo-Rhinol.* **89**, 63.
Gossare, P. (1967). Ph.D. Dissertation, Fac. Méd., University of Paris.
Gossens, N., and Durr, M. (1953). *Med. Monatsschr.* **7**, 225.
Gosset, A. (1933). *Compt. Rend.* **196**, 880.
Graatz, H. (1958). *Hippokrates* **29** (2), 54.
Graatz, H. (1962). *Z. Haut Geschlechtstkrankh.* **33**, 360.
Grundmann, I. (1950). *Med. Monatsschr.* **4**, 570.
Grünewald (1934). *Zentr. Landarzte* p. 465.
Gueytat, M. (1934). *Lyon Med.* **153**, 166.
Gürtler, J. Quoted by Rottman (1948).
Gürtler, J. (1964). *Therap. Umschau* **8**, 308.
Haag, F. E., and König, H. (1936). *Klin. Wochschr.* **15**, 1321.
Hadley, H. G. (1940). *Med. Record* **152**, 391.
Haevernick, I., and Brandner, R. (1956). *Therap. Gegenw.* **95**, 342.
Hagenmüller, F. (1956). *Med. Klin.* **51**, 1833.
Hance, J. B. (1937). *Indian Med. Gaz.* **72**, 76.
Hanut, C. J. (1936). *Compt. Rend. Soc. Biol.* **123**, 1232.
Hanzséros, J. (1934–1936). Private communications.
Hargreaves, A. B. (1943). *J. Am. Med. Assoc.* **122**, 855.
Heiss, H. (1954). *Wien. Med. Wochschr.* **104**, 833.
Heiss, H. (1956). *Wien. Med. Wochschr.* **106**, 512.
Helbig, G. (1961). *Med. Klin.* **56**, 1512.
Henderson, A. B., and Gray, H. O. (1959). *J. Natl. Med. Assoc.* **51**, 97.
Henriques, O. B., Fichmann, M., and Henriques, S. B. (1960). *Biochem. J.* **75**, 551.
Hesse, W. (1949). *Med. Monatsschr.* **3**, 688.
Hills, R. G., and Firor, W. M. (1952). *Am. Surgeon* **18**, 875.
Hofbauer, R. (1936). *Z. Aerztl. Fortbild.* **33**, 285.
Hohnen, H. W. (1957). *Z. Ges. Exptl. Med.* **128**, 427.
Holzknecht, F. (1958). *Abstr. Commun. 7th Congr. Intern. Soc. Hematol.*, p. 278.
Horányi-Hechst, B. (1941). *Wien. Med. Wochschr.* **91**, 470.
Horváth, M., and Ludwigh, K. (1962). *Magy. Belorv. Arch.* **15**, 239.
Hourand, V. (1938). *Fortschr. Therap.* **14**, 425.

Hubert, W. (1958). Ph.D. Dissertation, Fac. Med., University of Hannover.
Iriart, M. L. (1965). Ph.D. Dissertation, Fac. Méd., Bordeaux.
Iskhaki, Y. B. (1959). *Vestn. Oto-Rino-Laring.* **21**, 44.
Istomina, K. V., and Konevetseva, T. V. (1959). *Sov. Med.* **23** (2), 133.
Jackman, A. I. (1954). *Diseases Nervous System* **15**, 99.
Jelašić, F., Maretić, Z., and Radej, I. (1954). *Liyecn. Viesn.* **76**, 628.
Jelašić, F., Maretić, Z., and Paić, V. (1955). *Med. Klin.* **50**, 1336.
Jenkin, C. L., and Pendleton, A. S. (1913). *J. Am. Med. Assoc.* **63,** 1749.
Jeschek, S. (1938). *Klin. Wochschr.* **17**, 583.
Joedicke, P. (1913).
Johne, H. O., and Sperr, A. (1956). *Medizinische* p. 1145.
Jorns, G. (1957). *Hippokrates* **28**, 115.
Kajimoto, Y. (1937). *Proc. Japan. Pharmacol. Soc.* **11**, 23.
Karapata, A. P., and Shumakov, A. G. (1961). *Klinich. Med.* **39**, 142.
Kauer, G. L., Bird, R. M., and Reznikoff, P. (1943). *Am. J. Med. Sci.* **205**, 116.
Keiter, A. (1904).
Keiter, A. (1914).
Kellaway, C. H. (1934). *Australian J. Exptl. Biol. Med.* **12**, 17.
Keller, W. (1936). *Schweiz. Rundschau Med.* **25**, 751.
Kelly, R. J. (1938). *Arch. Dermatol. Syphilis* **38**, 599.
Keuth, U. (1959). *Monatsschr. Kinderheilk.* **107**, 353.
Keuth, U. (1968). *Pediat. Prat.* **7**, 233.
Kirschen, M. (1936). *Wien. Klin. Wochschr.* **49**, 648.
Kirchner, E. (1940). *Med. Welt* **14**, 528.
Kleinmann, A., Page, R. C., and Preisler, P. W. (1945). *J. Lab. Clin. Med.* **30**, 448.
Kock, N. G. (1957). *Nord. Med.* **57**, 331.
Körbler, J. (1934). *Klin. Wochschr.* **13**, 1185.
Koressios, N. T. (1935). "Le venin de cobra, son action physiologique, ses indications thérapeutiques." Maloine, Paris.
Kornalik, F. (1960). *Arch. Exptl. Pathol. Pharmakol.* **240**, 72.
Kots, I. A. L., and Barkagan, Z. S. (1957). *Vest. Oto-Rino-Laring.* **19** (5), 97.
Krebs, W. (1929). *Z. Aerztl. Fortbild.* **26**, 250.
Kretschy, F. (1928). *Z. Wisse. Baderk.* **3**, 112.
Kroner, J., Lintz, R. M., Tyndall, M., Andersen, M., and Nichols, E. (1938). *Ann. Internal. Med.* [N.S.] **11**, 1077.
Kürn, K. G. (1955). Ph.D. Dissertation, Fac. Med., University of Munich.
Kuthy, J., Rivero, O., and Ruiz, C. (1960). Unid. Neumologia, Hopital Gen., Mexico, D.F.
Lagomarsino, A. (1939). *Ann. Ottalmol. Clin. Ocul.* **67**, 143.
Laignel-Lavastine, M., and Koressios, N. T. (1933). *Bull. Soc. Med. Hop. Paris* **49**, 274.
Laignel-Lavastine, N., and Koressios, N. T. (1934). *Bull. Soc. Med. Hop.* **50**, 487.
Laignel-Lavastine, M., Wurmser, L., and Koressios, N. T. (1934). *Bull. Soc. Med. Hop.* **50**, 494.
Lange, E. (1931). *Rheuma-Jahrb.* p. 82.
Langer, J. (1897). *Arch. Exptl. Pathol. Pharmakol.* **38**, 381.
Langer, J. (1899). *Arch. Intern. Pharmacodyn.* **6**, 181.
Langer, J. (1915). *Jahrb. Kinderheilk.* **81**, 234.
Langlade, P. (1936). Ph.D. Dissertation Fac. Med and Pharm., University of Marseille.
Larget-Hue, E. (1963). Ph.D. Dissertation, Fac. Med., University of Paris.
Larralde, C., and Labat, M. (1965). *J. Med. Bordeaux, Sud-Ouest* **1**, 101.

Laterza, G., and De Serio, A. (1967). *Minerva Pediat.* **19**, 16.
Lavedan, J. (1935). *Bull. Acad. Med. (Paris)* **113**, 195.
Lee, C. Y., Johnson, S. A., and Seegers, W. H. (1955). *J. Michigan State Med. Soc.* **54**, 801.
Lémery, N. (1716). "Dictionnaire des drogues simples." 3rd ed. Paris.
Lhermitte, J., Porsin, H., and Bétourné, C. (1954). *Presse Med.* **62**, 1017.
Liashenko, M. S., and Bolshekova, V. P. (1963). *Pediat. Akush. Ginekol.* **5**, 51.
Lindenberg, W. (1953). *Muench Med. Wochschr.* **95**, 292.
Löebel, R., and Simo, A. (1930). *Med. Klin. (Munich)* **26**, 512.
Löwenstein, W. (1938). *Wien. Klin. Wochschr.* **51**, 10.
Lumpkin, W. R., and Firor, W. M. (1954). *Am. Surgeon* **20**, 756.
Macfarlane, R. G., and Barnett, B. (1934). *Lancet* **II**, 985.
Machon, F. (1944). *Rev. Med. Suisse Romande* **64**, 212.
Macht, D., and Macht, M. (1939). *Arch. Intern. Pharmacodyn.* **63**, 179.
Macht, D. I. (1936). *Proc. Natl. Acad. Sci. U. S.* **22**, 61.
Macht, D. I. (1938). *Ann. Internal Med.* [N.S.] **11**, 1824.
Macht, D. I. (1940). *Urol. Cutaneous Rev.* **44**, 119.
Macht, D. I. (1943). *Ann. Internal Med.* [N.S.] **18**, 772.
Mackenna, F. S. (1939). *Med. Press Circ.* **201**, 386.
MacLagan, N. F., Billimoria, J. D., and Curtis, C. (1958). *Lancet* **II**, 865.
McLane, E. G. (1943). *Minn. Med.* **26**, 1061.
Malmierca, R. M. (1947). *Dia Med.* **19**, 2198.
Maretić, Z. and Jelašić, F. (1953). *Acta Trop.* **10**, 209.
Marsten, J. L., Chan, K. E., Ankeney, L. J., and Botti, R. (1966). *Circulation Res.* **19**, 514.
Marucci, L. (1958). *Atti Soc. Oftalmol. Lombardia* [N.S.] **13**, 115.
Marucci, L. (1960). *Atti Soc. Oftalmol. Lombardia* [N.S.] Fasc. II.
Mashbits, M. G. (1969). *Akush. Ginek. (Moscow)* **45**, 66.
Mauro, C. (1950). *Giorn. Ital. Chir.* **6**, 686.
Meiselas, L. E., and Schlecker, A. A. (1957). *N. Y. State J. Med.* **57**, 2067.
Meneghetti, M. (1934). *Rev. Radiol. Clin.* **3**, 828.
Moacyr, A. E. (1939). *Am. J. Opthalmol.* **22**, 1130.
Moch, B. (1942). Ph.D. Dissertation, Fac. Med., Gen. Colon. Pharm., Marseille.
Much, V. (1948). *Ophthalmologia* **115**, 89.
Natale, P. (1935). *Tumori* **9**, 324.
Nekula, J. (1936). *Zentr. Gynaekol.* **60**, 328.
Neumann, W., and Stracke, A. (1951). *Arch. Exptl. Pathol. Pharmakol.* **213**, 8.
Noc, F. (1904).
Noguera Toledo, J., Noguera Martins, J. R.; and Fonseca Herreiro, G. (1965). *Gac. Med.* **39**, 17.
Nowotny, H. (1932). *Muench. Med. Wochschr.* **79**, 1159.
Oaks, L. W., and Quinn, J. H. (1954). *Trans. Pacif. Coast Oto-Ophthalmol. Soc.* **35**, 71.
Ohlenschlaeger, G., and Schwalbe, J. (1966). *Med. Monatsschr.* **20**(4), 179.
Orticoni, A. (1934). *Presse Med.* **42**, 112.
Osako, S., Sogabe, R., and Yoshida, K. (1964). *Otolaringol. (Tokyo)* **36**, 359.
Ottoboni, A., and Ciurlo, E. (1957). *Minerva Otorinolaringol.* **7**, 173.
Page, R. C., and De Beer, E. J. (1943). *Am. J. Med. Sci.* **205**, 336.
Passow, A. (1927). *Klin. Monatsbl. Augenheilk.* **79**, 814.
Paula Souza, G. H. (1914). *Ann. Paulist. Med. Cir.* **2**, 33.
Peck, S. M. (1932). *Proc. Soc. Exptl. Biol. Med.* **29**, 579.

Peck, S. M., Crimmins, M. L., and Erf, L. A. (1935). *Proc. Soc. Exptl. Biol. Med.* **32,** 1525.
Peck, S. M., and Goldberger, M. A. (1933). *Am. J. Obstet. Gynecol.* **25**, 887.
Peck, S. M., and Rosenthal, N. (1935). *J. Am. Med. Assoc.* **104,** 1066.
Peck, S. M., and Sobotka, H. (1931). *J. Exptl. Med.* **54,** 407.
Pengler, R. S., and Pribert, G. (1935). *Deut. Med. Wochschr.* **61**, 962.
Penido Monteiro, J. (1936). *Biol. Med.* **3,** 21.
Pepeu, F. (1938). *Deut. Med. Wochschr.* **64,** 1109.
Perrin, M., and Cuenot, A. (1933). *Rev. Med. Est.* **61**, 261.
Petsulenko, V. A. (1961). *Sov. Med.* **25,** 94.
Pfannenstiel, W. (1952). *Med. Klin.* (*Munich*) **35**, 1144.
Phisalix, M. (1936). *Synthèse* **1**, Oct. 10.
Picha, E., Rockenschaub, A., and Weghaupt, K. (1957). *Wien. Med. Wochschr.* **107,** 74.
Pierotti, F. G. (1906). *Arch. Ital. Biol.* **46,** 97.
Plath, P., and Salchi, E. (1968). *Z. Laryng. Rhinol. Otol.* **47**(2), 117.
Pollack, H. (1928). *Klin. Monatsbl. Augenheilk.* **81**, 668.
Pollack, H. (1932). "Zum Thema Apicosan in der Behandlung der *Iritis rheumatica.*" Enke, Stuttgart.
Poriadin, V. T. (1962). *Kazansk. Med. Zh.* **4,** 73.
Poulain, J. (1937). *Presse Med.* **45,** 1156.
Prikhod'ko, V. I. (1965). *Vrad Delo.* **5**, 148.
Ravina, A. (1934). *Presse Med.* **42,** 4.
Redi, F. (1664). "Osservazioni intorno alli vipere." Firenze.
Regoeczi, E., Gergely, J., and McFarlane, A. S. (1966). *J. Clin. Invest.* **45**, 1202.
Reid, H. A., and Chan, K. E. (1968). *Lancet* **I,** 485.
Reid, H. A., Chan, K. E., and Thean, P. C. (1963a). *Lancet* **II**, 617.
Reid, H. A., Thean, P. C., Chan, K. E., and Baharon, A. R. (1963b). *Lancet* **II**, 621.
Reinert, M. (1936). *Festsch. Emil Barrell* p. 407.
Riseman, F. (1945). *New Engl. J. Med.* **233,** 462.
Rosenfeld, S., and Lenke, S. E. (1935). *Am. J. Med. Sci.* **190,** 779.
Rossi, R. (1933). *Rev. Asoc. Med. Arg.* **47,** 3317.
Rottmann, A. (1948). *Wien. Klin. Wochschr.* **60,** 826.
Rudat, K. D. (1952). *Psych. Neurol. Med. Psychol.* **4,** 311.
Rutenbeck, R. (1934). *Muench. Med. Wochschr.* **81,** 957.
Sachs, J. J. (1837). "Mediz. Almanach für das Jahr 1837," 3rd ed., p. 64. Heymann. Berlin.
Sacristán Alonso, T., and Gómez López, F. (1960). *Rev. Clin. Espan.* **76,** 103.
Sanchez Fayos, J., Outeirino, J., Paniagua, G., Serrano, J., and Martinez, M. T. (1960), *Rev. Clin. Espan.* **77**, 215.
Schill, R. (1956). *Wien. Klin. Wochschr.* **68,** 576.
Schmidt, W. (1952). *Med. Klin.* (*Munich*) **47**, 307.
Schoger, G. A. (1960). *Med. Welt.* **5**, 267.
Schübel, K. (1942). *Aerztl. Fortbild.* **39**, 145.
Schwab, R. (1934). *Muench. Med. Wochschr.* **81**, 734.
Schwab, R. (1937). *Fortschr. Therap.* **13**, 560 and 615.
Schwab, R. (1938). "Bienengift als Heilmittel." Thieme, Leipzig.
Scola, E. (1959). *Commun. 9th Ann. Meeting Soc. Espanola Oto-Rinolaring.*
Seiler, J. (1936). *Muench. Med. Wochschr.* **83**, 527.
Self, L. E. (1908).
Seliger, H. (1951). *Med. Welt* **20,** 638.
Sergent, E. (1937). *Arch. Inst. Pasteur Algeier* **15**, 20.

Sharp, A. A., Warren, B. A., Paxton, A. M., and Allington, M. J. (1968). *Lancet* **I**, 493.
Shershevskaya, O. I. (1949). *Vestn. Oftalmol.* **28**, 43.
Skursky, J. (1955). *Wien. Med. Wochschr.* **105**, 887.
Sommerville, J. (1948). *Arch. Belges Dermatol. Syph.* **4**, 129.
Spangler, R. H. (1910). *N. Y. Med. J.* **92**, 462.
Spangler, R. H. (1911). *N. Y. Med. J.* **94**, 517.
Spangler, R. H. (1912). *N. Y. Med. J.* **96**, 520.
Spangler, R. H. (1913). *N. Y. Med. J.* **97**, 699.
Spangler, R. H. (1918). *N. Y. Med. J.* **107**, 727.
Spangler, R. H. (1924). *Atlant. Med. J.* **28**, 138.
Spengler, R., and Pribert, G. (1935). *Deut. Med. Wochschr.* **61**, 962.
Stacher, A., and Böhnel, J. (1959). *Wien. Klin. Wochschr.* **71**, 333.
Stefan, H. (1960). *Konstitutionelle Med.* **8**.
Steigerwaldt, F. Mathies, H., and Damrai, F. (1966). *Ind. Med. Surg.* **35**, 1045.
Steinbrocker, O., McEachern, G. C., La Motta, E. P., and Brooks, F. (1940). *J. Am. Med. Assoc.* **114**, 318.
Steinhoff, H. E. (1960). *Deut. Med. J.* **11**, 415.
Stockton, M. R., and Franklin, G. H. C. (1931). *J. Am. Med. Assoc.* **96**, 677.
Stöger, H. (1944). *Deut. Zahnaerztl. Wochschr.* **47**, 50.
Strebel, J. (1931). *Klin. Monatsbl. Augenheilk.* **86**, 657.
Ströder, J. (1956). *Med. Klin. (Munich)* **51**, 1351.
Ströder, J. and Ambs, E. (1967). *Arch. Kinderheilk.* **175**, 157.
Sugano, T. and Maruyama, R. (1962). *Proc. 119th Meeting Jap. Otorhinolar. Assoc. (Osaka)*
Süttinger, H. (1955a). *Med. Klin. (Munich)* **50**, 138.
Süttinger, H. (1955b). *Aerztl. Forsch.* **9**, 256.
Taguet, C. (1933a). *Bull. Mem. Soc. Med. Paris* **137**, 310.
Taguet, C. (1933b). *Bull. Mem. Soc. Med. Paris* **137**, 651.
Taguet, C., and Monaclesser, A. (1933). *Bull. Acad. Med. Paris* **109**, 371.
Tan, M. G., and Ines-Tan, A. R. (1947). *J. Philippine Med. Assoc.* **23**, 593.
Taren, J. A. (1953). *Univ. Mich. Med. Bull.* **19**, 206.
Taylor, K. P. A. (1929). *Bull. Antivenin Inst. Am.* **3**, 42.
Terč, F. (1888). *Wien. Med. Presse* **29**, 1261, 1295, 1326, 1359, 1393, and 1424.
Terč, R. (1912). Quoted by Walch (1943).
Tetsch, C., and Wolff, K. (1936). *Biochem. Z.* **288**, 126.
Thom, D. A. (1914). *Boston Med. Surg. J.* **171**, 933.
Törteli, A., and Hattyasy, D. (1965). *Schweiz. Monatsschr. Zahnheilk.* 75, 1214.
Turner, W. A. (1910).
Utz, W. (1968). *Deut. Zahnaerztl. Z.* **23**, 464.
Velvo, A. (1968). *Ann. Ottal.* **94**, 1011.
van Esveld, L. W. (1936). *Biochem. Z.* **283**, 343.
Vaz, E., and Pereira, M. (1944). *Anais Inst. Pinheiros (Sao Paulo)* **7**, 45.
Velázquez, B. L., Varela de Carvalho, F., Blanco Gallego, G., and Lucea Martinez, A. (1960). *Arch. Inst. Farmacol. Exptl. (Madrid)* **12**, 1.
Vigne, P., and Bouyala, J. (1923). Quoted by Langlade (1936).
Visetti, M. (1958). *Minerva Dermatol.* **33**, 375.
Vladimorova, K. F. (1959). *Klin. Med. (Moscow)* **37**, 139.
Voitik, V. F. (1958). *Klin. Med. (Moscow)* **36**, 131.
von der Trappen, P. (1938). *Therap. Gegenw.* **79**, 44.
Vulpian, E. F. A. (1854). *Compt. Rend. Soc. Biol.* **6**, 133.

Wagner, W. A., and Hubbel, A. O. (1958). *Oral Surg., Oral Med., Oral Pathol.* **11,** 1118.
Walch, J. (1943). *Rev. Med. Suisse Romande* **63,** 403.
Wasserbrenner, K. (1928). *Wien. Klin. Wochschr.* **41,** 1255.
Watson, G. I. (1941). *Lancet* **I,** 301.
Wazen, A., and de Moura, R., (1942). *Hospital* (*Rio de Janeiro*) **22,** 83.
Wiese, E. (1956). *Landarzt.* **32,** 860.
Williams, E. Y. (1960). *J. Natl. Med. Assoc.* **52,** 320.
Wilms, H. (1960). *Zentr. Chir.* **85,** 1575.
Wolpe, G. (1937). *Wien. Klin. Wochschr.* **50,** 1257.
Yannowitch, M. G., and Chabowitch, X. (1932).
Yawger, N. S. (1914). *J. Am. Med. Assoc.* **62,** 1533.
Zachariae, G. (1934). *Fortschr. Med.* **53,** 215.
Zalla, M. (1914). *Epilepsia* **5,** 81.
Zimmer, O. (1931). *Rheuma-Jahrb.* p. 122.
Zürn, H. (1957). *Folia Haematol.* **74,** 402.
Zürn, H. (1959). *Aertzl. Forsch.* **13,** 117.

Author Index

Numbers in italics refer to the pages on which the complete references are listed.

A

Abalos, J. W., *273*
Abbud, L., 431, *441*
Abonnenc, E., 121, 129, 131, *155*, 165, *167*
Abrahams, G., 64, *89*
Abreu, L. C., 109, *118*
Ackermann, D., 63, *89*, 412, *416*
Ackermann, J., 466, *470*
Adam, K. R., *346*, 355, 357, 360, *368*
Adams, A., 377, *391*
Adriano, J. A., 363, 365, *368*
Adrouny, G. A., 31, *56*
Aguilar, P. G., *273*
Albee, F. H., 451, *470*
Albl, F., 63, 64, 70, *89*
Alesker, E. M., 466, *470*
Alexander, A. J., 316, 453, *470*
Allard, H. A., 121, 126, 129, 130, *155*, 165, *167*
Allard, H. F., 121, 126, 129, 130, *155*, 165, *167*
Allington, M. J., 468, *476*
Almeida, R. F., 158, *167*
Altenkirch, G., 24, 25, *56*
Alvarenga, Z., 161, *167*
Ambrose, R. B., 454, *472*
Ambs, E., 455, *477*
Amend, G., 63, 67, 79, 83, *92*
Amitai, P., *347*
Amoreux, M., 326, *346*
Anchiera, J., 157, *167*
Ancona, L., *273*
Andersen, M., 463, *474*
Anderson, J. F., 39, *56*, 102, *117*, 457, *470*
Andrade, C., 162, *167*
Andrysek, K., 67, *89*
Anguiano, G. L., 345, *346*, 363, 365, 367, *368*
Ankeney, L. J., 467, *475*
Anson, M. L., 114, *117*
Anton, H., 87, *89*
Antonio, A., *314*, 345, *346*
Aravindakshan, I., 84, *89*
Archey, G., *195*
Ardao, M. I., 108, *117*, 140, 144, 145, 149, *155*
Aristotle, VII, *368*
Artault, 148
Arustamyan, A. T., 307, *309*
Arvy, L., 272, *274*
Attems, C. G., 171, 186, *195*
Attinger, E., 464, *470*
Aubry, M., 453, *471*
Audouin, V., *274*, 301, 302
Auerbach, St. I., *195*
Ausserer, A., 212, 213, 214, 215, *274*
Autrum, H.-J., 18, 20, 29, *56*
Azevedo, M. P., 453, *471*

B

Badcock, H. D., 218, 219, *274*
Báer, E. C., *273*
Baharon, A. R., 467, *476*

Bahr, G. F., 83, *92*
Baird, W., *440*
Baker, G. A., 452, *471*
Baldermann, G., 463, *471*
Baldone, J. A., 458, *472*
Baliña, P. L., *155*
Balozet, L., 352, 354, 355, 357, 358, 359, 360, 362, 365, 366, *368*, *369*
Balt, C. S., *297*
Bangham, A. D., 85, *90*
Banks, N., *274*, 328
Barbosa, A., *453*, *471*
Barkagan, Z. S., 454, *474*
Barker, S. A., 74, *90*
Barnes, J. H., 414, *416*
Barnett, B., 452, *475*
Barrio, A., *274*, *297*
Barron, E., 75, *91*
Barros, E. F., 362, *368*
Barsoun, G. S., 364, 367, *368*
Barth, R., 105, *117*, 250, *274*
Barthmeyer, A., 193, *195*
Barú, R. J., 447, *471*
Baumgartner, H., 454, *471*
Bayne, R., 465, *471*
Bayyuk, S. I., 74, *90*
Beard, R. L., 28, 30, 32, *56*
Becher, J., 465, *471*
Beck, B. F., 463, *471*
Becker, L., 268, 269, *274*, 304, 323
Becker, S., 463, *471*
Behrman, H. T., 414, 416, *416*
Behrmann, W., 451, *471*
Beille, L., 102, 142
Bell, W. R., 468, *471*
Benitte, A. C., 464, *471*
Beraldo, W. T., 64, *92*
Bercowitz, S., 102, *117*
Berg, C., *155*
Bergamo, P., 444, *471*
Bergenhem, B., 79, *90*
Bergh, R., *391*
Bergström, G., 29, *56*
Berland, L., 9, 28, *56*
Bernard, F., 9, 28, *56*
Bernardi, R., 453, *471*
Bernstein, St., 35, *58*
Berntrop, J. C., 412, 414, *416*
Berry, S. S., 385, 386, *391*
Bert, P., 351, 363, *368*
Bertaggia, A., 454, *471*
Bertkau, P., 214, 221, *274*
Bessler, H., *472*
Bétourné, C., 464, *475*
Bettini, S., 302, 306, 307, *309*
Betts, A., 5, 10, *56*
Beyer, O. W., 5, 17, *56*
Bhattacharya, K. L., 83, *90*
Bhoola, K. D., 63, 64, 65, *90*
Biasi, P., *274*
Billimoria, J. D., 454, *475*
Billiotet, J. C., 451, *471*
Binet, L., 414, *416*
Biosca Florensa, M., 306, *309*
Bird, R. M., 453, *474*
Birula, A., 325, 326, 327, 329, *346*
Bischoff, H., 2, *56*
Bishopp, F. C., 102, 154
Blackwell, 219
Blagodarny, Y. A., 307, *309*
Blanchard, E., *346*
Blanco Gallego, G., 454, *477*
Blasiu, A. P., 449, *471*
Bleech, D. M., 454, *471*
Bleyer, G. A. C., 102, 105, 108
Blombäck, B., *471*
Blombäck, M., *471*
Blondeau, P., *471*
Blum, M. S., 10, 11, 16, 18, 21, 22, 23, 30, 31, 32, 33, 34, 38, *56*, *57*, *58*, *59*, *101*
Boardman, W., 432, *441*
Böhnel, J., 454, *477*
Boerner, A., 165, 166, *168*
Bogen, E., 269, *274*, 304, *309*
Bohn, H., *275*, 287, *297*, 355, *369*
Boinet, E., 465, *471*
Bolivar, J. I., 355, *368*
Bollmann, C., 170, 171, 185, 187, *195*
Bolshekova, V. P., 466, *475*
Bonner, P., 218, 223, *274*
Boquet, P., 366, *369*, 452, *471*
Bordas, 143, 148
Bordas, L., 18, 22, 23, 24, *57*
Borelli, 325
Bornet, J., 464, *471*
Borowitz, J. L., *347*, 365, *371*
Boscaino, N., 454, *473*
Boso, J. M., *155*

Botti, R., 467, *475*
Bouché, G., 457, *471*
Bouisset, L., 307, *309*
Bourquin, F., *155*
Bouvier, E-L., 121, 391
Bouyala, J., 465, *477*
Bovet-Nitti, F., 78, *90*
Boyé, H., 165, *167*
Boyé, R., 121, 122, 123, 125, 126, 127, 129, 130, 132, 149, *155*
Bradbury, A. F., 66, *93*
Brady, W. H., 432, 440, *441*
Braganca, B. M., 83, 84, *89*, *90*
Brand, I. M., 35, *57*
Brander, J., 29, *59*
Brandner, R., 466
Brandt, J. F., 170, *195*
Brazil, V., *274*, 280, *297*, 311, *315*, *346*, 446, 450, *471*
Brecher, I., 445, *471*
Breda, R., 453, *471*
Breithaupt, H., 69, 73, *90*
Brethes, J., *274*
Breyer, A., *155*
Brignoli, P. M., *274*
Brimacombe, J. S., 74, *90*
Broadbent, J. L., 111, *117*
Brockerhoff, H., 75, *91*
Brölemann, H. W., 187, *195*
Brooks, F., 450, *477*
Brown, W. L., Jr., 17, *57*
Bruck, H., 454, *471*
Brünner-Ornstein, M., 450, *471*
Brues, C. T., 120, *155*
Brulle, 328
Bryson, K. D., 451, 458, *471*
Bucher, G. E., 11, *57*
Buchert, 453, *471*
Büchel, K. H., 97, *101*
Bücherl, W., 194, *195*, 214, 216, 223, *274*, *275*, 279, *297*, 312, *315*, 334, *346*, 365, *369*
Bulard, M., 351, *369*
Bulbrich, R. A., 458, *471*
Burkardt, A., 450, 463, 465, *471*
Burke, D. C., 454, *471*
Burn, J. H., 314, *315*
Burnet, 143
Burnstein, M., 414, *416*
Burstein, 148
Butler, C. G., 24, *57*
Buttler, 219
Byrne, K., 416, *416*

C

Cadrobbi, P., 454, *471*
Caffrey, 142
Caius, J. F., 358, *369*
Calderón, H. R., 446, *471*
Cali, G., 454, *472*
Callahan, P. S., 10, 11, 16, 21, 22, 23, 30, 31, 32, 33, *56*, *57*, *101*
Calle, J., 63, 64, 65, *90*
Callewaert, G. L., 66, *93*
Calmette, A., 457, *471*
Calvi-Zampetti, A., 454, *471*
Cambridge, F., 214, 215, 222, 223, *274*
Cambridge, O. P., 211, 218, 223, *274*, 302
Caporiacco, L., 218, 219, *274*, 325, *346*
Carlet, G., 7, 23, *57*
Caro, M. R., 33, *57*
Carpenter, F., 120, *155*
Carpito, 126
Carthy, J. P., 35, *57*
Carvalho, P., 311, 313, *315*
Castellani, A., 105, *117*
Castelli, 269
Castro, B., 193
Castro, Araujo, S. G., 447, *471*
Castro Escalada, P., 447, *471*
Causse, J., 453, *471*
Cavill, G. W. K., 21, 23, 30, 31, 35, 36, *57*, 96, 97, 98, *101*
Cazzola, D., 454, *471*
Chabowitch, X., 465, *478*
Chain, E., 73, *90*
Chaix, E., *346*
Chalmers, A. J., 102, 105, *117*, 193
Chamb, 187
Chamberlain, 218, 219
Chamberlin, R., 212, 214, *274*
Chamberlin, R. V., 170, 186, 187, *195*
Chan, K. E., 467, 468, 469, *471*, *475*, *476*
Charamis, I., 465, *471*
Charitonov, 219
Charnot, A., *346*

Chatterjee, A. K., 83, *90*
Chellini, T., 306
Cheverton, R. L., 102, *117*, 150, *155*
Chidester, J. C., 31, *57*
Chopra, R. N., 452, *472*
Chowhan, J. S., 452, *472*
Christensen, P. A., *346*, 367, *369*
Chu, 143
Ciurlo, E., 454, *475*
Clare Rudkin, 374, 375, *391*
Clark, A. H., 432, *440*
Clark, W. B., 458, *472*
Clark, W. G., 454, *471*
Clausen, C. P., 32, *57*
Clauser, F., 453, *472*
Cleland, J. B., 376, *391*, 414, *416*, 432, *440*
Clench, W. J., 374, *391*
Clerck, C., *274*
Clower, D. F., *101*
Cobb, M. C., 414, 423, 424, 432
Code, C. F., 112, *117*
Cohen, V. L., 458, *472*
Cohic, F., 10, *57*
Collier, H., 79, *90*
Comstock, J. H., *274*
Condrea, E., 80, *90*, *91*
Constant, Y., 121, 126, 129, 130, 131, 132, *155*
Cook, 171
Coombs, F. P., 458, *472*
Cordero, A. D., 18, 34, *56*
Cordova, C. A., 367, *368*
Corney, B. G., 376, *391*
Cornilleau, R., 459, *472*
Cornwall, 193
Corrado, A. P., 313, 314, *315*, 345, *346*
Corteggiani, E., 80, *90*, 446, *473*
Costa, A. T., *274*
Cotton, B. C., 384, 387, 388, *391*
Cottone, M. A., 83, 84, *93*
Courville, D. A., 410, *416*
Cox, J. C., 376, *392*
Crabill, R. E., *195*
Craig, C. F., 104, *117*
Cravarezza, F., 454, *472*
Creighton, W. S., *101*
Crimmins, M. L., 452, *476*
Crome, W., *274*
Crosse, H., 374, 377, *391*
Croxatto, M., 453, *472*
Cuenot, A., *476*
Curley, A., 38, *56*
Curtis, C., 454, *475*

D

da Costa Lima, 105, *117*, *155*
da Fonseca, F., 105, *117*, 140, 142, *155*, 158, *167*
Dahl, F., 218, 219, *275*
Dakhil, T., 81, *90*
Dallas, E. D., 102, *117*, 121, 125, 128, 129, *155*, 165, *167*
da Matta, A., *155*
Dameski, D., 300, *309*
D'Amous, 268, 269
Damrai, F., 464, *477*
Dance, S. P., 375
Darwin, C., 35, *57*
Da Silva, D. F., 458, *472*
Davin, E. J., 453, *472*
Dawson, R. M., 85, *90*
Dawson, R. M. C., 77, 85, *90*
Dax, G., 304
De Beer, E. J., 453, *475*
de Haas, G. H., 75, 76, 77, *90*, *93*
de Hurtado, I., 115, *117*
de Hurtado, L. M., 115
Deichmann, W. B., 82, *92*
de Klobusitzky, D., 454, *472*
De la Lande, I. S., 30, 33, *57*
De Lello, E., 20, 21, 22, *58*
de Leon, W., *472*
Delesalle, D., 308, *309*
Delezenne, C., 79, *90*
Delgado, A., 150, 152, *156*, 275
Del Pozo, E., 345, *346*, 363, 365, *369*
de Magalhães, O., 352, 358, 363, *369*
de Maupertuis, P. L. M., 351, *369*
de Mello, C. O., *346*
de Mello-Leitão, C., 211, 214, 219, 223, *275*
de Moura, R., 453, *478*
Denéchau, D., 459, *472*
Dennis, C. C., 458, *472*
Deoras, P. J., *346*, 360, 365, *369*
De Oreo, G. A., 414, *416*
Derbes, V. J., 31, 33, *56*, *57*
Derbez, J., 363, 365, *369*
Descoeudres, P., 463, *472*

De Serio, A., 458, *475*
Desimoni von Eickstedt, V., *275*
des Lignieris, M., 448, *472*
de Varigny, H., 353, 359, *370*
de Vries, A., 80, *90*, *91*, *472*
Dewitz, H., 5, 16, 17, *57*
Dianzani, M. U., 83, *92*
Diaz, M. O., *275*
Di Gunta, E., 454, *472*
Diniz, C. R., *275*, 283, 284, 287, *297*, 312, 313, 314, *315*, 345, *346*, 355, 357, 364, *369*
Dinstman, M., *472*
Dix, J., 454, *472*
Djaldetti, M., *472*
Dodge, E., 410, *416*
Dörfel, H., 67, *90*
Doerr, R., *369*
Doery, H. M., 79, *90*
Dold, H., 87, *90*
Donisthorpe, H., 29, 31, *57*
Dor, F., 350, *369*
Doro, D. S., 358, *369*
Dubosq, 193
Dufour, L., 22, 23, *57*, 216, 219, *275*, 301
Duncan, W. Y., 454, *472*
Durante, I. J., 454, *472*
Durr, M., 454, *473*
Duthie, E. S., 73, *90*
Dyckerhoff, H., 82, *90*

E

Earle, K. V., 432, *440*
Edman, P., 357, *370*
Egger, F., 458, *472*
Ehrenb, 324, 325
Ehrenberg, C. G., *346*
El Karemi, M. M. A., 86, *91*, *92*, 367, *370*
Eltringham, H., 39, *57*
Emdin, W., 358, *369*
Emmelin, N., 108, *117*
Endean, R., 374, 375, *391*, *393*
Engels, E., 32, *59*
Epstein, B., 74, 77, *90*
Erbland, J., 75, *92*
Ercoli, A., 74, 85, *90*
Erf, L. A., 452, *476*
Esable, 55
Esnouf, M. P., 457, 468, *472*
Espiñoza, N. C., *275*
Essex, H. E., 414, *416*
Estable, C., 108, *117*, 140, 144, 145, 149, *155*
Etzenperger, P., 353, 354, 359, *369*
Evans, H. E., 32, 33, *57*

F

Fabre, J. H., 32, *57*, 104, *117*, 333, *346*
Fabricius, J. C., 218, *275*
Fackenheim-Cassel, S., 455, 457, *472*
Fahraeus, R., 79, *90*
Fairbairn, D., 78, *90*
Faure, L., *346*
Faust, B. C., 104, *117*
Faust, E. S., 143, *155*, 193, *195*, 460, *472*
Favilli, G., 358, *369*
Fehlow, W., 463, *472*
Feldberg, W., 69, 70, 72, 81, 82, *90*, 108, 112, *117*
Fellinger, K., 465, *472*
Fergus, J. O'Rourke, *275*
Ferrari, J. A., 458, *471*
Ferreira-Berruti, P. E., 108, *117*, 140, 144, 145, 149, *155*
Ferreira da Silva, D., 449, *472*
Ferreira, M. C., 312, *315*
Ferri, G., 450, *472*
Ferriére, F., 463, 464, *472*
Fichmann, M., 454, *472*, *473*
Fiehrer, A., 458, 459, 460, *472*
Finlayson, M. H., *275*
Finnegan, 326
Firor, W. M., 449, 451, *475*
Fischenko, L. Y., 460, *472*
Fischer, F. G., 67, *90*, *275*, 287, *297*, 355, *369*
Fish, C. J., 414, *416*, 423, 424, 432, *440*
Fisher, A. A., *472*
Fishkov, E. L., 466, *472*
Fleckenstein, A., 83, *90*
Flecker, H., 376, 384, 387, 388, 413, *416*
Fleming, W., *472*
Flemming, H., 5, 9, 10, 11, 13, *57*
Fletscher, D. J., 35, *57*
Floch, H., 121, 125, 126, 129, 130, 131, 132, *155*, 165, *167*
Flury, F., 462, *472*

Foerster, E., 2, 10, 16, 17, 37, *57*
Fonseca Herreiro, G., 455, *475*
Fonssagrives, J. B., 432, *440*
Fontana, F., 452, 465, *472*
Fontana, G., 465, *473*
Foot, 52, *57*, 104, 140, 142, 144, 149
Forattini, O. P., 121, 123, 124, 126, 129, 130, 131, 132, *155*, 159, 162, 163, 164, *167*
Ford, D. L., 31, *57*, 96, *101*
Forel, A., 2, 16, 17, 18, 20, 21, 22, 23, 25, 28, *57*
Forster, K. A., 462, *473*
Foubert, E. L., 89, *90*
Fourneau, E., 79, *90*
Fraenkel-Conrat, H. L., 62, *93*
Frandoli, G., 454, *473*
Franganillo, B. P., 216, 218, *275*
Frank, R. T., 452, *473*
Franklin, G. H. C., 452, *477*
Frauchiger, E., 463, *473*
Fredholm, 73
Free, J., 36, *57*
Freire, R., 453, *473*
Freire-Maia, L., 312, *315*
Frenkler-Bleissner, I., 466, *473*
Freund, E., 463, *473*
Frias, D., *156*
Friede, K. H., 466, *473*
Fröhlich, K. O., 22, *57*
Fuhn, I. E., 307, *309*
Fujiwara, T., 431, 432, *440*
Furlanetto, R. S., *275*
Fusco, R., 96, *101*

G

Gajardo-Tobar, R., *275*
Gallai-Hatchard, J., 74, 77, *90*, *92*
Gaminara, A., 108, 139, 142, 144, 145, *155*
Garrett, A., 377, *391*
Gary, N. E., 24, *58*
Garza de los Santos, A., 150, *155*
Gascoin, H., 459, *473*
Gastpar, H., 86, *90*
Gaud, J., 308, *309*
Gautrelet, J., 80, *90*, 446, *473*
Gergely, J., 468, *476*
Gerlich, N., 71, *90*
Gerschmann, B., 214, *277*
Gerstaecker, A., 2, *58*
Gertsch, W. J., 216, *275*
Gervais, 187, 324, 327
Geske, H., 85, *90*
Ghent, L. R., 18, 24, *58*
Ghezzi, M., 453, *473*
Ghiretti, F., 390, *391*
Ghost, B. N., 83, *90*
Giardiello, A., 454, *473*
Gibbons, A. J., 84, *90*
Gibson, P. C., 452, *471*
Giffo, E., 454, 458, *473*
Gill, W., 375, 376, *392*
Gillett, K., 420, 440, *440*
Gilmer, P. M., 42, 45, 50, *58*, 106, *117*, 139, 140, 142, *155*
Ginsburg, H., 360, *371*
Giovanni, F., 453, *473*
Girard, A., 184, *195*
Girard, G., *275*
Githins, T. S., 453, *473*
Gitter, S., 80, *90*
Giunio, P., 414, *416*
Glaninger, J., 454, *473*
Goeldi, E. A., *155*
Goldberger, M. A., 452, *476*
Goldman, J., 144, 145, 149, 153, *155*
Goldman, S., 144, 145, 149, 153, *155*
Goldmann, L., *58*, 144, 145, 149, 153, *155*
Gomez, M. V., 312, *315*, 345, *346*
Gómez López, F., 454, *476*
Goncalves, J. M., 312, *315*, 345, *346*, 355, 357, 364, *369*
Gonzalez, C., 219, *275*
Gonzalez, Q. J., 345, *346*
Goodwin, J. F., 468, *471*
Goossens, T., 102, 108, *117*
Gossare, P., 454, *473*
Gossens, N., 454, *473*
Gosset, A., *473*
Gorka, V., 102
Gorter, E., 80, *90*
Gottfried, E. L., 76, *90*
Goubert, 333
Graatz, H., 466, *473*
Graham, S. A., 415, *416*
Grandi, G., 42, *58*
Grasser, E., *346*, 352, 353, 358, 360, *369*, 448, *472*
Gray, 269
Gray, G. M., 76, *92*
Gray, H. O., 454, *473*

Greig, M. E., 84, *90*
Grevinus, J., 304, *309*
Grosse, A., 78, *90*
Grünanger, P., 31, *59*, 95, *101*
Grünewald, 463, *473*
Grujic, I., 307, *309*
Grundmann, I., 451, 453, *473*
Guérin-Menéville, E. F., *275*
Gürtler, J., 454, *473*
Gueytat, M., 448, *473*
Guggenheim, M., 88, *90*
Guibert, H., 304
Gusmão, H. H., 121, 123, 124, 126, 129, 130, 131, 132, *156*, 159, 162, 163, 164, *167*
Guyon, D., *346*

H

Haag, F. E., 465, *473*
Haase, E., 170, 171, 185, 187, 193, *195*
Habermann, E., 30, *58*, 63, 64, 66, 67, 68, 69, 70, 71, 72, 73, 74, 77, 78, 79, 80, 81, 82, 83, 84, 86, *90*, *91*, *92*
Hackstein, F. G., 72, *91*
Hadjissarantos, H., 307, *309*
Hadley, H. G., 453, *473*
Haevernick, I., 466, *473*
Hagenmüller, F., 466, *473*
Hahn, G., 67, *91*
Hallen, A. H., 376, *392*
Halpern, N., 446, *473*
Halstead, B. W., 385, 386, 388, *391*, *392*, 410, *416*, 440, *440*, *441*
Hanahan, D. J., 75, 77, 83, *91*, *92*
Hance, J. B., 452, *473*
Hangartner, W., 35, *58*
Hanna, A. D., 11, *58*
Hansen, H., 63, 68, *92*
Hanut, C. J., 453, *473*
Hanzséros, J., 450, *473*
Hargreaves, A. B., 452, *473*
Hartree, E. F., 78, *91*
Hase, A., 28, *58*, *156*
Hashimoto, T., 431, *441*
Hassan, A., 367, *369*
Hastings, N., *437*
Hattyasy, D., 454, *477*
Haux, P., 66, *91*
Hawk, R. E., 18, 38, *56*
Hawthorne, J. N., 76, *91*
Hecht, E., 83, *91*
Hedden, 268
Heemskert, C. T. T., 75, 76, 77, *93*
Heinecken, H., *275*
Heiss, H., 454, *473*
Helbig, G., 466, *473*
Hemprich, F. G., 324, 325, *346*
Hemstedt, H., 24, *58*
Henderson, A. B., 454, *473*
Henderson, M. P., 467, *471*
Henri, V., 431, *441*
Henriques, O. B., 454, *472*, *473*
Henriques, S. B., 454, *473*
Hentz, L., 218, 219, 220, 221, *275*
Henze, M., *392*
Herbst, 324, 327
Hermann, H. R., 10, 18, 21, 30, 33, *57*, *58*, *101*
Hermans, J., 80, *90*
Hermitte, L. C. D., 376, 383, *392*
Herms, 269
Hesse, W., 464, *473*
Heselhaus, F., 18, 20, 22, *58*
Hessel, D. W., 410, *416*
Hewitt, 326
Heydenreich, H., 70, 72, 80, *91*
Heymons, R., *195*
Hickmann, V. V., 211, *275*
Higuchi, K., 121, 130, 133, *156*
Hills, R. G., 449, *473*
Hinde, B., 376, *392*
Hinegardner, Ralph T., 382, *392*
Hinterberger, H., 31, *57*, 97, *101*
Hirst, S., 325, 326, *346*
Hiyana, T., *392*
Hodgson, J. A., 352, 353, 358, 360, *369*
Hodson, J. A., *346*
Hofbauer, R., 463, *473*
Hofmann, H. T., 70, *91*
Hofshi, E., 142, 148, 149, *156*
Högberg, B., 80, 81, *91*
Hohnen, H. W., 454, *473*
Holden, H. F., 81, 82, *90*
Hölldobler, K., 37, *58*
Hollow, K., *275*
Holmgren, E., *58*, 102, *117*
Holmberg, E. L., 217, 219, 220, *275*
Holtz, F., 412, *416*
Holzknecht, F., 454, *473*
Homma, M., 82, *92*
Hope, 143, *156*

Horányi-Hechst, B., 451, *473*
Horen, W. P., *275*
Horváth, M., 454, *473*
Hoshi, S., 82, *92*
Hougie, C., 83, *92*, 455
Hourand, V., 450, *473*
Houssay, B. A., *156*, 311, *315*, 357, 358, 359, 363, *369*
Hubbel, A. O., 454, *478*
Hubert, W., 454, *474*
Hübscher, G., 76, *91*
Huidobro, H., *156*
Humbert, S., 186, *195*
Hustin, A., 457, *471*
Hyman, L. H., 396, *416*

I

Ibarra-Grasso, A., *274*
Ibn Khaldun, 351, *369*
Ibrahim, H. H., 364, *370*
Illingworth, J. F., 149, *156*
Immhoff, L., *195*
Imms, A. D., 2, 11, *58*
Ines-Tan, A. R., 458, *477*
Ingenitzky, J., 105, *117*
Ingram, W. W., *275*
Iriart, M. L., 454, *474*
Isgrove, A., 388, *392*
Iskhaki, Y. B., 454, *474*
Istomina, K. V., 464, *474*
Ivie, 218, 219

J

Jackman, A. I., 451, *474*
Jager, J., 86, *90*
James, H. C., 11, *58*
Janet, C., 9, 11, 16, 21, 23, *58*
Jaques, R., 63, 64, 74, *91*, 358, *369*
Jelasić, F., 461, *474*, *475*
Jenkin, C. L., 457, *474*
Jensen, O. M., 85, 87, *91*
Jentsch, J., *58*, 66, 68, *91*
Jeschek, S., 458, *474*
Joedicke, P., 457, *474*
Johannsen, O. A., 121, 129, 139, 143, 149, *156*
Johne, H. O., 466, *474*
Johnson, R. M., 358, *369*
Johnson, S. A., 454, *475*
Jones, D. L., *58*, 149, *156*
Jörg, M. E., 102, *117*, 121, 123, 124, 125, 126, 127, 129, 130, 131, 132, 140, 142, 144, 145, 149, 150, 153, *156*, 165, *167*
Jorns, G., 466, *474*
Jousset de Bellesme, 351, 353, *369*
Joyeux-Laffuie, J., 351, *369*
Juminer, B., 365, *370*
Jung, F., 80, 85, *90*, *91*
Jung, R. C., 31, 33, *37*, *56*

K

Kachler, H., 69, *91*
Kadish, U., 306, 307, *309*
Kahlenberg, H., 5, *58*
Kaire, G. H., 270, *275*
Kaiser, E., *275*, 287, *297*, *351*, 410, *416*
Kajimoto, Y., 459, *474*
Kaku, I., 352, *369*
Kaku, T., 352, 355, 360, *369*
Kannowski, P. B., 18, *57*
Karapata, A. P., 466, *474*
Karsch, F., 211, 218, *275*, 323, 324, 326, 328, 329, *346*
Kaston, B. J., 219, *275*
Katzenellenbogen, I., *156*
Kauer, G. L., 453, *474*
Kayalof, E., 431, *441*
Keegan, H. L., 268, *347*, 365, *371*
Keiter, A., 462, 463, *474*
Kellaway, C. H., 69. 70 72, 81, 82, *90*, *91*, 112, *117*, 305, *309*, 452, *474*
Keller, C., 102, *117*
Keller, W., 464, *474*
Kellog, 269
Kelly, R. J., 458, *474*
Kemp, P., 76, *91*
Kemper, H., 39, 44, 47, 48, *58*
Kephart, C. F., 46, *58*, 104, 106, 142
Kerr, W. E., 20, 21, 22, *58*
Kessler, 326
Keuth, U., 454, 455, *474*
Kew, 333
Keyserling, 279, 326
King, P. E., 18, 19, *59*
Kingston, C. W., 412, 414, *416*
Kirchner, E., 464, *474*

Kirschen, M., 447, *474*
Kirschmann, C., 80, *90*
Kleinmann, A., 453, *474*
Klibansky, C., 80, *90*, *91*
Klobusitzky, D. V., 453, 456
Kloft, W., 29, 37, *59*
Kneitz, H., 18, 20, 29, *56*
Knight, B. C. J. G., 62, *92*
Knight, E., 139, *156*
Knight, H. H., 148
Kobert, 269, 301, 302, 304, 306, 307, 308, *309*
Koch, C. L., 185, *195*, 211, 212, 213, 214, 215, 218, 219, 221, 223, *275*, 302, 311, 324, 325, 326, 327, 328, 329, *346*
Kock, N. G., 454, 455, *474*
Koehler, P., *156*
König, H., 456, 465, *473*
König, P., 454, *472*
Körbler, J., 447, *474*
Kohlrausch, E., *196*
Kohn, A. J., 374, 375, 377, 378, 379, 382, *392*
Kondo, Y., 374, *391*
Konevetseva, T. V., 464, *474*
Koob, K., 33, *59*
Kopstein, F., 432, *441*
Koressios, N. T., 445, 446, 449, 452, *474*
Kornalik, E., 454, *474*
Korte, F., 97, *101*
Koshewnikow, G. A., 23, *58*
Koszalka, M. F., 86, 87, *91*
Kots, I. A. L., 454, *474*
Kowaleck, H., 68, *91*
Kraepelin, K., 171, 185, 186, 324, 326, 327, 328, 329
Kraepelin, R., 5, 7, 10, 15, 23, 26, *58*
Krag, P., *346*
Kraus, O., *196*
Kraus, R., 353, *369*
Krebs, W., 463, *474*
Kretschy, F., 462, 463, *474*
Kreuzinger, R., 403, 407, 408, 409, 439, *428*
Kroner, H., 83, *90*
Kroner, J., 463, *474*
Krusche, B., 80, 82, *91*
Krynicki, J., 219, *275*
Krynicky, 302
Kubota, S., 358, *369*
Kübels, 463, *474*
Kürn, K. G., 454, *474*
Kürschner, I., 22, *57*
Kupeyan, C., 356, 357, *370*
Kuthy, J., 454, *474*

L

Labat, M., 454, *474*
Lacaze-Duthiers, H., 2, 5, 7, 9, 10, *58*
Lagomarsino, A., 465, *474*
Laignel-Lavastine, M., 445, 446, *474*
Lalesque, F., *58*
Lamarck, J. B., 212, *276*
La Motta, E. P., 450, *477*
Lands, W. E. M., 84, *91*
Lane, C. E., 410, *416*
Lane, F. W., 384, 387, 388, *392*
Lange, E., 463, *474*
Langer, J., 85, *91*, 462, *474*
Langlade, P., *474*
Lapie, G., 102
Larget-Hue, E., 454, *474*
Larralde, C., 454, *474*
Larroy, G., 307, *309*
Larson, O. A., *156*
Laterza, G., 458, *475*
Latreille, P. A., 170, *196*, 211, 214, 218, 221, *276*, 327
Laudon, 102, 108, 120, 126
Lavedan, J., 458, *475*
Lawrence, R. F., *196*, 215, 219
Layrisse, 115
Layrisse, M., 115, *117*
Leach, W. E., 170, 182, 186, 326, *346*
Lecomte, J., 36, *58*
Leditschke, H., 67, *91*
Lee, C. Y., 454, *475*
Leffkowitz, M., 306, 307, *309*
Legér, 123, 124, 126, 129
Leger, M., 39, *58*, 165, *167*
Lemaire, R., 414, *416*
LeMare, D. W., 440, *441*
Léméry, N., 459, *475*
Lenke, S. E., 452, *476*
Leod, 247
Léon l'Africain, 351, *369*
Lepolova, A. S., 307, *309*
Leuenberger, F., 7, *58*
Leuthold, R. H., 35, *58*

Levedan, 448, *475*
Levene, P. A., 78, *91*
Levi, H. W., 218, *276*, 301, 302, 307, *309*
Levin, O. L., 414, 416, *416*
Levine, A., 144, 145, 149, 153, *155*
Levrat, E., 432, 440, *441*
Lévy, R., 358, *369*
Lewis, J. G. E., *196*
Leydig, F., 18, 19, *58*
Lhermitte, J. P., 464, *475*
Liashenko, M. S., 466, *475*
Liepelt, W., 35, *58*
Light, S. F. 413, 414, *416*
Lima, A. C., 158, *168*
Lindauer, M., 18, 29, *58*
Lindenberg, W., *475*
Linnaeus, C., 170, 182, *196*, 212, *276*, 325, 329, *346*
Lintz, R. M., 463, *474*
Lissitzky, S., 353, 354, 355, 356, 357, 359, *369*, *370*
Locksley, H. D., 31, *57*, 96
Löebel, R., 463, *475*
Löfquist, J., 29, *56*
Löwenstein, W., 450, *475*
Long, C., 75, 76, *92*
Loomis, R. N., 269, *274*
Lowe, R. T., *275*
Lucas, H., 183, *196*, 216, 221, *276*, 324, *347*
Lucas, S., *275*
Lucas, T. A., 102, *117*, 152, *156*
Lucea Martinez, A., 454, *477*
Ludwigh, K., 454, *473*
Lukoschus, F., 19, 24, *58*
Lumpkin, W. R., 451, *475*
Lutz e Mello, 311

M

Maccary, A., 333, 351, *369*
MacClean, D., 358, *370*
McCrone, J. D., *276*
McEachern, G. C., 450, *477*
McFarlane, A. S., 468, *476*
Mac Farlane, R. G., 62, *92*, 452, *475*
Macgillavray, J., 374, 377, *392*
Machado, O., 194, *196*
Machon, F., 460, *475*
Macht, D. I., 447, 452, 458, *475*
Macht, M., 453, *475*
McIvor, 269
Mackenna, F. S., 463, *475*
MacLagan, N. E., 454, *475*
McLane, E. G., 466, *475*
MacLeod, J., *196*
McMichael, D. F., 384, 386, 387, *392*
McNeill, F. A., 413, *416*, 420, 440, *440*
Maeno, H., 82, *92*
Magalhães, O., 316, 312, 313, *315*, 343, *347*
Magee, W. L., 74, 76, 77, 78, *90*, *92*
Maidl, F., 18, *58*
Maimonides, M., 350, *370*
Mair, E., 451, *470*
Maksianowich, M. I., 307, *309*
Malmierca, R. M., 453, 454, *475*
Mann, T., 78, *91*
Marder, C., *58*
Maretic, Z., *276*, 300, 307, 308, *309*, 461, *474*, *475*
Marie, E., 374, 377, *391*
Marinetti, G. V., 75, *92*
Marmocchi, F., 304, *309*
Marples, E. A., 84, 85, *92*
Marshall, A., 102, *117*
Marsten, J. L., 467, *475*
Martinez, M. T., 454, *476*
Martini, E., *58*
Marucci, L., 454, *475*
Maruyama, R., 454, *477*
Marx, R., 82, *90*, 327
Marzan, B., 268, 301, *309*
Maschwitz, U., 15, 20, 24, 33, 34, 35, 36, *59*
Mashbits, M. G., *475*
Masin, G., 300, *309*
Matheson, 143
Mathias, A. P., 411, *416*
Mathies, H., 464, *477*
Matta, A., 161, *168*
Matthiesen, F. A., 335, *347*
Mauer, H., 63, *89*
Maurano, H. R., *347*, 353, *370*
Mauro, C., 453, *475*
Mazza, S., *156*
Mazzella, H., 108, *117*, 144, 149, *156*
Meinert, E., 16, *59*, 170, 185
Meise, 327
Meiselas, L. E., 451, *475*

Melander, A. L. 120, *155*
Mello, 327
Mello-Leitão, C. de, 327, 329
Mendes, E. G., 431, *441*
Meneghetti, M., 446, *475*
Mengin, 148
Mercier, J., 353, 354, 359, *369*
Mercier, P., 308, *309*
Mezie, A., 457, *471*
Mhaskar, K. S., 358, *369*
Michl, H., *369*, 410, *416*
Miller, J. H., *58*, 149, *156*
Millot, J., *276*
Mills, R. G., 102, 129, 140, 142, 149
Minini, F., 454, *471*
Miranda, F., 353, 354, 355, 356, 357, 359, *369*, *370*
Mitsuhashi, S., 82, *92*
Miyauchi, Y., 431, *441*
Moacyr, A. E., 453, *475*
Moch, B., 459, *475*
Mochulsky, V., 304
Mölbert, E., 69, 78, *91*
Moeschlin, S., 89, *92*
Mohammed, A. H., 86, *92*, 358, 364, 367, *369*, *370*
Monaelesser, A., 444, *477*
Monroe, R. S., 31, *57*
Montano, C. R., 365, *368*
Monte, O., 158, *168*
Mootrouzier, X., *392*
Moore, J. H., 75, *92*
Moorhead, M., 411, *417*
Morgan, D. B., 458, *472*
Mori, H., 352, 358, 360, *370*
Morisita, T., 121, 149, *156*
Morison, G. D., 5, 10, *59*
Morren, C. F. A., 102, *117*
Morrison, A., 83, 84, *93*
Mortensen, T., 426, 427, 428, 429, 432, *441*
Moser, J. C., 34, *56*, *59*
Moutsos, A., 454, *472*
Mouzcls, 123, 124, 126, 129
Mouzels, P., 39, *58*, 165, *167*
Much, V., 465, *475*
Mueller, H. L., 89, *92*
Muench, H., 284, *297*
Mulaik, C., 216, *275*
Mulder, I., 75, *90*
Müller, P. S., 153, 219, *276*
Mullin, C. A., 440, *441*
Musgrave, A., *275*, *276*

N

Nabawy, M., 364, 367, *368*
Nashbits, 458, *475*
Natale, P., 465, *475*
Neiko, E. M., 460, *472*
Nekula, J., 448, *475*
Netzlof, M. L., *276*
Neumann, W., 63, 64, 67, 68, 70, 71, 73, 77, 78, 79, 83, *92*, 465, *475*
Neumann, W. P., 67, *90*
Neveu-Lemaire, M., 147, 154, *156*
Newport, G., 170, 184, 185, *196*
Ng, K. Y., 81, *92*
Nichols, E., 463, *474*
Nicoler, A. C., 216, 219, 220, *276*
Nikander, Theriaka, *309*
Nilsson, I. M., *471*
Nobre, A. F., 412, *416*
Noc, F., 456, *475*
Noguera Martins, J. R., 455, *475*
Noguera Toledo, J., 455, *475*
Novak, A. F., 31, 32, 33, *57*
Nowotny, H., 463, *475*
Nygaard, A. P., 83, *92*

O

Oaks, L. W., 458, *475*
O'Connor, R., 30, *59*
Odavič, R., 75, *92*
Oeser, R., 2, 3, 5, 7, 8, 10, 11, 16, *59*
Ohlenschlaeger, G., 455, *475*
Okada, K., 431, *441*
Okonogi, T., 82, *92*
Olberg, G., 28, 33, *59*
Olessi, L. C., *347*
Oliver, 325
Orfila, R. N., *155*
Ornelas, E. S., 358, *369*
Ornelas, V., 355, *368*
O'Roke, E. C., 415, *416*
Orticoni, A., 446, *475*
Osaka, S., 454, *475*
Oshima, H., 432, *441*

Osman, M. F. H., 29, 37, *59*
Otto, D., 29, 37, *59*
Ottoboni, A., 454, *475*
Ouljanin, 5, 10, *59*
Outeirino, J., 454, *476*
Owano, 193
Oytun, H. S., 307, *309*

P

Packard, A. S., 102, 105, *117*, 142, 143, *156*
Padovani, F., 18, 38, *56*, *57*
Page, R. C., 453, *474*, *475*
Pallary, P., 325, *347*, *370*
Palmer, D. J., 74, *90*
Paniagua, G., 454, *476*
Papahadjopoulos, D., 83, *92*
Paradice, W. E., 416, 432, 440, *441*
Paseiro, P., 149, *156*
Passow, A., 463, *475*
Patetta, M. A., 108, *117*, 144, 149, *156*
Paula Souza, G. H., 456, *475*
Pauly, A., 350, *370*
Pavan, M., 16, 18, 25, 29, 31, 33, 35, 37, *59*, *60*, 95, 96, *101*, 358, *370*
Pawlowsky, E. N., 18, 19, 20, 22, 25, 28, 29, 33, *59*, 108, *117*, 127, *156*, 193, *196*, 307, 309, 333, 340, *347*, 412, *416*
Paxton, A. M., 468, *476*
Pearson, J. E., 79, *90*
Peck, S. M., 452, *475*, *476*
Pellaton, M. G., 108, *117*
Pendleton, A. S., 457, *474*
Pengler, R. S., 463, *476*
Penido Monteiro, J., 447, *476*
Pennino, P., 454, *473*
Penny, I. F., 76, *92*
Penther, 327, 329
Pope, E. C., 413, 414, *416*
Popescu, V., 307, *309*
Pepeu, F., 450, *476*
Perdomo, C. S., 108, *117*
Pereira, M., 453, *477*
Pereira Lima, F. A., *276*, 280, 281, 282, 284, 287, 288, 289, 290, 291, 293, 296, *297*
Perin, M., *476*
Perotti, N., 304, *309*
Perty, M., 221, 223, *276*, 311, 327
Pesce, H., 150, 152, *156*
Petagna, V., *276*
Peters, H., 323, 324, 326, 328
Petrauskas, L. E., 378, 384, *392*
Petrunkevitch, A., *276*, *347*
Petsulenko, V. A., 464, *476*
Pfannenstiel, W., 466, *476*
Pflugfelder, O., *347*
Phillips, C., 432, 440, *441*
Phillips, L. L., 454, *472*
Phisalix, C., 353, 359, *370*
Phisalix, M., 120, 121, 125, 128, 143, 147, 148, *156*, *347*, 353, 354, *370*, 463, *476*
Piaggio Blanco, R., 149, *156*
Picarelli, Z. P., 52, 54, 55, *60*, 109, *118*, 142, 144, 145, *156*, 158, *168*
Picha, E., 454, *476*
Piek, T., 32, *59*
Pierotti, F. G., 459, *476*
Pietruszko, R., 76, *92*
Pirano, J. J., 65, *92*
Pitney, W. R., 468, *471*
Piza, S. D. T., *276*, *347*
Plath, P., 454, *476*
Pocock, R. I., 185, 211, 212, 213, 214, 215, *276*, 323, 324, 325, 326, 327, 328, 329, *347*
Pocock, R. J., 170, 171, 181, 185, 186, *196*
Podleski, T. R., 70, 81, *92*
Pogrund, R. S., 454, *471*
Pollack, H., 462, 463, *476*
Pope, E. C., 424, 440, *441*
Porat, C., 186
Poriadin, V. T., 464, *476*
Porsin, H., 464, *475*
Porter, C., 193, *196*
Portier, P., 414, *416*
Portocarrero, C. A., 35, *57*
Poulain, J., 458, *476*
Poulton, 143
Powell, 219
Pradinaud, R., 166, *168*
Prado, J. E., 158, *168*, 327
Prado, J. L., 52, 54, 55, *60*, 109, *118*, 142, 144, 145, *156*
Preisler, P. W., 453, *474*
Preston, F. S., 416, *416*
Preto, 343
Pribert, G., 463, *476*
Prikhod'ko, V. I., 464, *476*

Prock, P. B., 411, *417*
Pugh, W. S., 431, 432, *441*
Pulleine, R. H., 211, *276*
Purcell, 215, 324

Q

Quastel, I. H., 83, *90*
Quilico, A., 31, *59*, 95, *101*
Quinn, J. H., 458, *475*

R

Radaody-Ralarosy, 459
Radej, I., 461, *474*
Radomski, J. L., 82, *92*
Railey, 269
Raillier, A., 105, *117*
Rainbow, W. J., 211, *276*
Ramos, H., 313, *315*
Ramzin, S., 307, *309*
Rand, M. J., 314, *315*
Rapport, M. M., 76, *90*
Ratcliffe, N. A., 18, 19, *59*
Rathbone, L., 76, 78, *92*
Rathmayer, W., 11, 20, 24, 30, 32, 37, *59*
Ravina, A., 446, *476*
Réaumur, R. A. F. de., 2, *59*
Redelberger, W., 83, *92*
Redi, F., 351, 443, *476*
Reed, C. F., *276*
Reed, L. J., 284, *297*
Reese, A. M., *276*
Regino, *309*
Regnier, F. E., 29, *59*
Regoeczi, E., 468, *476*
Rehm, E., 10, *59*
Reid, H. A., 467, *471*, *476*
Reinert, M., 63, *92*, 462, *476*
Reinwein, H., 412, *416*
Reiz, K.-G., 63, 66, 67, 73, 74, 79, *91*
Remane, A., *59*
Renoux, G., 365, *370*
Reyes, H., *156*
Reznikoff, P., 453, *474*
Ribaut, 181
Ribeiro, B. L., *156*
Richet, C., 414, *416*, *417*
Rietschel, P., 5, 9, 10, 13, 14, 26, 35, *59*
Riley, W. A., 121, 129, 139, 143, 149, *156*
Rimon, A., 76, *92*
Rios Patiño, 327
Riseman, F., 458, *476*
Rivero, O., 454, *474*
Rizza, C. R., 467, *471*
Roberts, J. E., 31, 33, *57*
Robertson, P. L., 18, 19, 20, 21, 22, 24, 30, 35, 36, *57*, *59*, 96, 98, *101*
Robins, D. C., 76, *92*
Robinson, N., 78, 80, *92*
Robson, G. C., 384, *392*
Rocci, 128
Rocha e Silva, M., 64, *92*, *276*
Rochat, C., *347*, 355, 356, 357, *370*
Rochat, H., 356, 357, *370*
Rockenschaub, A., 454, *476*
Rodriguez, G., 355, *368*
Röthel, M., 70, *92*
Roewer, K. F., 223, *276*
Rohayem, H., 358, 367, *370*
Roholt, O. A., 75, 77, *92*
Rolf, I. P., 78, *91*
Romanini, M. G., 390, *392*
Ronchetti, G., 16, 25, *59*
Rosen, A. J., 453, *472*
Rosenberg, P., 70, 81, *92*, *276*
Rosenbrook, W., 30, *59*
Rosenfeld, G., 64, *92*
Rosenfeld, S., 452, *476*
Rosenthal, N., 452, *476*
Rosi, R., 446, *476*
Rosin, E., *347*
Rosin, M., *472*
Ross, D. M., 411, *416*
Ross, G. N., 34, *57*
Ross, H. H., 2, *59*
Rossbach, F., 71, *92*
Rossi, B., 218, 219, 221, *276*, 301, 302, 304
Rossikow, K. N., 304, *309*
Rotberg, A., 121, 123, 124, 126, 129, 130, 131, 132, *156*, 159, 162, 163, 164, 165, 166, *167*, *168*
Rothschild, A. M., 80, *93*, *276*
Rottmann, A., 450, *476*
Roughley, T. C., 440, *441*
Rouville, E. de, 390, *392*
Rudat, K. D., 451, *476*
Ruiz, C., 454, *474*

Rummelhardt, K., 465, *472*
Rutenbeck, R., 450, *476*
Ruttner, F., 10, *59*

S

Sachs, J. J., 450, *476*
Sacristán Alonso, T., 454, *476*
Saer, F. A., *275*
Sala, F., 453, *471*
Salama, S., 364, 367, *368*
Salchi, E., 454, *476*
Salem, G., 454, *471*
Sales Vásquez, M., 306, *309*
Salzmann, G., 63, 72, *93*
Samaan, A., 364, *370*
Sampayo, R. L., 248, *276*, 283, *297*
Sanchez Fayos, J., 454, *476*
Sandbank, U., *472*
Sanders, H., 77, *92*
Sargent, E. J., 75, *92*
Sauerländer, S., 33, *59*
Saunders, P. R., 374, 375, 382, *392*
Saussure, H., 185, 186
Savoura, A., 307, *309*
Sawerthal, H., 66, *91*
Say, 327
Sayer, F., 144, 145, 149, 153, *155*
Schaafsma, A., *346*, 352, 353, 358, 360, *369*
Schachter, M., 63, 64, 65, *90*, *91*, *93*, 411, *416*
Schade, F., 149, *156*
Scheidter, F., 46, 47, *59*
Schenberg, S., *276*, 280, 281, 282, 284, 287, 288, 289, 290, 291, 293, 296, *297*
Schenone, H., *276*
Schiapelli, R., 214, *277*
Schildknecht, H., 33, *59*
Schill, R., 454, *476*
Schimer, K., 465, *472*
Schlamowitz, M., 75, 77, *92*
Schlecker, A. A., 451, *475*
Schlusche, M., 14, 22, 23, 28, *59*
Schmalfuss, H., 193, *195*
Schmid, T., 465, *472*
Schmidt, G., 223
Schmidt, W., 453, *476*
Schmitz, 153
Schöttler, W. H. A., 354, *370*
Schoger, G. A., 454, 466, *476*
Schröder, C., 42, *59*
Schübel, K., 451, 457, *476*
Schütz, R. M., 81, *93*
Schulze, W., *347*
Schwab, R., 463, *476*
Schwalbe, J., 455, *475*
Schwartz, L., 414, *417*
Schweinitz, 149, 150
Scola, E., 454, *476*
Scott, H. H., 412, 414, *417*
Seegers, W. H., 454, *475*
Seifert, E., 81, *93*
Seiler, J., *476*
Seitz, W., 86, *90*
Self, L. E., 444, *476*
Seliger, H., 449, *476*
Sellach, J., 454, *472*
Sergent, E., *347*, 359, 362, 365, 366, 367, *370*, *371*, 452, *476*
Serrano, J., 454, *476*
Shaffer, J. H., 89, *93*
Shafrir, E., 355, *371*
Shapiro, B., 74, 76, 77, 78, 85, *90*, *92*, *93*
Shapiro, M., 403, 404
Sharp, A. A., 468, *477*
Shaw, H. O. N., *393*
Shchensnowich, M., 304, *309*
Shcherbina, A., 304, *309*
Shepardson, G. R., 83, 84, *93*
Shershevskaya, O. I., 465, *476*
Shewchenko, M. S., 307, *309*
Shipoline, R., 66, *93*
Shulov, A., 219, *276*, 302, 334, *347*, 354, 355, 359, 360, *371*
Shumakov, A. G., 466, *474*
Shumway, 149, 150
Sibler, B., 71, *93*
Silva, S. G., 312, *315*
Simms, H. S., 78, *91*
Simo, A., 463, *475*
Simon, E., 211, 212, 214, 215, 216, 219, 221, 222, *276*, 323, 324, 326, 329, *347*
Simon Thomas, R. T., 32, *59*
Simpson, Y., 24, *57*
Skin, A. K., *59*
Skursky, J., 466, *476*
Slagle, T. D., 414, *417*
Slotta, K. H., 62, 74, 83, *91*, *93*

Smithers, R., 219, *275*
Snodgrass, R. E., 5, 6, 7, 10, 13, 16, 24, 26, *59*, *60*
Soares, B., *276*
Sobotka, H., 452, *476*
Sogabe, R., 454, *475*
Sohier, 459
Soliman, H. S., 10, 11, *60*
Sollmann, A., 7, *60*
Sommerville, J., 458, *476*
Southcott, R. V., 334, 412, 413, 414, 415, *416*, *417*
Späth, W., 67, *93*
Spangler, R. H., 444, 455, 456, *477*
Spengler, R., *476*
Sperr, A., 466, *474*
Spielmann, F., 453, *472*
Spinanger, J., 144, 145, 149, 153, *155*
Springer, H., 69, 81, *91*
Stacher, A., 454, *477*
Stahnke, H. L., *347*, 367, *371*
Stanic, M., 300, 307, 308, *309*
Stefan, H., 454, *477*
Steigerwaldt, F., 464, *477*
Stein, A. K., 108, *117*, 127, *156*, *196*, 412, *416*
Steinbrocker, O., 450, *477*
Steinhoff, H. E., 454, 464, *477*
Stern, I., 78, *93*
Stern, J., 306, 307, *309*
Sticka, R., 223, *277*
Stier, R. A., 89, *90*
Stockebrand, P., 69, 79, 80, 82, 83, *93*
Stockton, M. R., 452, *477*
Stöger, H., 453, *477*
Stotz, E., 75, *92*
Stracke, A., 465, *475*
Strand, E., 219, 223, *276*
Strauss, W. T., 89, *93*
Strebel, J., 465, *477*
Striebeck, C., 69, *93*
Ströder, J., 455, 466, *477*
Stuart, M. A., 414, *417*
Stumper, R., 17, 29, 31, 33, 36, *60*
Stura, L., 454, *472*
Süttinger, H., 460, *477*
Sugano, T., 454, *477*
Sumner, J. B., 83, *92*
Sundara Rajulu, 6, *196*
Suzuki, T., *276*
Swammerdam, J., 2, 18, *60*

T

Taft, C. H., 432, *441*
Taguer, C., 444, 445, *477*
Tahori, A. S., 142, 148, 149, *156*
Talaat, M., 363, *371*
Tamashiro, M., 32, *60*
Tan, M. G., 458, *477*
Tarabini-Castellani, G., 358, *371*
Taren, J. A., 451, *477*
Tattrie, N. H., 75, *93*
Tauböck, K., 78, *90*
Taylor, K. P. A., 452, *477*
Terč, F., 462, 463, *477*
Tetsch, C., 63, *93*, 355, 360, *371*, 462, *477*
Thain, E. M., 63, 64, 65, *93*
Thean, P. C., 467, *476*
Thom, D. A., 456, *477*
Thomas, C. I., 458, *472*
Thomas, D. W., 30, 33, *57*
Thompson, J. L. C., *347*
Thompson, R. H., 74, 76, 77, 78, 84, 85, *90*, *92*
Thorell, T., 214, 217, 219, *276*, 325, 326, 327, 328, 329
Thorp, R. W., *277*, *309*
Tinkham, E. R. *277*
Tippelt H., 83, *90*
Tisseuil, J., 102, *118*, 123, 125, 126, 128, 129, *156*
Todd, C., 357, 359, 363, *371*
Törteli, A., 454, *477*
Tonkes, P. R., 45, 46, *60*, 104, 108, *118*, *156*
Torres, J. M., 313, *315*
Trajan, E., 5, 7, 9, 10, 11, 16, 22, 23, 28, *60*
Trave, R., 31, *60*, 96, *101*
Trethewie, E. R., 81, *91*, 112, *117*
Tulgat, T., 360, 365, *371*
Tulipan, L., 414, *417*
Tullgreen, 215
Tunnah, G. W., 467, 468, *472*
Turner, A. W., 112, *117*
Turner, W. A., 456, *477*
Tweedie, M. W., 432, *440*
Tyler, M., 30, 33, *57*
Tyndall, M., 463, *474*
Typl, H., 454, *471*
Tyzzer, E. E., 102, 104, 105, 108, 142, 144, 148, 149

U

Uexküll, J., 430, *441*
Umiji, S., 431, *441*
Urabe, H., 121, 131, 133, *156*
Urquhart, 219, 221
Utz, W., 455, *477*
Uvnäs, B., 80, 81, *91*, 412, *417*

V

Vachon, M., 325, 326, 333, *347*
Valentin, 351, *371*
Valeri, V., 313, *315*
Valetta, G., *156*
Valle, J. R., 52, 54, 55, *60*, 109, *118*, 142, 144, 145, *156*, 158, *168*
Van Campen, W. G., *392*
van Deenen, L. L. M., 75, 76, 77, *90*, *93*
Vanela de Carvalho, F., 454, *477*
van Esveld, L. W., 448, *477*
Vanovski, B., 307, *309*
Van Ripes, 268, 269
Vaz, E., 453, *477*
Velázquez, B. L., 454, *477*
Vellard, J., *156*, 211, 214, 215, 248, 270, *274*, *277*, 280, *297*, 304, 305, 306, 307, *309*, 324
Velvo, A., 454, *477*
Venzmer, G., 193, *196*
Vercellone, A., 96, *101*
Verhoeff, K. W., 170, 171, *196*
Vernon, C. A., 66, *93*
Verrill, A. E., 432, *441*
Vigne, P., 465, *477*
Vinson, A., 219, *277*
Vintilă, I., 307, *309*
Vintilă, P., 307, *309*
Visetti, M., 454, *477*
Vladimurova, K. F., 466, *477*
Vogel, W. C., 74, *93*
Vogt, W., 81, *90*, *93*
Voitik, V. F., 466, *477*
von Bruchhausen, F., 70, 71, 81, *93*
von der Trappen, P., 465, *477*
von Hasselt, W. M., 219, *277*
Von Ihering, H., 2, *60*
von Ihering, R., 102, 105, *118*, *156*, 157, 158, *168*
von Keyserling, E., 216, 221, 222, *277*
Vulpian, E. F. A. 460, *477*

W

Wacker, T., 463, *472*
Wade, H. W., 412, 413, 414, *417*
Wagner, W. A., 454, *478*
Waite, C. L., 414, *417*
Walch, J., 464, *478*
Walckenaer, C. A., 211, 212, 214, 217, 218, 219, 221, 222, *277*, 301, 302
Walker, I. R., 31, 32, 33, *57*, *101*
Wallace, A. L., 225, *277*
Walter, J. R., 10, 11, 16, 21, 22, 23, *57*, *101*
Warren, B. A., 468, *476*
Warter, S. L., 31, *57*
Wasserbrenner, K., 463, *478*
Watermann, J. A., 343, *347*
Watson, G. I., 465, *478*
Wazen, A., 453, *478*
Weaver, C. S., 386, *393*
Weber, H., 5, 6, 11, 42, *60*
Webster, G. R., 74, 78, 84, *90*, *92*, *93*
Weckering, R., 96, *101*
Weghaupt, K., 454, *476*
Weidner, H., 39, 42, 46, 47, 51, 53, *60*
Weinert, H., 11, 35, *60*
Weiss, B., *346*
Weiss, C., 355, 357, 360, *368*
Weiss, Ch., *346*
Weissmann, A., *347*, 355, 359, 360, *371*
Welsh, J. H., *297*, 411, 412, *417*
Werner, F., 326, 327, 329, *347*
Whelden, R. M., 21, *60*
Whitfield, F. B., 21, 30, *57*, 96, 98, *101*
Whitley, G. P., 432, *441*
Whittemore, F. W., 268, *347*, 365, *371*
Whitwell, G. P. B., 104, *118*
Whyte, J. M., *393*
Widdicombe, M., 367, *369*
Wiener, S., *277*, 374, 375, 382, *392*
Wiese, E., 466, *478*
Will, F., 102, *118*
Williams, D. L., 75, *92*
Williams, E. Y., 451, *478*

Williams, P. J., 23, *57*
Wilms, H., 454, *478*
Wilson, E. O., 29, 33, 35, *57*, *59*, *60*, *101*
Wilson, W. H., 343, 354, 363, *371*
Witt, P. N., *277*
Witter, R. F., 83, 84, *93*
Wloszyk, L., 73, *93*
Wolff, K., 63, *93*, 355, 360, *371*, 462, *477*
Wolpe, G., 465, *478*
Wood, H. C., 186, 193, 327, 329
Woodson, W. D., 269, *277*, *309*
Wurmser, L., 446, *474*

Y

Yannowitch, M. G., 465, *478*
Yarovoy, L. V., 307, *309*
Yasiro, H., 376, *393*
Yawger, N. S., 456, *478*
Yoshida, K., 454, *475*

Z

Zachariae, G., *478*
Zaky, O., 358, 367, *370*
Zalla, M., 456, 457, *478*
Zander, E., 5, 9, 11, 14, 15, 16, 24, *60*
Zeller, E. A., 62, 74, *93*
Zeruos, S. G., 414, 417
Zieve, L., 74, *93*
Zimmer, O., 463, *478*
Zinsser, H. H., 454, *472*
Ziprkowsky, L., 142, 148, 149, *156*
Zolessi, L., 334,
Zumpt, F., 50, *60*
Zürn, H., 454, *478*

Subject Index

A

Abyssinia,
  scorpions of, 352
  spiders of, 219, 236
Acanthaster ellisi, range of, 420
Acanthaster planci,
  range of, 420
  venom apparatus, 421–423
  wounds,
    clinical characteristics, 423–424
    prevention, 424
Acanthoscurria,
  bites, wounds and, 198
  description of, 231
  distribution, 236
  key for, 214
  mating behavior, 239
  number of eggs, 241
  venom,
    extraction of, 263–264
    toxicity, 270
Acanthoscurria antillensis, 215
Acanthoscurria atrox, 215, 251
  adolescent phase, 244
  habits, 231
  venom,
    amounts, 266
    toxicity, 268
  venom apparatus,
    glands, 260
    inoculating apparatus, 255
Acanthoscurria brocklehursti, 215
Acanthoscurria convexa, 215
Acanthoscurria ferina, 215
Acanthoscurria geniculata, 215, 251
  venom apparatus,
    glands, 260
    inoculating apparatus, 255
Acanthoscurria gigantea, 215
Acanthoscurria insubtilis, 215
Acanthoscurria juruenicola, 251
  bites, importance of, 231
  venom apparatus,
    glands, 260
    inoculating apparatus, 255
Acanthoscurria minor, 215
Acanthoscurria musculosa, 215
  venom,
    amounts, 266
    toxicity, 268
Acanthoscurria rhodothele,
  venom,
    amounts, 266
    toxicity, 168
Acanthoscurria sternalis, 215, 251
  adolescent phase, 244
  habits, 231
  venom,
    amounts, 266
    toxicity, 268
  venom apparatus
    glands, 260
    inoculating apparatus, 255
Acanthoscurria suina, 215
Acanthoscurria tarda, 215
Acanthoscurria theraphosoides, 215
Acanthoscurria violacea, 251
  habits, 231
  venom,
    amounts, 266
    toxicity, 268

venom apparatus,
glands, 260
inoculating apparatus, 255
Acanthothraustes,
venom, toxicity, 342–343
Acanthothraustes brasiliensis, 329
Acetic acid, ant venom and, 96
Acetylcholine,
hymenopteran venoms and, 30, 109, 113, 145
nettle sting and, 108
scorpion venom effects and, 313–314, 363–365
sea urchin venom and, 431
Acids, *Phoneutria* venom and, 289, 290
Acronicta, urticating caterpillars, 41
Acropora palmata, 405
range, 406
Acroporidae, venomous representatives, 406
Actinia, 397
stings, anaphylaxis and, 414
Actinia equina,
range of, 406
stings, skin changes, 412
Actiniidae, venomous representatives, 406
Actinopus,
leg claws of, 208
metatarsus and tarsus of, 208
sternum and lip, 206
venom, extraction of, 263–264
venom apparatus, 247
venom glands, 203
venom jaws, 204
Actinopus crassipes, 251
morphology, 200
venom, amounts, 266
venom apparatus,
glands, 260
inoculating apparatus, 255
musculature, 252–253
Aden, spiders of, 219
Adenyl compounds, caterpillar venom and, 112
Adolia, urticating caterpillars, 40
Adoneta,
urticating caterpillars, 134
venom apparatus, 142
Adrenaline, scorpion venom and, 363, 364, 365, 368
Adrenocorticotropic hormone, lepidopterism and, 133
Adriatic Sea, venomous sea urchins of, 425
Afibrinogenemia, snake venom and, 455
Africa, *see also* East Africa, *etc.*
venomous animals,
caterpillars, 105
centipedes, 185, 186, 187
coelenterates, 406
scorpions, 332
sea urchins, 424, 427
Age, scorpion venom and, 359, 362
Agkistrodon piscivorus, hemostasis and, 452
Agkistrodon rhodostoma,
venom, anticoagulant effect, 467–470
Aglaophenia cupressina, 398
range, 401
Ahriman, scorpions and, 349–350
Aidos, caterpillars of, 158
Aiptasia, tentacular extracts, 411–412
Alabama, centipedes of, 184
Alarm, hymenopteran venoms and, 33–34
Algeria,
latrodectism in, 307
scorpions of, 333, 343, 351, 362
Alipes,
classification, 171
occurrence of, 186
Alkali, *Phoneutria* venom and, 289, 290
Allergic syndrome, lepidopterism and, 130–131, 152–153
Alluropus,
classification, 171
occurrence of, 186
Amazonia, spiders of, 212
Amines,
hymenopteran venoms and, 31
injurious effects, 29, 32
Amino acids, *Phoneutria* venom and, 287, 297
Ammonium hydroxide, coelenterate stings and, 415
Ammonium sulfate, *Phoneutria* venom and, 290, 291
Ammophila,
Bordas gland, 24
venom, effects, 32
Amphinomidae, venomous representatives, 434–435

Analgesia,
snake venoms and, 445–452
toad venoms and, 459
Anal gland, 18
structure, 25
toxic secretions, 37
Anal tufts, hairs of, 48
Anamorpha, suborders, 170
Ananteris,
venom, toxicity, 342–343
Anaphe, urticating caterpillars, 41, 134
Anaphylaxis, coelenterate stings and, 414
Andrena, sheath gland, 25
Androctonus,
distribution of, 332–333
venom, toxicity, 343
Androctonus aeneas, 325
distribution, 351
venom, 358
toxicity, 343
Androctonus aeneas liouvillei, venom, 358
Androctonus amoreuxi, 325
distribution, 351, 352
venom, 357, 358, 359
toxicity, 343
Androctonus australis, 325, 351
distribution, 351–352
venom, 357, 358, 359
amount, 354
antivenin and, 355, 365–366
electrophoresis, 355
hyaluronidase in, 358
purification, 356
toxicity, 343, 344, 359, 362
Androctonus australis hector,
venom, purification of, 357
Androctonus crassicauda, 325
distribution, 352
venom,
antivenins and, 366
toxicity, 343, 360
Androctonus hoggarensis, 325
distribution, 352
Androctonus maroccanus,
venom, toxicity, 343
Androctonus mauretanicus, 325
Androctonus sergenti, 325
Anemonia, 397
Anemonia sulcata, range of, 406
Anesthesia, phoneutrism and, 281, 282
Anethops,
classification, 171
occurrence of, 186
Annelida,
characteristics of, 433
stings,
clinical characteristics, 440
prevention, 440
treatment, 440
venom apparatus, 436–438
venomous representatives, 434–435
Anomalobuthus rickmersi, 326
Anoplobuthus parvus, 325
Ant(s),
sting reduction in, 36–38
sting, uses of, 95
venom, composition of, 95–96
venom apparatus, 95
venom glands, 18, 19, 21–22
Antarctia fusca, caterpillars of, 159
Anthophora,
Dufour gland, 22, 23
sheath gland, 25
Anthophora acervorum, venom glands, 20
Anthozoa,
characteristics and habitats, 397–398
families and venomous representatives, 406–407
Antibiotics, scorpion venom and, 359
Anticoagulant therapy, snake venoms and, 467–470
Antigenicity, *Phoneutria* venom, 287–288
Antigua, centipedes of, 185
Antihistaminics,
caterpillar venom and, 114, 154
coelenterate stings and, 415
lepidopterism and, 133
Antisera, spider venoms, 273
Antivenin,
latrodectism and, 308
local pain and, 280
Phoneutria venom and, 290
scorpion venom and, 345, 355, 365–366
Anuroctonus phaeodactylus, 329
Apatela,
poisonous setae, 139
urticating caterpillars, 41, 134
Apatela americana, caterpillars of, 105
Aphaengaster, sting apparatus, 37

Aphonopelma, 212
Apicosan, nature of, 462
Apidae,
  Dufour gland, 22
  further sting glands, 25
  sting apparatus,
    special peculiarities, 14–16
    structure of, 9
  venoms,
    alarm and, 34
    composition, 30
    effects, 33, 35
  venom glands, 18, 20, 21
Apis,
  alarm substance, 34
  Dufour gland, 22–23
  Koshewnikow gland, 23–24
  sheath gland, 24–25
  sting apparatus, 17
  sting retraction in, 27
  venom ejection by, 28
  workers, sting loss in, 35–36
Apis mellifera,
  venom, tumors and, 445
Apis mellifica,
  central nervous system, 11
  sting apparatus,
    muscles, 10, 12–13
    position, 3–4
    structure, 8
  venom composition, 30
Apistobuthus pterygocerius, 326
Arabia,
  scorpions of, 332, 339, 343, 352
  spiders of, 219, 236
Arachnidism, definition of, 267
Araña de los rincones, see *Loxosceles*
Araña homicida, see *Loxosceles*
Araneae, key for families, 209–210
Aranea tredecimguttata, 219
Araneida,
  chelicerae, 203
  key for families, 209–210
Araneidae,
  description and venomous species, 220
  key for, 210
Araneinae, venomous species, 220
Araneus,
  bites, wounds and, 198
Araneus diadematus, venom of, 272
Arbacia lixula,
  range, 424
  venom, pharmacology, 431
Arbaciidae,
  pedicellariae of, 430
  venomous representative, 424
Arctia, urticating caterpillars, 40
Arctia caja, hair of, 43
Arctiidae,
  erucism, symptoms, 149
  hairs of, 44
  host plants, 159
  poisonous setae, 139
  urticating caterpillars, 40, 104, 133
    distribution, 134
Argentina,
  centipedes of, 184
  scorpions of, 323
  spiders of, 219, 223, 236, 237, 269, 271
Argiope,
  bites, wounds and, 198
Argiopidae, key for, 210
Arizona,
  centipedes of, 184
  scorpions of, 332, 343
Armadeiras, *see Phoneutria*
Arrhabdotini, 171
  description, 185
Arrhabdotus,
  classification, 171
  dental plates and sternites, 180
Arrhabdotus octosulcatus, occurrence of, 185
Artace cribraria, caterpillars of, 159
Artemis, scorpions and, 350
Arthritis,
  bee venom and, 463–465
  snake venoms and, 450, 451
Arthropods, urticating caterpillars and, 56
Arthrorhabdus,
  classification, 171
  key and distribution, 181
Arvin,
  clinical indications for use, 468–470
  unit of, 468
Asageninae, description and venomous species, 217
Asanada,
  classification, 171
  distribution of, 185

Asanadini,
  distribution of, 185
  members of, 171
Ascorbic acid, bee venom therapy and, 465
Asia, spiders of, 216, 219
Asia Minor, scorpions of, 333
Asteroidea, venomous members, 420–421
Asthenosoma ijimai, range, 426
Asthenosoma varium,
  range of, 426
  spines of, 428, 432
Atlantic Ocean,
  venomous animals,
    annelids, 435
    coelenterates, 398, 401, 405, 406
    sea urchins, 427
Atrax, 251
  aggressiveness of, 198, 225
  bites, importance of, 198, 225
  description, 225–226
  distribution of, 236
  habits of, 224–225
  maternal behavior, 240
  mating behavior, 239
  venom,
    antisera, 273
    extraction of, 263–264
  venom apparatus, 247
  venom glands, 203, 258
Atrax, formidabilis, 211
  bites, importance of, 225
Atrax modesta, 211
Atrax pulvinator, 211
Atrax robustus, 211
  bites, importance of, 225
  spinnerets, 202
  venom,
    amounts, 265
    toxicity, 270–271
Atrax tibialis, 211
Atrax valida, 211
Atrax venenatus, 211
Atrax versuta, 211
Atraxini, key and venomous members, 211
Atropine,
  caterpillar venom and, 109, 113
  scorpion venom effects and, 314, 364, 367
  venom induced salivation and, 281
Atta,
  sting apparatus, 37
  trail marking by, 34
Attidae, *see also* Salticidae
  key for, 210
Australia,
  venomous animals,
    centipedes, 181, 185, 186
    coelenterates, 401, 402, 405
    octopuses, 386–387, 388
    sea urchins, 425
    spiders, 211, 219, 221, 225, 236, 269
Automeris,
  urticating caterpillars, 41, 134
    venom apparatus, 142
Automeris acuminata, caterpillars of, 158
Automeris aurantiaca, caterpillars of, 158
Automeris complicata, caterpillars of, 158
Automeris coroesus,
  venom, toxicology, 144–145
Automeris illustris,
  caterpillars of, 158
  poisonous spines of, 54
Automeris incisa, caterpillars of, 158
Automeris io,
  caterpillars of, 105
  venom, toxicology, 144–145
Automeris melanops, caterpillars of, 158
Automeris viridescens, caterpillars of, 158
Autonomic nervous system, scorpion venom and, 313–314
Avicularia,
  bites, wounds and, 198
  description of, 225
  habits of, 225
  leg claws of, 208
  mating bulbs, 207
  metatarsus and tarsus of, 208
Avicularia avicularia, 212, 251
  morphology, 199
  venom, amounts, 266
  venom apparatus,
    glands, 260
    inoculating apparatus, 255
Avicularia caesia, 212
Avicularia detrita, 212
Avicularia laeta, 212
Avicularia metallica, 212
Avicularia rutilans, 212
Avicularia velutina, 212
Avicularia versicolor, 212
Aviculariidae, *see* Theraphosidae

Aviculariinae,
  key and venomous members, 212
  maternal behavior, 240
Avicularioidea, *see* Orthognatha
Azores, venomous sea urchins of, 424, 427

B

Babycurus, 326
Bacillus thuringensis, paralyzed insects and, 32
Bacteria, hymenopteran venoms and, 33
Bahamas, venomous coelenterates of, 406
Balkans, scorpions of, 351
Baltic Sea, venomous coelenterates of, 405
Baptista,
  extracts, venom combinations and, 466
Barbados, scorpions of, 332
Banana spiders, *see* Ctenidae
Barbiturates, scorpionism and, 368
Barychelidae,
  key for, 209
  subfamilies and venomous species, 215
Barychelinae, description, 215
Bee(s), *see also Apis*
  venom, 461–462
    antirheumatic and antiarthritic uses, 463–465
    combinations, 466
    other therapeutic applications, 465–466
    pharmaceutical preparations, 462–463
Belisarius xambeui, 339
Bellafoline, scorpionism and, 367
Bermuda Islands, venomous coelenterates of, 401
Bethylidae,
  ovipositor, 28
  venom effects, 33
Bezoar, scorpion stings and, 350
Birds, urticating caterpillars and, 35, 55
Bird spiders, *see* Theraphosidae
Bitis arietans,
  venom, pain relief and, 448
Black Sea, venomous coelenterates of, 406
Black widow spider, *see Latrodectus*
Blood,
  changes, latrodectism and, 300
Blood pressure,
  caterpillar venom and, 109, 111, 114, 144
  Phoneutria venom and, 293
  scorpionism and, 361, 364
Blood worm, *see Glycera*
Bolivia,
  centipedes of, 183
  scorpions of, 331, 343
  spiders of, 217, 223, 236, 237
Bombus distinguendus, venom glands, 20
Bombycidae, urticating caterpillars of, 105, 133, 134
Bombyx, urticating caterpillars, 134
Bonin Islands, venomous sea urchins of, 425
Bordas gland, 18
  structure of, 24
Borneo, centipedes of, 185, 186
Bothriocyrtum,
  venom, toxicity, 270
Bothriomyrmex, anal gland, 25
Bothriomyrmex meridionalis, anal gland secretion, 25
Bothriuridae,
  key to, 322
  sternum, 320
  subfamilies, 323
Bothriurus,
  development, 337
  venom, toxicity, 342–343
Bothriurus bonariensis, mating behavior, 334
Bothroponera, venom glands, 21
Bothrops,
  venom,
    combinations of, 466
    pain and, 447
Bothrops alternatus,
  venom, pain, relief and, 446, 447
Bothrops atrox,
  venom, hemostasis and, 452, 453, 454, 455
Bothrops cotiara,
  venom, pain relief and, 447
Bothrops jararaca,
  venom,
    arthritis and, 450
    chicken pox and, 458
    hemostasis and, 452, 453, 454
    pain relief and, 447
Bothrops jararacussu,
  venom, pain relief and, 447
Brachypelma, 212
Braconidae,
  venom composition, 30

venom glands, 19
Bradykinin, caterpillar venom and, 110
Brazil,
lepidopterism,
clinical symptoms, 159–166
history, 157
pathology, 167
species, 158–159
treatment, 167
scorpion stings in, 311
venomous animals,
annelids, 435
caterpillars, 105
centipedes, 180, 181, 183, 184, 185, 186
scorpions, 331–332, 339, 343
sea urchins, 424
spiders, 219, 220, 223, 225, 232, 236, 263
Bristle worms, *see* Amphinomidae
British Columbia, scorpions of, 333
British Guiana, centipedes of, 186, 187
Broteas, 329
Broteochactas, 329
Brown spiders, *see Loxosceles*
Brown-tail moth, *see Euproctis*
Brush-hairs, structure of, 47–48
Bufo bufo bufo,
venom, active ingredient, 460
Bufogenins, heart disease and, 460
Bulgaria, latrodectism in, 307
Burma,
centipedes of, 185
spiders of, 219
Buthacus,
distribution of, 332, 333
venom, toxicity, 343
Buthacus arenicola, 326
distribution, 351
venom, 358
amount, 354
antivenin and, 365
coagulant activity, 359
toxicity, 343, 344, 360
Buthacus foleyi, 326
Buthacus leptochelys, 326
Buthacus villiersi, 326
Butheoloides milloti, 325
Butheolus bicolor, 326
Butheolus ferrugineus, 326
Butheolus melanurus, 326
Butheolus pallidus, 326
Butheolus thalassinus, 326
Buthidae,
distribution of, 339
key to, 322
sternum, 320
subfamilies, key to, 324–325
Buthinae,
genera of, 325–326
key to, 325
venom glands, 340
Buthiscus bicalcaratus, 325
Buthotus,
distribution of, 332, 333
venom, toxicity, 342–343
Buthotus franzwerneri, 326
Buthotus judaicus, 326
distribution, 352
mating behavior, 334
venom, 358, 359
electrophoresis, 355
protein content, 356
toxicity, 343, 344, 360
Buthotus minax,
venom, electrophoresis, 355
Buthus,
distribution of, 332, 333, 351
venom toxicity, 343
Buthus atlantis, 326
Buthus barbouri, 326
Buthus famulus,
distribution, 352
venom, toxicity, 360
Buthus maroccanus, 326
Buthus martensis,
distribution, 352
venom,
chemical composition, 355
hemolytic properties, 358
toxicity, 360
Buthus occitans, 326
distribution, 351
mating behavior, 333
venom, 357, 358, 359
amount, 354
antivenin and, 365, 366
electrophoresis, 355
hyaluronidase in, 358
purification, 356
toxicity, 343, 344, 360
venom glands, 340

Buthus occitanus tunetanus,
venom, purification of, 357

C

Calchas nordmanni, 329
Calcium, latrodectism and, 308
Calcium gluconate, coelenterate stings and, 414
California,
octopus bites in, 385, 388
venomous animals,
annelids, 434
centipedes, 186
scorpions, 332
starfish, 420
Callimorpha, urticating caterpillars, 40
Camphor, scorpionism and, 368
Camponotinae,
alarm substances, 34
Dufour gland, 23
sting apparatus, 17, 37
trail marking by, 34–35
venoms,
composition, 28, 29, 31
effects, 35
emission, 28
venom glands, 22, 37
Camponotus,
sting apparatus, 17
venom glands, 22
Camponotus ligniperda,
venom, composition, 29
Campsomeris, venom glands, 20
Campylostigmus,
classification, 171
key and occurrence, 181
Canada, spiders of, 219, 221, 236
Canary Islands, venomous sea urchins of, 424, 427
Cannibalism,
scorpions and, 338, 353–354
spiders, 243
Cantharidin, caterpillar venom and, 108
Caraboctoninae, key to, 328
Caraboctonus,
venom, toxicity, 342–343
Caraboctonus keyserlingi, 328
Carama, urticating caterpillars, 40
Caribbean Sea, venomous coelenterates of, 398
Carybdea alata, range, 401
Carybdea marsupialis, range, 401
Carybdea rastonii, range of, 401
Carybdeidae, venomous representatives, 401
Cataglyphis, venom glands, 22
Catecholamines, scorpion venom and, 369–370
Caterpillars,
poisonous, 104–105
classification, 133–134
distribution, 134
morphology of venom apparatus, 138–143
pharmacology, 108–116
somatic morphology, 135–138
toxicology, 143–145
venom apparatus, 105–107
synecology, 145–147
Catocola,
caterpillars, 41, 105, 134
venom apparatus, 142
Catostylidae, venomous representatives, 402
Catostylus,
stings, effects, 413
Catostylus mosaicus, range of, 402
Caucasus, scorpions of, 333
Cauterization, scorpion stings and, 367
Central Africa, spiders of, 216
Central America,
centipedes of, 185, 187
dangerous scorpions of, 331, 343
spiders of, 211, 216, 218, 236, 237
Centruroides,
description, 327
distribution of, 331, 332
mating behavior, 334
venom,
electrophoresis, 355
toxicity, 343, 345
Centruroides bertholdi, 327
Centruroides bicolor, 327
Centruroides danieli, 327
Centruroides elegans, 327
Centruroides exilicauda, 327
Centruroides exsul, 327
Centruroides flavopictus, 327
Centruroides fulvipes, 327

Centruroides gertschi, 327
venom, toxicity, 343
Centruroides gracilis, 327
Centruroides granosus, 327
Centruroides insulanus, 327
Centruroides limbatus, 327
Centruroides limpidus,
venom, toxicity, 343, 344
Centruroides margaritatus, 327
Centruroides nigrescens, 327
Centruroides nigrimanus, 327
Centruroides nitidus, 327
Centruroides noxius,
venom, toxicity, 343
Centruroides ochraceus, 327
Centruroides ornatus, 327
Centruroides pantheriensis, 327
Centruroides rubricauda, 327
Centruroides sculpturatus, 327
venom, 358
antivenins and, 366
toxicity, 343
Centruroides subgranosus, 327
Centruroides suffusus,
venom, toxicity, 343
Centruroides testaceus, 327
Centruroides thorelli, 327
Centruroides vittatus, 327
venom, toxicity, 343
Cephalonomia,
sting apparatus, position of, 4
Cephalothorax, spider, 202–203
Cephalotoxin, nature of, 390
Cephus pygmaeus, ovipositor, 3
Cerastes cornutus,
venom, therapeutic use, 451, 452
Cerceris, stinging by, 32
Cermatobiidae, 170
Cerura,
cryptotoxic caterpillars, 134
erucism and, 148
toxic apparatus, 143
Ceylon,
urticating caterpillars of, 105
Chacoana antherata, 219
Chacoana carteri, 219
Chacoana cretaceus, 219
Chacoana distincta, 219
Chacoana flavodorsata, 219
Chacopsis isignis, 329
Chactas, 329
Chactidae,
key to, 323
sternum, 320
subfamilies, key to, 329
Chactinae, genera and species, 329
Chaerilidae,
genera and species, 329
key to, 323
Chaerilus, 329
Charmus indicus, 326
Charmus laneus, 326
Chaunopelma, 212
Cheiracanthium, see *Chiracanthium*
Chelicerae, spider, 202–204
Chicken pox, snake venom and, 458
Chile,
centipedes of, 183
spiders of, 212, 217, 219, 236, 269, 271
Chilopoda,
biology of, 189–190
classification and distribution, 180–189
key to families, 180
nomenclature, history of, 170–171
subclasses, 170–171
China,
venomous animals,
annelids, 434
centipedes, 185, 186, 187
coelenterates, 405
sea urchins, 425
Chiracanthium,
bites, importance of, 198
description of, 236–237
distribution, 236
maternal behavior, 240
venom,
extraction of, 264
toxicity, 272
Chiracanthium brevicalcaratum, occurrence, 221
Chiracanthium diversum, occurrence, 221
Chiracanthium inclusum, distribution, 221
Chiracanthium punctorium, distribution, 221
Chironex fleckeri, 402
range, 405
stings, effects of, 413
Chiropsalmus quadrigatus,
range of, 405
stings, effects of, 413

Chloeia,
  envenomation, mechanism, 435
  stings, 440
Chloeia flava, range of, 434
Chloeia viridis, range of, 434
Chlorpromazine, scorpionism and, 368
Cholinesterase, scorpion venom and, 364, 365
Chrysaora quinquecirrha, 404
  range of, 405
Chrysididae,
  ovipositor, 28
  sting apparatus, position of, 3–4
Chymotrypsin,
  caterpillar venom and, 116
  Phoneutria venom and, 289, 290, 296
  scorpion venom and, 357
Cicileus exilis, 325
Citharacanthus, 212
Cities, scorpions in, 339
Clarion Islands, venomous starfish of, 420
Clavopelma, 212
Cleopatra, scorpions and, 350
Cleptidae,
  ovipositor, 28
  sting apparatus,
    position of, 3–4
    structure of, 7
Clubiona terrestris, venom of, 272
Clubionidae,
  description and dangerous species, 221
  key for, 210
Clubioninae, description and dangerous species, 221
Cnethocampa pitycampa, vesicating properties, 103
Coagulant unit, determination of, 453
Cobra, *see also Naja*
  venom,
    arthritis and, 450, 451
    neuralgia and, 450
    tabes and, 451
    tumors and, 445, 448
Cocaine, scorpion venom effects and, 364
Cochlidiidae,
  poisonous spines of, 50, 51, 52, 53, 54
  urticating caterpillars, 40
Cocoons, urticating hairs and, 39
Coelenterates,
  characteristics of, 395–396
  mechanism of intoxication, 407–409
  stings,
    clinical characteristics, 413–414
    pathology, 412–413
    prevention, 415–416
    treatment, 414–415
  venom,
    chemistry of, 410–411
    pharmacology, 411–412
  venom apparatus, 409–410
Colletes, sheath gland, 25
Colombia,
  centipedes of, 183, 185, 186
  scorpions of, 332
  spiders of, 223, 236
Coloradia, urticating caterpillars, 134
Communication,
  hymenopteran venoms,
    alarm, 33–34
    trail marking, 34–35
Condylactis, tentacular extracts, 411–412
Cone shells,
  foods of, 375, 380
  poisonous,
    classification, 374–375, 380
    distribution, 376–379, 381
    symptomatology of stings, 383–384
    venom apparatus and venom, 381–383
Congo, centipedes of, 181, 186
Conus abbreviatus, food of, 375
Conus aulicus, 380
  stings, 375, 377
    symptoms, 383
Conus capitaneus, food of, 375
Conus catus, 380
  food of, 375
  stings, 375, 379
    symptoms, 383
Conus ceytanensis, food of, 380
Conus chaldus, food of, 380
Conus clavus, food of, 380
Conus coronatus, food of, 375, 380
Conus distans, food of, 380
Conus ebraeus, food of, 380
Conus emaciatus, food of, 380
Conus figulinus, food of, 380
Conus flavidus, food of, 380

Conus geographus, 380
food of, 375
stings,
fatalities and, 374, 376–377
symptoms, 383–384
venom, pharmacology 383
Conus glans, food of, 380
Conus imperialis,
food of, 380
stings, 375, 379
symptoms, 383
Conus leopardus, food of, 380
Conus litteratus, 380
stings, 375, 379
symptoms, 383
Conus lividus,
food of, 375, 380
stings by, 375, 379
Conus magus, food of, 375
Conus marmoreus, 380
food of, 375
stings, 375
symptoms, 383
venom duct, 382
Conus miles, food of, 380
Conus nanus,
stings, 375, 378
symptoms, 383
Conus obscurus, 380
food of, 375
stings, 375, 378
symptoms, 383
Conus omaria, 380
stings, 375, 378
symptoms, 383–384
Conus pennaceus, food of, 375
Conus pertusus, food of, 380
Conus pulicarius,
food of, 380
stings, 375, 378
symptoms, 383
Conus quercinus,
food of, 375, 380
stings by, 375, 379
Conus rattus, food of, 380
Conus sponsalis, food of, 380
Conus stercusmuscarium, food of, 375
Conus striatus, 380
food of, 375
preying by, 382
stings by, 375
Conus textile, 380
food of, 375
stings, 374–375, 377
symptoms, 383–384
venom of, 383
venom duct, 382
Conus tiaratus, food of, 380
Conus tulipa, 380
food of, 375
stings, 375, 377–378
symptoms, 383
Conus vexillum, food of, 380
Conus virgo, food of, 380
Conus vitulinus, food of, 380
Coral,
cuts, treatment of, 415–416
Cormocephalus,
classification, 171
key and occurrence, 181
Corticoids, scorpionism and, 367
Corticosterone, erucism and, 154, 167
Cosmotriche, urticating caterpillars, 41
Cossidae, cryptotoxic species, 134
Cossus, cryptotoxic caterpillars, 134
Cossus cossus, toxic apparatus, 143
Cossus ligniperda,
erucism and, 148
toxic apparatus, 143
Craterostigmidae, 170
Craterostigmomorpha, 170
Cremastogaster,
sting apparatus, 37
trail marking by, 35
Cremastogaster scutellaris, Dufour gland, 22, 23, 38
Cricula, urticating caterpillars, 41
Crossocerus elongatus, sting apparatus, 33
Crotalin, epilepsy and, 455–457
Crotalas,
venom, combinations of, 466
Crotalus adamanteus,
venom,
epilepsy and, 444
neuralgia and, 449–450
tumors and, 445
Crotalus durissus terrificus,
venom,
epilepsy and, 456
neuralgia and, 450

pain relief and, 446, 447, 448, 449
Cryotherapy, scorpion stings and, 367
Cryptarachne bisaccata, 220
Cryptopidae,
key and distribution, 186–187
members of, 171
Cryptopinae,
key and distribution, 186–187
members of, 171
Cryptops,
classification, 171
distribution of, 186–187
habits of, 190
legs of, 178
maxillae, 174
prey capture by, 189–190
Cryptops iheringi,
distribution, 187
venom, toxicity, 195
venom apparatus, 191
Ctenidae,
description, 222–223
key for, 210
Ctenus, description of, 222
Cuba,
centipedes of, 185
scorpions of, 332, 338
spiders of, 219
Cupiennius, description, 222
Curaçao,
scorpions of, 332
spiders of, 219
Curare, scorpion venom effects and, 363–364
Cuticle, ant venoms and, 38
Cuy dorado, *see Megalopyge*
Cuy rojizo, *see Megalopyge*
Cyanea,
stings, effects, 413
Cyanea capillata, 403
extracts, histamine and, 412
range of, 405
size of, 397
Cyaneidae, venomous representatives, 405

D

Dactylometra,
stings, effects, 413
Dalmatia, centipedes of, 185
Dasychira, urticating caterpillars, 40
Dasychira pudibunda, brush-hairs of, 47–48, 49
Delopelma, 212
Dendrolasius fuliginosus, venom of, 96
Dendrolimus, urticating caterpillars, 41
Dendrolimus pini, hairs of, 45
Dendryphantes, 237
bites, wounds and, 198
distribution, 236
mating behavior, 239
venom, toxicity, 272
Dendryphantes noxiosus, description, 222
Deoxycorticosterone glycosides, bee venom therapy and, 465
Dermatitis, lepidopterism and, 129–130, 149–150
Desert, scorpions of, 338–339
Diadema, spines of, 428, 432
Diadema antillarum, range of, 424
Diadema setosum, range of, 425
Diadematidae, poisonous representative, 424
Dialysis, *Phoneutria* venom and, 290, 291
Diamma, venom glands, 20
Digitipes,
classification, 171
occurrence of, 186
Dinocryptopinae, key and distribution, 187
Dinocryptopini, key and distribution, 187
Dinocryptops, 171
distribution of, 187
Dinoponera,
venom, effects, 33
Diplocentrinae,
genera and species, 323–324
key to, 323
Diplocentrus,
venom, toxicity, 342–343
Diplocentrus antillanus, classification, 323
Diplocentrus grundlachi, classification, 323
Diplocentrus hasethi, classification of, 324
Diplocentrus keyserlingi, classification, 324
Diplocentrus scaber, classification, 324
Diplocentrus whitei, classification, 324

Diplothelinae, description, 215
Diplura bicolor, 251
venom apparatus,
glands, 260
inoculating apparatus, 255
Dipluridae,
key for, 209
subfamilies with venomous members, 211
Diplurinae, key for, 211
Dirphia,
extracts, pharmacology of, 109–112
poisonous spines of, 52–53, 54
urticating caterpillars, 41, 134
venom apparatus, 107, 142
venom, toxicology, 144–145
Dirphia multicolor, caterpillars of, 105
Dirphia sabina, caterpillars of, 105
Dogs, *Phoneutria* venom effects, 280–283
Dolichoderinae,
alarm substance, 34
anal gland, 25, 37
sting apparatus, 17, 37
trail marking by, 35
venoms,
composition, 28, 31
effects, 33, 35
emission, 28
venom glands, 18, 21
Dolicherus, anal gland, 25
Dolichoderus clarki,
venom, composition, 31
Dolichoderus dentata,
venom, composition, 31
Dolichoderus scabridus,
venom, composition, 31
Dolichodial,
isolation of, 97
venoms and, 29, 31
Dorylinae,
sting apparatus, 17, 37
sting reduction in, 36
trail marking by, 34–35
venom glands, 21
Dryinidae, ovipositor, 28
Dufour gland, 18
structure and function, 22–23, 34, 38
Dugesiella, 212
Dutch Guiana, spiders of, 219
Dysdera,
venom, toxicity, 271
Dysdera crocata, venom of, 272

E

Eacles, urticating caterpillars, 134
East Africa, *see also* Africa
centipedes of, 181, 184
spiders of, 219, 236
Eberthella typhosa, scorpion venom and, 359
Echinacea,
extracts, venom combinations and, 466
Echinidae,
pedicellariae of, 430
venomous representative, 425
Echinoderms,
characteristics and venomous classes, 419
venom apparatus,
sea urchins, 428–430
starfish, 421–423
Echinoidea, characteristics and venomous representatives, 424–427
Echinothrix calamaris, range of, 425
Echinothuridae, venomous representatives, 426
Echis coloratus,
venom, anticoagulant effect, 468
Ecpantheria orsa, caterpillars of, 159
Ecuador,
centipedes of, 183
scorpions of, 331
Edema,
bee venom and, 465
lepidopterism and, 161
snake venom and, 446
Egg(s),
spider, development of, 242
Egypt,
scorpions of, 333, 343, 352
spiders of, 219
Ejaculation, phoneutrism and, 282
Electrocardiogram, latrodectism and, 300
Electrophoresis,
Phoneutria venom, 290, 291
scorpion venoms, 355
Eledone, venom apparatus, 388, 389
Elk horn coral, *see* Acroporidae

Empretia,
urticating caterpillars, 134
venom apparatus, 142
Enteritis, caterpillars and, 147
Epeira cornigera, 220
Epeira gasteracanthoides, 220
Epilepsy, snake venom and, 444, 455–457
Epimorpha, suborders of, 171
Ergotamine, scorpionism and, 367
Ergotoxine, scorpion stings and, 367
Erucism, *see also* Lepidopterism
etiopathogenesis, 147–149
prevention,
humoral changes, 154
individual measures, 154
sanitation, 153–154
symptomatology,
allergic syndrome, 152–153
circumscribed, 150–152
general, 152
histopathology, 149–150
treatment, 154
types of, 133
Erythria, scorpions of, 352
Esberitox, components, 466
Eserine,
scorpion venom effects and, 314, 364
venom induced salivation and, 281
Ethanol, *Phoneutria* venom and, 289, 290
Ethiopia,
centipedes of, 185, 186
scorpions of, 333
Ethmostigmus,
classification, 171
distribution of, 186
Euchaetia, distribution of, 134
Euchaetis egle,
caterpillars of, 104
poisonous setae, 139
Eucleidae,
habitats of, 158
urticating caterpillars, 104, 133, 134
venom apparatus, 142
Euglyphes ornata, caterpillars of, 159
Eunice aphroditois, bite of, 433
Eupalaestrus,
description of, 225
key for, 213
number of eggs, 241
venom, extraction of, 263–264
Eupalaestrus campestratus, 214
Eupalaestrus pugilator, 214
Eupalaestrus rubropilosum,
venom,
amounts, 266
toxicity, 268
Eupalaestrus spinosissimus, 214
Eupalaestrus tarsicrassus, 214
Eupalaestrus tenuitarsus, 214
venom,
amounts, 266
toxicity, 268
Euproctis, urticating caterpillars, 40, 134
Euproctis chrysorrhoea, 105
cocoons, urticating hairs and, 39, 129
eggs, urtication by, 127
hairs of, 45–46, 48
venom apparatus, 104, 106
Euproctis flava,
adults, toxicity, 149
cocoons, toxicity, 129
occurrence of, 121
Euproctis phaerrhoea,
adults, toxicity, 148–149
venom, toxicology, 144–145
venom apparatus, 142
Eupseudosoma aberrans, caterpillars of, 159
Eupseudosoma involutum, caterpillars of, 159
Euptoieta, urticating caterpillars, 134
Europe,
centipedes of, 185
spiders of, 216, 221, 269
urticating caterpillars of, 105
Euryda variolarus, caterpillars of, 158
Eurypelma, 212
distribution, 237
venom, toxicity, 270
Eurythoë,
envenomation, mechanism, 435
stings, 440
Eurythoë brasiliensis, range, 434–435
Eurythoë complanata, 434
range, 435
setae, 435, 436–437
Euscorpioninae, genera and species, 329
Euscorpius, distribution of, 333, 351
Euscorpius carpathicus, 329
venom glands, 340

Euscorpius flavicaudus, 329
Euscorpius germanus, 329
Euscorpius italicus, 329
  mating behavior, 334
  venom,
    antibiotic in, 359
    hyaluronidase in, 358
Euvanessa, urticating caterpillars, 40
Euvanessa antiopa, caterpillars, 104
Evan's blue, caterpillar venom and, 113–114, 115
Eye(s),
  diseases,
    bee venom and, 465–466
    snake venom and, 458
  spider, 205, 207

F

False coral, *see Millepora*
Feather hydroid, *see Lytocarpus*
Fibrinogenopenia, snake venom and, 455
Fiji Islands, spiders of, 221
Filistata, distribution of, 237
Filistata hibernalis, description and distribution, 271
Fire caterpillar, *see Megalopyge*
Fire coral, *see Millepora*
Florida,
  centipedes of, 185
  octopus bites in, 386, 388
  spiders of, 219
  venomous annelids of, 435
Food,
  cone shells, 375, 380
  venomous spiders, 246
Forest, scorpions of, 338
Formaldehyde, *Phoneutria* venom and, 289, 290
Formic acid,
  aculeate venoms and, 28, 29, 31, 37
  ant venom and, 96
  caterpillar venom and, 108
  injurious effects, 29
Formica,
  sting apparatus, special peculiarities, 14, 17
  venom,
    alarm and, 34
    effects, 33
  venom glands, 22
Formica polyctena, venom glands, 20
Formica pratensis, sting apparatus, 17
Formica rufa,
  venom, 96
    composition, 29
    emission, 28
  venom glands, 22
Formicidae,
  Koshewnikow gland, 24
  sting reduction in, 36
  venom,
    alarm and, 34
    composition, 30–31
  venom glands, 20
Formicoidea,
  venoms, composition, 30–31
Formiga do fôgo, *see Solenopsis*
France,
  latrodectism in, 306
  scorpions of, 333, 351
Fungi, hymenopteran venoms and, 33
Funnel-web spiders, *see Dipluridae*

G

Galapagos Islands,
  venomous animals,
    annelids, 434
    centipedes, 185
    spiders, 219
    starfish, 420
Gel filtration,
  Phoneutria venom, 290, 291–293
  scorpion venom, 356–357
Georgia,
  centipedes of, 184
  spiders of, 218
Glutamate, *Phoneutria* venom and, 287, 297
Glycera dibranchiata,
  bite of, 434, 435–436, 440
  jaws of, 437–439
  range of, 435
Glycera ovigera, range, 435
Glyceridae, venomous species, 435
Glyptocranium,
  distribution, 236
  venom, extraction of, 264

Glyptocranium gasteracanthoides, 220
Gompsobuthus acutecarinatus, 326
Gompsobuthus werneri werneri, 326
Grammostola, 212
  life history, 244–245
  mating bulbs, 207
  number of eggs, 241
  venom, toxicity, 270
Grammostola actaeon, 213, 251
  adolescent phase, 244
  venom,
    amounts, 266
    toxicity, 268
  venom apparatus,
    glands, 260
    inoculating apparatus, 255
Grammostola gossei, 213
Grammostola iheringi, 251
  adolescent phase, 244
  venom,
    amounts, 266
    toxicity, 268
  venom apparatus,
    glands, 260
    inoculating apparatus, 255
Grammostola mollicoma, 213, 251
  adolescent phase, 244
  description of, 225
  mating behavior, 239
  venom,
    amounts, 266
    toxicity, 268
  venom apparatus,
    glands, 260
    inoculating apparatus, 255
Grammostola pulchripes, 251
  venom,
    amount, 266
    toxicity, 268
  venom apparatus,
    glands, 260
    inoculating apparatus, 255
Grammostolinae,
  key and venomous members, 212–213
  venom, extraction of, 263–264
Greece,
  latrodectism in, 307
  scorpions of, 351
  spiders of, 219
Grenada, scorpions of, 331
Gryllus domesticus, scorpion maintenance and, 354
Guadelupe, centipedes of, 185
Guatemala, scorpions of, 332
Guianas, scorpions of, 331
Guinea pig,
  ileum, *Phoneutria* venom and, 283–284, 290, 291, 292, 295
  scorpion venom effects, 358, 360
  phoneutrism in, 287
Gulf of Suez, venomous sea urchins of, 426
Guyana, spiders of, 236

H

Hadogenes,
  distribution of, 332, 352
  venom,
    collection of, 353
    hemolytic properties, 358
    toxicity, 343, 344
Hadogenes opisthocanthoides, 324
Hadogenes paucidens, 324
Hadogenes trichiurus, 324
Hadogenes troglodytes, 324
Hadruroides,
  venom toxicity, 342–343
Hadruroides lunatus, 328
Hadrurus, distribution of, 333
Hadrurus hirsutus, 329
Hair(s),
  urticating, types of, 42–55
Hairy caterpillar, *see Megalopyge*
Hairy stinger, *see* Cyaneidae
Haiti,
  centipedes of, 185
  scorpions of, 332
  spiders of, 219
Halisdota, urticating caterpillars, 134
Halistemma, nematocysts of, 410
Hapalochlaena, bites, symptoms, 390–391
Hapalochlaena lunulata, bites by, 384, 387
Hapalochlaena maculosa,
    bites by, 384, 386–387
  venom apparatus, 389–390
Harpactirella, 251
  aggressiveness of, 198
  bites, importance of, 198, 230, 315
  distribution of, 236

venom,
antisera, 273
extraction of, 263–264
venom apparatus, 247
venom glands, 203, 258
Harpactirella domicola, 215
Harpactirella flavipilosa, 215
Harpactirella helenae, 215
Harpactirella karrooica, 215
Harpactirella lapidaria, 215
Harpactirella lightfooti, 215
description of, 230–231
spinnerets, 200, 202
Harpactirella longipes, 215
Harpactirella magna, 215
Harpactirella schwarzi, 215
Harpactirella spinosa, 215
Harpactirella treleaveni, 215
Harpagoxenus, sting apparatus, 37
Harpya,
erucism and, 148
toxic apparatus, 143
Hawaii,
octupus bites in, 385, 386, 388
spiders of, 221, 236, 237, 270, 272
venomous coelenterates of, 398, 401
Heart disease, toad venoms and, 460
Heat,
latrodectism and, 308
Phoneutria venom and, 289, 290, 293
scorpion venom and, 355
Hemerocampa, urticating caterpillars, 40, 134
Hemerocampa leucostigma,
behavior of, 55–56
poisonous setae, 139, 148–149
Hemibuthus crassimanus, 326
Hemicholinium, scorpion venom and, 314
Hemileuca,
urticating caterpillars, 41, 134
venom apparatus, 142
Hemileuca maia, caterpillars of, 105
Hemileuca oliviae, caterpillars of, 105
Hemileucidae,
host plants, 158
urticating caterpillars of, 105
Hemiscorpion lepturus, 324
Hemiscorpioninae, 324
key to, 323
Hemolysis,
caterpillar venom and, 108, 115, 144
scorpion venoms and, 357–358
Hemolytic factor, ant venom and, 96
Hemophilia,
epilepsy and, 456
snake venom and, 455
Hemorrhage,
toad venom and, 460
treatment, snake venom and, 452–455
Hemostasis, snake venoms and, 452–455
Heptadecane, Dufour gland and, 23
cis-Heptadec-8-ene, Dufour gland and, 23
Hermodice,
envenomation, mechanism, 435
stings, 440
Hermodice carunculata,
range of, 435
setae, 436–437
Herpes zoster, snake venom and, 450, 458
Heterometrus,
distribution of, 332
stridulating apparatus, 321
venom,
hemolytic properties, 358
toxicity, 342–343
Heterometrus bengalensis, 324
Heterometrus caesar, 324
Heterometrus cyaneus, 324
Heterometrus fulvipes, 324
Heterometrus indus, 324
Heterometrus latimanus, 324
Heterometrus liphysa, 324
Heterometrus liurus, 324
Heterometrus longimanus, 324
Heterometrus phipsoni, 324
Heterometrus scaber, 324
Heterometrus swammerdami, 324
Heterometrus xanthopus, 324
Heteropoda venatoria,
distribution, 237
venom, toxicity, 272
Hexamethonium, venom induced salivation and, 281
Histamine,
ant venom and, 96
bee venom therapy and, 463, 465
hymenopteran venoms and, 30, 31, 54–55, 109, 111–112, 113, 144, 145
nettle sting and, 108

Phoneutria venom and, 287, 289, 292, 294
release, coelenterate toxin and, 411, 412
Hodgkin's disease, toad venom and, 459–460
Honduras, centipedes of, 185
Honeybee, *see Apis*
Hormiga de fuego, *see Solenopsis*
Horse, scorpion venom effects, 360
Huberia, trail marking by, 34
Humans,
ecology of poisonous lepidoptera and, 125–126,146–147
Phoneutria bites in, 287
scorpion envenomation in, 360-362
Hyaluronidase,
antivenin efficacy and, 366
ant venom and, 96
hymenopteran venoms and, 30, 31, 108
octopus venom and, 390
Phoneutria venom and, 287
scorpion venom and, 312, 358
Hydergine, scorpionism and, 367
Hydrocorallina, characteristics and habitats, 396
Hydroxy acids, hymenopteran venoms and, 30
5-Hydroxytryptamine, *see* Serotonin
Hydrozoa,
characteristics and habitats, 396–397
families and venomous representatives, 398–401
Hylesia,
adults, poisonous, 121
allergic reactions to, 130–131
ecology,
humans and, 125–126
moths and, 125
epidemic dermatitis and, 126
morphology,
general, 121–122
sting weapon, 122–124
poisonous setae, 163, 164
urticating caterpillars, 41
venomous phase, 158
Hylesia canitia, occurrence of, 159
Hylesia continua, distribution, 121, 159
Hylesia fulviventris,
coloration, 122
nests, toxicity, 127, 149
occurrence of, 121, 159
venom of, 125
Hylesia linda,
coloration, 122
distribution, 121
Hylesia nigricans
adults, sting weapons, 123
coloration, 121–122
distribution, 121, 159
males, toxicity, 128
nests, toxicity, 127, 149
venom, 124
Hylesia urticans,
adult, sting weapon, 123
distribution, 121, 159
males, toxicity, 128
nests, toxicity, 127–128
toxic substance of, 124, 166
Hylesia volvex,
coloration, 122
distribution, 121, 159
Hymenoptera,
venom apparatus,
evolution, 35–38
function, 1–2, 25–35
morphology, 3–25
Hyperglycemia, scorpionism and, 362, 364
Hypotension, phoneutrism and, 283
Hypothalamus, scorpion venom and, 363, 365

I

Ichneumon,
ovipositor, function, 28
Ichneumonoidea, venom composition, 30
Imaginal tufts, hairs of, 48
Immunization, lepidopterism and, 132
Immunoelectrophoresis, *Phoneutria* venom, 288
Imported fire-ant, *see Solenopsis*
Inderal, scorpion venom and, 314
India,
centipedes of, 181, 185, 186
scorpions of, 332, 333, 342
spiders of, 219, 236
urticating caterpillars of, 105
venomous octopuses near, 388
Indian Ocean,
venomous animals,
annelids, 434

coelenterates, 398, 401, 405
mollusks, 381
sea urchins, 426
Indies, spiders of, 219
Indonesia, venomous sea urchins of, 426
Insect(s), hymenopteran venom effects, 29, 32–33
Insecticides,
contact, spiders and, 247
scorpions and, 339
urticating lepidoptera and, 131–132
Instituto Butanan, spiders sent to, 263
Ion-exchange chromatography,
Phoneutria venom, 290, 293–297
scorpion venom, 356–357
Iran,
centipedes of, 185
scorpions of, 352
spiders of, 219, 220, 236
Iridodial,
isolation of, 97
venoms and, 29, 31
Iridolactone, isolation of, 97
Iridomyrmecin,
hymenopteran venoms and, 31
injurious effects, 29, 32
isolation of, 96, 97
Iridomyrmex, anal gland, 25
Iridomyrmex conifer,
venom, composition, 31
Irodomyrmex detectus,
venom, composition, 31
Iridomyrmex humilis,
anal gland secretion, 25
venom, 96, 97–98
composition, 31
Iridomyrmex myrmecodiae,
venom, composition, 31
Iridomyrmex nitidiceps,
venom, composition, 31
Iridomyrmex nitidus,
venom, 96
composition, 31
Iridomyrmex pruinosus,
venom, composition, 31
Iridomyrmex rufoniger,
venom, composition, 31
Ischnurinae,
genera and species of, 324
key to, 323
Isobutyric acid, ant venom and, 96
Isoiridiomyrmecin,
hymenopteran venoms and, 31
isolation of, 96
Isometrinae, key to, 325
Isometroides angusticaudus, 326
Isometroides vescus, 326
Isometrus,
venom, toxicity, 342–343
Isovalerianic acid, ant venom and, 96
Israel,
centipedes of, 181
latrodectism in, 306
scorpions of, 352
Italy,
latrodectism in, 306
scorpions of, 351
spiders of, 219

J

Jamaica,
centipedes of, 183
scorpions of, 332
Japan,
venomous animals,
annelids, 434
centipedes, 187
coelenterates, 398, 401, 405
sea urchins, 425, 426, 427
Java, centipedes of, 185
Jordan, scorpions of, 352
Jumping spiders, *see* Salticidae
Jurinae, key to, 328
Jurus,
venom, toxicity, 342–343
Jurus dufoureius, 328

K

Kansas, centipedes of, 184
Karasbergia methueni, 326
Kartops, classification, 171
Kartops guiannae, occurrence, 187
Kenya, spiders of, 218
Kethops, classification, 171
Kethops utahensis, occurrence of, 187
Khohe caterpillars, ingestion of, 147

Kinin(s), caterpillar venom and, 114, 145
Kininlike factor, ant venom and, 96
Korea,
  centipedes of, 187
  scorpions of, 352
Koshewnikow gland, 18
  structure and function, 23–24

L

Labidognatha,
  chelicerae, 204
  distribution, 236
  families with venomous species, 215
  key for families, 209–210
  morphology, 201
  venom apparatus, 247, 250
Lachesis muta,
  venom, therapeutic use of, 451
Lacrimation,
  Phoneutria envenomation and, 280, 284
Lagarta aranha, *see* Eucleidae
Lagarata-capedula, *see Megalopyge*
Lagarta de fogo, *see Megalopyge*
Lagoa, urticating caterpillars, 40, 134
Lagoa crispata, venom apparatus, 105
Lancets,
  honeybee sting, 35
  protrusion of, 26–27
  retraction of, 27
Larra,
  venom, effects, 32
Larvae,
  scorpion, 337
  spider, development of, 242–243
Lasiocampa, urticating caterpillars, 41, 134
Lasiocampa quercus, nettling hair of, 44, 45
Lasiocampa trifolii, hairs of, 45
Lasiocampidae,
  caterpillars of, 41, 133, 134, 159
  erucism, symptoms, 149
  hairs of, 44–45
  host plants, 159
Lasiodora,
  description of, 227
  distribution, 236
  key for, 213
  leg claws of, 208
  mating behavior, 239
  metatarsus and tarsus of, 208
  number of eggs, 241
  venom,
    extraction of, 263
    toxicity, 270
  venom jaws, 204
Lasiodora klugi, 214, 251
  adolescent phase, 244
  description of, 227
  venom,
    amounts, 266
    toxicity, 268
  venom apparatus,
    glands, 260
    inoculating apparatus, 255
Lasiodora saeva, 214
Lasiodora spinipes, 214
Lasiodora striatipes, 214
Lasius,
  alarm substances, 34
  sting apparatus, 17
  venom glands, 22
La temible or huayruro, see Loxosceles
Latrodectinae, description and venomous species, 217–220
Latrodectism,
  clinical symptoms of, 299–301
  epidemiology of, 304–306
  folklore treatment of, 308
  history, 302, 304
  spread in Mediterranean countries, 306–308
  symptoms, 267, 268–270
Latrodectus,
  adolescent phase, 244
  aggressiveness of, 198
  bites, importance of, 198
  classification of, 218
  common names, 217–218
  description of, 232–233
  food of, 246
  habits, 232–233
  longevity, 246
  maternal behavior, 240–242
  mating behavior, 239
  mature life, 244
  received by Instituto Butanan, 263
  venom,
    antisera, 273
    extraction, 264

toxicity, 267–270
venom apparatus, 247, 253
glands, 258
musculature, 253–254
young, behavior of, 243
Latrodectus albomaculatus, 219
Latrodectus ancorifer, 219
Latrodectus apicalis, 219
Latrodectus argus, 219
Latrodectus carolinum, 219
Latrodectus concinnus, 218, 219
Latrodectus conglobatus, 219
Latrodectus curacaviensis, 252
distribution, 218, 219, 236
egg cases, 240
insecticides and, 247
mating behavior, 239
migration of, 243–244
preying by, 232
venom,
amounts, 265, 266
toxicity, 268
venom apparatus,
glands, 249, 250–251, 260, 261, 262
inoculating apparatus, 255–256, 257
musculature, 249, 250
Latrodectus dahli, occurrence, 220
Latrodectus erebus, 219
distribution of, 301
Latrodectus foliatus, 219
Latrodectus formidabilis, 218
Latrodectus geographicus, 219
Latrodectus geometricus, 252
description of, 232
distribution, 218, 236
egg cases, 240
venom,
amounts, 266
toxicity, 268
venom apparatus,
glands, 249, 260
inoculating apparatus, 255
musculature, 249
Latrodectus geometricus modestus, 218
Latrodectus geometricus obscuratus, 218
Latrodectus geometricus subalbicans, 218
Latrodectus dahli, 219
Latrodectus hasselti, 219
Latrodectus hasselti aruensis, 219
Latrodectus hasselti elegans, 219
Latrodectus hystrix, distribution, 219
Latrodectus incertus, 219
Latrodectus indistinctus, 219
Latrodectus indistinctus karooensis, 219
Latrodectus insularis insularis, 219
Latrodectus insularis lunulifer, 219
Latrodectus intersector, 218, 219
Latrodectus katipo, 219
Latrodectus luzonicus, 219
Latrodectus mactans,
distribution, 218–219
egg cases, 240
molting of, 245
subspecies, venom toxicity, 272
venom, tachyphylaxis and, 283
venom apparatus,
glands, 248–249
musculature, 249
Latrodectus mactans bishopi, 219
Latrodectus mactans cinctus,
collection of, 233
distribution, 219, 236
Latrodectus mactans curacaviensis, occurrence of, 305
Latrodectus mactans hasselti,
collection of, 233
distribution, 219, 236, 305
Latrodectus mactans mactans,
description of, 232
distribution, 218, 236
habitats, 304–305
venom,
amounts, 266
toxicity, 268, 269
venom apparatus,
glands, 260
inoculating apparatus, 255–256
Latrodectus mactans menavodi,
bites, 269
collection of, 233
occurrence, 219, 236
Latrodectus mactans mexicanus, 219
Latrodectus mactans texanus, 219
Latrodectus mactans tredecimguttatus,
description of, 233, 301–302
distribution, 219, 236, 301–302
habitats, 305, 306
venom, toxicity, 269, 270
Latrodectus malmignathus, 219
Latrodectus malmignathus tropica, 219

Latrodectus martius, 219
Latrodectus obscurior, 218, 219
Latrodectus occulatus, 219
Latrodectus pallidus, distribution and description of, 219, 236, 358
Latrodectus pallidus immaculatus, 219
Latrodectus pallidus pavlovskii, 219
Latrodectus perfidus, 218
Latrodectus quinqueguttatus, 219
Latrodectus renivulvatus, 219
Latrodectus revivensis, 219
Latrodectus sagittifer, 219
Latrodectus scelio, 219
Latrodectus scelio indica, 219
Latrodectus stuhlmanni, 219
Latrodectus thoracicus, 219
Latrodectus tredecimguttatus, 218
  venom, therapeutic use, 461
  venom apparatus, glands, 247–248
Latrodectus variegatus, 219
Latrodectus variolus, 218
Latrodectus venator, 219
Latrodectus zickzack, 218
Lava-pé, *see Solenopsis*
Lebanon, scorpions of, 352
Lecithin, hymenopteran venoms and, 30
Lecithinase, scorpion venoms and, 357–358
Legs, spider, 207, 209
Legends, scorpions and, 349–351
Leiurus, distribution of, 332, 333
Leiurus quinquestriatus, 326
  distribution, 352
  mating behavior, 334
  venom, 357, 358, 359
    chemical composition, 355
    electrophoresis, 355
    serotonin, 357
    toxicity, 343, 344, 360
Leiurus quinquestriatus quinquestriatus,
  venom, purification of, 357
Lepidoptera,
  poisonous,
    classification, 120–121
    distribution, 121
    morphology, 121–124
    toxicology, 124–125
  urticating hairs of larvae,
    ecological importance, 55–56
    scope of problem, 38–42
    types of, 42–55
Lepidopterism, *see also* Erucism
  clinical symptoms, 159
    caterpillars, 160–162
    moths, 162–166
  etiopathogenesis, 126–129
  historical data, 103–104, 120
  occurrence of, 120
  pathology of, 167
  prevention,
    humoral modifications, 132
    individual measures, 132
    sanitation, 131–132
  species involved in Brazil, 158–159
  symptomatology,
    allergic, 130–131
    toxic, 129–130
  treatment, 167
    general, 133
    local, 132
Leprosy,
  bee venom and, 465
  snake venoms and, 444, 452
  spider venom and, 444
Leptopelmatinae, description and venomous genera, 215
Lestrimelitta limao, venom glands, 21
Libya,
  scorpions of, 339, 343, 352
  spiders of, 219, 236
Lights, lepidopterism prevention and, 132
Limacodidae,
  poisonous spines of, 50
  urticating caterpillars, 40, 42, 133
Limonene,
  ant venom and, 95
  hymenopteran venoms and, 31
Liobuthus kessleri, 326
Liometopum, anal gland, 25
Lion's mane, *see* Cyaneidae
Liparidae, urticating caterpillars of, 40, 42, 45, 104, 105, 133
Liparis, urticating caterpillars, 134
Liparis chrysorrhoea,
  setae, ingestion of, 148
Liris, stinging by, 32
Lissothus bernardi, 325
Lithobiidae, 170
Lithobiomorpha, 170

Lithosia, urticating caterpillars, 134
Lithosiidae, urticating caterpillars of, 133, 134
Lithyphantes,
bites, importance of, 198
distribution, 236
venom, toxicity, 271
Lithyphantes anchoratus, 217
Lithyphantes andinus, 217
Louisiana, centipedes of, 184
Loxosceles, 251
adolescent phase, 244
bites, importance of, 198
chelicerae, 203
classification of, 216
description of, 231
distribution, 236
eyes of, 207
food of, 246
habits, 231
longevity, 246
maternal behavior, 240–242
mating behavior, 239
mature life, 244
received by Instituto Butanan, 263
venom,
antisera, 273
extraction, 264
toxicity, 267–268
venom apparatus, 247, 253
glands, 258
musculature, 253–254
young, behavior of, 243
Loxoscelles decemdentata, 216
Loxosceles distincta, 216
Loxosceles laeta, synonym, 216
Loxosceles reclusa, 216
Loxosceles rufescens, 216
venom glands, 260
Loxosceles rufipes, 216
male palps, 206
venom,
amounts, 266
toxicity, 268
venom apparatus,
glands, 260, 262
inoculating apparatus, 255
Loxosceles similis,
description of, 231
male palps, 206
Loxosceles similis,
venom,
amount, 265, 266
toxicity, 268
venom apparatus,
glands, 259, 262
inoculating apparatus, 255, 256
Loxosceles spadicea, 216
Loxosceles unicolor, 216
Loxoscelids, venom glands, 203
Loxoscelinae, description and venomous members, 216
Loxoscelism, symptoms, 267
Ludia, urticating caterpillars, 134
Lupus, bee venom and, 465
Lycaena,
cryptotoxic caterpillars, 134
toxic apparatus, 143
Lycaenidae, cryptotoxic species, 134
Lychas, 308
Lycosa,
bites, wounds and, 198
chelicerae, 203
cocoons of, 240
description of, 234, 236
distribution, 236
food of, 246
habits of, 234, 236
longevity, 246
maternal behavior, 240–242
mating behavior, 239
mature life, 244
venom, antisera, 273
venom apparatus, 247, 253
glands, 258
musculature, 253–254
young, behavior of, 243
Lycosa albopilata, 221
Lycosa auroguttata, 221
Lycosa capensis, 221
Lycosa carolinensis, habits of, 234
Lycosa erythrognatha, 221, 252
aggressiveness of, 198
description of, 234
molting in, 245
received by Instituto Butanan, 263
venom,
amounts, 264–265, 266
toxicity, 267, 268

venom apparatus,
glands, 259, 260, 262
inoculating apparatus, 255, 256, 258
Lycosa hilaris, 221
Lycosa narbonensis, 221
Lycosa nordenskioldi, habits of, 234, 236
Lycosa nychthemera, 221
fangs, 256
Lycosa ornata, 221
Lycosa pampeana,
fangs, 256
habits of, 234, 236
Lycosa poliostoma, 221
Lycosa punctulata, habits of, 234, 236
Lycosa tarentula, 221
Lycosa thorelli, 221
fangs, 256
habits of, 234, 236
Lycosidae,
description and venomous species, 220–221
key for, 210
Lycosinae, description and poisonous species, 220–221
Lymantria, urticating caterpillars, 40
Lymantria dispar, urticating hairs of, 39
Lymantriidae,
caterpillars, 104, 133, 134
venom apparatus, 142
urticating hairs of, 45–46, 47–48, 139
Lysergic acid diethylamide, caterpillar venom and, 109, 113, 114
Lytechinus variegatus,
venom, pharmacology, 431
Lytocarpus philippinus, range, 401

M

Macromphalia lignosa, caterpillars of, 159
Macrothelinae, key and venomous members, 211
Macrothylacia, urticating caterpillars, 41
Macrothylacia rubi, hairs of, 45
Macrurocampa, toxic apparatus, 143
Madagascar,
centipedes of, 185
scorpions of, 332
spiders of, 218, 219, 236, 269
Madeira, venomous sea urchins of, 424
Magula symmetrica, 251
venom apparatus,
glands, 260
inoculating apparatus, 255
Malacosoma, urticating caterpillars, 41, 134
Malayan Archipelago, venomous coelenterates of, 401, 405
Malayan pit viper, *see Agkistrodon rhodostoma*
Mammals, urticating caterpillars and, 55
Manchuria, scorpions of, 352
Mandibular gland, 18
function of, 38
Mastophora,
bites, importance of, 198
description of, 233
distribution, 236
habits of, 233
maternal behavior, 240
mating behavior, 239
venom, toxicity, 271
Mastophora bisaccata, occurrence, 220
Mastophora cornifera, occurrence, 220
Mastophora cornigera, occurrence, 220
Mastophora extraordinaria, occurrence, 220
Mastophora gasteracanthoides, occurrence of, 220
Mauritania, scorpions of, 352
Mediterranean region,
centipedes of, 181, 184, 185, 187
scorpions of, 333, 343, 351
species of *Latrodectus* in, 301–302
spiders of, 219, 236
Mediterranean Sea,
venomous coelenterates of, 401, 406
venomous sea urchins of, 424, 425, 427
Megachile, venom glands, 20
Megacorminae, genera and species, 329
Megacormus,
venom, toxicity, 342–343
Megacormus granosus, 329
Megalopyge,
extracts, pharmacology of, 112–116
hairs and bristles of, 140
nests, urticaria and, 127, 149
Peruvian, 135–136
description of, 136–138
poisonous setae of, 142–143
urticating caterpillars, 40, 134
venom apparatus, 106, 141, 142
Megalopyge crispata, caterpillars of, 104

Megalopyge lanata,
caterpillars of, 104, 158
Megalopyge opercularis,
poisonous spines of, 51, 52
venom, toxicology, 144–145
venom apparatus, 104, 142
Megalopyge radiata, caterpillars of, 158
Megalopyge superba, caterpillars of, 158
Megalopyge undulata, caterpillars of, 158
Megalopyge urens,
caterpillars, 104
venom apparatus, 142
venom,
nature of, 55, 108
toxicology, 144–145
Megalopygidae,
erucism, symptoms, 149, 151, 152
habitats of, 158
poisonous spines of, 50, 51, 52
urticating caterpillars of, 40, 42, 104, 133, 134
Megaphobema,
bites, wounds and, 198
distribution, 236
key for, 213
venom, 270
extraction of, 263–264
Megaphobema robusta, 214
description of, 226–227
Megascolia, stinging by, 32
Melipona marginata, venom, glands, 21
Melipona quadrifasciata, venom glands, 21
Meliponinae,
Dufour gland, 22
sting reduction, 38
venom glands, 18, 21
Meliponula bocandei, venom glands, 21
Membrane cell, insect hairs and, 43, 45, 47
Mesobuthus eupeus, 326
Mesopotamia, centipedes of, 181
Messor, sting apparatus, 37
Meta hispida, 219
Meta schuchii, 219
Methyl-*n*-amylketone, hymenopteran venoms and, 31
Methyl heptenone,
ant venom and, 97
hymenopteran venoms and, 31
Methyl hexanone,
ant venom and, 97
hymenopteran venoms and, 31
Mexico,
centipedes of, 181, 184, 185, 187
dangerous scorpions of, 331, 332, 333, 339, 343, 344
spiders of, 219, 221
venomous annelids of, 434
Mice,
phoneutrism in, 284–286
scorpion venom effects, 344–345, 359–360
antivenin and, 366
Microbracon hebetor,
venom,
composition, 30
effects, 32
Microbuthus litoralis, 326
Microbuthus pusillus, 326
Micrurus,
venom, pain relief and, 447
Millepora, 396
Millepora alcicornis, 399
range, 398
Milleporidae, venomous representative, 398, 399
Mimops,
classification, 171
sternites of, 186
Mimops occidentalis, occurrence of, 186
Mimops orientalis, occurrence of, 186
Mithras, scorpions and, 349–350
Molting,
scorpions, 337
spiders, 245
urticating hairs and, 56
Monoamine oxidase, scorpion venom and, 365
Monomorium, trail marking by, 34
Morocco,
latrodectism in, 307–308
scorpions of, 333, 343, 351
Morphidae,
caterpillars, host plants, 159
Morphine,
coelenterate stings and, 414
scorpionism and, 368
Morpho, urticating caterpillars, 134
Morpho achillaena achillaena, caterpillars of, 159
Morpho cypri, caterpillars of, 159
Morpho hercules, caterpillars of, 159

Morpho laertes, caterpillars of, 159
Morpho menelaus, caterpillars of, 159
Morpho rhetenor, caterpillars of, 159
Morphoidae, urticating caterpillars of, 133
Moths, urticating, 159
Mouth, spider, 204–205
Mucin, scorpion venom, 354–355
Mucous membranes, lepidopterism and, 130
Muscles,
lancet retraction and, 27
spider venom apparatus, 252–254
sting apparatus, 10, 12–13
sting protrusion and, 26
tone, scorpion envenomation and, 361, 363–364
Mutilla maura, venom glands, 20
Mutillida,
sting apparatus, structure of, 9
Mutillidae,
venom, effects, 33
venom glands, 19, 20
Mydriasis,
phoneutrism and, 280, 284, 287
scorpionism and, 361, 364
Mygale, see *Trechona*
Mygalomorpha, *see also* Orthognatha
key, 209
Myrmecia,
Dufour gland, 23
venom effects, 33
venom glands, 21
Myrmecia gulosa,
venom, 96, 98
composition, 30
Myrmecia pyriformis,
venom, composition, 30
Myrmeciinae,
sting apparatus, 37
venom,
composition, 30
effects, 33
Myrmica,
Dufour gland, 23
sting apparatus, 17, 37
sensory organs, 11
venom, effects, 33
venom glands, 21
Myrmica laevinodis, venom glands, 20
Myrmicaria natalensis,
venom, composition, 31, 95–96
Myrmica ruginodis,
venom, composition, 31
Myrmicinae,
alarm substance, 34
sting apparatus, 37
special peculiarities, 16, 17
venoms,
alarm and, 34
composition, 31
ejection, 28
venom glands, 21

N

Naja flava,
venom, pain relief and, 448
Naja haje,
venom, purification of, 357
Naja naja,
venom,
arthritis and, 450, 451
hemostasis and, 452
neuralgia and, 451
ophthalmology and, 458
pain relief and, 444, 445, 446, 447, 448 449
tabes and, 449–450
Nanobuthus andersoni, 326
Nasonia vitripennis, venom gland, 18–19
Natada, urticating caterpillars, 40
Nebo, classification, 323
Nebo hierochonticus, mating behavior, 334
Necrosis, latrodectism and, 268
Nematocysts,
coelenterate stings and, 407–409
development of, 410
structure of, 410
Neobuthus berberensis, 326
Neomyrma rubida,
venom, effects, 33
Neostothis gigas, 251
venom apparatus,
glands, 260
inoculating apparatus, 255
musculature, 252
Nephila,
bites, wounds and, 198
venom, toxicity, 271
Nephila brasiliensis,
venom, amounts, 266

Nephila clavipes,
venom, amounts, 266
Nerves, sting apparatus, 10, 11
Neuralgia,
bee venom and, 463, 464, 465
snake venoms and, 449, 451
Neuromuscular synapse, hymenopteran venom and, 32
Neurotoxicity,
centipede venom and, 194
scorpion venom, 362–363
New Caledonia, centipedes of, 181
New Guinea,
centipedes of, 185, 186
venomous coelenterates of, 402
New Mexico, centipedes of, 187
Newportia,
classification, 171
habits of, 190
occurrence of, 187
tarsi of, 178
venom apparatus, 188, 190
Newportia ernsti,
head and postcephalic segment, 188
last sternite, 189
New Zealand,
octopus bites in, 386
spiders of, 211, 219, 221, 236
venomous annelids of, 435
Nicaragua, scorpions of, 332
Nikethimide, scorpionism and, 368
Noctuidae,
hairs of, 44, 47
poisonous setae, 139, 142
poisonous spines of, 50
urticating caterpillars of, 41, 105, 133, 134
Norape, urticating caterpillars, 134
North Africa, *see also* Africa
centipedes of, 184
latrodectism in, 307–308
scorpions of, 332, 333, 343, 351–352
spiders of, 216
North America,
spiders of, 216, 218, 236
urticating caterpillars of, 105
Notechis scutatus,
venom, hemostasis and, 452
Notodontidae,
cryptotoxic caterpillars, 134, 148
erucism, symptoms, 149, 152
hairs of, 45
toxic apparatus, 143
urticating caterpillars, 104, 133, 134
venom apparatus, 142
Notostigmorpha, members of, 170
Nucleoprotein, scorpion venom, 354
Nudanriella cytherea, ingestion of, 147
Nymphalidae,
hairs of, 44
urticating caterpillars of, 40, 104, 133, 134

O

Ocneria, urticating caterpillars, 40
Octopus(es),
classification, 384, 388
stings, symptomatology, 390–391
venomous, distribution, 388
Octopus apollyon,
bites, 384, 385
symptoms, 390
Octopus fitchi,
bites, 384, 385
symptoms, 390
venom apparatus, 388–389
Octopus rubescens, bites by, 384, 385
Octopus rugosus, bites by, 384, 387
Octopus vulgaris, venom of, 390
Odonturus baroni, 326
Odonturus dentatus, 326
Odontobuthus doriae, 326
Oeclus purvesi, classification, 323
Ointment, bee venom and, 462–463
Olindiadidae, range and venomous representatives, 398
Olindioides formosa, range, 398
Olios rapidus,
bites, importance of, 271
Onuphis teres, bite of, 433
Ophion,
ovipositor, function, 28
Opisthacanthus,
venom, toxicity, 342–343
Opisthacanthus africanus, 324
Opisthacanthus asper, 324
Opisthacanthus cayaporum, 324
Opisthacanthus elatus, 324
Opisthacanthus lecomtei, 324
Opisthacanthus madagascariensis, 324

Opisthacanthus punctulatus, 324
Opisthacanthus rugulosus, 324
Opisthacanthus validus, 324
Opisthophthalmus,
distribution of, 332, 352
venom,
collection of, 353
toxicity, 343, 344
Opisthophthalmus ater, 324
Opisthophthalmus austerus, 324
Opisthophthalmus breviceps, 324
Opisthophthalmus capensis, 324
Opisthophthalmus carinatus, 324
Opisthophthalmus flavescens, 324
Opisthophthalmus fossor, 324
Opisthophthalmus fusciceps, 324
Opisthophthalmus glabrifrons, 324
Opisthophthalmus granicauda, 324
Opisthophthalmus granifrons, 324
Opisthopthalmus intermedius, 324
Opisthophthalmus laticauda, 324
Opisthophthalmus latimanus, mating behavior, 334
Opisthophthalmus macer, 324
Opisthophthalmus nitidiceps, 324
Opisthophthalmus opinatus, 324
Opisthophthalmus palcalvus, 324
Opisthophthalmus pictus, 324
Opisthophthalmus pugnax, 324
Opisthophthalmus schlechteri, 324
Opisthophthalmus wahlbergi, 324
Orb-weaving spiders, *see* Araneidae
Orgyia, urticating caterpillars, 40, 134
Orion, scorpion and, 350
Orthochirus,
venom, toxicity, 342–343
Orthochirus innesi, 326
Orthognatha,
chelicerae, 203, 204
distribution, 236
food of, 246
key for families, 209
molting in, 245
morphology, 201, 202, 210–211
sternum of, 209
venom apparatus, 250
glands, 203
Oryga, toxic apparatus, 143
Osmia spinulosa, venom glands, 20
Otocryptops, classification, 171
Otostigminae,
description, 185
members of, 171
Otostigmini,
description and distribution, 185–186
members of, 171
Otostigmus,
classification, 171
habits of, 190
prey capture by, 189–190
secondary sex characters, 189
Otostigmus caudatus, venom apparatus, 190
Otostigmus scabricauda,
description and range, 186
eggs of, 189
venom, toxicity, 195
venom apparatus, 191
ventral sternites, 178
Otostigmus tibialis,
body segments, 176
eggs of, 189
Ovipostor,
sting apparatus and, 2, 28
structure of, 4–5
Oxybelus, sting apparatus of, 33
Oxytrigona, venom glands, 18

P

Pachystopelma, 212
Pacific Ocean,
venomous animals,
coelenterates, 398, 401
mollusks, 381
sea urchins, 425, 427
starfish, 420
Pain,
latrodectism and, 356
local, *Phoneutria* venom and, 280, 284, 287
snake venoms and, 444–452
Palestine,
centipedes of, 185
scorpions of, 333, 339, 343
spiders of, 219, 236
Pale tussock moth,
caterpillar, venom apparatus, 104
Paltothyreus,
venom, effects, 33

Pamphobeteus,
bites, wounds and, 198
description of, 227
distribution, 236
key for, 213
mating behavior, 239
mating bulbs, 207
number of eggs, 241
venom,
extraction of, 263–264
toxicity, 270
Pamphobeteus antinous, 214
Pamphobeteus augusti, 214
Pamphobeteus benedenii, 214
Pamphobeteus ferox, 214
Pamphobeteus fortis, 214
Pamphobeteus insignis, 214
Pamphobeteus isabellinus, 214
Pamphobeteus platyomma,
venom,
amounts, 266
toxicity, 268
Pamphobeteus roseus, 214, 251
adolescent phase, 244
venom,
amounts, 266
toxicity, 268
venom apparatus,
glands, 260
inoculating apparatus, 255
Pamphobeteus soracabae, 251
adolescent phase, 244
venom,
amounts, 266
toxicity, 268
venom apparatus,
glands, 260
inoculating apparatus, 255
Pamphobeteus tetracanthus, 214, 251
adolescent phase, 244
aggressiveness, 227
venom,
amounts, 266
toxicity, 268
venom apparatus,
glands, 260
inoculating apparatus, 255
Pamphobeteus vespertinus, 214
Panama,
scorpions of, 331, 332
spiders of, 236, 237
venomous annelids of, 434
Pandinus,
distribution of, 332
stridulating apparatus, 321
venom, toxicity, 342–343
Pandinus arabicus, 324
Pandinus exitialis, 324
Pandinus bellicosus, 324
Pandinus cavimanus, 324
Pandinus colei, 324
Pandinus dictator, 324
Pandinus imperator, 324
Pandinus meidensis, 324
distribution, 352
Pandinus pallidus, 324
Pandinus phillipsi, 324
Pandinus viatoris, 324
Papilio, cryptotoxic caterpillars, 134
Papilio asteria, toxic apparatus of, 143
Papilio cresphontes, toxic apparatus, 143
Papilionidae, cryptotoxic species, 134, 148
Parabroteas montezuma, 329
Parabuthus,
distribution of, 332, 333
venom,
collection of, 353
hemolytic properties, 358
toxicity, 343, 344
Parabuthus brevimanus, 326
Parabuthus capensis, 326
venom, toxicity, 343
Parabuthus fulvipes, 326
Parabuthus granulatus, 326
venom, toxicity, 343
Parabuthus liosoma, 326
distribution, 352
Parabuthus transvalicus, distribution, 352
Parabuthus triradulatus, distribution, 352
Parabuthus villosus, 326
Paracentrotus lividus,
venom, pharmacology, 431
Paracryptops,
classification, 171
distribution, 186
Paraguay,
centipedes of, 184
scorpions of, 332, 343
spiders of, 217, 219, 237

Paralysis,
coelenterate toxins and, 411–412
distensive, phoneutrism and, 286, 293, 294, 295
flaccid, *Phoneutria* venom fraction and, 288–289, 290, 291, 293, 294, 295
Paraponera,
venom, effects, 33
Paraponera clavata,
venom, composition, 30
Parasa, urticating caterpillars, 40, 134
Parasa chloris, caterpillars of, 104
Parasa hilarata, venom apparatus, 142
Parasa latistriga, caterpillars of, 104
Parasa vivida, poisonous spines of, 50, 51
Parascorpiops montanus, 328
Parasemia, urticating caterpillars, 40
Paruroctonus gracilior, 329
Patagonia, spiders of, 219, 236
Pedicellariae, structure of, 429–430
Pelagia noctiluca, range of, 405
Pelagiidae, venomous representatives, 405
Pentadecane, Dufour gland and, 23
Pepsin,
caterpillar venom and, 116
Phoneutria venom and, 289, 290, 296
Peptides, hymenopteran venoms and, 30, 145
Persia, scorpions of, 333, 343
Peru,
caterpillars of, 135–136
erucism in, 146–147
scorpions of, 331
spiders of, 236, 269
Pheidole, sting apparatus, 37
Pheidole fallax, trail marking by, 34
Phentolamine, scorpionism and, 367
Philanthus,
Bordas gland, 24
venom, effects, 32
Philanthus triangulum,
sting apparatus, sensory organs, 11, 14
venom,
composition, 30
effects, 32
Philippines,
centipedes of, 185, 187
spiders of, 219
urticating caterpillars of, 105
venomous coelenterates of, 401, 402, 405
Phobetron, urticating caterpillars, 134
Phobetron hipparchia, caterpillars of, 158
Pholcus phalangioides, venom of, 272
Phoneutria,
adolescent phase, 244
aggressiveness of, 198
bites, importance of, 198
chelicerae, 203
cocoons of, 240
description, 222–223
distribution, 236
food of, 246
longevity, 246
maternal behavior, 240–242
mating behavior, 239
mature life, 244
molting in, 245
venom,
antisera, 273
extraction, 264
venom apparatus, 247, 253
fangs, 256
glands, 248, 258, 259
musculature, 253–254
young, behavior of, 243
Phoneutria andrewsi, 223
Phoneutria boliviensis, occurrence, 223
Phoneutria colombiana, 223
Phoneutria fera,
occurrence, 223
venom, toxicity, 283
venom glands, 260
Phoneutria nigriventer,
description of, 237–238
female with egg sac, 241
habits of, 237–238
occurrence, 223
received by Instituto Butanan, 263
venom,
amounts, 264–265, 266, 280
characteristics of components, 287–291
fractionation, 291, 297
toxicity, 267, 268, 280
venom apparatus, inoculating apparatus, 255
Phoneutria nigriventroides, occurrence, 223
Phoneutria reidyi, occurrence, 223
Phoneutria rufibarbis, occurrence, 223
Phoneutria sus, occurrence, 223
Phoneutriinae, description of, 222–223

Phoneutrism, symptoms, 267, 280–287
Phormictopus,
 bites, wounds and, 198
 description of, 229
 distribution, 236
 key for, 214
 venom, toxicity, 270
Phormictopus cancerides, 214
Phormictopus cubensis, 214
Phormictopus hirsutus, 214
Phormictopus meloderma, 214
Phospholipase, hymenopteran venoms and, 30
Physalia,
 stings, effects, 413, 414
 structure, 396–397
 toxin, properties of, 410–411
Physalia physalis, range, 398
Physalia utriculus, 400
 range, 398
Physaliidae, venomous representatives, 398
Physoctonus physurus, 329
Pieris, cryptotoxic caterpillars, 134
Pieris brassicae, toxity, 143
Pieris rapae, toxic apparatus, 143
Pieridae, cryptotoxic species, 134
Pimpla,
 ovipositor, function, 28
Plagiolepis, sting apparatus, 17
Plesiochactas dilutus, 329
Plesiochactas dugesii, 329
Pleurostigmorpha, members of, 170–171
Plumulariidae, venomous representatives, 398
Plutonium zweierleinii, occurrence of, 187
Podalia,
 Peruvian, 135–136
 urticating caterpillars, 134
Podalia albescens, caterpillars of, 158
Podalia chryscoma, caterpillars of, 158
Podalia orsilochus, caterpillars of, 158
Podalia radiata, caterpillars of, 158
Polybetes maculatus,
 venom, toxicity, 272
Polybetes pythagoricus,
 venom, amounts, 266
Polychaeta, characteristics of, 433
Polyergus,
 sting apparatus, 17
 venom glands, 22
Polyergus rufescens,
 venom, quantity, 29
Polypeptides,
 Phoneutria venom, 283–284, 287
  dialysis, 291
  heat resistance, 289, 290,
Pompilidae,
 sting apparatus, structure, 7, 8, 9
 venom,
  effects, 33
  ejection of, 28
 venom glands, 19
Pompilus biguttatus,
 venom, effects, 32
Ponerinae,
 sting apparatus, 16, 37
 trail marking by, 34–35
 venom,
  composition, 30
  effects, 33
 venom glands, 21
Porthesia, urticating caterpillars, 40, 134
Porthesia similis, caterpillars of, 104
Porthetria, urticating caterpillars, 134
Portugal venomous coelenterates of, 401
Portuguese man-o-war, *see Physalia*
Priapism,
 phoneutrism and, 281–282, 284–286, 287, 289, 290, 292, 294, 295
 scorpion envenomism and, 361
Prionoxystus, cryptotoxic caterpillars, 134
Prionoxystus robinae, toxic apparatus, 143
Proconvertin,
 deficiency, snake venom and, 455
Proctotrupoidea, ovipositor, 3
Propionic acid, ant venom and, 96
Propyl isobutyl ketone,
 ant venom and, 97
 hymenopteran venoms and, 31
Prostigmine, scorpion venom and, 314
Protein(s),
 hymenopteran venoms and, 30–31, 145
 scorpion venom, 354, 355–356
Proteinase(s),
 caterpillar venom and, 108, 114, 145
 Phoneutria venom and, 287, 289, 292, 293, 296–297
Prothrombin,
 deficiency, snake venom and, 455
Psalmopoeus, 212

Psammechinus microtuberculatus, range, 425
Psammobuthus zarudnyi, 326
Pseudocryptops,
  classification, 171
  distribution of, 185
Pseudohazis, urticating caterpillars, 134
Pseudolychas, 326
Pseudomyrmex, venom, effects, 33
Pseudomyrmex pallidus,
  venom, composition, 30
Pseudomyrmicinae,
  sting apparatus, 16–17
  venom, composition, 30
  venom glands, 21
Pterinopelma, 212
Pterinopelma vellutinum,
  venom, amounts, 266
Pterombrus,
  venom, effects, 32
Puerto Rico,
  centipedes of, 185
  scorpions of, 331, 332
Purpura haemorrhagica, snake antivenin and, 452
Puss caterpillar, *see Megalopyge*
Pyrilamine, caterpillar venom and, 109, 110

R

Rabbits,
  Phoneutria venom and, 287
  scorpion venom effects, 360
Rabies, viper venom and, 444
Rats,
  Phoneutria venom in, 287
  scorpion venom effects, 360
Red Sea, venomous sea urchins of, 425
Regeneration, spiders, 245–246
Reptilase, hemostasis and, 455
Reserpine, scorpion venom and, 314
Respiration,
  caterpillar venom and, 114
  scorpionism and, 361, 362, 364, 368
Respiratory infections, Esberitox and, 466–467
Respiratory tract, lepidopterism and, 130, 148
Rheumatic diseases, spider venom and, 461
Rheumatism,
  bee venom and, 461–462, 463–465
  snake venoms and, 450, 451–452
Rhoda,
  classification, 171
  key and occurrence, 181
Rhopalurinae,
  genera and species of, 327
  key to, 325
Rhopalurus,
  description of, 327
  venom, toxicity, 342–343
Rhopalurus acromelas, 327
Rhopalurus agamemnon, 327
Rhopalurus borelli, 327
Rhopalurus debilis, 327
Rhopalurus iglesiasi, 327
Rhopalurus intermedius, 327
Rhopalurus junceus, 327
Rhopalurus laticauda, 327
Rhopalurus pintoi, 327
Rhopalurus rochai, 327
Rhopalurus stenochirus, 327
Rhysida,
  classification, 171
  distribution of, 186
Rhysida brasiliensis, head, 173
Romania, latrodectism in, 307
Rosy anemone, *see* Sagartiidae
Rotschildia, urticating caterpillars, 134
Russia,
  latrodectism in, 307
  spiders of, 219, 236, 270

S

Sagartia, 397
  lesions produced, 413–414
Sagartia elegans, 407
  range of, 406
Sagartiidae, venomous representatives, 406–407
Sahara, scorpions of, 339, 351–352
Saint Croix, centipedes of, 185
Saint Thomas,
  centipedes of, 185
  scorpions of, 332

Saint Vincent, scorpions of, 331
Saline,
  injection, scorpion stings and, 367
Salivation,
  Phoneutria envenomation and, 281, 284, 286, 287, 289, 290, 292, 293–294, 295
  scorpion envenomation and, 360, 361, 364
Salticidae,
  description and dangerous species, 221–222
  key for, 210
Samia, urticating caterpillars, 41
Santa Lucia, scorpions of, 331
Sapygidae,
  ovipositor, 28
  sting apparatus, structure of, 8
Sardinia, latrodectism, 270
Sarohmę blanco amarillento, *see Podalia*
Sasoninae, description, 215
Saturniidae,
  adults, poisonous, 121
  erucism, symptoms, 149
  poisonous spines of, 50, 52, 53
  urticating caterpillars of, 41, 105, 133, 134,
  venom apparatus, 142
Sceliphron caementarium,
  venom, composition, 30
Schizura,
  cryptotoxic caterpillars, 134
  erucism and, 148
  toxic apparatus, 143
Schwarze Witwe, see *Latrodectus*
Scleroderma,
  venom, effects, 33
Scolia, venom glands, 21
Scolia villosa, venom glands, 20
Scoliidae,
  venoms, effects, 33
  venom glands, 19, 20
Scolopendra,
  classification, 171
  description of, 181–182
  key, 181
  legs of, 178
  life span, 190
  prey capture by, 189–190
  secondary sex characters, 189
  venom, toxicity, 193
  venom apparatus, 173, 190
Scolopendra alternans, distribution, 185
Scolopendra angulata, description and distribution, 183
Scolopendra armata, distribution, 185
Scolopendra calcarata, occurrence of, 185
Scolopendra canidens, occurrence of, 185
Scolopendra cingulata,
  distribution of, 185
  venom, toxicity, 193
Scolopendra clavipes, occurrence of, 185
Scolopendra crudelis, occurrence of, 185
Scolopendra dalmatica, occurrence of, 185
Scolopendra explorans, occurrence, 185
Scolopendra galapagoensis, occurrence of, 185
Scolopendra gigantea,
  description and distribution, 182–183
  venom, toxicity, 194
Scolopendra gordulana, occurrence of, 185
Scolopendra gracillima, occurrence, 185
Scolopendra hardwickei, occurrence, 185
Scolopendra heros,
  description and range, 184
  venom, toxicity, 193
Scolopendra hirsutipes, occurrence, 185
Scolopendra laeta, occurrence of, 185
Scolopendra madagascariensis, occurrence of, 185
Scolopendra metuenda, occurrence of, 185
Scolopendra morsitans,
  description and distribution, 182
  legs of, 179
  venom, toxicity, 193
Scolopendra pinguis, occurrence of, 185
Scolopendra robusta, distribution of, 185
Scolopendra spinosissima, occurrence, 185
Scolopendra subspinipes, 177
  description and distribution, 182
  maxillae, 175
  venom, toxicity, 193, 194
Scolopendra sumichrasti, distribution, 185
Scolopendra valida, description and range, 183–184
Scolopendra viridicornis, 172
  description and range, 184
  exuvia of, 190
  venom, toxicity, 194
  venom apparatus, 191–192
  ventral view, 173
Scolopendra viridicornis viridicornis, reproduction in, 189

Scolopendra viridis, distribution, 185
Scolopendridae,
key for subfamilies, 180
members of, 171
molting by, 190
Scolopendrinae, key for tribes, 180
Scolopendrini,
key to important genera, 180–181
members of, 171
Scolopendromorpha,
description of, 172–180
families in, 171
Scolopendropsis,
classification, 171
key and occurrence, 180
Scolopocryptopinae, members of, 171
Scolopocryptopini, key and distribution, 187
Scolopocryptops,
classification, 171
distribution of, 187
Scolopocryptops ferrugineus,
distribution of, 187
eggs of, 189
Scolopocryptops ferrugineus ferrugineus,
venom, toxicity, 195
venom apparatus, 190–191
Scorpamines, properties of, 356
Scorpio,
venom, toxicity, 342–343
Scorpio boehmi, 324
Scorpio maurus, 324
venom,
anticoagulant activity, 359
hemolytic properties, 358
hyaluronidase in, 358
Scorpio testaceus, 324
Scorpion, *see Androctonus*, *Buthos*, *Tityus*, etc.
ambulatory leg, 321
dangerous neotropical, 329–330
development of, 337
distribution, 317
old world, 332–333, 351–352
South American, 331–332
habitats, 338–339
maternal behavior, 335–337
morphology, 318–322
pairing of, 333–335
preying by, 337–338
stings, treatment of, 345–346
venom,
amounts, 342
chemical properties, 354–357
enzymes in, 357–358
extraction, 341–342, 352–354
humans and, 360–362
pathological physiology, 362–365
toxicity, 342–345, 359–360
venom apparatus, 339-341
worldwide, keys to families and subfamilies, 322–329
Scorpionidae,
key to, 321
sternum, 320
subfamilies, key to, 323
Scorpioninae,
genera and species of, 324
key to, 323
Scorpionism,
deaths and, 343
history, 349–351
treatment,
nonspecific, 367–368
specific, 365–366
Scorpiops, 328
Scorpiopsinae, key to, 328
Scutigeridae, 170
Scutigeromorpha, 170
Scyphozoa,
characteristics and habitat, 397
families and venomous representatives, 401–405
Scythodidae, *see* Sicariidae
Sea anemone, *see* Actiniidae
Sea blubbers, *see* Catostylidae, Cyaneidae, Pelagiidae
Sea mouse, *see Chloeia*
Sea nettle, *see* Cyanaidae, Pelagiidae
Sea of Azov, venomous coelenterates of, 406
Sea urchins, *see also* Echinoidea
stings,
clinical characteristics, 431–432
prevention, 432–433
treatment, 432
venom, pharmacology, 430–431
Sea wasps, *see* Carybdeidae, Chirodropidae
Segestria, distribution of, 237

Segestria perfida,
venom, toxicity, 271
Segestria ruficeps,
venom, toxicity, 271
Senegal, scorpions of, 333
Sensory organs, sting apparatus, 10–11, 14
Sepia officinalis, venom of, 390
Sericopelma,
description of, 227
distribution, 237
key for, 213
venom, toxicity, 270
Sericopelma communis, 214
Sericopelma fallax, 214
venom, amounts, 266
Sericopelma rubronitens, 214
Serotonin,
coelenterate toxin and, 411
hymenopteran venoms and, 30, 113
Phoneutria venom and, 287, 289, 292, 294
scorpion venom and, 357, 363
Setae,
annelid, 436–437
structure of, 42–43, 139–143
Shock, coelenterate stings and, 414
Sibine, urticating caterpillars, 40, 134
Sibine stimulea,
caterpillars, 104
venom apparatus, 142
poisonous spines of, 50
Sicariidae,
description and subfamily with venomous members, 215–216
key for, 210
Sicily, centipedes of, 187
Siphonophora, characteristics and habitats, 396–397
Sisyrosea, urticating caterpillars, 134
Skeleton, sting apparatus, 4–9
Skin,
changes, jellyfish stings and, 412–413
Smooth muscle, scorpion venom and, 312
Snake(s),
venoms, 444–445
analgesic applications, 445–452
anticoagulant therapy, 467–470
hemostatic applications, 452–455
other therapeutic uses, 455–458
Sneezing,
Phoneutria venom and, 280, 284
scorpions and, 361
Sodium hyposulfite, lepidopterism and, 132
Solenopsis,
Dufour gland, 23
trail marking by, 34
venom, effects, 33
Solenopsis fugax, prey of, 37
Solenopsis saevissima,
systematics, 98–100
venom, composition, 31
venom glands, 21
Solenopsis saevissima richteri,
adult worker, 97
biological cycle, 100–101
distribution of, 98, 99
eggs of, 98
larvae of, 100–101
pupa of, 100, 101
sexed forms of, 101
distribution of, 99
nest of, 96
venom, composition, 31
Solomon Islands, centipedes of, 185
Somalia,
centipedes of, 181
scorpions of, 333
South Africa,
centipedes of, 181, 185, 186
scorpions of, 332, 333, 339, 343, 344, 352
spiders of, 211, 218, 219, 230, 236, 269, 271
South America,
centipedes of, 187
dangerous scorpions of, 331–332, 338, 343, 344
poisonous lepidoptera of, 105, 121
spiders of, 211, 216, 218, 234, 236, 237
South West Africa, spiders of, 219
Spain,
latrodectism in, 270, 306
scorpions of, 333, 351
Sparteine, scorpionism and, 368
Spatangidae, pedicellariae of, 430
Spatangus purpureus,
venom, pharmacology, 431
Sphaerechinus granularis,
range of, 426–427
venom, pharmacology, 430–431
Sphecidae,
Bordas gland, 24
Dufour gland, 23

prey specificity of, 32–33
venom,
composition, 30
effects, 33
venom glands, 19, 20
Sphex flavipennis, sting gland, 25
Sphingidae, urticating caterpillars of, 133, 134
Sphinx, urticating caterpillars, 134
Spider(s), *see also Latrodectus, Phoneutria, etc.*
bites,
medical importance of, 197–199
treatment of, 273
classification, 209–222
dangerous,
development, 242–245
food, 246
longevity, 246
molting, 245
motherhood, 240–242
pairing, 238–240
regeneration, 245–246
descriptions of, 222–234, 237–238
distribution of, 236–237
morphology of, 199–209
protection against, 246–247
received by Instituto Butanan, 263
venom,
extraction, 263–265
leprosy and, 444
therapeutic uses, 461
toxicity, 265–273
treatment of bites, 273
Spiegelhaare, types of, 45–47
Spilosoma, urticating caterpillars, 134
Spines,
poisonous, 50–55
types of, 42–55
Spinnerets, spider, 200–202
Sponge fishermen's disease, coelenterates and, 413–414
Staphylococcus aureus, scorpion venom and, 359
Starfish, *see Acanthaster*
Status thymolymphaticus, coelenterate stings and, 414
Stenochirus politus, 326
Stenochirus sarasinorum, 326
Stilpnotia, urticating caterpillars, 134
Sting apparatus,
aculeate, special peculiarities of, 14–17
adult lepidoptera, 122–124
hymenopteran, 2
evolution, 35–38
function, 1–2, 25–35
morphology, 3–25
mobility of, 36–37
nonglandular parts,
muscles and innervation, 10, 12–13
position, 3–4
sensory organs, 10, 14
skeleton, 4–9
protrusion of, 26
retraction of, 27–28
Stinging coral, *see Millepora*
Stinging process, venom emission and, 25–28
Sting sheath gland, 18
structure, 24–25
Stomatitis, ingestion of setae and, 148
Strongylocentrotidae, pedicellariae of, 430
Strongylognathus, sting apparatus, 37
Sucuarana, *see Megalopyge*
Sudan, scorpions of, 333, 343, 352
Sumatra, centipedes of, 185
Sunda Islands, centipedes of, 185
Sympathetic nervous system, scorpion venom and, 363, 364
Synagris, stinging by, 32
Syntropinae, key to, 328
Syntropis macrura, 328
Syria,
centipedes of, 181
scorpions of, 333, 339, 343, 352
spiders of, 219, 236

T

Tachyphylaxis, spider venoms and, 283
Tachysphex, Bordas gland, 24
Tachysphex nitidus, venom glands, 20
Tannin, lepidopterism and, 167
Tapinauchenius, 212
Tapinoma, anal gland, 25
Tapinoma nigerrimum,
venom, composition, 31
Taragama, urticating caterpillars, 41

Taragania igniflua, caterpillars of, 105
Tarantism, latrodectism and, 308
Tarantulas, *see also* Orthognatha
  received by Instituto Butanan, 263
  sternum and lip, 205
  venom, extraction of, 263–264
  venom glands, 203
Tarantulinae, *see* Lycosinae
Tarentula apulia, 221
Tarentula melanogaster, 221
Tarentula poliostoma, 221
Tarentula thorelli, 221
Tasmania,
  spiders of, 221
Tatorana, *see Megalopyge*
Tegenatria,
  bites, wounds and, 198
Tegenaria derhami, venom of, 272
Tegenaria parientina, venom of, 272
Temnopleuridae, pedicellariae of, 430
Terebrantia,
  Bordas gland, 24
  ovipositor venom gland, 28
  venom composition, 30
Tetramethylammonium chloride, coelenterates and, 412
Tetramorium, trail marking by, 34
Teutana grossa, venom of, 272
Teuthraustes, 329
  venom, toxicity, 342–343
Texas,
  centipedes of, 184
  latrodectism in, 270
  scorpions of, 343
  spiders of, 219
Thais,
  cryptotoxic caterpillars, 134
  toxic apparatus, 143
Thaumetopoea, erucism and, 41, 126, 134
Thaumetopoea pinivora,
  setae, ingestion of, 148
  urticating hairs of, 46–47
Thaumetopoea pityocampa, urticating hairs of, 46–47, 142
Thaumetopoea processionaria, urticating hairs of, 47, 48
Thaumetopoea processionea,
  caterpillars of, 104
  urticating hairs of, 46–47
Thaumetopoea wilkinsoni,
  caterpillars, 104
    venom apparatus, 142
Thaumetopoeidae,
  hairs of, 46–47
  urticating caterpillars of, 41, 42, 104
Theatops, classification, 171
Theatops spinicauda, occurrence of, 187
Theatopsinae, 171
  distribution of, 187
Theraphosa,
  bites, wounds and, 198
  distribution, 236
  key for, 214
  venom, 270
    extraction of, 263–264
Theraphosa blondi, 214
  description of, 227–229
  food of, 246
Theraphosa leblondi,
  venom apparatus,
    glands, 260
    inoculating apparatus, 255
Theraphosidae,
  cocoons of, 240
  key for, 209
    subfamilies with venomous members, 212–215
  leg claws of, 209
  longevity, 246
  maternal behavior, 240
  molting in, 245
  morphology, 211–212
  regeneration in, 245–246
  venom apparatus, 247
  venom glands, 203
  young, behavior of, 243
Theraphosinae,
  key for, 212
    genera, 213–215
  mating behavior, 239
Theridiidae,
  description and subfamilies, 216–217
  key for, 210
Theridion atritus, 219
Theridion katipo, 219
Theridion lineatum, 219
Theridion lugubre, 219
Theridion melanoxantha, 219
Theridion mirabilr, 219

Theridion verecundum, 218
Theridion zebrina, 219
Thrombin, snake venom and, 454
Thromboplastin, snake venom and, 455
Thrombosis, venom therapy of, 468–470
Thuja occidentalis,
  extracts, venom combinations and, 466
Thynnidae, venom glands, 20
Tidops,
  occurrence of, 187
  tarsi of, 178
Tiphia,
  venom, effects, 32
Tiphiidae,
  sting apparatus, structure of, 9
Titya proxima, caterpillars of, 159
Titya undulosa, caterpillars of, 159
Tityinae,
  genera and species of, 327–328
  key to, 325
Tityus,
  venom,
    antivenin and, 365
    electrophoresis, 355
    pharmacological properties, 313–314
    serotonin in, 357
    toxicity, 342–343
Tityus androcottoides, 327
Tityus antillanus, 327
Tityus atriventer, 327
Tityus bahiensis, 327
  cities and, 339
  distribution, 331–332
  life story, 337
  mating behavior, 333, 334, 335
  morphology, 319, 330
  oocytes of, 336
  venom,
    amount, 342
    antivenins and, 366
    chemistry, 312–313
    extraction, 341
    toxicity, 343–345, 312
  venom glands, 340–341
Tityus bolivianus, 327
Tityus cambridgei, 328
  habitat, 338
Tityus championi, 327
Tityus clathratus, 327
Tityus colombianus, 327
Tityus costatus, 327
Tityus discrepans, 328
Tityus ecuadorensis, 327
Tityus forcipula, 327
Tityus insignis, 327
Tityus macrochirus, 328
Tityus magnimanus, 328
Tityus melanostictus, 327
Tityus metuendus, 327
Tityus obtusus, 328
Tityus pachyurus, 327
Tityus paraensis, 327
Tityus paraguayensis, 327
Tityus pictus, 327
Tityus pugilator, 327
Tityus pusillus, 327
Tityus serrulatus, 327
  cannibalism in, 338
  cities and, 339
  description of, 330
  development, 337
  distribution, 331
  parthenogenesis in, 335
  venom, 358
    amount, 342
    antivenins and, 366
    chemistry, 312–313
    extraction, 341
    pharmacological properties, 313–314
    symptoms of intoxication, 312
    toxicity, 343–345, 312
  venom glands, 340–341
Tityus silvestris, 327
Tityus stigmurus, 327
Tityus timendus, 327
Tityus trinitatis, 327
  chelicerae, 320
  distribution, 332
  morphology, 330
  venom, toxicity, 343, 344
Tityus trivittatus, 327
  description, 332
  mating behavior, 334, 335
  morphology, 330
  oocytes of, 336
  venom, toxicity, 343
Toad(s),
  venoms, therapeutic uses of, 458–460
Tokyo Bay, venomous coelenterates of, 398
Tolype, urticating caterpillars, 134

Tourniquets, scorpion stings and, 367
Toxicity, *see under name of organism*
Toxopneustes elegans, range of, 427
Toxopneustes pileolus,
  range, 427
  venom, pharmacology, 431, 432
Toxopneustidae, 433
  pedicellariae of, 430
  venomous representatives, 426–427
Trachoma, bee venom and, 465
Trachycormocephalus,
  classification, 171
  key and distribution, 181
Trail marking, hymenopteran venoms and, 34–35
Tranquilizers, scorpionism and, 368
Trasiphoberus,
  key for, 214
  musical apparatus, 229–230
Trasiphoberus parvitarsis, 215, 229–230
  spermatic bridge, 229, 238
Trechona,
  bites, importance of, 198
  description of, 223 224
  distribution, 236
  habits, 224
  leg claws of, 208
  mating behavior, 239
  maternal behavior, 240
  metatarsus and tarsus of, 208
  venom, extraction of, 263–264
  venom apparatus, 247, 260
  venom glands, 203
  venom jaws, 204
Trechona venosa, 251
  cephalothorax, 202, 203
Trechona venosa, 251
  distribution, 211
  spinnerets, 201, 202
  stridulating apparatus, 207
  venom,
    amounts, 266
    toxicity, 268, 270
  venom apparatus,
    glands, 260
    inoculating apparatus, 255
    musculature, 252–253
Trechonini, key and venomous members, 211
Trichogen, insect hairs and, 43, 45, 47
Trigona droryana, venom glands, 21
Trigona freiremaiai, venom glands, 21
Trigona postica, venom glands, 21
Trigona spinipes, venom glands, 21
Trigona tataira, venom glands, 21
Trigona xanthotricha, venom glands, 21
Trinidad,
  centipedes of, 183
  dangerous scorpions of, 331, 332, 338, 343
Tripolitania, spiders of, 219, 236
Trosia, caterpillars of, 158
Trypsin,
  caterpillar venom and, 116
  Phoneutria venom and, 289, 290, 296
Tumors,
  bee venom and, 465
  pain, snake venoms and, 444–449
  toad venom and, 459
Tunisia, scorpions of, 333, 351
Turkestan, spiders of, 221
Turkey,
  latrodectism in, 307
  scorpions of, 352
Tyramine, octopus venom and, 390

U

Uca mordax, coelenterate toxins and, 411–412
Ulcers, coral cuts and, 415
United States,
  centipedes of, 184, 185, 187
  dangerous scorpions of, 331, 332, 333, 343
  latrodectism in, 269, 270
  spiders of, 218, 219, 220, 236
Uroctoninae,
  genera and species, 329
  key to, 328
Uroctonoides fractus, 329
Uroctonus mordax, 329
Urodacinae,
  genus and species of, 324
  key to, 323
Urodacus, mating behavior, 334
Urodacus abruptus, 324
Urodacus armatus, 324
Urodacus darwini, 324

Urodacus excellens, 324
Urodacus granifrons, 324
Urodacus hoplurus, 324
Urodacus novaehollandiae, 324
Urodacus planimanus, 324
Urodacus woodwardi, 324
Uroplectes, 326
Urtica urens, sting of, 108–109
Uruguay,
  scorpions of, 332
  spiders of, 220, 223, 269
Utah, centipedes of, 187
Utcu bayuca, see *Megalopyge*
Utheteisa ornatrix, caterpillars of, 159
Utheteisa pulchella, caterpillars of, 159

V

Vanessa, urticating caterpillars of, 40, 134
Vanessa io, caterpillars of, 104
Vanessa urticae, dorsal spine, 43
Vascular permeability,
  caterpillar venom and, 114, 115
  smooth muscle and, 312
Vejovidae,
  key to, 322
  subfamilies, key to, 328
Vejovinae,
  genera and species of, 328–329
  key to, 328
Vejovis,
  distribution of, 333
  venom, toxicity, 342–343
Vejovis boreus, 329
Vejovis carolinus, 328
Vejovis cristimanus, 328
Vejovis flavus, 329
Vejovis granulatus, 328
Vejovis intrepidus, 329
Vejovis mexicanus, 328
Vejovis nitidulus, 329
Vejobis punctatus, 329
Vejovis pusillus, 328
Vejovis spinigerus, 329
Vejovis subcristatus, 328–329
Vejovis variegatus, 329
Venezuela,
  centipedes of, 183, 185, 186
  scorpions of, 331, 332
  spiders of, 236
Venom,
  aculeate,
    composition, 28–29, 30–31
    effects of, 29, 32–33
    ejection of, 27–28
  coelenterate,
    chemistry, 410–411
    pharmacology, 411–412
  combinations, therapeutic use of, 466–467
  scolopendromorph, toxicity, 193–195
  snake, 444–445
    analgesia and, 445–452
    hemostasis and, 452–455
    other therapeutic uses, 455–458
Venom apparatus,
  annelid,
    jaws, 437–438
    setae, 436–437
  caterpillar, morphology, 138–143
  coelenterate, 409–410
  cone shells, 381–383
  octopus, 388–390
  scolopendromorph, 175, 190–193
  scorpions, 339–341
  sea urchin,
    pedicellariae, 429–430
    spines, 428
  spiders, 247–252
    glands, 257–262
    inoculating apparatus, 254–257
    musculature, 252–254
  starfish, 421–423
  urticating caterpillars, 105–107
Venom glands,
  aculeate,
    general view, 18
    structure, 18–21
  ant, 21–22, 37
  Meliponinae, 18, 21
  spider, 203
Veromessor, venom gland, 38
Vertebrates, hymenopteran venom effects, 33
Vespa,
  Dufour gland, 22

Koshewnikow gland, 24
sting apparatus,
nerves of, 10
special peculiarities, 14
sting retraction in, 27
Vespa crabro,
venom, composition, 30
Vespa germanica, venom glands, 20
Vespa vulgaris,
venom, composition, 30
Vespidae,
Dufour gland, 22, 23
sting apparatus, structure of, 8, 9
venoms,
alarm and, 34
composition, 30
ejection, 28
venom glands, 19, 20
Vibrio comma, scorpion venom and, 359
Vipera ammodytes,
venom,
pain relief and, 447
rheumatism and, 450
Vipera aspis,
venom, rheumatism and, 450
Vipera berus,
venom,
pain relief and, 451
rabies and, 444
tumors and, 445
Vipera russelli,
venom, hemostasis and, 452, 453, 455
Visceral congestion, jellyfish stings and, 412
Viuda negra, see *Latrodectus*
Viuva negra, see *Latrodectus*
Vomiting, scorpionism and, 362

W

Wandering spiders, *see* Ctenidae
Water, scorpions and, 338
Weather, scorpion envenomation and, 360–361
West Africa, venomous coelenterates of, 401, 405
West Asia, centipedes of, 184
West Indies,
venomous animals,
annelids, 435
centipedes, 183, 184, 185, 187
coelenterates, 401, 405, 406
scorpions, 331
sea urchins, 424
spiders, 212, 216, 221
White-marked Tussock moth, *see Hemerocampa leucostigma*
Wolf spiders, *see Lycosa*
Wound healing, toad venom and, 460

X

Xenesthis,
bites, wounds and, 198
description of, 226
distribution, 236
key for, 213
venom, 270
extraction of, 263–264
Xenesthis immanis, 214
Xenesthis intermedius, 214
Xenesthis monstruosus, 214

Y

Yemen,
scorpions of, 333, 339
spiders of, 219
Yucatan, scorpions of, 332
Yugoslavia, latrodectism in, 305–306, 307

Z

Zabius, 327
venom, toxicity, 342–343
Zanzibar, spiders of, 218
Zygoena,
adults, poisonous, 121
occurrence of, 121
toxic secretion, 125, 128